SUPERSYMMETRY OR CHAOS

A JUDEO-CHRISTIAN COSMOLOGICAL MODEL
OF THE ORIGIN OF THE UNIVERSE

Book 2 of The Machine or Man Apologetics Series

BY HENRY PATIÑO

Isaiah 44:6; Psalm 24:1
(Gebo Wunjo Othala—Chi Rho Owns Earth)

Areli Media

Supersymmetry or Chaos
A Judeo-Christian Cosmological Model of the Origin of the Universe
Book 2 of The Machine or Man Apologetics Series
Copyright © 2018 by Henry Patiño
Published by Areli Media

Photo credits: Henry Patino (567); NASA (218, 351, 389, 523, 532, 706, 709, 725); Sarah Vega (48, 279); Image copyright 2018 R. Jay GaBany, Cosmotography.com (143, 214); ESA/Herschel/PACS & SPIRE Consortium, O. Krause, HSC, H. Linz (383); BICEP2 Collaboration, Harvard University, Cambridge, MA. (404); Image credit DEA/G. DAGLI ORTI via Getty Images (492).

ISBN-10: 0-9962441-4-X
ISBN-13: 978-0-9962441-4-5
eISBN-10: 0-9962441-8-2
eISBN-13: 978-0-9962441-8-3

TABLE OF CONTENTS

INTRODUCTION

● ● ●

The year was 1957; I was five and a half years old. In those days, that half was important. Having been raised on a farm, I had never seen the ocean. That summer, my family decided to visit my grandparents in Havana, Cuba. On a typical hot, balmy day, they took us to Varadero Beach, perhaps one of the most celebrated beaches in the Caribbean. The vastness of the blue ocean made a deep impression on me. As I walked through the hot sand to the edge of the water, waves were curling over, capped with foam and pushing the salt water until it nearly reached my feet. The sound of crashing waves was both terrifying and beautiful. I decided to explore those waves, so when the water rushed back into the sea, I ran forward toward them, but as the waves got closer to me, I turned and ran back to the dry sand.

My intrepid sister Nena was a year older. She stayed put and allowed the wave to go by her. "You see," she said turning to me, "there is nothing to be afraid of."

"I'm not afraid," I shouted back indignantly.

"Yes you are," she said. "Come on!"

I was afraid, but I would never let my sister know that, so I said, "No, I want to collect some shells."

I had never seen shells before. I was fascinated and intrigued by their many shapes and colors. So I began to walk up the shore and dig up beautiful shells. I don't know how long I meandered along the shore, but at a certain point, my pockets were full of shells and I had no more room to carry them. Yes, in those days bathing suits had pockets.

I turned around and began walking back, but I could not find my sister or my brother Robert playing in the water. I scanned the shore but could not find my mom in the midst of the crowd. After some time, I decided to recruit some help and walked up to the lifeguard stand.

"Have you seen my mom? She seems to be lost," I said.

The two lifeguards began to laugh. I did not think there was anything funny about my mother being lost.

"What is her name?" said one of the lifeguards.

"Her name is Isabel Patiño." I raised my hand above my head and continued, "She is about this tall and has light brown hair. Her hair is very long, all the way to her waist, but she has it in a bun behind her head, and she is very beautiful."

The lifeguards laughed again, and I became a bit irritated that they found it amusing that my mother was lost.

"And what is your name, young man?"

"I'm Kike, but my real name is Enrique."

"Oh, I see. Let me see if we can help you."

The lifeguard picked up the phone and called headquarters. In a few moments he answered, "I think we found your mom."

It was not long before my family came running over to us, and I saw my mother frantic and crying. She hugged me and said, "We were so worried about you."

"I was worried about you too, Mom."

The lifeguard then piped in, "He says he was not lost but that you were lost."

"I'm not lost," I protested. "I know exactly where I am. I am right here."

From my perspective, I knew exactly where I was. It was my mother who was lost. Looking back at this childhood experience after I became an adult, I realized that if we are to get to the bedrock of truth, we must be willing to look at reality through every possible perspective before coming to any conclusions. Sometimes, our personal perceptions may be erroneous. Sometimes, we must be willing to view other perspectives if we really want to see reality. Only after analyzing all possible alternatives can we rationally choose what best fits the data.

I have lived now more than six decades, and it has become apparent to me that humans as finite beings can never presuppose that we have come to a perfect and complete understanding of reality. Paradigms of science are doomed to be eventually overturned by future discoveries. Arrogance is a blinding curse that keeps people from understanding how little they really know about reality. The seeker of truth must be willing to humbly admit when his or her presuppositions have been shown to be scientifically false. But the problem is that our egos are such that we are prone to bias for psychological and moral reasons.

I came to the United States when I was in the fourth grade. By the time I was in the eighth grade, I had rejected the formalism of the Catholic Church. The nuns at St. Michael's did not like my rebellious attitude very much, and the feeling was mutual. By the time I was in high school, I considered myself an atheist and an ardent evolutionist. Science had always been my favorite subject. In fact, when I was in Auburndale Elementary School in Miami, Florida, I was a guest of the television science teacher, Mr. Greenburg. In those days, all the students piled into the cafeteria, and several televisions were positioned around us as we learned science and history. I marveled at the miracle of television as I watched myself on the screen with Mr. Greenburg teaching about the differences between mixtures and compounds.

It took some time before I accepted the fact that our universe could not have evolved by random ordering. I invite you to be a seeker of reality. Do not limit your mind to only one particular perspective.

Listen to all the possible alternatives and then make an informed choice. Question everything. Do not just accept that things are as you have been taught. Learn to read between the lines. True science is the pursuit of truth, wherever the evidence leads.

It is my purpose to record the evolution of the evolutionary theory. I will trace back to the beginning of the evolutionary cosmological model, consider the evidence posited by naturalists who reject the biblical Genesis narrative as historical, and continue all the way to our present time. You will find that the supposed empirical evidence that has been proffered to support the naturalistic argument in antithesis to the Genesis record has time and time again been shown to not be empirical data at all, but in fact a false interpretation of the true data. But sometimes paradigms persist, not because of empirical data but because of personal biases in the echelons of authoritative academicians. All true scientific breakthroughs had to break through these entrenched biases through patient and rational discourse.

As science has progressed, the Genesis record continues to be vindicated. If you are a stalwart evolutionist, you may think you are on the side of science and that the Judeo-Christian worldview is nothing more than mythology. I urge you to analyze the historical evolution of the naturalistic argument and decide for yourself which model best fits the true empirical evidence that science provides us. If you are willing, you will find that the evolutionary dogma that was presented as scientific empirical fact has consistently been proved erroneous as science and technology have progressed and more empirical data have been gathered. I invite you to be an objective and open-minded scientist in search of the true and historically real interpretation of the empirical data so reason may guide your choices.

From the onset, we must also acknowledge that many Christians have adopted the gradualist evolutionary theory for the creation of our universe and life. Many add God as the initial spark to the evolutionary theory and simply reinterpret the Genesis account as a poetic book that was not meant to be a true historical account of the creation of our universe. Others have tried with sundry hermeneutical gymnastics to force-fit the evolutionary scheme and timetable into the Genesis narrative.

Our position offers a Judeo-Christian cosmological model that is an antithesis to these views and regards the Genesis narrative as a true space-time historical account of the creation of our universe. We intend to show evidence that God does not need random chemical chance reactions to accomplish His will. He is not an impotent God who needs the constructs of human folly to achieve His will. Although the outskirts of our universe may be billions of years old, He does not need gradualism and billions of years to create humankind or the universe that He has so magnificently designed for us to inhabit. I therefore intend to show the scientific evidence for this marvelously designed habitat that supports the literal interpretation of the Genesis narrative. Reason and science form the path that leads us to our true historical space-time reality.

.

CHAPTER 1

• • •

THE NATURALIST ARGUMENT FOR THE ORIGIN OF THE UNIVERSE

God, who gives life to the dead and calls into being that which does not exist.

—Romans 4:17 NASB

Today, the universe is, but science has discovered that once the universe was not, and then somehow the universe became. You might think that the previous simple statement is an obvious observation, but it is not. Not too long ago—in fact, less than fifty years ago—naturalists believed that the universe had no beginning and no end in both time and space. That is, naturalists believed that the universe was infinite in size and had existed forever.

The naturalist presupposes that all events in the universe are explained only by the operation of the natural evolution of chemicals, through purely random or chance processes, devoid of any divine intervention. The idea that the universe had a beginning was quite abhorrent to the naturalist who preferred to think that "thus has it always been."

7

The notion that our material universe was static and eternal was infinitely more palatable to those with an atheistic proclivity. It removed from consideration any discussion relating to the very origin of the universe and avoided the possibility of a creator. They understood that there could be no logical scientific mechanism to traverse from nothing to something. So they simply avoided the subject altogether by declaring that the universe was eternal in time and size. The concept of a beginning was considered an illogical choice, for how could something begin from nothing in a naturalistic system?

> *Nobody can imagine how nothing could turn into something. Nobody can get an inch nearer to it by explaining how something could turn into something else. It is really far more logical to start by saying, "In the beginning God created heaven and earth" even if you only mean "In the beginning some unthinkable power began some unthinkable process." For God is by its nature a name of mystery, and nobody ever supposed that man could imagine how a world was created any more than he could create one. But evolution really is mistaken for explanation. It has the fatal quality of leaving on many minds the impression that they do understand it and everything else; just as many of them live under a sort of illusion that they have read the Origin of Species (Chesterton 1955, 23).*

If nothing existed at first, then nothing should exist now, yet evolutionists ask us to believe not only that nothing could become something but even more preposterously that nothing created everything. Such a proposition has nothing to do with empirical data. It is a metaphysical choice made purely because of psychological reasons and metaphysical preferences.

Someone said to me once that if there was nothing in the beginning, then there was something, because nothing is something. However, that is nothing but semantic deception, which ignores the very meaning of the words utilized. Nothing is the complete absence of something, and by the way, space is not an absolute void or nothing. Everywhere in space, there is at least a three-dimensional spatial reality intertwined in time. If the string theory is correct, then there may be ten spatial dimensions intertwined in time. If our understanding of the Higgs boson and the Higgs field is correct, then there is a pervasive non-zero energy everywhere in the universe. That is, every square inch of our entire universe contains a certain amount of energy.

In other words, space is not an absolute vacuum. It is not a void of nothing. It has a definitive shape and can be stretched or contracted. Moreover, there is an enormous amount of debris in the form of space dust as well as energy waves, distributed in minutely varying degrees of concentration throughout the entire spatial dimensions in our universe. As a matter of fact, there is good reason to believe, as Albert Einstein did, that the very matrix of space is more like an invisible fabric. If space can be bent, then it is not just a void.

If there really was nothing to begin with, then even the word or concept of nothing was non-existent, and there can be no mechanism to evolve into something from nothing. Thus, within the materialistic presupposition, it was much more palatable to accept a universe that had no beginning.

For years, evolutionary scientists argued that matter, as observed today, existed for all eternity. This Steady State theory and the idea that our universe was infinite in space and time as proposed by the naturalists monopolized the scientific consensus for centuries after the Enlightenment. The notion of a special creation by a divinity was considered a blind leap of faith in the unobservable past and was ridiculed as superstitious, unscientific mythology and naïve at the very least.

The possibility of a divine creation was summarily rejected as unscientific nonsense amounting to nothing more than mythological superstitions. But if science is really the pursuit of truth and facts, how can we discount what we have not been able to rule out with facts? The idea that God could have created the universe was flatly declared to be a subject that was outside the parameters of science, which only dealt with sensible, observable reality. The idea of the divine was adamantly ruled to be out of the purview of science and a consideration that belonged only in the subjective category of mythology.

Christians countered that if by chance there is a God and He created the universe, then science would have supporting evidence of this previsioned design. Naturalists, however, insisted and continue to insist that such consideration is beyond the scope of science because they presuppose that true science cannot delve into areas outside of our space-time continuum, which can be testable and observable. But this limitation is nothing more than an arbitrarily assigned qualification to the scientific process by those who choose to begin from a naturalistic bias. Such a limitation is not required by any scientific law; it is simply a metaphysical choice made a priori.

Theoretical physicist Lawrence M. Krauss said in an ad for the group Freedom from Religion Foundation in the November 2016 edition of *Scientific American*, "Lack of understanding is not evidence for God. It is evidence of lack of understanding, and a call to reason to try and change that" (Krause 19). What Krauss fails to understand is that it is not "lack of understanding" that leads us to this conclusion; it is, in fact, *understanding* the profound and rational reality that to believe blindly that nothing could become everything is irrational. The organization's motto "In Reason We Trust" is an oxymoron, since the quote begins with a metaphysical choice in the negative and then moves forward to try to substantiate their fundamental belief. That is not reasoning; it is rationalizing. The former objectively considers all possible alternative explanations. The latter subjectively begins with a preconceived choice and attempts to correlate it with the facts.

This atheistic view of the evolution of the natural world did not, however, begin with modern science or even the Enlightenment. As far back as the first century BC, a Roman naturalist named Lucretius, building on the concepts of "atoms" as delineated by Democritus (fourth century BC), established a cosmological model that formed the basis for most of today's atheistic cosmological models. Lucretius theorized an infinite number of universes processing through an infinite number of cycles of formation and destruction, leading to reformation. Lucretius was well aware of the improbability of ordered life arising by mere chance from a disordered state. But he rationalized that by the sheer number of cycles, the improbability problem would be resolved.

Lucretius made three major assumptions that formed the very bedrock of the naturalist position for quite some time:

1. The universe is infinite.
2. Time is infinite.
3. There is an infinite number of universes.

But what is the underlying motivation to begin with these three arbitrary a priori assumptions? Ilya Prigogine, winner of the 1977 Nobel Prize in Chemistry, and Isabelle Stengers admit in their book *Order Out of Chaos* that the motivation behind the concept is antipathy toward the notion of a deity.

> *The urge to reduce the diversity of nature to a web of illusions has been present in Western thought since the time of Greek atomists. Lucretius, following his masters Democritus and Epicurus, writes that the world is "just" atoms and void and urges us to look for the hidden behind the obvious: "Still, lest you happen to mistrust my words, because the eye cannot*

perceive prime bodies, hear now of particles you must admit exist in the world and yet cannot be seen."

Yet it is well known that the driving force behind the Greek atomists was not to debase nature but to free men from fear, the fear of any supernatural being, of any order that would transcend that of men and nature. Again and again Lucretius repeats that we have nothing to fear, that the essence of the world is the ever-changing associations of atoms in the void.

Modern science transmuted this fundamentally ethical stance into what seemed to be an established truth (emphasis added) (3).

In other words, Lucretius described the naturalistic idea that our universe is a closed system. All that exists is what exists within that material universe that humankind can see, touch, taste, or hear. No consideration of a supernatural being or of anything that is not sensible and palpable is considered plausible. The obvious angst, which preoccupied Lucretius, is shared by all "deophobes" who reject and rebel against the creator. The idea of a closed system predicated on these three fundamental assumptions made by Lucretius has appealed to those wishing to discard the notion of a deity, which would have, as a consequence, moral boundaries—an element that is quite repugnant to the naturalist. These boundaries are viewed as the cause of underlying fear and are therefore despised by many. It is therefore quite ironic that Prigogine and Stengers phrase their atheistic metaphysical presupposition as an "ethical stance." The real meaning of their ethical stance is that there is no divine force and all ethics are therefore relativistic.

However, modern science has not corroborated those three "ethical" assumptions. Quite the contrary, it has refuted them concretely. Today, modern naturalists can no longer consider the idea of an eternally existent universe as scientifically possible. The mathematics of

Einstein has proved beyond all doubt that our universe began at a finite time. But naturalists' dislike for the ethical ramifications of a creator is no less intense. Hence, attempts are still being made to explain the origin of our universe by way of a naturalist cause. We will discuss those in "Chapter 12, The Genesis Singularity."

The fact is that since Einstein's general theory of relativity, the first two assumptions made by Lucretius and followed by naturalists have been proven to be scientifically false. Time and space had a beginning and are therefore not infinite. Moreover, Lucretius's third assumption requires us to believe that something outside our universe exists. This third naturalistic dogma that other universes exist means that their other assumption that the universe is a closed system is also wrong.

Ironically, their assumption that there are other universes forces them to take shelter in doing what they claimed was unscientific for Jews and Christians. The Judeo-Christian claim that an infinite God exists outside our space-time universe was ridiculed as unscientific because it was outside the purview of science, which according to them was limited to testable and observable reality within our physical universe. Anything outside our universe could not be directly tested and therefore could not be considered within the purview of science. I find it a bit ironic that naturalists now have no choice but to look outside our universe for the answer to origins. Somehow, it is scientific if naturalists do it but not if Jews or Christians do.

Modern humankind, in a quest to avoid moral accountability, has created a naturalistic worldview that regards living things as nothing more than the accidental components of stardust, a biological machine void of purpose and transcendental significance. Gazing outwardly from this third rock hurtling through space and spinning around our sun, naturalists look at the cold light of stars and imagine that they are the masters of their own destinies. Standing defiantly in this lonely vantage point, the naturalist peers into the inky black stillness of a cold and forbidding space inhabited by violent, hostile black holes and forbidding distances and wonders if we are truly alone in the universe. Something inside wishes desperately to believe that we are not alone. Naturalists want to find life, any kind of life out there, not just to assuage their loneliness, but also to claim that their fundamental presupposition that life is but an accident could be supported.

For all their optimism, the realistic statistics of being able to truly search out our universe are quite impossible. It is rather like scooping a glass of water from the ocean and studying it to determine what exists in every other area of the vast oceans of our earth. In their arrogance, naturalists believe that their finite minds are the measure of all reality—the driving motive that spurs them to disregard the obvious symmetry and order in every level of reality throughout the universe and irrationally declare that the universal symmetry and amazing complexity we observe is simply a very fortuitous accident of nature.

That forced and illogical interpretation of the pervasive symmetry in our universe is not a scientific conclusion but rather a metaphysical square peg being forcibly pounded into a round hole by their personal psychological needs. The battle is not between science and faith; it is between two faiths. One faith declares that there is no God and all we are is the product of the survival of the fittest, and the other says that our universe could not have achieved such symmetry without a master designer.

The fact is that for years, most scientists involved in astronomy and physics were devout people of faith, and there was no dichotomy between their understanding of God and nature. As a matter of fact, it is reasonable to conclude that the Judeo-Christian worldview is responsible for the development of modern science as we know it. It was commonly maintained that the universe must have had a creator who set it in motion in an ordered fashion and maintained it thus. It is this underlying presupposition of the Judeo-Christian concept of reality that allowed the very concept of the scientific process in the modern sciences to develop.

Because there was an ordered universe, humankind could, through reason, learn about it. Science was possible because the universe was created for us to explore and manage. (And that is why we are responsible to take care of our environment; we were placed here, not to exploit it but to manage it.)

Alfred North Whitehead succinctly distilled it thus:

> *I mean the inexpugnable belief that every detailed occurrence can be correlated with its antecedents in a perfectly definite manner, exemplifying general principles. Without this belief the incredible labours of scientists would be without hope. It is this instinctive conviction, vividly poised before the imagination, which is the motive power of research:—that there is a secret, a secret which can be unveiled. How has this conviction been so vividly implanted in the European mind?*
>
> *When we compare this tone of thought in Europe with the attitude of other civilisations, when left to themselves, there seems but one source for its origin. It must come from the medieval insistence on the rationality of God, conceived as with the personal energy of Jehovah and with the rationality of a Greek philosopher. Every detail was supervised and ordered:* the search into nature could only result in the vindication of the faith in rationality. *Remember that I am not talking of the explicit beliefs of a few individuals. What I mean is the impress on the European mind arising from the unquestioned faith of centuries. By this I mean the instinctive tone of thought and not a mere creed of words* (emphasis added) (12).

There is no dichotomy between true science and the Judeo-Christian faith. On the contrary, the Judeo-Christian faith is the mother of rational science in this sense. The idea that humankind could through reason and observation discover the underlying rational principles that govern our universe could not have risen from an ideology that declared our universe to be guided by random order. A universe birthed by chaos has no justification for the orderly pursuit of truth through a scientific methodology that can discover universal laws. How can random chaos produce universal laws? How can random chaos legitimize an orderly pursuit of truth through the scientific process?

Prigogine and Stengers, in contrast, cannot bring themselves to quite admit this, stating:

> *It is not our intention to state, nor are we in any position to affirm, that religious discourse in any way determined the birth of theoretical science, or of the "world view" that happened to develop in conjunction with experimental activity (46).*

They go on to present the disclaimer that they do not know whether it was theological discourse or the "scientific myth" that came first and triggered the other. They are correct, however, in stating that in a universe birthed by chaos, scientific methodology is nothing more than a myth. Absolute truths do not exist outside of a creator. Naturalists are caught in a dichotomy that no universals can be absolute without God as the integration point; hence, they intellectually accede that science is but a mythology, yet they cannot function in their respective fields of endeavor without acknowledging that there are universal laws of physics.

Perceiving the Christian position as analogous to the watch, Prigogine and Stengers state, "A watch is a *contrivance* governed by a rationality that lies outside itself, by a plan that is blindly executed by its inner workings. The clock world is a metaphor suggestive of God the Watchmaker, the rational master of a robotlike nature" (46).

This, however, may be true of clocks, but it is flawed as a comparison to the Judeo-Christian concept of the cosmos. While it is true that the contrivance of the watch is governed by a rationality that obviously lies outside itself, since the very mechanism could not have caused itself to be, it is not an accurate representation of the Judeo-Christian position.

It was the evolutionists and deists who ardently pontificated that the universe was a clock set in motion by the laws inherent in it, since infinity. Furthermore, they stipulated that there was no interplay or interference from an outside source so that all things moved in a deterministic, robot-like fashion. The planets revolved around the sun because of natural laws, and there was no need to invoke God. In their minds, the universe was self-created and self-sufficient, needing no divine cause or interference.

The Judeo-Christian position, on the other hand, is that our choices within that framework or reality are meaningful and do have real consequences. Moreover, from our perspective, the designer of the watch (God) is also in continuous interface with the inner workings of its mechanism. Thus, the watch does not move in a robot-like fashion since the creator and designer here and there allows our choices to impact the movement. Neither do our lives go on in robot-like fashion, for our choices have real and significant consequences. This is an absolutely important distinction. We do not believe in a deterministic fate that is unchangeable.

As for who came first, it is rather naïve to assert that modern science could have risen in a naturalistic community that would regard all truths as relative and would not have had a foundation or underlying premise to even attempt to question anything, seeing that no real concrete truth could be ascertained in such a system. The natural outworking of such a philosophy would be the pantheistic concept of reality perceiving all to be Maya—an illusion. To state anything else is mere self-deception or perhaps even outright deceit. Therefore, Prigogine and Stengers are forced to refer to the scientific methodology as the "scientific myth" (46). It can only be called a myth because in a relativistic system, there is no foundation for truth and no platform from which the scientific process could be deemed an absolute maxim.

It must be clearly noted that although the naturalistic scientists of their day interpreted Newton's world in a structured and mechanistic fashion, the biblical concept of the cosmos is not mechanistic, as implied by Prigogine and Stengers. It is the biblical concept of humans as free within the providence of God that exemplifies the proper cosmology intimated by nature in the interplay with the macro world and the subatomic world. Here is the interplay between freedom and form, between free will and sovereignty, between the freedom exemplified by the seeming anarchy of subatomic particles and the Newtonian adherence of the macro world.

Thus, the Judeo-Christian faith is more in line with Einstein's theory and with quantum mechanics than any other philosophical framework. It provides for true freedom while maintaining structure. Here, and only here, the universals and the particulars are in a unified field (I will treat this matter more completely later).

Absolute Chaos Does Not Exist

Is our entire universe and all that is contained within it birthed by randomly generated physical reactions in absolute chaos? Those who hold to the atheistic/Darwinist worldview believe that is so. I say "believe" because this interpretation is strictly a matter of faith. The scientific process is not devoid of faith, as the naturalists are prone to claim. Faith is an integral component even of the scientific methodology. The scientific process of experimentation, observation, and accumulation of data is a fairly linear process. But once the data are assembled, they must be correlated into a theory that provides a conclusion for the way the data relate to our physical reality. That interpretation is a matter of faith created by a person's imagination.

Now the faith interpretation made by the scientist is subjected to rigorous testing in order to find out if it stands the litmus test of reality. More data are used to either disprove the faith interpretation or prove it. After some time, a consensus is generally reached when the scientific community either accepts the consensus as fact or rejects it. But that consensus is, in fact, also a faith interpretation. History has shown time and time again that an accepted consensus is not always correct if future evidence disqualifies the accepted faith interpretation. Such was the case for the Steady State theory that claimed our universe was infinite in size and had no beginning.

The naturalist continues to insist that the matrix of reality is absolute chaos and that the seeming design observed in the universe is simply accidentally generated by random chemical processes. But in doing so, naturalists must turn a blind eye to the pervasive symmetry that dominates our entire universe, to all the natural components residing within it. The very process of scientific methodology is the science of discovering the symmetries between the separate components of the data accumulated in order to find the designed purpose. In fact, the key component that ratifies the correctness of the faith interpretation is called repeatability. In other words, other scientists must be able to repeat the process that accumulated the data in order to verify the conclusion. That leads us to the next step of the scientific methodology called predictability. If the faith interpretation presented is true to reality, then it is possible to make logical predictions that would result in the verification of that faith interpretation. This, for example, was accomplished by Arthur Eddington's famous photograph that verified Einstein's prediction from his equations that claimed gravity could bend light. The photographs of a solar eclipse showed that a star that ought to have been behind the sun was seen at the very edge of the penumbra as light bent by the gravity of the giant mass of the sun caused gravitational lensing.

But we must understand that predictability is completely dependent on repeatability and would be impossible in a universe created by absolute chaos and random ordering. A completely randomly generated reality could never achieve any degree of repeatability. What naturalists fail to understand is that repeatability and predictability are deeply rooted in teleology. In other words, our universe and all of its components evidence a purposeful design. Had it not been intentionally designed, nothing in our universe could be either repeatable or predictable. By their very nature, repeatability and predictability require order and point to a purpose.

In fact, even the very process of the scientific methodology stands as proof that our universe could not have been birthed out of absolute chaos. Absolute chaos cannot create order anymore than a scientific methodology could produce constructive faith interpretations or conclusions through random ordering. Without the ordered process of scientific investigation, nothing could be concluded of any significant value.

That does not mean that chaos does not exist at all. Our Judeo-Christian cosmological model maintains that chaos is real. It may surprise you to find out that God created chaos for a very specific purpose. Genesis 1 tells us that chaos existed first at the Alpha Point before He created light. Nothing that God does is without purpose and significance. There is a reason for this, but let us first discuss the true nature of chaos.

What kind of chaos did God create? The answer is deterministic chaos. You might ask as you scratch your head: What on God's green earth is deterministic chaos? The answer begins with the story of Edward Lorenz who served as a meteorologist in the United States Army Air Corps during World War II. After the war, Lorenz earned advanced degrees from MIT (Massachusetts Institute of Technology) and worked on creating equations that he hoped would eventually allow him to predict the weather with unlimited accuracy. By 1961, he had compiled all the weather simulation programs for about a dozen equations. Confident that he would succeed in distilling the essential data necessary to perform such accurate weather predictions, he labored on.

But on a fateful winter day in 1961, Lorenz discovered something that would turn out to be exceedingly more consequential than his lifelong goal of accurately predicting the weather. Something that would alter radically and forever the way he and others viewed the personality of our universe.

It began with his deciding to rerun his weather predicting program to double check his most recent results. It seemed like a little thing. But as he stood watching the output of his trusty, 740 pound Royal McBee LGP-30 computer, he couldn't believe his eyes. Rather than replicating the previous data set—which is what he expected, given that he hadn't changed any of the meteorological parameters— his program was spitting out results that were outlandishly different.

The inner workings of the LGP-30 relied on 113 vacuum tubes, so Lorenz's first hunch was that the computer had blown a tube. But he checked things out and couldn't find anything wrong. Eventually, he came upon a possible culprit. When reentering the data for the second run, he had inadvertently entered numbers that were rounded off. They were extremely close to the first set of numbers but not identical.

If that was indeed the explanation, Lorenz was now more puzzled than ever. Weather was a linear system—or presumed to be—so the tiny discrepancies in the input data should have produced only tiny discrepancies in the output data. But that hadn't happened. Instead, the tiny input changes had produced results vastly different from those of the first run. . . . Clearly, trivial aberrations in the initial conditions could have unforeseen unbridled consequences (Guillen 122–123).

Lorenz realized that the processes that ruled meteorology were not linear, and in 1963, he published a paper titled "Deterministic Nonperiodic Flow" in the *Journal of the Atmospheric Sciences*, which became the foundation for the deterministic chaos theory. His description of what was then termed the "butterfly effect" followed in 1969. At the time, few scientists paid much attention to his discovery.

In 1972, however, at a conference in Washington, DC, of the American Association for the Advancement of Science, Lorenz delivered a talk whose very title made his discovery hard to ignore. It was stated in the form of a provocative question: "Predictability: Does the Flap of a Butterfly's Wings in Brazil Set Off a Tornado in Texas?"

The answer was yes. The slightest disturbance could set the world aflutter. Chaos was real.

From then on, not only other meteorologists, but scientists of every stripe—biologists, astronomers, physicians, you name it—hopped aboard Lorenz's "chaos" bandwagon. Even those in the humanities joined in (Guillen 123–124).

For this reason, Lorenz has been generally recognized as the father of the chaos theory. Lorenz eventually condensed three major cardinal features that defined deterministic chaos:

1. Chaos does not arise only out of complexity; it also arises out of simplicity.
2. Chaos does not emerge slowly; it leaps almost instantaneously.
3. Although chaotic behavior appears to be purely random, it is not; it is deterministic.

In other words, chaos is not purely patterns of absolute randomness. It has parameters that cause it to be deterministic. Those parameters are evidence of design rather than a universe generated through pure, absolute randomness. But for what purpose would God produce this deterministic chaos? The answer absolutely blew my mind when He revealed it to me. It is because without deterministic chaos, there could be no free will and individuality.

A universe with absolute randomness cannot have order. But a universe with absolute order cannot have individual choices that are meaningful. Chaos is real, but it is not unrestrained. It is not absolutely random. It is constrained by God's sovereignty, and thus we can trust in His providential will to bring the Alpha Point to the Omega Point. He did not make us as automatons, but He also disallows absolute anarchy.

What this means is that our choices are real, and they impact the course of history in ways unimaginable to our human eyes. One act of kindness, faithfulness, or courage can have global effects. The smallest person can change the course of history because our choices have real consequences. Frodo also lives in God's mind.

Without deterministic chaos, there could be no individuality in our personhood and no free will. On the other hand, unrestrained chaos is anarchy and the domain of the rebellious angels. Those who pander this deception are doing the will of the enemy of humanity. Absolute chaos does not exist. It is a religious myth promoted by those who rebel against the creator and claim godhood for themselves.

Providence is the evidence of God's sovereignty to restrain chaos deterministically toward the Omega Point, regardless of our personal choices. The evidence is all about us, for those who have eyes to see and ears to hear. It was deterministic chaos that brought jealousy and bullying by the brothers of a young boy named Joseph in Israel. It was deterministic chaos that tossed him in a well and sold him as a slave to a caravan that went to Egypt. It was deterministic chaos that unjustly cast him into jail. But that chaos was not unfettered. It was God's providence that brought him to the right hand of the pharaoh. It was providence that allowed him to prepare for the seven years of famine. It was providence that brought the brothers who had so wronged him to his feet. It was providence that allowed Joseph through his grace and mercy to save the nation of Israel. If those personal choices had not been made, there would have been no lineage for the Messiah to be born. Rest, my brothers and sisters, on this truth—chaos is not, never was, and will never be unrestrained. God's providence can take what Satan meant for evil and turn it into good.

Freedom in Form: Free Will and Sovereignty

> *God is clever, but He is not malicious.*
>
> —Albert Einstein

The idea that the universe had a beginning was an irrefutable, mathematical reality, but it was not accepted by the naturalists for a very long time, simply because it was evidence against their underlying metaphysical worldview. Einstein in time gave grudging acceptance to the necessity for a beginning and eventually to "the presence of a superior reasoning power" (Ross 58–59). Some have claimed that he never did quite completely accept the doctrine of a personal God, although in some of his writings, as in the above quote, he clearly spoke of God as a person who acted with purpose and design and with a distinct character in His act of creation. While it is true that there were some deep-seated subjective motives that Einstein could not quite shake due to some unpleasant experiences in his childhood and in Nazi Germany, his later writings seem to indicate that he moved past these initial stumbling blocks.

> *Two specific obstacles blocked his way. According to his journal writings Einstein wrestled with a deep felt bitterness toward the clergy, towards priests in particular, and with his inability to be able to resolve the paradox of God's omnipotence and man's responsibility for his choices (Ross 58–59).*

This misunderstanding of the interplay between man's free will and God's sovereignty has been used as an excuse to deny the reality of God by a great many people. It is ironic that this tension exists in much the same way, even in the subatomic level, which Einstein had a difficult time correlating under a single mathematical theory. Einstein's equations dealt with the macroworld and quite accurately described the physics of the movements of large bodies such as stars, galaxies, and planets. But the quantum equations describe the microworld. Because of the nature of matter at the quantum level, their actions are not as predictable as in the macroworld.

I am speaking of the Heisenberg Uncertainty Principle, which asserts that the position and momentum of a subatomic particle cannot be simultaneously measured with high precision. The reason for this is the fact that as one proceeds downward in size to the atomic level, the particles can no longer be considered like a hard or solid sphere. The smaller the dimension, the more wave-like the particle becomes. This wave-particle duality shows us that at the fundamental level, solid matter is, in fact, an expression of energy. The net effect of this wave-particle duality is that in this quantum world, particles have a sort of free will that is unpredictable. We can come up with probability graphs but not exact measurements, which proves that chaos is deterministic—it has parameters.

Yet in spite of this seeming anarchy in the microworld, there are parameters that maintain a deterministic outcome so the macroworld continues to function as it does. As we have begun to understand the nature of subatomic particles, a curious relationship between the macrocosm and the microcosm, reflective of our relationship with the creator, emerges. Initially, the relationship between the microcosm and the macrocosm was not perceived. The two seemed to be separate and acting distinctly without any congruence. Quantum mechanics explained a microcosm that seemed to be in complete anarchy, while the general theory of relativity explained the macrocosm acting in elegant order. The microworld, explained by quantum mechanics, seemed to be bizarre and ever fluctuating. Attempts to merge the mathematics of general relativity and quantum mechanics failed. They could not be correlated mathematically.

> *The experiments conducted by Davisson . . . were the first direct observation of the astonishing new principle at work. As an introduction to the new theory, let us re-examine the idea of the law of motion.*
>
> *Suppose a ball is projected from a place A and moves along a path to some other place B.*
>
> *If the procedure is repeated, we expect the ball to follow in exactly the same path (so long as the initial conditions are identical). This property was also expected of atoms, and their constituent particles, electrons and nuclei. The shattering discovery of the quantum theory was that this is not so.*
>
> *A thousand different electrons will travel from A to B along a thousand different paths. The rule*

of mathematical law over the behavior of matter appears at first to be finished, presenting the specter of subatomic anarchy. It is hard to over emphasize the immense implications of this discovery, forever since Newton had found that matter behaves according to definite rules, it was supposed that some sort of rules would apply on all scales, from the atom to the cosmos. But now it seems that the orderly discipline in the macroscopic world of our experience collapses into chaos with the atom (emphasis added) *(Davies 29–30).*

This subatomic anarchy is quite attractive to evolutionists, because it provides a scientific underpinning for their existentialist philosophical presupposition and to their hypothesis of a randomly generated universe. Curiously, it seems as though God has programmed an intrinsic sort of free will even in matter within the subatomic level.

At first glance, the entire subatomic world may seem to be in complete chaos and anarchy, but it is not. The very fact that the fabric of the universe does not rip apart is evidence that somehow it is not complete, unrestrained anarchy. If it did not have some overriding restraining force, there would be an extremely high degree of probability that matter would come in contact with antimatter and destroy itself. Chaos is not purely random; it is deterministic. Otherwise, we could not produce probability charts. Several observations immediately stand out in my mind against Paul Davies's arguments.

We live in a universe in which reality includes dimensions where invisible antiparticles of antimatter coexist with us. The result of the union of antimatter and matter is, of course, instant annihilation. Yet here we are. How can a chance and chaotic matrix of reality account for the fact that we have not annihilated? How can a chance and chaotic matrix account for the symmetry that is prevalent even in the subatomic world?

What is antimatter? In 1928, Paul Dirac came up with a theory that was consistent with both quantum mechanics and the special theory of relativity. Dirac's theory predicted that the electron should have a partner, a particle that was then named the positron (an anti-electron). In 1932, the positron was discovered, confirming Dirac's theory. We now know that every particle of matter has an antiparticle of antimatter corresponding to it. Contact between the two means instant annihilation.

Some scientists claim that there are whole antiworlds with antipeople made of antiparticles. This is stretching it quite a bit. But if you ever meet your antiself, as Michio Kaku said, "Whatever you do, make sure you don't shake hands!"

Obviously, there is a force that maintains order within the seeming anarchy that keeps matter from contacting antimatter and annihilating altogether. This is quite a feat when one considers that there is just as much antimatter in the universe as there is matter.

If the macroworld is but an extension of the microworld, and that is all that it could be in a naturalistic system, then the very same laws that govern the microuniverse would apply in the macroworld. The naturalistic paradigm would predict that the macroworld would be simply an extension of the microworld. The idea that traditional science has counted on in order to be able to build upon previous knowledge and technological progress is based on the premise that any given experiment, if performed accurately, has one and only one correct outcome, which can be testable and repeatable. In quantum mechanics, scientists dispense with this notion:

> *The mathematics underlying quantum mechanics—*
> *or at least, one perspective on the math—suggests*
> *that all possible outcomes happen, each inhabiting*
> *its own separate universe. If a quantum calculation*

predicts that a particle might be here, or it might be there, then in one universe it is here, and in another it is there. And in each such universe, there's a copy of you witnessing one or the other outcome, thinking—incorrectly—that your reality is the only reality (Greene 2011, 6).

I have a deep problem with the logic of that claim. If the quantum event originated in our universe, then each of the potential outcomes should be causally connected to our universe and would not be truly separate universes. If they originated from ours, then we should be able to observe it in theirs. But that is not what happens, is it?

After decades of closely studying quantum mechanics, and after having accumulated a wealth of data confirming its probabilistic predictions, no one has been able to explain why only one of the many possible outcomes in any given situation actually happens. When we do experiments, when we examine the world, we all agree that we encounter a single definite reality. Yet, more than a century after the quantum revolution began, there is no consensus among the world's physicists as to how this basic fact is compatible with the theory's mathematical expression (Greene 2011, 6).

Perhaps the failure to match this mathematical expression with reality tells us that not all potential mathematical expressions can match with reality. Is not the search for true science the connecting of our mathematical expressions and experiments with actual reality? If the theoretical constructs cannot be supported by predictions that can be physically verified, then they must be labeled speculation and not hard science. If there is a physical causal connection between the microworld and the macroworld, then there must be a unified theory that connects them. Quantum mechanics and the general theory of relativity cannot form that bridge between Jiffyland and our macroworld.

There must be a deeper fullness, a richer quality to the nature of reality that allows the macroworld to exist precisely because of the way the microworld functions. There must be a theory that, when tested, does not produce mathematical expressions that are in conflict with reality. There must be a theory that can encompass the verifiable aspects of both quantum mechanics and the general theory of relativity, which can explain all of reality.

Elegance and Symmetry

The title of this book is *Supersymmetry or Chaos*. When a physicist speaks of symmetry, he or she means that the subatomic particles have certain other particles that are very similar to one another, and these symmetries are also related to the symmetries of the four forces in nature. These four forces were once unified at the very high heat of the Big Bang. As the universe cooled down, the four forces condensed from that one force. Hence, there is a unity, intrinsic to all of matter and all forces, that indicates previsioned design. A universe guided by random chaos could not be expected to form an elegant and symmetrical design in all aspects of reality.

When a physicist speaks of supersymmetry, he or she means that all particles of the Standard Model have completely new superpartners to which they are related by supersymmetry. All of these have not yet been discovered. The title of this book is not referring only to such supersymmetry. What I am also referring to is the deep harmony that exists in all matter and forces that seem to indicate an intelligent designer. We can point to the fine-tuned parameters that were necessary for life to exist. We can point to the elegance of these symmetries between the forces of nature. We can even go beyond that to show the numerical symmetries that are pervasive in nature. It seems that these symmetries fall upon precise groups of two, three, four, and seven.

For those who are not familiar with the scientific terminology that describes the subatomic particles, the following section may be a bit laborious. But do not worry about remembering the names and understanding what properties they exhibit; instead, try to see the symmetry they express. This is the vital element that allows us to see that our universe is not the product of happenstance but of an intended design and order.

Number 2

The numbers 2, 3, 4, and 7 have special significance in the Judeo-Christian worldview, which was established thousands of years before the discovery of these subatomic particles and their properties. The number 2 symbolizes the dual nature of the Messiah, who, being God, was made into the likeness of a man to bring salvation to humankind. This salvation is first spiritual, accomplished by His death on the cross in His first coming, and second physical, which will be accomplished in His second coming to restore our broken world. In the first coming, He came in the flesh, but in the second coming, He will come in His resurrected body, displaying the fullness and glory of God. Hence, the number 2 speaks of the two aspects of reality, the physical visible form and the spiritual invisible form, which is just as real. They are not two separate realities, but two expressions of reality.

The number 2 also speaks of His body of believers, who are both Jew and Gentile, and is represented by the two prophets who will come at the end of this age to usher in His kingdom. It is reflected in the duality of solid matter, which we spoke of earlier. Matter in its fundamental particles exhibits a wave-particle duality that evidences its origin from the very energy of the initial explosion in the Big Bang—the energy of the creator.

The supersymmetry I speak of is the harmony of every aspect of God's creation that reinforces the unity and interrelatedness between Him and His creation. A universe created by chaotic processes could hardly be expected to manifest itself in repeated symmetrical patterns that are universally pervasive. Reason dictates that such supersymmetry must be regarded as the fingerprint of God in His creation. Judge for yourself.

The number *2* is represented by the fermions and bosons. According to the Standard Model of this quantum mechanic theory, particles come in two types: (1) *fermions*, or particles with mass that make up matter, and (2) *bosons*, or massless particles that carry forces.

We can also see this duality in tardyons and tachyons. Tardyons are ordinary particles of matter that are constrained below the speed of light. Tachyons are theoretical particles hypothesized through mathematics that are supposed to travel faster than light.

We can further see this duality in hadrons. All of the subatomic particles are composed of quarks. All particles composed of quarks are called hadrons. But quarks are never found alone. There are two kinds of hadrons: baryons and mesons.

Hadrons made of three quarks are called baryons. These form ordinary visible matter, like protons and neutrons (also called nucleons). Hadrons made of one quark and one antiquark are called mesons. These hadrons do not form ordinary matter. Because a meson consists of a particle (a quark) and an antiparticle (an antiquark), it is very unstable. In other words, its lifespan is extremely short. The one exception is the K meson, which lives longer than most mesons, which is precisely why it was called *strange*. And for this reason, scientists gave this name to the strange quark, which is one of its components.

So we see here that the hadrons made up of two quarks are strange and almost ephemeral particles, but the ones made up of three quarks comprise the building blocks or nucleons of tangible matter in our universe.

Number 3

The number *3* speaks of the ministry of Christ, the visible member of the triune Godhead, who began His ministry at the age of 30, served for three years, died on the 33rd year of His life, and rose from the dead on the third day. It should be noted that no human can become God, but God can become a human if He so chooses. The number *3* represents the visible part of God when He entered into physical form. It is therefore not surprising that the visible portion of our universe is divided into three spatial dimensions in one overarching dimension of time. There are three constants in nature: constant h; the speed of light (denoted by c), which is constant under all conditions; and Newton's constant G that measures the strength of the gravitational force. The Planck constant is expressed in three forms: (1) Planck length is $(Gh/c^3)^{1/2}$, which is about equal to 10^{-35} meters; (2) Planck time, which is $(hG/c^5)^{1/2}$ and translates to about 10^{-44} seconds; and (3) Planck mass $(hc/G)^{1/2}$, which is about 10^{-8} kilograms.

There are *three* fermions: protons, neutrons, and electrons. As we already stated, protons and neutrons are composed of *three* quarks. In our macroworld, we generally think of an object with greater mass as being larger than one with lesser mass, although that is not always so. Denser objects often have less volume than less denser materials. For example, a pound of cotton has greater volume than a pound of iron. Of course, the weight of one pound is not the measure of mass, but rather the interplay of gravity upon that mass. In the microworld, particles with greater mass have smaller sizes. Protons and neutrons are more massive than electrons and are therefore smaller. Electrons have considerably less mass and are therefore larger than protons and neutrons.

The reason is that a more massive particle is, in fact, a more energetic particle ($E=MC^2$ — energy and mass are interchangeable). High-energy vibrations (high frequency) have a smaller wavelength; that is, they are more tightly packed and hence take up less space.

Those *three* (protons, neutrons, and electrons) are, in fact, made up of three particles called quarks: (1) up quarks; (2) down quarks; and (3) electrons. These are held together by three specific forces, which decoupled from one single force that existed during the plasma stage of the Big Bang. The three forces that rule the subatomic world are (1) strong nuclear force; (2) weak nuclear force; and (3) electromagnetic force. The force of gravity does not rule in the quantum world. On the other hand, the force of gravity and the electromagnetic force rule the macroworld.

Number 4

The number *4* represents Earth. The symbol of Earth is a cross with *four* points, which represent the *four* cardinal points and the *four* seasons, which govern our life on this planet.

There are *four* forces in nature: (1) strong nuclear force; (2) weak nuclear force; (3) electromagnetic force; and (4) gravity.

The interplay between matter (fermions) and forces (bosons) is what produces our observed reality.

1. Strong nuclear force: What force can keep positively charged protons tightly packed into the nucleus of an atom when like poles repel? The force is so powerful that when it is released, we have the enormous release of energy we call nuclear power. Scientists using the Standard Model of Particle Theory call these gluons. According to this theory, gluons, a vibration in the strong nuclear force field, are responsible for keeping protons packed tightly in the nucleus in spite of the fact that they are repelling one another with incredible force.

2. Weak nuclear force: What force holds the quarks together in a nucleon (proton or neutron), maintaining the integrity of that particle? Scientists call these W bosons and Z bosons. According to this theory, these bosons are a vibration in the weak nuclear force field. When these bonds are broken, we observe radioactive decay.

3. Electromagnetic force: What force holds an electron to the nucleus? According to the Standard Model, it is the exchange of photons. Photons are a vibration in the electromagnetic field. By exchanging photons with the nucleon, the electrons are kept within their given shells, zipping around the nucleus.

4. Gravity: When the moon causes ocean tides through its gravitational influence, it is because gravitons are passing back and forth between the two bodies. A graviton is a vibration in the gravitational field.

Hence, all forces are, in fact, vibrations in fields. Actually, all particles are the result of a vibrating wave in their particular field. Just like sound waves propagate through the medium of air, vibrations that propagate through quantum fields create our observable reality. How does this understanding of the foundational properties of reality harmonize with the Judeo-Christian worldview?

The scriptures say that God spoke, and the universe came to be. Sound is but a vibration in the air. But vibrations in space (matter or energy, for they are interchangeable) are simply the voice of God. Hence, even the Standard Model of quantum physics speaks of God's vibrating energy causing the universe not only to exist, but also to be held together.

The Interconnection between 3 and 4

Three of the four forces are carried by one boson respective to that force: (1) strong nuclear force: gluons; (2) gravity: gravitons; (3) electromagnetic force: photons. But one of the forces is carried by *three* bosons. The weak nuclear force is carried by (1) Z bosons that are neutral; (2) the positively charged W boson (W^+); and (3) the negatively charged W boson (^-W).

Matter particles come in *four* groups of *three. Two* of the groups of *three* are quarks, and two of the groups of three are leptons. Quarks are particles that "feel" the strong nuclear force, and leptons are particles that do not.

The first group of *three* quarks is the up-type quark with a charge of +2/3: (1) up quark; (2) charm quark; and (3) top quark.

The second group of *three* quarks is the down-type quark with a −1/3 charge: (1) down quark; (2) strange quark; and (3) bottom quark.

The first group of three leptons is the charged lepton with a −1 charge: (1) electron; (2) muon; and (3) tau.

The second group of *three* leptons is the neutrino with a 0 charge: (1) electron neutrino; (2) muon neutrino; and (3) tau neutrino.

Everything you have ever seen with your eyes, touched with your hands, and heard with your ears is composed of some combination of the *three* particles we know as protons, neutrons, and electrons (the *three* "Lego blocks" of visible reality), along with the *four* forces that interact between them: gravity, electromagnetism, and the strong and weak nuclear forces that hold protons and neutrons together. That is all visible reality in the universe that exists within our *three* visible spatial dimensions in time and displays this uncanny symmetrical order that screams out "previsioned design" to the rational and objective mind.

The interplay between these particles and the four forces is what makes our universe possible. This interplay is quite distinctly balanced within a very narrow parameter in order for our universe to be possible. If the values were to change only by slight numbers, it would not be possible to have a universe as we know it.

Protons and neutrons are massive in comparison to electrons. A nucleus is about 99.9 percent of the mass of an atom. The positively charged proton is about 1,836 times more massive than an electron. The neutron is even more massive, about 1,842 times more than an electron.

As we noted earlier, it is the strong nuclear force that holds the nucleus together in spite of the fact that protons are packed tightly together (remember, like charges repel). But this nuclear force has a limit. If the nucleus has too many protons, it becomes unstable and tends to decay in a process we know as radioactivity. It literally sheds a given combination of particles until it reaches a stable configuration.

But the fact that the proton is positively charged is what holds the electrons zipping about the nucleus (remember, unlike charges attract). The electromagnetic attraction between the proton and the electron is about 1039 times stronger than the gravitational attraction between them. If the gravitational strength were stronger, we would literally sink through the surface of the earth into the core. It is the strength of the electromagnetic force that maintains the integrity of matter in balance to the force of gravity. The interplay between these forces must be finely tuned within a specific parameter in order for molecules to exist intact. Hence, the balance between the electromagnetic force and gravity had to be exactly as it is in order for us to be able to function as we do. It is the force that creates the integrity of the shape of atoms and molecules. That force must be exactly what it is in order to accomplish this. There is no other possibility in the magnitude of that force that would allow atoms to stay intact and not be sucked down by gravity in striating levels according to their respective mass.

The Elegance of Symmetry

There is an undisputed elegance to the symmetry, even of the microworld. A cursory examination of the Standard Model reveals that anarchy does not rule the subatomic world unrestrained. Subatomic particles are constrained by what physicists call *conservation laws*. These are unbreakable rules that govern what interaction among particles is possible. It is not ruled by chaos. When a particle decays into its subcomponents, there is a conservation of energy. That is, if we add up the energies/mass of the subcomponents, we end up with the energy/mass of the original particle. Moreover, there is also a law of conservation of (1) the electric charge; (2) the number of quarks; and (3) the number of leptons. There is an indisputable structural order even within the seeming anarchy of the microworld.

When I was a child, it was popular to play with tops. I would tie a string around my blue top and throw it spinning to the ground.

It would twirl around perfectly upright until it slowed down and began to wobble before falling on its side and coming to a stop. We intuitively expect spinning objects to run out of energy, slow down, and eventually stop. Not so for subatomic particles that have an exact value of spin that is eternally invariable. They have perpetual motion. What gives this perpetual motion that eternal energy?

The most amazing of all things is that both quantum theory and M-theory predict the existence of another particle called the Higgs boson. The Higgs boson is a vibration in the Higgs field. The Higgs field is unlike any other field. It is, almost like God, literally omnipresent; that is, it is literally all around us. I say almost like God because God extends beyond our universe, while the Higgs field is restrained within our universe.

The Higgs field extends to every point in the universe and has everywhere a resting non-0 energy. That is a fancy term for saying the Higgs field has a positive steady value at every point in the universe—quite a provocative thought. When space is completely empty of matter, most fields are set to zero; that is, they are turned off with zero energy. Not so for the Higgs field; its lowest energy state is universally at 246 GeV (giga electron volts).

But even more incredible, all other fields interact with the Higgs field, and it is this interaction that gives different masses to different particles. It is what gives all particles their unique qualities. The more a particle reacts with the Higgs field, the more mass it has. It is that constant interaction that creates the mass of all the particles. Hence, the intrinsic energy of the Higgs field, pervasive throughout the entire breadth of spatial dimensions, is what gives rise to our visible reality. That is an amazing thing to consider!

A particle according to the Standard Model is simply a little vibration in a field. It is a quantum or bundle of energy produced when the field is nudged away from its original value. The Higgs field is different. It has a constant positive value even in empty space. Einstein was right: space is not an empty void.

The Higgs boson is a vibration around that steady universal value. That is what makes it so unique. Thus far, every particle scientists have found is either a fermionic (matter particle) or a boson (force particle) derived from the connection fields associated with a symmetry. But the Higgs boson seems to be a horse of a different color.

If it were not for the Higgs field, the universe would be a featureless collection of uniform, massless particles zipping around at the speed of light. Reality as we know it would not exist. Because of these godlike qualities, some have dubbed it, much to the naturalist's chagrin, the "God particle." It is indeed a powerful symbol of the interplay between God and His creation. It is His fingerprint, if you will, left for those who seek to see it.

If the Higgs field did not exist at exactly the value that it has in its resting state, the universe as we know it could not be possible. It is that precise calibration that gives the electron the exact and necessary mass, slight as it might be compared to protons and neutrons. It turns out that the size of atoms is determined by a fundamental parameter in our universe—the mass of the electron. That precise calibration is absolutely essential in order for the universe that we know to exist.

It is important to understand that unlike planetary movements in solar systems, the electron seeks to always move closer to the nucleus due to the attractive force between its negative charge and the proton's positive charge. If the electron is excited with energy, it begins to zip around faster and move away from the nucleus. It will therefore give off a photon in order to release that excess energy and return to its place near the nucleus within its prescribed shell. That exact place is determined by the mass of the electron.

If the mass of the electron were smaller, it would be larger in size. But this does not translate into a bigger universe. Molecules are made because atoms share electrons. Any slight change to the mass of the electron would impede this process created by electromagnetic connections. A very tiny change in mass would allow only simple compounds such as water (H_2O) or methane (CH_4) to form. But larger complicated molecules such as proteins and DNA could not conglomerate if the atom's size were even slightly different. Life as we know it could simply not exist.

If the mass of the electron were changed considerably, the changes to our physical universe would be radical. Atoms would become macroscopic in size and perhaps even astronomical in size. If an atom were as big as a solar system, there would be very little interaction between atoms, since they would simply fly between each other without the subatomic particles making electromagnetic connections. Change the mass enough, and the electrons could not even stick to the nucleus.

In fact, if it were not for the precise charge of the background Higgs field, quarks and leptons would also be identical. Matter as we know it could not coalesce and form the familiar compounds that make up our physical world. Our world would be a monotonous and invariable world of sameness. In effect, in order for our world to exist as it is, there had to be symmetry at first, and yet this symmetry had to be broken at precisely the right amount to allow our universe to exist with its precise and limited variations that consequently allow for life to thrive. That is an elegant description of deterministic chaos.

Do you see random chaos, or do you see a predefined, previsioned pattern in the magnificent design of matter? Consider the following biblical passage written in the first century. It speaks of Christ:

> *For by Him all things were created, both in the heavens and on earth,* visible and invisible, *whether thrones or dominions or rulers or authorities—all things have been created through Him and for Him. He is before all things, and* in Him all things hold together (emphasis added) *(Colossians 1:16–17 NASB).*

In Him—Christ—all things hold together. It is the energy of God vibrating throughout the entire breadth of the universe that holds reality together and allows our universe to exist.

Do you think random ordering could design this multivariate pattern of *threes*? The deophobes with a subjective predilection to substantiate their Hegelian, naturalistic worldviews gravitate toward anything that could substantiate their dysteleological presupposition. In other words, they wish to see the universe as purposeless in order to avoid God. They abhor the idea of God, and thus the label *deophobes*. But in spite of some quantum weirdness, the microworld is anything but chaotic.

Number 7

The number 7 is the number that represents God the Father (invisible). He created the world in *seven* days. His archangels about Him are *seven* in number. His judgment upon the world at the end of this age shall come in three groups of *seven* judgments, indicating the complicity of God the Father and Christ. It is therefore not surprising that according to the M-theory (string theory), the invisible portion of our universe would be divided into *seven* invisible spatial dimensions, as we shall soon see, and three visible spatial dimensions. (We will address the M-theory more in "Chapter 18, The Grand Unification Theory.")

Quantum Weirdness

What is this "quantum weirdness"? The calculations of the Standard Model provided some ammunition for the naturalist's arguments against a designer God. The quantum weirdness comes from the apparent paradoxes created through the mathematical calculations. For example, it insinuates that particles can be in two places at once, that information can travel instantaneously (faster than the speed of light), and that cats can be alive and dead at the same time.

Say what? That's right, you read correctly. The famous example of Schrödinger's cat, which illustrates the way quantum mechanics works, says to imagine that a cat and a vial of poison are sealed inside a box. Before the observer looks inside the box, the wave function describing the system is in a superposition of both states. That means the cat is both alive and dead. It is the observation of the cat that collapses that wave function into one state or another. The observer then is what gives reality to the thing observed.

The obvious inference proposed is that the universe is not a reality independent and outside of a person's mind. That person is what gives realness to reality by the act of observation. The act of observation is then instantaneously transmitted to the thing observed and causes it to either make the cat dead or alive. It is this relativistic appearance that has enamored deophobes to the quantum explanation of our universe.

But it has two serious scientific flaws. First, it has the same flaw that marred Newton's idea of gravity. Information cannot travel faster than the speed of light. Gravity is not instantaneous, which Einstein showed us. Nothing can outrun a photon. Our observation cannot instantaneously change anything.

Second, the assertion that two antithetical things are both true is irrational. Of course, it is the endearing aspect for deophobes, but it is irrational nonetheless. How could science proceed if we did not have absolute answers for mathematical calculations? A physician presented with a patient suffering from a bacterial infection may give the patient antibiotics or arsenic. The two choices are not equal. One is absolutely right, and the other is absolutely wrong. To ignore that reality would be quite catastrophic.

I know that a watched pot seems to boil more slowly, but measure it, and you will find that it does not. Our observation has nothing whatsoever to do with the speed of boiling water. The assertion that Schrödinger's cat is truly both alive and dead at the same time is a scientific and logical absurdity. It is the delusion of deophobic megalomaniacs that have by faith subjectively chosen to believe that their own personal state of mind is what makes the world come into being. That is what Einstein rejected and called "spooky action at a distance."

The Copenhagen ideology of Neils Bohr does not reflect true reality. There is a seeming underlying anarchy in the subatomic world, yet an obvious structural order remains. Obviously, quantum mechanics provides an imperfect view of reality, although it contains some truth. It is an incomplete view of reality.

An objective analysis of our universe, both in the microworld and the macroworld, reveals the symmetry between reality and the Judeo-Christian worldview. The voice of God not only brought forth all forces and all forms of matter, but it is the energy that keeps all subatomic particles spinning and all forces vibrating within their respective fields. It is the power that brings forth and maintains the reality about us. By Him, all things consist.

The four fundamental forces in nature govern all reality, but quantum mechanics can only explain three of those forces. The Standard Model has given us great understanding of the subatomic world, but it fails mathematically when dealing with gravity; hence, it is an incomplete view of true reality. Quantum weirdness is the result of this missing information in the equation. We see that there is symmetry between all four forces, thus hinting at the fact that maybe we just have not discovered the deep overarching law that governs our entire reality.

This is the all-encompassing equation that Einstein so fervently sought toward the end of his life (unfortunately to no avail)—the theory of everything that can mathematically join all four forces into one fundamental and elegant equation. Unfortunately, he passed away before finding that all-encompassing equation that could bring the mathematics of gravity in the macroworld together with the other three fundamental forces that rule the microworld in our universe.

It is plain to see that somehow the microscopic world is not complete chaos, for it holds together and allows us to function in Newtonian fashion in the macroscopic world—in the supraworld. That is the paradox that rules out the assertion by naturalists that the microcosm proves that the universe runs in complete anarchy and pure chaos. It contradicts the notion that our universe is purposeless and guided only by random ordering. For how could the processes of random chance have succeeded in maintaining order from a supposedly disordered origin without the aid of an external directed energy to guide it? How could the universe begin in the Big Bang with such symmetry? How could that symmetry be broken in just the right quantity to make possible our intricately structured material universe unless it was so designed?

While it is true that the subatomic world functions in what could be termed a controlled unpredictability, it is not pure chaos. There is a sort of controlled microscopic anarchy; that is, the laws of probability govern things in a seeming state of randomness, yet it is not pure and absolute randomness. Quantum mechanics does give us probability graphs that allow us to predict outcomes. It is deterministic chaos.

The laws of conservation do rule the microworld. Moreover, the fluctuating and frothing randomness in the microworld is somehow tamed in order to reflect a steady, smooth, ordered, and elegant macroworld. Hence, that quantum description of reality is not quite accurate. It is at odds with the way the macroworld functions and must therefore be an incomplete view of reality. In other words, it is at odds with observable science in the macroworld, and if we are simply the product of random ordering, the macroworld could only be a reflection of the microworld. It is therefore at odds with the fundamental principles of science and logic. Something is obviously missing in the quantum equations that can link it to the macroworld.

Ironically, when the Newtonian view of the universe was vogue, the Enlightenment naturalists claimed that this clockwork mechanistic universe was eternal and self-sufficient; there was no room for God as espoused in the Steady State theory. The universe was viewed as a finely tuned machine that needed no outside interference to run and that had existed eternally. With the advent of relativism, this view of the universe as a smoothly running machine gave way to a seemingly random subatomic world. And the naturalist immediately adapts to this view and supports his or her relativistic presupposition with this one observation within the microworld.

But although most of the predictions provided by quantum mechanics are true, they fail to explain all of reality. It turns out that although quantum mechanics elegantly describes three of the forces of nature, it does not mathematically reconcile with the properties of gravity. Einstein's general theory of relativity does explain gravity, but not the other three forces. Hence, it is also an incomplete view of reality. Both these marvelous equations are missing a piece of the puzzle that can put reality into a continuum.

Because the general theory of relativity forces upon us a universe that had a beginning, a concept toward which a naturalist is quite antipathetic, naturalists tend to gravitate toward quantum mechanics, and for theological motives, many of them effectively act as though gravity does not exist in the world of physics. They are quite content to ignore the fact that all the forces do have symmetry and that therefore there must be some underlying theory that will bind the two equations together.

It has taken some 90 years since Einstein unveiled the general theory of relativity for scientists to begin to accept the incredible implications of this equation. The idea that evolution did not have infinity to function in its gradualistic model was a serious blow to naturalism and not well received. Naturalists thus gravitate to this randomness feature in the subatomic world as evidence that there is no God who designed and organized the universe in order and purposeful prevision. But in fact, the very opposite is what the quantum theory proves.

The very fact that the universe does not annihilate in the midst of this seeming randomness is proof of God's dynamic interrelationship with our space-time continuum. The pervasive symmetry observed in subatomic particles does not speak of randomness but of order, design, and purpose, as we have already noted. The plain fact remains that we know that three of the forces in the universe can be explained well by the quantum particle theory (strong nuclear force, weak nuclear force, and electromagnetic force), and they describe the microcosm but cannot explain the macrocosm. On the other hand, the general theory of relativity explains the fourth force in the universe that rules the macrocosm (gravity) but cannot explain the microcosm. So it is obvious that both theories are true and yet incomplete in and of themselves.

We know that there is an intrinsic connection between the two, for how can a macroworld of order evolve from a microworld of chaos? How is it that the macroscopic universe can function in a Newtonian fashion when supposedly purely random processes rule the subatomic world? Obviously, Einstein was right, and scientists had not yet stumbled across the right equation that could neatly fit these two magnificent equations in resonant continuity in order to explain their interrelationship.

The bottom line is that neither quantum mechanics nor the general theory of relativity can give us a comprehensive view of all reality. A more comprehensive theory is needed to unite both aspects of the supraworld (gravity) and the subatomic world (strong nuclear force, electromagnetic force, weak nuclear force). Perhaps it already has been discovered, and the much-celebrated string theory may soon be vindicated by the new and more powerful particle accelerators.

The Great Divorce

Unfortunately, when the Christian faith abandoned its proper place and became a political government in Rome, much harm was done to the Christian cause. Absolute power corrupted absolutely, and leaders with overblown egos and no spiritual discernment drifted to legalism and quenched those who sought truth.

Buffoons dressed in clerical garb and armed with their narrow view of scripture harassed and persecuted good men of faith who simply sought truth. It was a sad day for Christianity, for this gave Satan the opportunity to set a wedge between science and religion. It was the birth of the great divorce of science and faith. But that was an unnatural division created not by true faith but by religious autocratic despots.

Hopefully, objective students will recognize this as an unnatural wedge, since true truth is not self-contradictory. Since the state of the church at that time was marked by certain peculiar conditions, it led to the development of a clergy that was, by and large and with a few exceptions, less than sincere in their faith. I am speaking of the extravagant opulence and political power that could be attained only through service to the Mother Church and which acted as a magnet that gravitated to the clergy every scoundrel and rascal who was looking for power and wealth. It was, in fact, a very poor representation of what the scriptures stand for with their Christian principles exemplified by the humble and disenfranchised members of the early church.

Sadly, the upshot of this debacle was that sincere Christians were made to choose between the authority of the church and science. It never occurred to these narrow-minded men of the clergy that perhaps their interpretation of the scriptures was fallible. True truth cannot contradict itself.

If science brings to light facts and these facts are a seeming contradiction to a biblical tenet, then one of two things is true: Either our understanding of the scripture is in error and further study will elucidate the correct interpretation, or the scientific postulate is in error and further study will also bring that to light. But no one should ever be prevented from freely exercising the scientific process in search of truth. Truth shall eventually—always—rise to the top, for it is more than able to compete in the free market of ideas.

And for this reason, I consider it a scandal of equal magnitude that the proponents of intelligent design are not allowed to present their scientific model for the existence of our universe in the public education arena. It is an unconscionable repression of the scientific process and smacks of the repressive spirit of the Inquisition. What hypocrisy!

It is fair to say that as a result of Newton's laws and the subsequent bifurcation of the church and science, most scientists began to drift toward a concept of the universe that, apart from God, ran in a clockwork fashion. The natural resentment scientists developed toward their persecutors helped seal the divorce of science and religion.

Sincere and honest people, justifiably rejected the subjective and authoritarian clergy that imposed upon them irrational restrictions from their misguided notions of the biblical mandate. A great many souls became convinced that Lucretius was right. Had I lived in that time and been exposed to those pressures, without the benefit of hindsight, I might also have gone the way of Lucretius. In fact, I did in my youth and for many of the same reasons.

(By permission of Exploring Philosophy by Peter A. French)

CHAPTER 2

• • •

NEWTON AND THE CLOCKWORK UNIVERSE

We have already documented in the first book of this series the influence that Platonic thinking had and still has in our culture. But I think it is important to reiterate this development that was concurrent with the great divorce of science and religion. Plato viewed reality in two separate and ununifiable halves. He believed and taught that our physical world is but a shadow of the reality behind it, which he called immaterial "forms." These forms, such as goodness, truth, and beauty, could be attained by the enlightened, only when a person realized that the physical forms of matter were but an illusion of the idealized and real forms behind them.

$$\frac{\text{Form} - \text{(True reality)}}{\text{Matter} - \text{(Shadow} - \text{Illusion)}}$$

Plato believed that both form and matter were eternal. Hence, in this dualism, matter represents chaos and evil, while form represents order and goodness. Today postmodernists have also divided reality into two separate and unconnected spheres. Francis A. Schaeffer likens this dualistic worldview to a two-story home in which there is no connection between the first and second floors, a home without a staircase. In the lower story, modern humanity has placed science as the product of objective and rational thought. In the upper story, they have placed religion and any metaphysical ideology as subjective, irrational, and non-cognitive faith.

(Upper Story)

Metaphysical Realm–Irrational, subjective, non-cognitive faith

Physical Realm–Science, Objective, rational, cognitive data

(Lower Story)

The illusion created by this dualistic worldview is that Darwinism or naturalism is squarely in the lower story of objective reality, while theism is in the upper story of subjective and irrational knowledge. The total acceptance of this schism in epistemology by our academia can be traced to our Western universities when they divided the disciplines of humanities and natural sciences into two separate schools.

The natural sciences have since been seen as objective verifiable truths, while the humanities are seen as relativistic, mythological, unverifiable, and subjective truths that are essentially boiled down to personal preferences. Sadly, even Christians and Jews have been duped into accepting this unnatural division of truth, which is in complete antithesis to the teachings of the scriptures.

That dualistic framework has served to create a divided field of knowledge in which science has been placed in a realm that is completely divorced from the metaphysical. All considerations in the metaphysical are deemed to be subjective and non-rational choices and not part of rational empirical inquiry. Darwinism, on the other hand, is considered an empirical scientific fact belonging in the lower story.

Most of you reading this see no problem with this artificial division. It seems natural. That is because it has been inculcated in us from the time we began school. I invite you to reason with me, and I promise you will, in thinking more deeply, see the grave inconsistency of this naturalistic doctrine. The truth of the matter is that the naturalist automatically presupposes an atheistic metaphysical worldview when he or she insists that the realm of science by definition cannot evidence a designer of this universe. This choice is an arbitrary one not predicated by scientific necessity but by an underlying atheistic proclivity.

Darwinism is not a scientific model. It is a metaphysical model that uses science to support it. But that same scientific methodology can also be used to support the idea that our universe was designed with intelligence.

It is up to the reader to consider the scientific evidence posited for both worldviews in order to obtain an objective understanding of reality. The arbitrary division that moves metaphysical choices to the irrational upper story is nothing more than smoke and mirrors designed to sideline any consideration of a master designer. The insistence that natural science cannot come to a rational conclusion about the existence of God is by definition an arbitrary subjective designation, which is nothing more than an atheistic doctrine and amounts to a metaphysical choice. There is no scientific reason to automatically discard this possibility. It is nothing more than a metaphysical choice to deny the investigation of the possibility of an intelligent designer.

It is akin to restricting the scientific investigation of a computer whereby the investigators are prohibited from concluding or even considering that human intelligence imagined it, designed it, engineered it, and ultimately created it. The paltry rationalization used to reject anything that cannot be sensed with our finite human senses must be laid bare as a cheap conjurer's trick to deny the reality of most of our universe.

But there is a great inconsistency in their schizophrenic attempt to avoid the reality of God. We cannot sense, hear, or see dark energy or dark matter, but naturalistic scientists are convinced that both are there nonetheless. They claim that dark energy is an invisible antigravity force that causes the acceleration of our universe to increase with distance. They claim that dark matter is an invisible and exotic form of matter that causes the distant galaxies in our universe to maintain their integrity in spite of the enormously high speed of rotations observed for these galaxies. In fact, according to their most recent calculations, only about 4 percent of the universe can be seen heard or sensed. Thus their dogma that only the sensible can be discussed by science is just a ruse to discredit the existence of the divine. The plain truth is that the Darwinist in the lower story is equally making a metaphysical choice, albeit a choice in the negative. Having said that, let us continue with the historical development of the naturalistic model.

The Historical Development of the Naturalistic Model

In 1734, Emanuel Swedenborg, a Swedish mystic, published his theory of the development of galaxies and stars in a mechanistic framework. As galaxies were brought into view through the advancement of optics, people peered more deeply into space. The universe seemed endless, and the concept of an infinite universe became entrenched in science.

It was at this time that another Scandinavian, a theologian named Imanuel Kant, published his Universal *Natural History and Theory of the Heavens*. Kant proposed that all the galaxies emerged from primal nebulae. These nebulae with molecules in constant motion, he theorized, began to aggregate into cores of mass, which eventually began to rotate.

In this manner, then, proto-stars and planets were formed. By the beginning of the eighteenth century, the vast majority of scientists still believed that the universe was infinite and eternal in time. Some time later, Kant published his much-celebrated *Critique of Pure Reason*. It was in this seminal work that Kant divided the realm of faith from the realm of the rational, ushering in the divided field of knowledge and the era of postmodernism. He introduced two dominating concepts that have deeply affected both science and theology to this very day. In science, he is considered the "father of modern cosmology." In theology, he is considered the "father of the New Theology."

The New Theology then hoped to cohabitate with the naturalist paradigm by divorcing the two fields of thought. Faith no longer needed rational explanation, and therefore it no longer needed to rationally correlate with the scientific paradigm of the age. The Judeo-Christian concept of the unity of truth was challenged by the postmodern concept of the divided field of knowledge. In other words, religious truths were placed in a separate category from scientific truths. Scientific truths were understood through reason, and religious truths were accepted by blind faith, outside of reason.

This artificial division has segregated religious thought into the mythical and the realm of mysticism, which cannot be substantiated. But the Judeo-Christian worldview holds that all truth is unified because it emanates from the creator. Hence, what is scientifically really true is also theologically really true. Reason is the highway to truth, but in the end, after accumulating all the empirical data, reason must bring us to the door of faith before we can conclude in truth. Faith is the key that opens that door. That process is not much different; in fact, it is exactly the same in scientific experimentation. Once the experiments are concluded and the evidence is amassed, the scientist by faith accepts the conclusions, which are rational.

Now evolutionists will tell you that their worldview is arrived not by faith, but by reason. That is an absolute lie. Their model uses scientific facts, but their conclusion is a matter of faith in the construction of the interpretation of those facts. It is no different than our model, which also relies on scientific facts to provide a rational interpretation of the data that are consistent with reality. In fact, it takes less faith to believe in the Judeo-Christian model than it does to believe in the naturalist model for reality, as we shall see.

Nevertheless, during this historical transition when religious truth was first segregated to the mythological, Charles Lyell's uniformitarian theory in geologic processes gained favor against the previous idea of catastrophism. All geologic processes were now seen to undergo very small changes for very long periods of time. The seemingly large changes were simply understood as the accumulated product of immense times. Since they perceived the universe to be eternal in time and space, time was their most cherished ally. Enormous spans of time were needed to make small and gradual changes that eventually accumulated into drastic changes. This became the foundation for the naturalistic doctrine of gradualism that then dominated all scientific disciplines.

In the area of cosmology, the idea of an infinite and static universe was held by the vast majority of scientists, who subsequently rejected the Judeo-Christian position that God had created the universe and our planet a finite time ago. At this point in time, the naturalists viewed the universe as a vast machine running for all infinity, needing no control or input from an exterior force that existed outside the universe. All processes were selected by random choices that naturally propelled the more successful choices in a gradual fashion.

The stage had been perfectly set for Charles Darwin to step forth. The Darwinian Theory of Evolution by Natural Selection fell right into place to explain the biological evolution from one species into another through long periods of very small changes due to selective, adaptive pressures upon all life forms. The origin of life itself was understood as the fortuitous product of chemical recombinations.

The scientific consensus that developed from the time of Newton forward by the holistic effect of these theories as a unified system created a formidable bastion that seemed impregnable. The idea that

God had created our universe was relegated to the realm of mythology and outside the jurisdiction of the scientific process.

The New Theology provided an escape for those who wished to have faith, which did not need to correlate with the dominant scientific paradigm of the day. Those who rejected the scientific model proposed by the naturalistic paradigm were considered irrational, ignorant, and superstitious. The naturalist, armed with such scientific armaments provided by the combination of these theories, believed that the machinelike universe was self-perpetuating, self-reliant, and eternal. No God was necessary to explain any part of our universe.

They not only believed that the universe had existed from eternity past, but that it inhabited a three-dimensional space that was infinite in size. As recently as the middle of the twentieth century, naturalists believed that somehow, inanimate matter had existed for eternity with its intrinsic order and complexity. There was no need to consider origins.

They simply accepted that the laws of physics that govern our universe have done so for eternity. The machinelike universe needed no divine help. It was a perpetual machine. One of the key elements used to promote this view was Newton's law of universal gravitation. Stars and planets were not moving by the finger of God, but because of gravity. God was no longer necessary to explain how the universe functioned. The light of science illuminated the darkness of superstitions. There was no gray area anymore; everything was black and white with science, but this was a simplistic boast that would prove to be erroneous.

We can see an example of this naïve and arrogant position in Newton's intellectual successor, French physicist Pierre-Simon, Marquis de Laplace, who was convinced that through calculus and the laws of physics, one could know everything about the future and the past in the natural world. Science could provide the knowledge of everything, given that the universe was seen as simply a machine.

An intelligence knowing all the forces acting in nature at a given instant, as well as the momentary positions of all things in the universe, . . . would be able to comprehend in one single formula the motions of the largest bodies as well as the lightest atoms in the world. . . . To it nothing would be uncertain, the future as well as the past would be present to its eyes (Guillen 121).

But this simplistic view of the universe was on a collision course with reality. Newton's incredible insight that a planetary body is ruled by the same laws of gravity that an object on earth experiences was an extraordinary accomplishment; nevertheless, the very nature of gravity remained a mystery to him, and he sensed that it was not yet completely accurate in explaining the phenomenon. Although it could be used to make highly accurate predictions about planetary movements, the very mechanism of gravity in his theory meant that objects millions of miles apart would have an instantaneous effect on each another. He found that disturbingly unlikely. Things were not as black and white as many of the naturalists supposed and boasted.

It is inconceivable, that inanimate brute matter, should, without the mediation of something else, which is not material, operate upon and affect other matter without mutual contact, as it must be, if gravitation, in the sense of Epicurus, be essential and inherent in it. And this is one reason why I desired you would not ascribe innate gravity to me. That gravity should be innate, inherent, and essential to matter so that one body may act upon another at a distance through a vacuum, without the mediation of any thing else, by and through which their action and force may be conveyed from one to another, is to

> me so great an absurdity, that I believe no man, who
> has in philosophical matters a competent faculty
> of thinking, can ever fall into it. Gravity must be
> caused by an agent acting constantly according to
> certain laws; but whether this agent be material or
> immaterial, I have left to the consideration of my
> readers (Newton 634).

According to Newton's law of universal gravitation, a celestial body exerts a gravitational pull on another body with the strength determined by the mass of the two objects involved and the distance that separates them. For example, if the sun were to suddenly explode and cease to exist (change in mass), Earth would instantaneously feel a change in the gravitational mutual attraction. Earth is roughly 93 million miles from the sun. Yet, according to Newton's law of universal gravitation, the change in mass would incite an immediate response from Earth's point of view. That would require gravity to act upon Earth faster than the speed of light.

And then along came Einstein with his theory of special relativity. He showed us mathematically that no signal could outrun the speed of light. Nothing in this entire universe runs faster than a photon. The loss of mass from the sun could not affect Earth instantaneously. Hence, he sought a new theory of gravity and came up with the general theory of relativity. Gravity was explained by the contortions of space-time, which created a force that channeled the objects such as planets upon a specific trajectory. Space-time was now understood as a fabric that could be molded into shapes that affected the trajectory of physical objects interacting with it. Instead of an attractive force, gravity was seen as a force that made space denser and pushed down on matter to bring two bodies closer together. Einstein's equations showed that gravity from the mass of matter warps space and dilates time so that all three (space, time, and matter/energy) are interacting and intertwined in a continuum.

In addition, the equations proved mathematically that the universe was dynamic and not static. The universe had to be either expanding or contracting. In either case, the possibility of an infinite universe was irreversibly destroyed. In other words, the general theory of relativity forced upon scientists the mathematical conclusion that the universe began at a finite point in time. The first two assumptions of Lucretius were proved wrong. Not only was the universe created a finite time ago, but if it began at a finite point in space, it was also not infinite in size. The two major pillars of evolution were crushed in a single blow. The underpinnings of the naturalistic paradigm were no longer as black and white as they had believed and boasted.

CHAPTER 3

• • •

THE IMPACT OF EINSTEIN'S THEORY OF RELATIVITY

The old idea of an essentially unchanging universe that could have existed, and could continue to exist, forever was replaced by the notion of a dynamic, expanding universe that seemed to have begun a finite time ago, and that might end at a finite time in the future.

—Stephen Hawking

In the early part of the twentieth century, two major scientific accomplishments shattered the notion that our universe was eternal in existence and infinite in size. The first was Einstein's general theory of relativity, and the second was the discovery of the Doppler effect.

However, Einstein was not the first to conclude this from his equations. In fact, when Belgian physicist Georges Lemaître informed Einstein that his equations inferred a paradigm shift in the age of the universe by mathematically showing that it must have had a beginning, Einstein was not convinced. His answer to the young physicist was, "Your math is correct, but your physics is abominable" (Greene 12). This exchange took place in the 1927 Solvay Conference on Physics. Einstein had previously read Lemaître's theory and could find no faults with his mathematical manipulations of Einstein's equation, but he could not accept the idea that the universe was not eternal in time and space.

It was not the first time that Einstein had been presented with this idea. Earlier, in 1921, Alexander Friedmann had presented something similar.

> In 1921, the Russian mathematician and meteorologist Alexander Friedmann had come upon a variety of solutions to Einstein's equations in which space would stretch, causing the universe to expand. Einstein balked at those solutions, at first suggesting that Friedmann's calculations were marred by errors. In this, Einstein was wrong; he later retracted the claim. But Einstein refused to be mathematics' pawn. He bucked the equations in favor of his intuition about how the cosmos should be, his deep-seated belief that the universe was eternal and, on the largest of scales, fixed and unchanging. The universe, Einstein admonished Lemaître, is not now expanding and never was.

Six years later, in a seminar room in Mount Wilson Observatory in California. Einstein focused intently as Lemaître laid out a more detailed version of his theory that the universe began in a primordial flash and that the galaxies were burning embers floating on a swelling sea of space. When the seminar concluded, Einstein stood up and declared Lemaître's theory to be "the most beautiful and satisfactory explanation of creation to which I have ever listened." The world's most famous physicist had been persuaded to change his mind about one of the world's most challenging mysteries. While still largely unknown to the general public, Lemaître would come to be known among scientists as the father of the big bang (Greene 2011, 12–13).

But not all scientists came face-to-face with reality as quickly as Einstein did. Most scientists continued to balk at the solutions of the equations as they held on to their cherished naturalistic bias for an eternal and static universe. The nail in the coffin of a static universe came with the discovery of the Doppler effect.

What is the Doppler effect? It is the measuring of wavelengths to determine if an object is moving toward you or away from you. But before we can understand the Doppler effect, we must know a little bit about the electromagnetic spectrum and its corresponding wavelengths.

Much earlier, Galileo had discovered that light going through a prism divided into the color spectrum. The colors corresponding to the shortest visible wavelengths (blue) were divided in one end of this rainbow spectrum. The colors with the longest visible wavelengths (red) were seen on the opposite end, creating the familiar rainbow pattern. But this visible portion of the entire spectrum is only a small part (one-third, to be exact) of the electromagnetic spectrum. There are three invisible long waves beyond red, and there are three invisible short waves before blue. All nine of these electromagnetic waves are created by photons. (Notice again the three patterns of three, which speak of prevision and design and not of random ordering.)

At the long wavelength end of the spectrum, we have, for example, radio waves, which are invisible to the human eye. These are followed by microwaves and then infrared waves before coming to the visible spectrum where red is the longest visible ray. Yellow lies in the middle, and at the other end, the shortest wavelength we can see is blue light. This is followed by invisible ultraviolet light, then x-rays, and finally the shortest wavelength, the gamma rays. These pack a bigger punch in energy, because the photons are more energetic, and there are more waves packed together (higher frequency).

Hence, we see that the electromagnetic spectrum forms another symmetrical pattern of three threes. There are three invisible electromagnetic rays in the long wavelength. There are three invisible electromagnetic rays in the short wavelength. And there are three visible wavelengths in the middle of the spectrum. All of humanity possesses trichromatic vision. In other words, the entire rainbow of colors visible to humanity is merely a combination of three primary colors of light: red, green, and blue. Once again, God's creation leaves behind His fingerprint for all with eyes to see—literally. Artists would tell you that the three primary colors are magenta, yellow, and cyan, but that is because with pigments, the colors are subtracted; whereas with light, they are added.

You may think that these rays are composed of different particles, since their appearance and their impact are so vastly different; after all, radio waves are harmless to us, but gamma rays are deadly. Yet this is not so. All electromagnetic waves are, in fact, streams of photons (massless particles) traveling in wavelike patterns at the speed of light. The higher the frequency of the waves, the more powerful and energetic are the photons. Hence, the shorter the wavelength, the more harm it does to our cells.

Let's return to the Doppler effect. If you have ever visited a racetrack, then you have noticed that as the vehicles approach, the sound of the motors seems to change in pitch. The closer the race car comes as it approaches the observer, the higher the pitch of the sound of the motor. And conversely, the further the car drives away from the observer, the lower the pitch becomes. This is known as the Doppler effect. As the object is moving toward the observer, the wavelengths of sound are pushed together into a higher frequency. When they are moving away, the wavelengths are pulled apart and lengthened, thus sounding at a lower pitch. This phenomenon applies to light waves as well as sound waves. As a matter of fact, it also applies to all electromagnetic waves. By studying the color of the light of stars, we can determine if they are lengthening or moving away (red-shifted) or if the light waves are shortening or coming toward us (blue-shifted).

In the 1920s, astronomers were studying the spectra of light in the stars when they began to notice that almost all stars were shifted to the red spectra (longer wave compared to the blue-shifted shorter wave), signifying that they were actually moving away from us. This red-shift analysis revolutionized astronomers' understanding of our universe. The universe was not static. It was now understood that the universe was both dynamic and finite.

Einstein's equation told us that the universe had a beginning, and the observation of the Doppler effect on stars told us that stars were almost exclusively moving away from us. Obviously, all stars were moving away from a central point, a point of origin, the point of the Genesis Singularity. If the universe were infinite in size and time, then we would expect to see either no shift to either end of the spectrum if it were static, or an equal number of red-shifted stars and blue-shifted stars if it were dynamic and infinite.

Scientists had fully expected the universe to have roughly the same number of blue-shifted stars as red-shifted stars, which would have been consistent with an eternal universe governed by randomness. Instead, the vast majority of stars were found to be red-shifted, signifying that almost all stars are moving away from us. And only a minute number of them are moving toward us.

In other words, all stars are moving away from a precise area in space, presumably the area where our universe began. And moreover, that point happens to be quite close to where we exist in the universe. The fact that almost all stars are red-shifted indicates that the universe is expanding from a starting point, somewhere close to our region of space in the universe (although this is still contested by naturalists, as we will discuss later). Nevertheless, fundamentally this means that the universe is neither eternal in existence nor infinite in size. In other words, modern scientists were forced to face the fact that the universe had a beginning and was finite in size. This mathematical revelation led to the eventual acceptance of the Big Bang theory as an explanation for our origins.

Two of the dogmatic scientific claims fundamental to the Enlightenment worldview regarding the origin of our universe were dramatically exposed as inconsistent with empirical science. The universe had a beginning, and it was finite in size. The biblical Genesis account was once again vindicated by science when the truth of reality was finally understood. But even more surprising is the fact that this red-shift shows us that the further the stars or galaxies are from us, the faster they are moving away from us. This is counterintuitive to the scientific presupposition that assumed that energy would dissipate with time as gravity dragged the stars to a halt. So unlike a normal explosion, the universe, instead of slowing down at its edge, is actually speeding up in direct proportion to its distance from us.

The fundamental base of the gradualist worldview is that everything in this universe is guided by random ordering. A randomly ordered universe would therefore predict randomness in the motions of stars, but that is not what the mathematical conclusions of special relativity showed, and it is not what observational empirical data showed.

> *At that time people expected the galaxies to be moving around quite randomly, and so expected to find as many blue shifted spectra as red shifted ones. It was quite a surprise therefore to find that most galaxies appeared red-shifted: nearly all were moving away from us! More surprising still was the finding that Hubbell published in 1929: even the size of a galaxy's red- shift is not random, but is directly proportional to the galaxy's distance from us. Or in other words, the farther a galaxy is, the faster it is moving away! And that meant that the universe could not be static, as everyone previously had thought, but is in fact expanding; the distance between the galaxies is growing all the time (Hawking 41).*

The evolutionary presupposition of randomness received another lethal blow. Initially, scientists had suspected that the farthest stars would be moving more slowly, as the pull of gravity would have caused them to eventually slow down through time. Once again, naturalistic science tried to come up with a scientific explanation that could do away with the Genesis account. Naturalists theorized that if the mass of the universe were great enough, eventually it would cause the universe to implode. The aesthetics of this proposition, from the naturalist point of view, was that one could hypothesize an oscillating universe and once again do away with special creation.

But this was found to also be in direct contradiction to observable data. If the stars are speeding up the farther they are, then it is hard to conceive that the universe could in any way implode. The fact that the farthest stars are moving faster than the closer stars was quite puzzling. In 1929, Edwin Hubble showed precisely that the speeds of these stars were actually increasing as they moved away from us, in exact proportion to their distance from us.

If the stars would have each been propelled at different speeds and varying directions, as would be expected in a randomly ordered universe in the initial explosion, then they would be moving at different speeds, varying among each other as they expanded outward. But the speeds were exactly uniform, relative to their distance. It is as if the very fabric of the space-time continuum were elastic and expanding and everything caught within it were expanding proportionally. The stars furthest away from us were therefore increasing in speed, as they ventured outward. In other words, they were accelerating. For an object to accelerate, it must have some kind of force acting upon it. What kind of energy could counter the drag of gravity that must have been gargantuan when all matter existed in a singularity and cause such acceleration for such a long time?

It is an unassailable fact that the space-time fabric of our universe is expanding in the matrix of hyperspace outside of it, whatever that is. No one can truly describe hyperspace outside our universe. Nevertheless, it implies an open reality rather than a closed reality, for our universe is within something other. That means conclusively that we live in a dynamic universe, not a static universe, and it implies an open system and not a closed system. Therefore, if one extrapolates backward in space-time, one will come to a primordial point, a beginning time and place for the universe, the much despised moment of creation that Enlightenment naturalists tried so hard to eradicate.

The Steady State dogma that provided the eons of years, which had been adamantly defended by evolutionists and which provided the very basis for the theory of evolution to be plausible, was now thoroughly discredited. The fact is that the red-shift discovery was not really necessary to have disproved this theory. Simple, common logic should have illuminated these evolutionary scientists. Let's back up a few years and examine the subjective bias of naturalistic science.

Common logic would dictate that if the number of stars were infinite and if they existed for all eternity, then the night sky should have been completely lit up a long time ago, with not an inch of darkness anywhere. Every inch of the night sky should have been as bright as the noonday sun. Light traveling from the infinite number of stars, which would have also been traveling for all eternity, should have from our planet's beginning arrived at our planet and filled every square inch of the sky long ago.

In other words, if the number of stars is infinite and the universe is also infinite, then the stars that are near enough to our solar system to be seen, as well as those in the deepest parts of the universe, would have been giving off light from their eternal past. And this light would have arrived at Earth and completely filled the night sky with their light from eternity's past. Our night sky would be so thick with visible stars that not one square inch of sky would be dark. Nighttime would be daylight.

As far back as 1576, Thomas Digges presented this seeming paradox. In 1715, the renowned Edmund Halley once again proposed this unexplainable paradox. Some time later, P. L. de Cheseaux in 1744 and Heinrich Olbers in 1823 came up with a hypothesis that seemed to resolve the problem at first glance. They theorized that there might be an interstellar medium, a dark matter, that could absorb this light and block it from view.

In the 1930s, Robert Trumpler seemingly provided the final nail in the coffin to this bothersome paradox for naturalists with the discovery of dark cloudlike structures that seemed to exist next to dense clusters of stars. Now a curious thing develops here; in spite of the fact that Josef Stefan's experiments in 1879 proved conclusively that this medium, or dark matter, must, in the process of absorption, reach a temperature at which the matter must radiate as much as the light it receives, modern scientists totally ignored Stefan's verifiable mathematical facts and doggedly held on to the dark matter concept to explain the paradox. They did so because they were predisposed to believe that the universe was eternal as a necessary component to their evolutionary hypothesis. It must be noted that the dark matter envisioned by the evolutionists at that time is not the same as the modern concept of dark matter.

Nevertheless, the evolutionists simply refused to accept Stefan's argument so they could avoid the paradox. Then along came Einstein, and the implications of the general theory of relativity showed mathematically that the universe must have had a beginning. It was not until 1960, over a half a century later, that mainstream scientists finally conceded that Halley's objections were not answered by this dark matter theory. Slowly and grudgingly, the implications of the general theory of relativity began to sink in to the emotional side of these naturalistic scientists, allowing them to face the brute facts.

It is a universal verity that old dogmas die slowly. Although they knew theoretically that, without question, Einstein's theory proved that the universe must be either slowing down or speeding up and, more importantly, that it had to have had at some point a beginning, they were unwilling to concede it as fact for more than half a century. No doubt, many hoped that some other explanation would spare them from the obvious inference that the universe must have had a beginning. Their cherished dogma that trivialized the Genesis account had been pulled out from under their feet like a cheap rug.

Einstein's equation proved that the fundamental assumptions of both Lucretius and Kant were utterly false. The universe was not eternal in time or space. Modern scientists reluctantly grappled with the startling notion that the universe, the space-time continuum, was irrefutably finite in time and in expanse. The idea that the universe had a beginning brought to center stage the repugnant idea that it must have had a "starter."

Some, like Einstein, gradually accepted the fact that a God had to exist. But others have refused to believe in a creator and have simply taken a leap of faith into the unverifiable and unsubstantiated in order to remain in the unfounded presupposition that they have chosen through their subjective biases.

The Universe Began at a Finite Time Ago

Sadly, it is ironic that scientists today are guilty of the same narrow-mindedness that afflicted the clergy of the Mother Church, which persecuted scientists during the Renaissance. In spite of the known facts, they hold tenaciously to their failed hypothesis, exhibiting the same dogma and narrow-minded subjectivity, which results in folly and ignorance. How sad! The same mindset that led people to believe the world was flat still exists.

The late Stephen Hawking, in his usual epigrammatical style, plainly stated with clarity the obvious conclusion, which many have tried so desperately to avoid:

> *The old idea of an essentially unchanging universe that could have existed, and could continue to exist, forever was replaced by the notion of a dynamic, expanding universe that seemed to have begun a finite time ago, and that might end at a finite time in the future (Hawking 35).*

Hawking, at that time, was working with another British physicist named Roger Penrose to try to understand black holes through the mathematics of Einstein's general theory of relativity. Many of their colleagues, who wished desperately to remain mired in the old paradigm, provided them with some stiff opposition. Speaking of his work with Penrose, his collaborator, Hawking stated:

> There was a lot of opposition to our work, partly from the Russians because of their Marxist beliefs in scientific determinism, and partly from people who felt that the whole idea of singularities was repugnant and spoiled the beauty of Einstein's theory. However, one cannot really argue with a mathematical theorem. So in the end our work became general-

ly accepted and nowadays nearly everyone assumes that the universe started with a big bang singularity. . . . As experimental and theoretical evidence mounted, it became more and more clear that the universe must have had a beginning in time, until in 1970 this was finally proved by Penrose and myself, on the basis of Einstein's general theory of relativity. *That proof showed that general relativity is only an incomplete theory: it cannot tell us how the universe started off, because it predicts that all physical theories, including itself, break down at the beginning of the universe* (emphasis added) *(Hawking 53–54).*

I believe that God has a sense of humor. He leaves His fingerprint on the universe for those who seek the truth. Mathematics shows us that at the precise moment of creation, no law of the universe or naturalistic cosmological theory can stand; everything breaks down. What that means to a mathematician is that the answers turn to infinities—the equations break down. Yet it is at this exact moment that all the parameters that govern the development of the universe are set, as if to plainly show us that only that which is above the energy and mass and space of the universe could control such an event. But the naturalists continue to insist that the king has clothes, in spite of the observable data. They cannot avoid the Big Bang but now insist that the universe was birthed infinite. We will examine this in more detail later.

This Big Bang, however, posed another formidable problem for naturalists. How could such an explosion create a universe that is homogeneous? The isotropic and uniform homogeneity of our universe is evidence of the creator's influence during its developmental phases. Although brave attempts have been made to explain this conundrum (the uniformly isotropic and homogeneous state in our present universe), no random processes have yet been shown to successfully execute the isotropic and uniform homogeneity observed in our universe. Their only candidate to date is the Rapid Inflationary Model, which we will address later. But let us first understand what scientists mean by this isotropic homogeneity that permeates our entire universe.

Interestingly, in 1922, several years before Hubbell's discovery, Russian physicist Alexander Friedmann predicted exactly what Hubble observed. Friedmann made two very simple assumptions about the universe. First, he observed that the universe looks identical in whichever direction we look. And second, he reasoned that if we were observing the universe from anywhere else in the world, this would also remain true. This uniformity could not be explained in a randomly constructed universe; therefore, the universe, he reasoned, could not be static.

Another important piece of evidence, which was accidentally discovered, lent even further proof to the homogenous nature of our uniform universe. In 1965, physicists working for Bell Telephone Laboratories accidentally stumbled on microwave radiation coming from space. You will remember that microwave radiation is one of the electromagnetic waves created by photons. American astronomists Arno Penzias and Robert Woodrow Wilson showed that microwave radiation is roughly the same temperature in every direction in which they turned their instruments into space.

That was an amazing physical confirmation that proved Friedmann right. Background radiation is believed to be the leftover radiation from the Big Bang singularity. It is believed to be the leftover energy of the annihilation of matter when it comes in contact with antimatter. That temperature in the black of space, the leftover microwave radiation, is uniformly 2.73 degrees Kelvin, which is approximately minus 272 degrees Celsius.

Here is the conundrum: The universe could not have been absolutely homogeneous in nature during the process of the explosion of the Big Bang. Some variation must have existed during the explosion; otherwise, galaxies could not have formed. But if there were too much variation, we would have a universe filled with black holes. How that explosion managed to find that absolutely small parameter in which the difference allows for the formation of stars and galaxies through random processes is so remotely probable that, for all practical purposes, it must be considered an impossibility.

We know that there are very minute variations because we have been able to measure them through a magnificently engineered satellite, the Wilkinson Microwave Anisotropy Probe (WMAP), which has on board extremely sophisticated and highly sensitive machinery that has enabled us to map the exact radiation density throughout the universe.

According to NASA, the differences in temperature concentrations are less than one part in 10,000 (0.00003 degrees Celsius). It is impossible to explain how a random explosion could achieve such a homogeneous expansion with such minute levels of fluctuations. How does the naturalist explain through pure random processes the accomplishment of this feat without any external guidance? It is statistically nothing less than a miracle of the first order. What external force could guide the explosion of the Big Bang in such a manner as to create this amazingly homogeneous universe? There is more to the story of the WMAP that was not publicly revealed regarding the mapping of our universe, but I will deal with that later. Nevertheless, for most of the universe, the homogeneous nature of the microwave background radiation is a remarkable feature that, once again, refutes a randomly guided explosion.

Once more, randomness had been given a severe blow by the observational data. The naturalists were backed into a corner. The race was on to create a mechanism that could explain this anomaly through random ordering. Friedmann had stipulated through his calculations that the speed at which the galaxies would be moving apart from one another was proportional to the distance between them. And that is exactly what Hubbell observed.

As a result of all these findings, there were three possible models for the Big Bang theory proposed by the naturalists:

1. The first was a model in which the universe is expanding sufficiently slowly so the gravitational pull between the expanding galaxies is able to slow it down enough to eventually stop it. If the force were great enough, it would then cause the galaxies to retract or contract as the gravitational pull overtakes the expanding force. The result of this implosion is the opposite of the Big Bang, namely the Big Crunch.

2. In the second model, the universe is expanding so rapidly that the gravitational pull is insufficient to stop it, although it does slow it down somewhat.

3. The third model envisions the universe expanding at a critical rate, only just fast enough to avoid recollapse.

The first model requires that space be bent on itself, like the surface of a planet (a sphere), causing it to be finite in extent. (According to this model, the interior of the sphere would not be considered part of the universe, only the surface.) The sphere could expand in diameter infinitely, but it would remain connected end-to-end, having finite form. However, those trapped within that spherical universe would observe it as infinite because they could travel around it for infinity, eventually returning to the place where they had departed.

The second model requires that space be bent the other way, such as the surface of a saddle, so the space is therefore able to expand to infinity. In the third model, space would be flat, and therefore, its expanse could also theoretically continue to infinity; however, those trapped inside would not be able to travel through it for infinity. If they could reach the edge of the universe, there would be a finite dimension to the universe.

Reason dictates that if there is a beginning, as proved by the general theory of relativity, then the universe began at a finite point in time and in a finite size; hence, reason dictates that it cannot be infinite in size within a finite range of time. There is no such thing as a singularity that is infinite in size. By definition, a singularity has a finite size and boundaries. And although most physicists today continue to prefer to think of our universe as infinite in space, there is absolutely no shred of evidence that proves or even demands that our universe must be infinite. It is only demanded by the metaphysical aversion they have to God as the creator. Some more honest naturalists admit that a finite universe is possible, but most simply state unequivocally that it is infinite because of their underlying metaphysical choice.

> *We don't know whether the universe is finite or infinite. The previous sections have laid out the case that both possibilities naturally emerge from our theoretical studies, and that both possibilities are consistent with the most refined astrophysical measurements and observations (Greene 2011, 29).*

According to Hawking, at the time he wrote *A Brief History of Time*, the initial evidence seemed to favor the first model. If the first model were correct, then that would mean that if a space traveler were able to travel in one direction, he or she would eventually reach, in a finite amount of time, the starting point. That is a remarkable feature.

But new evidence brought back from the WMAP seems to point out definitively that the universe is flat and accelerating at the edge, so not only could the space traveler not return to the beginning by traveling in a straight line, but the universe cannot recollapse. In fact, our stars and galaxies will not last for all eternity either. Cosmologists theorize that at some point, our universe will stretch so far that the fabric of galaxies, suns, and planets will be literally torn apart. Even molecules will fly apart, and our universe will end in a cold silence. That is called *heat death*. But I do not think that the ripping of the space-time fabric will result in a dead and cold universe, but rather it will burn up by the violence of giant quasars and their deadly beams of electromagnetic energy that will incinerate everything they do not consume into their black holes.

But we are getting ahead of ourselves. Let us return to the beginning.

CHAPTER 4

THE BIG BANG

Evolutionists have now been forced to grudgingly concede that, indeed, the universe had a starting point: possibly the Big Bang. The universe, much to their consternation, is therefore not eternal. Anything that has a beginning has an end. Hence, not only did our universe begin at a given point in time, but it will also have an end. Moreover, our universe has boundaries. It is finite. How does the finite bring forth itself from nothing? The implications of the general theory of relativity did not make the naturalists happy campers. Their long-cherished Steady State theory that falsely assumed the universe was infinite in space and time came crashing down like a lead balloon. Using Einstein's law of general relativity, many began to explore what the beginning could have been like.

Since Einstein's revelation, naturalists have posited that our universe began with a Big Bang. All that exists in our universe was, at the very beginning, believed to be concentrated in a single point of almost zero space within our visible, three-in-one, spatial, and time dimensions containing infinite density. That is called the *singularity* (for lack of a better word to explain it). The mother of all explosions ever then brought forth our universe as energy exploded outward stretching our visible spatial dimensions.

Not only did naturalists lose their cherished hope for a static, steady state universe, but their uniformitarian/gradualistic model was also broadsided since the universe was understood to begin in the most catastrophic explosion imaginable. It was not a long, slow, gradual, and almost imperceptible process (always the preferred evolutionary time scale), but an almost instantaneous explosion of such magnitude that our finite minds have trouble wrapping around it.

As stated earlier, evidence for the Big Bang theory comes from the pervasive cosmic background radiation (CBR). The detection of an extremely uniform radiation in every sector of our universe, whose temperature is measured at 2.73 Kelvin, is thought to be the leftover radiation from this mother of all explosions. In other words, that energy in the black of space is the afterglow of the explosion at the creation point. But the most amazing characteristic of this afterglow is that it is almost absolutely evenly distributed throughout the cosmos. That is, the texture of our energy gradient in the Big Bang was unimaginably smooth and not lumpy as would be expected in a randomly ordered explosion.

Somehow, an enormous unimaginable amount of energy, highly concentrated into infinite density, appeared in a finite space of such a very small size that it was almost zero volume in our visible dimensions. This was the Alpha Point. There was no matter yet in the universe at the Alpha Point; all was radiant energy. Radiant energy exists in a state that can be described as an "eternal now." Radiant energy is not trapped within the framework of time such as matter and space. According to the mathematics of the Big Bang, approximately 0.00001 seconds after the Alpha Point, that radiant energy turned into matter. Light began to create matter. Scientists call this the moment of quark confinement.

It might surprise you to find out that all matter has come from light. Science did not know about this until Einstein came along. But the biblical Genesis record, written more than 3,000 years ago, has uniquely stated that God's voice was the radiant energy that created the universe. Energy was turned into matter and the space-time dimensions we call our universe. This is not just some theoretical hypothesis. Modern scientists have not only turned matter into energy, as in atomic bombs, but they also have routinely turned light into matter.

> *In principle, light can be turned into ordinary matter—even flesh and blood.*
>
> *These two possibilities are not merely theoretical. Nowadays, scientists can and do make them happen repeatedly and without much effort. The two processes are called "pair annihilation" and "pair creation."*
>
> *In pair annihilation, an electron collides head on with a positron, its antiparticle. Result? The two particles—the two tardyons—annihilate and become light. In pair creation, light collides with light. Result? The light disappears and materializes in the form of two tardyons—an electron and positron.*
>
> *Taken together, the two discoveries science has made about light are truly extraordinary. The first—that light is in a realm by itself—is amazing enough. But the second—that light and ordinary matter are somehow interchangeable—is positively mind-blowing (Guillen 76).*

And so the moment that radiant energy turned to matter at the beginning of our universe, the temperature was approximately a million million times hotter than the afterglow of 2.73 degrees Kelvin we find today in the black of space. That means that since that first moment in time, our universe has cooled down and expanded a million million times from that Alpha Point, and yet the energy remains so homogeneous that there is only a 0.00003 degree Celsius difference that can be measured. That is absolutely incongruous with the naturalistic explanation of a randomly ordered reality. That the most powerful explosion in the entire universe should result in such ordered compliance is in direct contradiction to the naturalist hypothesis that randomness is the matrix of reality.

We can mathematically move backward in time up to that dramatic moment when it took place only 0.00001 second after the Alpha Point. But as we move forward in time, there is, however, little mathematical evidence that can substantiate the claim that the Big Bang could produce highly concentrated or rotating bodies, if it was a linear explosion without any external influence to cause it to spin.

> Galaxy rotation and how it got started is one of the great mysteries of astrophysics. In a Big Bang universe, linear motions are easy to explain: They result from the bang. But, what started the rotary motions? (Corliss 177).

The answer to that question may be explained if our initial Big Bang began the universe with a spin; that is, if the Genesis Singularity was purposefully spun by an outside source (God) at an exact accelerating rate from the moment of creation. We will return to that later. For now, whether or not the universe began in a Big Bang, the point is that it had to have a beginning. Naturalists were forced to face the dilemma of a beginning, because it has been mathematically shown to be an inescapable reality. The hard, empirical, scientific data show conclusively that the universe did, in fact, have a beginning. Our material universe began at precisely 0.0001 second after the Alpha Point when light was turned to matter in the form of quarks. From the point of view of our material world, time began at that moment.

However, from the point of view of the fabric of space, time began at the Alpha Point because space had to exist in order to contain the timeless radiant energy that exploded outward in a hot ball of plasma. It was at the Alpha Point that space-time came to be. There can be no matter without both a spatial dimension and time existing.

Reality initially existed in the form of timeless radiant energy in space-time prior to the condensation of physical matter. Not until the moment of quark confinement can we say that matter in its simplest form existed. Radiant energy is not bound by time. But Einstein taught us that space and time are intricately intertwined. We cannot have space without time, and we cannot have time without space. Time is, in fact, a property or function of space. Hence, the logical conclusion is that without an external input from a timeless source that facilitated the formation of the Alpha Point, there could not have been the radiant energy inside space-time that at the moment of quark confinement turned into matter. Something outside of our universe had to create our universe. Nothing inside our universe could cause itself to be created.

What naturalistic cause from a timeless and spaceless source can even be imagined that possessed that gargantuan radiant energy? And if naturalists now offer any materialistic explanation that requires reality to also exist outside our universe, then our universe is not a closed system, as they have so ardently pontificated for so long, when they ignorantly thought our universe was infinite.

If there is a beginning, then what is before the beginning? Technically, there is no before, because time was created in the beginning. Since all data, as we know it, are tied to time and place and there was no time and place before the beginning, modern scientists say that we cannot delve into that point. Much to-do is made about trying to evade their glaring problem.

We are all used to the idea of an absolute zero of temperature. It is impossible to cool anything below

−273.16° C, not because it is too hard or because no one has thought of a sufficiently clever refrigerator, but because temperatures lower than absolute zero just have no meaning—we cannot have less heat than no heat at all. In the same way, we may have to get used to the idea of an absolute zero of time—a moment in the past beyond which it is in principle impossible to trace any chain of cause and effect (Weinberg 149).

The problem with Steven Weinberg's argument is that it is an evasion that compares apples to oranges. When considering the meaning of a temperature below absolute zero, of course it is impossible to give an answer to that singular dimension when there is an absence of heat. But in the Big Bang, we are beginning not with zero energy and zero space; when the clock starts, there is a singularity with so much energy that our human minds can hardly wrap around it, and it is all crammed into a small, bounded spatial area we call a singularity. We are not trying to explain the absence of something; we are trying to explain the appearance of everything that is now contained in the universe, which is quite the opposite. The only thing absent is time. The logical conclusion, then, must be that all the energy contained in the Genesis Singularity must have originated from a timeless source, a source not trapped in space-time. What other rational conclusion can we make?

We say that the Big Bang occurred and not until 0.0001 second did matter appear. Space also started at that Alpha Point, which contained or enveloped that massive amount of radiant energy. What naturalistic explanation can rationally explain how nothing created space-time? We cannot say or even imagine what would be outside of space-time. What we can say with confidence is that only a timeless and spaceless entity abounding in timeless radiant energy could have brought our universe into existence.

But don't let that evasion fool you; they still try to come up with a materialistic cause before the Big Bang. The standard explanation that naturalists give is that our universe popped into existence from the quantum vacuum. But this is nothing more than a clever semantic smokescreen. The idea that the quantum vacuum existed prior to the moment of the Big Bang has absolutely no empirical evidence to support it. It is a subjectively chosen scenario that rather poorly attempts to evade the creator. What is a quantum vacuum? Michael Guillen, former ABC News science editor and Harvard physics instructor, explains it this way:

> According to one favored scenario, the universe abruptly grew out of the so-called quantum vacuum, which by definition is what is left over after all matter and energy have been removed from space. How can that be? How can an entire cosmos spring forth from something that is void of all matter and energy?
>
> The theory only makes sense when we realize that according to modern science, the quantum vacuum is actually far from being nothing. Rather, it is filled with all sorts of exotica—quantum fields, and virtual particles—that can't be detected by ordinary means. This makes the quantum vacuum the mother of all piñatas, the mother of all wombs, and, so says modern science, the mother of our universe.
>
> As to what provoked the quantum vacuum to cough up an entire universe, some proposals are clearly meant to studiously avoid any idea of a creator (Guillen 89–90).

This is nothing more than a smokescreen thrown to gullible people who have no idea what quantum vacuum means. Quantum vacuum does not exist in the absence of space-time. Quantum fields cannot exist outside of time and space. Virtual particles that pop up from other invisible spatial dimensions into our visible spatial dimensions cannot exist outside those spatial dimensions. It is a term used to describe a peculiar property of space itself. Moreover, space cannot exist without time. Space cannot even exist without energy. Space is not a vacuum. It is a fabric with energy, fields, and density that can be stretched or crunched or warped. Not until the moment of the Big Bang did space and time exist; hence, no quantum tunneling could exist prior to the moment that the Genesis Singularity appeared from outside of space-time altogether.

Since we know that space is expanding, there can be no doubt that it also had a beginning. The idea that random ordering through sheer serendipity caused our universe to pop up from the quantum vacuum is nothing more than a sleight-of-hand attempt to evade the unavoidable conundrum that nothing could not be the mother of our universe. It is not reverting to the unknown. It is reverting to the known—that massive radiant energy that was squeezed into a singularity of infinite density came from outside our universe to begin our universe.

That moment between the Alpha Point and the quark confinement point can only be measured in time if it is in reference to the creator's act of creation at the Alpha Point. Remember that Einstein told us that when we ask what time, we must also ask in reference to what? Like black holes, this primordial singularity contained an event horizon. That is the region that, when crossed, the immense gravity of the singularity does not allow anything to ever escape, not even light. The more mass inside the singularity, the greater the distance of the critical circumference of the event horizon. The primordial singularity, the Genesis Singularity, must have had an enormously large primordial critical circumference.

We know that the exact period between the Alpha Point and the moment of quark confinement existed, but from our viewpoint on the other side of the event horizon of the primordial singularity, we cannot really measure time. From the point of view of our material world, time was not created until matter/energy crossed the primordial critical circumference of the Genesis Singularity, since all observations from outside the primordial critical circumference show time to be standing still. All events that occurred prior to the crossing of the primordial critical circumference of the Genesis Singularity appear to us as timeless. But God is not caught within that finite framework of space-time.

Moreover, the moment of the creation of space-time and radiant energy cannot be measured without that precise point in which the creator spoke it into existence. It can only be understood if there is a creator from which that moment can be referenced. When Weinberg declares a timetable beginning at the Alpha Point, he is unknowingly referring to the timetable from God's perspective, because only God could see inside the critical circumference of the Genesis Singularity.

In fact, whether naturalists admit it or not, they have dealt with things prior to time and space, simply by speaking of the singularity of the Big Bang. Can inanimate matter and undirected energy cause itself to begin out of nothing? The obvious logical answer is, "Of course not." How, then, can our universe be a closed system? The only possible explanation for the genesis of the beginning singularity has to, by definition, exist outside the physical boundaries of our material existence and our space-time continuum. That we do know. It is not retreating into the unknown; it is accepting the known.

But the schizophrenia of naturalists regarding our origins is alarming. How is it that scientists routinely speak of and believe that there is a hyperspace beyond our universe, even though we cannot describe it through time and space descriptions? And yet the notion that a creator exists is immediately shot down on the basis that we cannot describe anything outside our space-time continuum.

According to the most widely accepted model of the Big Bang theory, the entire substance of the universe occupied a small volume in space at an extremely high temperature and at such an immense density that it is labeled *infinite density*. How does our finite universe through some finite naturalistic system overcome infinite density? And yet, somehow, this phenomenon called the Genesis Singularity exploded and expanded, supposedly seeding our universe with matter and eventually developing into the present stars and galaxies. Cosmologists believe that the initial interaction between matter and antimatter created the heat or energy, which fueled this massive explosion. But the problem is that this explosion is a completely choreographed ballet that delicately tiptoed around landmines that would have ended the entire dance. That was no ordinary explosion ruled by random ordering.

Here is the bottom line: the fact that the Big Bang existed is proof that we are not in a closed system; otherwise, there would not have been a Big Bang at all, not even a poof. The Genesis Singularity came from somewhere outside our closed system. It did not create itself from nothing. The naturalist, by faith, accepts that nothing created everything. How is that logical? The only logical conclusion is that some force caused this. What must this force be like in order to accomplish this?

Obviously, this creative force must be powerful enough to cause this incredible amount of radiant energy contained in the Genesis Singularity to be formed. Such power is quite difficult to fathom. Imagine the release of the atomic energy of every single atom in the universe, and we can begin to appreciate the enormity of this radiant energy that was packed into an extremely small space that scientists call an almost zero volume.

But what is even more spectacular is that the force must have carefully controlled and calibrated the explosion in such a way so as to create an almost homogeneous universe. This entity must have surely been a super genius to be able to design and engineer such a feat. Slice it however you want; the answer can only be explained as the direct action of an enormously powerful and intelligent entity that stands outside our reality. This is the very definition of the Judeo-Christian God.

This, of course, has resulted in an enigma to the naturalist, an enigma that is unanswerable within the materialistic framework of their philosophy. Hence, they have, in a pitifully futile attempt to mask the conundrum of the genesis of our universe, completely avoided the subject and placed it outside the field of scientific inquiry.

Of course, this decision to place the origin of the Genesis Singularity outside the field of inquiry is completely subjective in nature and based on their presupposition that we live in a closed system. It does not occur to them that if the concept of a closed system leaves such a glaring astronomical deficiency in explaining observable reality, then perhaps they should switch their cherished presupposition to another that is capable of explaining our observable phenomena.

The Oscillating Universe

As we briefly mentioned previously, some naturalists were initially unwilling to accept the obvious inference of a beginning, since it forces upon them the logical conclusion that it had to have a starter. In order to avoid this necessity, they consequently proposed that the universe has always been exploding and reimploding. They envisioned an eternally oscillating cycle of the process of creation over and over again.

They theorized that if the mass of the universe were big enough, at some point, the expansion process would stop due to the drag of gravity, in which case the universe would then begin to implode back on itself. This implosion would then, theoretically, consequently result in another singularity that could begin the process all over again and thus complete one cycle of their proposed eternally oscillating scenario.

For the naturalist, the aesthetic element of this proposition is that it effectively avoids a beginning altogether. But, unfortunately for them, we have since found this to be mathematically impossible and in direct contradiction to the physical evidence observed. Physical observations prove exactly the opposite.

The universe is not slowing down at the edges. In fact, it seems to be speeding up toward the outer edge. The physical evidence of a universe that would be able to contract would stipulate that the outer fringes would begin to slow down as they reached further away from the central Genesis Singularity from whence they came.

Direct physical evidence clearly shows that the edge of our universe is, instead, expanding at an ever-accelerating pace. More importantly and conclusively, the total mass of the universe is just not big enough to counter the expansion process, even with the addition of the dark matter and the exotic dark energy now hypothetically theorized to exist in outer space.

Although most of the initial evidence had favored a spherical model for the shape of our universe, the new evidence collected by the WMAP satellite favors a universe that is flat and expanding forever. Launched in June 2001, this $145-million orbiting satellite was a joint venture of NASA and Princeton University, designed to make accurate measurements of the background radiation as the observatory orbits Earth one million miles above us. Measurements between faraway stars showed that our universe was Euclidian or flat. In other words, the lines between these stars were straight and not curved.

Earth, of course, is not flat, but it looks like the universe is, according to WMAP. The evidence further indicates that the universe could never collapse, as those proposing the oscillating scenario had hoped. At any rate, Hawking's calculations and the findings of WMAP seem to indicate that there is just not enough mass in the universe to cause a contraction. And if the stars and galaxies in the outer fringe of the universe are accelerating exponentially as they move further away, how could they collapse?

We can determine the present rate of expansion by measuring the velocities at which other galaxies are moving away from us, using the Doppler effect. This can be done very accurately. However, the distances to the galaxies are not very well known because we can only measure them indirectly. So all we know is that the universe is expanding by between 5 percent and 10 percent every thousand million years. However, our uncertainty about the present average density of the universe is even greater. If we add up the masses of all the stars that we can see in our galaxy and other galaxies, the total is less than one hundredth of the amount required to halt the expansion of the universe, even for the lowest estimate of the rate of expansion. Our galaxy and other galaxies, however, must contain a large amount of "dark matter" that we cannot see directly, but which we know must be there because of the influence of its gravitational attraction on the orbits of stars in the galaxies. Moreover, most galaxies are found in clusters, and we can similarly infer the presence of yet more dark matter in between the galaxies in these clusters by its effect on the motion of the galaxies.

When we add up all this dark matter, we still get only about one tenth of the amount required to halt the expansion (emphasis added) *(Hawking 48)*.

According to the naturalists, the findings of WMAP indicate that only 4 percent of the universe is composed of atoms; about 23 percent is "cold dark matter," about which we know very little; and 73 percent is composed of "exotic dark energy," about which we know even less. In other words, 96 percent of reality is invisible according to their new findings. Hence, another foundational scientific myth that formed the ideology of the Enlightenment, which considers all things not sensible as superstitious, has been toppled by these recent claims. It always amazes me how their cherished foundational doctrines are quickly swept aside when convenient for the rationalization of their cherished Copernican principle.

The WMAP satellite, however, has not discovered dark matter or dark energy. The idea of dark energy, which works in the opposite way of gravity as an antigravity force that gets stronger with distance, is nothing more than a mathematical fix to explain away the accelerating expansion of our universe. The idea of dark matter is also a mathematical fix to explain away the speed of the rotation of galaxies in the outskirts of our universe, which, under Newton's law of universal gravitation, would disintegrate by the centrifugal forces created at these extremely fast rotational rates.

They have intentionally ignored the discrepancy of time dilation created by the stretching of space-time, simply because they want to make the age of Earth the same as the age of the outskirts of our universe. Without these enormous spans of time, evolution has no chance, and gradualism is completely debunked. But that is not the only thing they have intentionally avoided.

In fact, what WMAP and two other satellites sent to map the universe have discovered is that there are regions of space, aligned with our solar system, that form distinct areas of anomaly to the otherwise uniformly homogeneous and isotropic nature of our universe. These dipoles, quadrupoles, and octupoles are areas of anisotropy (opposite of isotropy) where the density of the microwave radiation is less than the mean gradient throughout the entire universe. But even more fantastic, these are aligned orthogonically with the sun-Earth ecliptic and with the equinoxes of our planet as it rotates around the sun.

What? You have not read that in popular science periodicals? Hmm, I wonder why. Maybe it is because they have tried to throw this information under the table in order to keep it from the public.

But why? Simply because it contradicts their cherished Copernican principle and shows that our solar system has a specific design that cannot be created by random processes, and that additionally shows that our solar system is exactly at the center of the universe. In other words, these three satellites have discovered that the design of our universe indicates that our Earth-sun ecliptic is at the center of our universe. How could random forces have engineered such a magnificent feat?

We will speak more on this and document it in the section about the center of our universe, but for now, let us complete our conversation regarding the proposition of an oscillating universe. The fact remains that no matter how we do the math, we can with fair certainty rule out the possibility of our universe imploding. There is just not enough mass in the universe to generate the gravitational pull to bring it back or contract it. We also know that for sure there is not a mechanism from which the material in the universe could re-explode. Even if it did go through the Big Crunch, all indications are that it would simply become a giant black hole forever.

Thus, the idea of an eternally oscillating universe made popular by Carl Sagan's *Cosmos* film series, is just not possible. It is nothing more than wish-projection stemming from his pantheistic presupposition, which has as its underlying premise the notion that the universe is bound in eternally oscillating cycles. Sadly, the public education system continues to promote this pantheistic illusion that has been completely debunked by true science.

Hawking had predicted, even before the findings of WMAP were made public, that there simply was not enough mass in the universe (gravity) to cause an implosion. But even if it did implode, Hawking believed that the universe could not re-explode, because much of its energy would be used up and would therefore be irretrievable.

John Bahcall of the Institute for Advanced Study in Princeton, New Jersey, referring to the information recently garnered by the WMAP satellite, also seems to agree with Hawking's conclusion.

Charles Bennett of NASA's Goddard Space Flight Center also concurs, "The universe will expand forever. It will not turn back on itself and collapse in a great crunch" (Suh).

However, although the mass is not sufficient to cause it to implode, and the experimental evidence does not support this actuality, it was possible, through the use of the general theory of relativity, to deduct that the universe could have decelerated. This mathematical possibility is what led some to propose the oscillating universe idea before the recent empirical findings ruled out that possibility.

> An interesting physical consequence results from merely subtracting (or canceling out) equation 5.4 from 5.3. The remainder is $2(d^2R/ d\ t^2)/R = -8\pi G(p+3p/c^2)/3$ (5.5). Since, the constant of gravity is a positive value, this remaining equation states that the universe is decelerating (Ross 48).

In the past, there were those scientists with pantheistic leanings, such as the late Carl Sagan, who favored the decelerating universe idea with the famous oscillating scenario. That followed naturally because it fit neatly into their pantheistic idea of a cyclical history. And more importantly, it featured the added bonus that it conveniently avoided the repugnant idea that the universe had a definite starting point.

Ironically, it has become increasingly popular for former atheists to turn to pantheism as a sort of compromise, where they can admit the notion of some form of spirituality without serious content while remaining steadfast in their naturalist framework. At first glance, the naturalistic presupposition of pantheism matches well with their evolutionary posture.

It allows them a semblance of some spirituality in an otherwise stark, machine universe. There is still the denial of a personal God, which by necessity dictates an absolute standard for morality. But through Eastern mysticism, they can adorn the outside of the machine with spiritual clothes and create an illusion of personality to the machine. The clever use of these spiritual, connotative words provides some semblance of spirituality while still remaining fast within their naturalistic framework.

Their bias, whether conscious or subconscious, is accepted in order to assuage their angst resulting from the cold, naked reality of a universe void of personhood. At the same time, they are able to remain free of guilt in their choices, for they still deny a personal God that could hold them accountable to an absolute moral standard.

Central to the pantheistic theology is the concept of cycles. According to their philosophical framework, the universe is undergoing movement that is always cyclical. The oscillating universe, then, fits in rather nicely with this concept. If the universe cycles, then there is, again, no beginning. The cycles can be deemed eternal, and thus the notion of an initial creation is avoided.

By turning the Big Bang into an ever-oscillating scenario (the collapsing model that leads to the Big Crunch), naturalists can again avoid an encounter with the primal cause. New Age gurus such as the late Carl Sagan have promoted and popularized this view extensively. As a result, even though it has been scientifically discredited, many who are ignorantly unaware still popularly hold this view. And the television science programs still abound with this defunct proposition.

There are several factors that contrapose the oscillating factor:

> *While it has* long been acknowledged that no known physical mechanism can ever reverse (rebound) a cosmic contraction, more fundamental limitations on a "bounce" have recently been discovered. Novikov and Zel'dovich have pointed out that uniform isotropic compression becomes violently unstable near the end of the collapse phase, and the collapsing medium breaks up into fragments. This problem, along with

some straightforward thermodynamics recognized in the last couple of years, rule out any possibility for a bouncing universe.

As noted earlier, the universe, with specific entropy of about a billion, ranks as the most entropic phenomenon known. Thus, even if the universe contained sufficient mass to force an eventual collapse, that collapse would not *produce a bounce. Far too much of the energy of the universe is dissipated in unreclaimable form to fuel a bounce. Like a lump of wet clay falling on a carpet, the universe, if it did collapse, would go "splat"* (emphasis added) *(Ross 104–105).*

Furthermore, even if there could be a bounce, the beginning would still be traceable. Science has concluded that if the universe had undergone cyclical expansion and contraction, it would result in an ever-increasing radius. Such an expansion in radius could then be traced backward to a beginning. There is, therefore, no way of avoiding a beginning.

Many naturalists have thus been forced to concede the repugnant idea that the universe had a beginning. Of course, they continue to attempt to reconcile the inescapable reality that our universe had a definite starting point with their cherished naturalistic paradigm, attempting to explain the formation of the universe devoid of any external influence by a deity.

As we previously stated, a universe sustained and guided by nothing outside its boundaries and controlled completely by the random chemical processes governed by pure chance, within its boundaries, is called, in their terms, a *closed system*. Somehow, evolutionists claim that between this initial explosion and our present time, by random chance recombination, the right chain of chemical combinations occurred, and galaxies and planets were created.

They propose that with the passage of a great deal of time, eventually life began its epic evolutionary ascent, presently culminating in humans, its most sophisticated achievement to date. This, of course, is a highly condensed sketch of the events that the evolutionists believe to have transpired. However, it serves to bring out a very crucial contradiction that even the casual observer cannot escape: the incredible assertion, which evolutionists must accept by blind faith, that order comes from disorder, without any external engineering, and that the complexity of the material universe (including life) is literally the product of an explosion *without directed energy.*

But this runs completely contrary to the laws of physics and the natural order of things observed. The idea that pure random chemical processes, without externally directed energy, through the chance recombination of chemicals could have created our material universe (as well as the wide variety of living creatures found therein) is not only scientifically unsubstantiated but infinitely improbable from a statistical standpoint. Yet this, in a nutshell, is the foundational maxim of the naturalist worldview: hydrogen evolved into human beings. And their cherished Copernican principle that claims our planet and solar system do not have any special significance in their positions within our universe is a reflection of their presupposition that there is no overt design to existence, but only a simple accident of random ordering. Simply stated, this dogma does not come from empirical data, but from their a priori metaphysical choice.

This foundational maxim, however, runs counter to one of the very basic laws of physics that dictates precisely that things are always running down instead of becoming more ordered. The second law of thermodynamics states that all things go from a low state of entropy (low degree of randomness or disorder) to a high degree of entropy (a high degree of randomness and disorder). Yet the big daddy of explosions, the Big Bang (no higher entropy is or ever will be possible in the universe) has, according to naturalistic presupposition, supposedly resulted in our present sophisticated state of order and complexity that permeates every part of our universe. And all this magically transpired through purely random processes.

The logical scientific conclusion can only be that without a mechanism to direct or funnel energy and create order out of disorder, there can be no physical evolution of our universe. This directive mechanism must be able to calibrate all the countless parameters necessary to create a universe, which is physically sustainable and can support life. As we shall see, these parameters are so infinitesimally small, compared to the infinite possibilities available should the whole process be relegated to blind random chemical recombinations, that the sheer odds of this happening through blind chance are literally insurmountable.

Some evolutionists attempt to evade the matter by pointing to living processes that do seem to go against the second law of thermodynamics. But living processes have a mechanism for directing energy, without which it could not make order out of disorder. Without a living metabolic "motor," which directs energy to perform specific tasks, there can be no escalation in order. Even so, in the end, even living mechanisms eventually cease to function when the organism dies, thus succumbing inevitably to the second law of thermodynamics. All things, whether living or nonliving, deteriorate with time.

There are no nonliving mechanisms that can duplicate this function of directing energy to create order out of disorder. Naturalists have made brave attempts to substantiate this claim, hoping to alleviate their embarrassment in this gaping inconsistency with their basic presupposition, but their assertions are infinitesimally irrelevant when compared to the overwhelming evidence to the contrary.

Hence, the creative process of the formation of our universe must have had some force outside our universe as a mechanism to counter the second law of thermodynamics in order to create the stars, galaxies, planets, and living organisms contained therein. The intricate design, which pervades our universe on every level and in every object, could not have developed without some intelligent designer to conceive and engineer it.

The sticky problem for naturalists, as we have been discussing, is that the theory does not really answer the question of origins because it begins with something already there. But it does so with no explanation as to how it got there in the beginning. Furthermore, the naturalist, in a closed system, has no way of being able to explain how it could have appeared out of nothing.

Even without considering the complexity of the universe and the negligible chance of its development through the random chemical processes insisted on by naturalists, one has no intelligible way of explaining why it is there or even how it is there. For how does one explain the passage from non-being to being in a natural or materialistic closed framework?

Since none of us were there at the origin of our universe, we must work backward with observable and verifiable facts and construct the best possible model that explains all the contingencies observed. The scientist is unable to go backward beyond the singularity. All science fails at that point. Yet the enigma is that there must have been something; otherwise, we would have nothing now. Hugh Ross explained it well in his book, *The Fingerprint of God*. Ross is an astronomer who was involved in research of quasars and galaxies for several years as a postdoctoral fellow at the California Institute of Technology. He currently heads up an organization in Pasadena, California, dedicated to providing scientific data in regard to the Christian faith.

> *While general relativity implies an age for the universe vastly beyond 6,000 years, it also implies that there is, indeed, a creation date. Expansion, coupled with deceleration, indicates a universe that is exploding outward from a point. In fact, through the equations of*

general relativity, we can trace that "explosion" back-ward to its origin, an instant when the entire physical universe burst forth from a single point of infinite density. That instant when the universe originated from a point of no size at all is called the singularity. No scientific model, no application of the laws of physics, can describe what happens before it. Somehow, from beyond itself the universe came to be. *It began. It be-gan a limited time ago. It is finite not infinite.*

The implications can only be described as monu-mental. Atheism, Darwinism, and virtually all the "isms" emanating from the eighteenth to twentieth century philosophies are built upon the assumption, the incorrect assumption, that the universe is infinite. The singularity has brought us face to face with the cause—or Causer—beyond/behind/before the uni-verse and all that it contains, including itself (empha-sis added) *(Ross 49–50).*

From a philosophical standpoint and by strict definition, no other entity than a supreme and transcendent God could be placed behind the singularity of the primordial white hole. A white hole is the opposite of a black hole. In a black hole, everything is being sucked in through the event horizon by the immense gravity of the singularity inside. In a white hole, the singularity is expelling everything outside of the event horizon. The Big Bang was probably a ginormous white hole.

Only a supreme being not bound by time and place could exist prior to the primordial white hole singularity. No other entity could possibly have caused to exist out of nothing the largest singularity ever, containing all the matter in the universe. Furthermore, what other force could cause the singularity to explode and allow matter to escape that immense gravitational grip?

Some have ridiculed this, saying that Christians are resorting to a "God of the gaps" reasoning. That is, they claim that God fills the gaps where there is no human knowledge. Nothing could be further from the truth. It is because of our knowledge that we understand that nothing, other than God, could accomplish the task of bringing order out of disorder. A mechanism for directing energy, such as what we observe within the mitochondria of the cell, must exist before disorder can be channeled into order. A genesis without such a mechanism could create nothing more than chaos and an overwhelming number of black holes instead of galaxies. God cannot be crunched into a finite box to become an object within our universe.

> For the classical theological tradition, God is not a being in the world, one object, however supreme, among many. *The maker of the entire universe cannot be, himself, an item within the universe, and the one who is responsible for the nexus of causal relations in its entirety could never be a missing link in an ordinary scientific schema. Thomas Aquinas makes the decisive point when he says that God is not ens* summum *(highest being) but rather* ipsum esse *(the sheer act of being itself). God is neither a thing in the world, nor the sum total of existing things; he is instead the unconditioned cause of the conditioned universe, the reason why there is something rather than nothing. Accordingly, God is not some good thing, but Goodness itself; not some true object but Truth itself; not some beautiful reality, but Beauty itself* (emphasis added) *(Barron)*.

The fact that galaxies exist shows that random chemical processes could not have been the guiding force in their creation. The fact that black holes exist in the center of most galaxies is evidence that these galaxies are succumbing to the second law of thermodynamics. The universe is moving toward greater chaos and not the other way around. Black holes did not turn into galaxies. Galaxies have deteriorated into black holes. Physicist Jacob Bekenstein has convincingly shown that the larger the black holes, the larger their entropy, so we can see that the universe is increasing in entropy from its time of creation (Susskind 2008, 149).

In fact, the second law of thermodynamics is the scientific term for the biblical theological doctrine of the fall. It is my suspicion that prior to the fall in the garden, the second law of thermodynamics did not exist. Creation may have been accomplished when the second law of thermodynamics did not exist. Therefore, the second law of thermodynamics did not complicate the construction of our universe. It impacted our universe after its creation. Moreover, the day will come when a new heaven and a new earth will be built by God, in which, once again, the second law of thermodynamics will no longer exist, a time in which, "No longer will there be anything accursed" (Revelation 22:3).

Had our genesis been purely chaotic, as the naturalists presuppose, the galaxies would simply not have been formed. It is amusing when I hear scientists speak of black holes as essential elements in the formation of galaxies. This subjective wish projection unmasks their fundamentally biased worldview. They desire to make chaos appear as the guiding force of our universe. It is not. It is the end result of our eventual decay. Black holes are the ghosts of dead stars and galaxies.

Black holes in our universe are not shrinking. They are growing, which means that the vector of time in the second law of thermodynamics demands that all systems are gaining greater entropy with the passage of time. As black holes swallow matter and energy, their mass inside within the singularity increases, and consequently their entropy increases. The size of the radius of their event horizon thus increases in direct proportion to their mass. In other words, the more they eat, the larger and more powerful they grow. The greater the mass in the singularity, the stronger and further their gravitational pull will stretch outward into the universe.

Through the general relativity equations, scientists have been able to learn about the nature of black holes and singularities as they search to understand the very nature of the Genesis Singularity. Although originally many doubted it, most modern astrophysicists have now come to accept the fact that black holes exist. The observational data are irrefutable. Although we cannot see the black hole, sometimes (if it is feeding) we are able to see the material spinning wildly around it as it is consumed. The friction created as it spins faster and faster before being consumed makes it radiate light.

How do black holes form? Some black holes are the result of the death of a star of an enormous size (>20 suns). As the fuel in the star is depleted, in the process of imploding, matter in the very core of the star increases in density as it collapses. The core consequently increases in gravitational power, sucking everything around it and crunching it into an ever-increasing density as it continues to implode. Eventually, the concentration of matter into such a small space results in a pull of gravity that becomes so great that not even light can escape its grip. Therefore, to us, from an external observation point, it looks like a giant black hole.

The more mass crunched into that central point, the greater the size of the singularity and the wider the critical circumference around it, which prevents light from escaping. The critical circumference is that area in the circumference around the singularity where the gravitational pull becomes so great that everything is sucked into the singularity and nothing can escape its pull. It is, literally, the no-turning-back point.

The existence of black holes is not only predicted by the general relativity equations, but astronomers have also produced some convincing observational evidence pointing to the existence of these black holes in our universe. In the very center of Messier 87, astronomers have discovered a black hole of enormous size. It is a gargantuan black hole, millions of times larger than ones previously observed, that was created by a single imploding star. These super-massive black holes are the result of millions of stars being sucked into its gravitational field.

Observations on the speed of stars circling the center of our sister galaxy, Andromeda, seem to support the view that it also has a giant black hole at its center. Many astronomers now theorize that at the very center of our own Milky Way galaxy, there is also a giant black hole. Black holes are more numerous than ever expected. The evidence of the second law of thermodynamics is everywhere in our universe.

In addition, astronomers now link unusual periodic outbursts of x-ray with a new class of intermediate-mass black holes, not as large as those within the core of galaxies but several thousand times that of the mass of our sun. These findings were published in the March 2005 issue of *The Astrophysical Journal Letters* and were based on observations of an ultraluminous x-ray source (ULX) in Messier 74, a spiral galaxy. This ULX emanates up to 1,000 times more x-ray energy than a neutron star or a small black hole and is now estimated to be a new class of black hole, about 10,000 solar masses in weight.

The periodic x-ray outbursts have a regular variation with a period of two hours, which physicists believe is caused by an accretion disc surrounding the black hole. These accretion discs are surprisingly turbulent, creating disturbances in density and magnetic field strengths, which may explain why they have random light variations. The suction into the singularity of the black hole causes the accretion disc to spin at great speeds. That, in turn, creates the friction that causes it to produce these outbursts of energy outside the event horizon of the black hole.

At one time, physicists believed that black holes were eternal in duration and that their internal temperature was at or near absolute zero. The temperature of an object is the measure of the sum of the energy contained in all its atoms. In a black hole, everything has been shrunk into a volume of near zero space; hence, it was not expected to have a temperature. As it turns out, Hawking discovered that black holes do radiate temperature. It has been dubbed *Hawking radiation.* It turns out that black holes do radiate some heat and therefore could not be eternal in duration. According to Hawking's calculations, pair annihilation of matter and antimatter particles near the event horizon may cause some slight loss of radiation. If heat (energy) is lost, then mass is lost, and eventually the black hole can evaporate. However, the time it would take for an ordinary, relatively small black hole created from an imploding star to evaporate is 10^{68} years, which is many times greater than the largest estimates by evolutionists of the age of our universe.

The measure of heat is, in fact, the measure of the movement of the atomic or subatomic constituents in a system; the higher the energy of atoms in a gas, the higher the heat, for example. The fact that black holes radiate heat tells us that matter is still present inside the black hole. It has not disappeared into another universe as some have conjectured. All matter swallowed by a black hole remains within our universe, even if all of it is not within the three visible spatial dimensions of our universe.

Black holes are what scientists call *black bodies.* Our ability to see objects comes from the molecular composition of the object that allows a certain wavelength of light to be reflected. Black bodies do not reflect any light. Our sun, believe it or not, is a black body. We can see the light that it is generating, but it does not reflect any light. In the same way, a black iron pan does not reflect light, but if we heat it, the pan will glow red as it emanates electromagnetic waves (photons). Apparently, black holes also radiate some heat. But the radiation or temperature is inversely related to their mass. In other words, the more massive the black hole, the less radiation it exudes. Hawking figured out that the temperature of a black hole is one-fourth of the horizon area measured in Planck units.

The coldest place in space within our universe is three degrees above absolute zero Kelvin, but the warmest (smallest) black holes are about a hundred million times colder. Since heat travels in order to find homeostasis, at this time in our universe, empty space is much warmer than any black hole, and energy then necessarily flows into the black hole. In order for black holes to evaporate, our universe has to be colder than the black holes. It would take a few hundred billion years for space to expand at its present rate to cool it below the temperature of black holes

But even the super-massive black holes in the center of galaxies having a mass of millions of stars are infinitesimally insignificant in relation to the immense mass/energy of the Genesis Singularity. If the gravitational pull of a singularity made from a collapsing sun only some 20 times larger than our own sun is so great that nothing can escape its pull, what would be the gravitational pull of a singularity containing hundreds of millions of stars at the center of a galaxy? What would be the gravitational pull of a singularity that contained all of the matter/energy in the universe?

If, in fact, all the matter and energy of the universe were concentrated into one singularity, how could anything escape its gargantuan gravitational pull and explode outward without some form of directed energy, outside of our space-time continuum to cause it to do so? Then, as if this were not impossible enough, what kind of power could control that immense explosion so it would remain almost exactly evenly distributed throughout the entire process until our present time? That this mother explosion of all explosions created a universe that is almost universally homogeneous and isotropic could do so by chance alone is an irrational belief bordering on madness.

The Big Bang begins with energy, a dimension of time intrinsically intertwined with space, which is a space-time dimensional universe, or what scientists refer to as the space-time continuum. These three spatial dimensions in time may not be the only dimensions in our universe, as we shall see later. But, where did this space and matter/energy/time come from?

The infinitely dense Genesis Singularity that contained all the matter that is presently found in our universe supposedly had a very small but, nonetheless, spatial dimension. How did it get that way? Naturalists have been searching for this missing piece of the puzzle with all fervor and resolve imaginable, but to no avail. Since they cannot explain it, through their naturalistic presupposition, they have instead conveniently placed the question out of the dialog.

Their rationalization is that no event in our universe is observable without time; therefore, no discussion of an event without time is sensible. But their assumption is predicated on an anthropocentric stance, which subjectively precludes the possibility of a being that is not bound to time and space, simply because we are bound to time and space. Their logic is circular and flawed. Since the first law of thermodynamics declares that energy cannot be created or destroyed within our universe, then something outside our universe had to begin the process of creation. What their rationalization actually admits is that there is no naturalistic explanation possible.

The very fact that the matter and space-time continuum came to be (before time and space) necessitates our concluding that at least one event has taken place without time, which is outside our spatial dimensions and therefore without any direct influence from our material universe. Whether naturalists like it or not, this is the proof that we live in an open system. This realization brings us the only possible alternative explanation that the immense radiant energy that came into existence simultaneously into a space-time continuum and could inhabit as the Genesis Singularity could not have magically appeared from nothing but was more rationally expressly created by a being outside our space-time limitations. Furthermore, since this Big Bang did not explode in a random manner but in a controlled and isotropic manner that developed rational universal laws to govern the matter and energy created, then this being had to be of superior intelligence and capable of controlling the explosion in such a way that life could have therefore existed within His creation.

It is a mathematical reality that the monstrously huge Genesis Singularity could not have remained in that primordial form from eternity past. Any black hole that could have existed for eternity would continue to exist for eternity, since there is no materialistic power that can undo the enormous entropy in its final state, going completely against the second law of thermodynamics. Therefore, it had to have come from somewhere or something, which not only created it but also caused it to explode outward. Some power of infinite proportion is necessary to overcome the immense gravity that created infinite density in the primordial Genesis Singularity. No finite power could accomplish that. It is a fact, not a speculation. Since there is no such infinite power in our finite universe, the logical and rational conclusion must be that it came from outside our universe. To exclude that answer on the grounds that it cannot be tested is nothing less than self-delusion, since all finite alternatives are incapable of either forming the Genesis Singularity from nothing or causing it to explode outward.

The problem for the naturalist is that to admit this is to admit that we are in an open system. And that brings upon them the conclusion that there is more to reality than the physically observable space-time continuum of our universe, and that is a heresy to the materialist religion. The foundational doctrine of materialists in their catechism of faith is that we live in a closed system that cannot be acted upon by exterior powers, and thus, there is no God.

But is this so difficult to understand when empirical data readily prove that there is an origin to our space-time continuum? If there is an origin to our space-time continuum, then there must exist something outside of it, from which our Genesis Singularity was brought forth to that point of origin in order to begin the process of the formation of our universe.

From Einstein's work on relativity came the recognition that there must be an origin for matter and energy. From Penrose, Hawking, and Ellis' work came the acknowledgment that there must be an origin for space and time, too.

With the knowledge that time has a beginning, and a relatively recent beginning at that, all age-lengthening attempts to push away the creation event, and thus the Creator, become absurd. Moreover, the common origin of matter, energy, space, and time proves that the act(s) of creation must transcend the dimensions and substance of the universe—a powerful argument for the biblical doctrine of a transcendent Creator (Ross 111).

Our entire space-time continuum (matter/energy, space, and time) had a beginning. None of the elements of our space-time continuum were eternal. All were created at a specific point in time. The problem for the naturalist is the inescapable fact that mathematics, a non-emotional and thus objective science, has proved that the very essence of the singularity (the beginning) must transcend the dimension and substance of our physical universe. That places the beginning square on the lap of a transcendent God. Physicists who hold to a naturalistic worldview dogmatically refuse to accept any view of reality that is not strictly materialistic and therefore irrationally continue to adamantly maintain that anything outside of our space-time continuum is out of the bounds of the study of science and therefore a moot point of discussion.

However, as previously stated, I find it quite ironic that this presupposition does not seem to stop them from imagining the possibility of separate universes, which would exist outside of the space-time continuum in our universe. And it serves only to prove the point that their arbitrary disqualification is predicated simply on their antitheistic bias stemming from their rabid deophobia.

It seems, then, to me equally valid to extrapolate from our present knowledge and theorize that since there is no possible mechanism capable of passing from a state of non-existence to a state of existence without the catalyst of an outside force, that our existence was, indeed, precipitated by an outside force, which not only had to be immensely powerful but unbelievably intelligent in order to control the process into such a fine-tuned chemical synthesis that allows our universe to exist as it does so that it can harbor life.

The universe is therefore an open system. Furthermore, our space-time continuum is rooted in hyperspace, and it is as much a part of reality as our universe, which is touched and bordered from the first moment of the Big Bang with a timeless and perhaps boundless hyperspace. No other explanation is either plausible, possible, or sensible. This, of course, has created a furor in the atheistic-evolutionary circles.

Hyperspace, the existence beyond our space-time horizon, is a commonly accepted reality by most naturalistic scientists, although we may not be able to describe or understand it within the confines of our dimensional space-time continuum. Yet many evolutionists see no problem in admitting the possible existence of hyperspace outside the bounds of our universe, which would not be bound to our space-time continuum. But ironically, evolutionists refuse, arbitrarily and illogically, to accept the notion of the possibility of the existence of anything of a divine nature outside the dimension of our space-time continuum. Is hyperspace testable? Can we even describe it?

Paul Davies, in his book *God and the New Physics*, tries desperately to avoid this conundrum. His reasoning is that all cause-and-effect phenomena in our universe are time-bound. That is, all events need time in order to occur. Since time had a beginning, then, he reasons, there could be no cause before time to begin time, since cause necessitates time to exist.

On the contrary, this is precisely the reason why only a transcendent creator, who stands outside of time, can be the primal cause. No other material entity in our universe that is time-bound (including quantum leaps) can escape being bound in the space-time continuum. Nothing except the concept of a transcendent God can stand outside of time and space to begin the whole process. The irrefutable fact is that the universe has been set in motion, and the notion that nothing set the universe into motion is illogical and ludicrous. But we must understand that it is science—and not theology—that has brought us to this inescapable conclusion!

And it is this beginning that causes the naturalist such consternation. Here, Davies takes his leap of faith into the irrational in order to avoid the unavoidable. As a matter of fact, some naturalists, including Davies, propose the existence of other universes that are also outside the confines of our space-time continuum. These cannot be observed or tested either. Yet the untestability of these universes does not seem to stop them from believing in them.

Curiously, influenced by the modern New Age movement and its emphasis on the human potential movement, many naturalists now claim that the possibility of the human mind is infinite. Yet when it comes to this primordial question in regard to the origin of the Genesis Singularity, they cop out and claim that our minds are not equipped to comprehend such phenomena occurring in either such large scales or such small scales.

The false and quite subjective assumption that humanity at its present state is the measure of all possible reality is, in fact, the epitome of our over-inflated egos and the zenith of anthropocentrism. Contrary to naturalist propaganda, the belief in God is not anthropocentric, but rather deocentric. God, and not humans, is at the center of reality. It is the naturalist worldview that makes humans the center of the universe.

While it is true that no human mind could truly grasp the enormity of the creator in all His glory, it does not naturally follow that we could not know Him at all if He chooses to reveal Himself to those He has created. Nor is it irrational to believe that He has chosen to reveal His character to those who seek to know Him. Our Judeo-Christian cosmological model maintains that the evidence of the creator is contained in the obvious design of the natural order of all things in the universe. From the smallest subatomic particle to the largest galaxies, the fingerprint of God is there for those who have eyes to see. But people who refuse to acknowledge the moral standards of a righteous God are prone to suppress the truth in order to assuage their willful spirits and avoid the guilt that comes from living in rebellion to God's moral standards. That is the fundamental driving force behind those who willfully blind themselves to the overwhelming evidence that surrounds them and causes them to subconsciously, or sometimes consciously, suppress the truth. The central sin of man, which first reared its head in the garden of Eden long ago, is voiced by the serpent in the day he tempted Eve: "You will be like God" (Genesis 3:5).

> *For the wrath of God is revealed from heaven against all ungodliness and unrighteousness of men, who by their unrighteousness suppress the truth. For what can be known about God is plain to them, because God has shown it to them. For his invisible attributes, namely, his eternal power and divine nature, have been clearly perceived, ever since the creation of the world, in the things that have been made. So they are without excuse (Romans 1:18–20).*

Modern people may not worship idols made of gold, silver, stone, or wood, but they have made humans and their finite minds their idol. It is people's willful rebellion against their creator and their deep desire to become their own gods that have wrought all of "man's inhumanity to man" since that day in the garden. But lest you ridicule this statement as superfluous, naturalist scientists who recognize the need for an intelligent mind to be behind the work of the creation of the universe have now posited that since there is no God, humanity must have been the cause. Yes, as incredible as that may seem, it is precisely what the new anthropic principle, a cosmological model, is promoting. Learned people of science who have understood the deep signs of intelligent design of the universe are now saying that humanity is the cause of creation.

The fact that humans, made in the image of God, are endowed with certain godlike attributes such as the ability to create and think about such lofty matters, does not give them the power to cause to happen anything before their existence, simply by observing it. We will deal with this more thoroughly in Chapter 16, which deals with the anthropic cosmological principle. Let us now explore the symmetry and unimaginably complex design of our universe, which we may be able to rationally judge as to whether or not random forces could have arranged it thus.

CHAPTER 5

• • •

DESIGN OR SERENDIPITY

In retrospect, the incredible number of processes that had to occur at precise and narrow parameters in order for carbon-based life to have evolved is so staggering that it requires an astronomical amount of blind faith to believe that these myriads of controls just simply happened by random chance processes, devoid of external control.

The irrationality of the naturalistic worldview is underscored by the extreme odds of ending up with the universe we have by mere serendipity. My college biology teacher would simply retreat to the safety of the harbor of irrationality and say, "I know the odds are infinitely small, but it just happened by blind chance. The fact that we are here proves that it happened. We won the lottery ticket." At that, I simply shook my head in sorrow at his defiant refusal to see the scientific evidence for design and simply said, "Then you are indeed not a man of science, but a man of faith!"

Science and technology have continued to verify that the limits or parameters necessary to allow life to exist on Earth are so minute, and the probability of this coming together by blind chance is so remote that to believe so requires an irresponsible leap of blind faith. Amazingly, this is not even argued by evolutionists, for they readily admit it. Davies wrote:

> An examination of life on Earth reveals just how delicately our existence is balanced on the scales of chance. There is a long list of indispensable prerequisites for the survival of our species. First there must be an abundant supply of the chemicals, which make up the raw materials of our bodies: carbon, nitrogen, hydrogen, oxygen, as well as some small but vital quantities of heavier elements such as calcium and phosphorous. Secondly, there must be no risk of contamination by other chemicals, which are poisonous: we would not want an atmosphere of methane or ammonia as found on many other planets.
>
> Thirdly, we require a rather narrow range of temperature so that our body temperature can proceed at the correct pace. . . . Fourthly a supply of free energy is needed, which in our case is supplied by the sun. It is important that this energy supply remains stable and is not subject to large fluctuations, which not only requires that the sun continues to burn with extraordinary uniformity, but that the Earth's orbit be nearly circular to avoid excursions toward and away from the solar surface. A fifth requirement is that the Earth's gravity is strong enough to restrain the atmosphere from evaporating into space, but weak enough so that we may move about easily and fall over once in a while without disastrous injury.
>
> Closer inspection shows that the Earth is endowed with still more amazing 'conveniences.' Without the

layer of ozone above the atmosphere, deadly ultravio-
let radiation from the sun would destroy us, and in the
absence of a magnetic shield, cosmic subatomic par-
ticles would deluge the Earth's surface (Davies 143).

The hard data are undisputed by naturalists. The astronomical statistical improbability of our universe forming through random processes is not denied. They simply choose to believe it in order to remain within their naturalistic paradigm. That is not reason. It is a metaphysical choice, spurred by emotional and psychological rationalizations. It is not the result of the scientific process of analysis and induction or deduction through reason, which chooses the most plausible and probable solution.

The design argument has been brilliantly put forth throughout the ages by such luminaries as Socrates, Plato, Xenophon, Cicero, Copernicus, Kepler, Linnaeus, Cuvier, Boyle, Newton, Kelvin, Faraday, and Rutherford. Efforts to explicate the marvelously engineered intricacies of our universe crested before ours in the Bridgewater treatise on the second century. Since that time, the design argument has been thoroughly attacked and ridiculed by proponents of the evolutionary paradigm. This attack was stimulated primarily by the singular development of the Darwinian concept of natural selection.

But the Darwinian concept did not come in a vacuum. Existentialist philosophy had been bantered about in Europe for some time as a philosophical progression stemming from the Enlightenment worldview. Many were attempting to find a way to unite the universals with the particulars in an all-encompassing worldview, without God at the center. That is, they were hoping to find a way to explain the particulars of reality with an atheistic universal worldview.

With the rise in popularity of the Darwinian model, the universe was no longer seen as a rational, ordered universe created by a creator. Instead, it was viewed as a fortuitous chance development of organisms guided by the principle of natural selection. Thus, the concept of natural selection slowly ousted the design argument from the scientific community. Eventually, most in the scientific community readily adopted the concept of natural selection as a way to explain the present order of the world within a naturalistic framework, thus avoiding God.

Even before Darwin, some such as David Hume had argued against the design argument, trying to make sense of a universe without a creator as the primal cause. After Darwin, Thomas Huxley and Ernst Haeckel catapulted this concept to the forefront, turning modern science on its head. It may be argued that without the efforts of Huxley, who referred to himself as "Darwin's Bulldog," Darwin's theory may not have caught on the way it did. Many also believe that Haeckel's efforts to promote his idea that "ontogeny recapitulates phylogeny" was even more productive in this process.

Huxley was one of the most powerful proponents of Darwinism. There was an exuberant and almost religious zeal to Huxley's denial of the design argument. Many who had existentialist leanings automatically embraced the concept of natural selection as the mechanism from which order evolved in a chaotic universe. It created an alternative to God and promoted the idea that science and faith in God could not be reconciled.

But now, almost two centuries later, with all the advancements in the areas of genetics, nuclear science, physics, and astronomy, more than ever the design of the universe is being made evident. And conversely, the process of natural selection, which had been the hope of the naturalists, as we shall see, has been found wanting as a means to explain this transition from chaos to order.

Several problems have arisen for the naturalist. First, their original assumption in the natural selection of the species required an infinite universe with an eternal time of existence in which the endless array of possible chance combinations were, by the eternal time period of existence, given the probability of developing through the long accumulated weight of tiny changes. Such was the thinking of Hume, as quoted by Barrow and Tipler.

> *Perhaps the development of the world is random but has had an infinite amount of time available to it so all possible configurations arise until eventually a stable self-perpetuating form is found.*
>
> *Let us suppose it [matter] finite. A finite number of particles is only susceptible of finite transpositions. And it must happen in an eternal duration, that every possible order or position must be tried an infinite number of times. . . . a chaos ensues; till finite through innumerable revolutions produce at last some forms, whose parts and organs are so adjusted as to support the forms amidst a continual succession of matter (Barrow and Tipler 70).*

The concept of an eternal duration of existence then made palatable what was statistically illogical. Thus, what would otherwise be impossible (the assumption that order could arise from disorder by mere chance, without directed energy) could be made plausible by the infinite amount of time available to create all the possible combinations of chemicals, "until eventually a stable self-perpetuating form is found" (Barrow and Tipler 70).

Hume realized that a finite universe with a finite amount of time would have only a finite number of transpositions allowable. The chance of such order arising from chaos in finite time is next to nil. He understood that. But it was the foundational premise from which the entire theory was constructed. The first assumption has now been completely disproved. The universe is not of an eternal duration. It is, against Hume's most ardent wish, finite.

What if we were able to calculate the total number of possible physical transitions that could take place in the most elementary particles throughout the entire 15 billion years of history of our universe using all the available matter within it? What would be the chances for simple random selection to create a single protein? I am not speaking about a living cell, but just a simple protein.

> *Due to the properties of gravity, matter, and electro-magnetic radiation, physicists have determined that there is a limit to the number of physical transitions that can occur from one state to another within a given unit of time. According to physicists, a physical transition from one state to another cannot take place faster than light can traverse the smallest physically significant unit of distance (an individual "quantum" of space). That unit of distance is the so-called Planck length of 10–33 centimeters. Therefore, the time it takes light to traverse this smallest distance determines the shortest time in which any physical state can occur. This unit of time is the Planck time of 10–43 seconds (Meyer 216).*

If we are to calculate the odds of proteins being formed from amino acids, then we must determine how fast these chemical transitions could take place. Instead of using the real time duration of such transitions, we will instead, in order to give chance every opportunity possible, use the smallest theoretically possible time that a transition could take place—the Planck time of 10-43 seconds. Using all the material existing in our universe and calculating each transition at the phenomenal speed of light through the Planck space can give us an idea if this is in any way even plausible.

> *Knowing this, Demski was able to calculate the largest number of opportunities that any material*

event had to occur in the observable universe since the big bang. Physically speaking, an event occurs when an elementary particle does something or interacts with other elementary particles. But since elementary particles can interact with each other only so many times per second (at most 10^{43} times), since there are a limited number (10^{80}) of elementary particles, and since there has been a limited amount of time since the big bang (10^{16} seconds), there are a limited number of opportunities for any given event to occur in the entire history of the universe.

Demski was able to calculate this number by simply multiplying the three relevant factors together: the number of elementary particles (10^{80}) times the number of seconds since the big bang (10^{16}) times the number of possible interactions per second (10^{43}). His calculations fixed the total number of events that could have taken place in the observable universe since the origin of the universe at 10^{139}. *This then provided the measure of the probabilistic resources of the entire observable universe* (emphasis added) (Meyer 216–217).

In other words, the upper limit, the largest possible number of physical reactions that could have taken place in the entire history of our universe, using every elementary particle in our universe, is 10^{139}. The probability of producing a single 150-amino-acid functional protein (the smallest theoretical protein possible, and no such protein actually exists) is 1 in 10^{164} (Meyer 217). That number documents that the universe does not contain the probabilistic resources necessary to allow chance random ordering to formulate even the most basic building block of life—the protein. Using all the matter in the universe at the fastest possible rate of recombination cannot even provide the resources to create a simple theoretical protein, much less a much larger real one.

The Missing Magical Recombination Machine

Second, there is no known mechanism in inanimate nature that can combine and recombine lifeless matter through these innumerable revolutions that would create the transpositions necessary to bring to pass this imagined self-perpetuating phenomenon. If such a mechanism existed, it would have reshuffled the components of the atom with an endless variety of possibilities in its fundamental composition. As Lord Kelvin so aptly described in his address to the British Association for the Advancement of Science in 1873:

> No theory of evolution can be formed to account for the similarity of molecules, for evolution necessarily implies continuous change, and the molecule is incapable of growth or decay, of generation or destruction. None of the processes of nature, since the time when nature began, have produced the slightest difference in the properties of any molecule. We are, therefore, unable to ascribe either the existence of the molecules or the identity of their properties to the operation of any of the causes which we call natural. . . . the molecules out of which these systems are built—the foundation stones of the material universe—remain unbroken and unknown (Barrow and Tipler 88).

We have certainly learned a great deal more about the molecule since that time. As a matter of fact, we can split an atom. But what Lord Kelvin said then is still true today. While it is true that we have been able to break down atoms artificially through particle accelerators, the building blocks of atoms remain the same throughout the universe. We have even brought into existence human-made elements by combining subatomic particles artificially, but the building blocks of atoms are nonetheless ever the same. The universe did not establish an innumerable number of designs to the building blocks of atoms, as would be predicted by a purely random chemical genesis.

The electron is the same anywhere you find it. The proton and the neutron are also the same in every molecule. The specific charge of the electron remains the same throughout the entire universe. Where are the traces of these innumerable transpositions in the design of the basic building blocks of all matter? The more we know, the more intricate and purposeful is every aspect of the atom. It is inconceivable and completely irrational to believe that this magnificently complex design was an accident of nature, if there are no alternatives from which the choice would have been made. There are no alternatives.

The stability of the atom is dependent on the juxtaposition of the charged particles within it. What force could have united these particles in such an ordered and stable fashion through random chance processes? The naturalist, by necessity, begins with quarks turning into atoms as a unit, but in a naturalistic universe, it is only logical that the elemental particles of an atom would have existed before the atom and would have gone through many trial-and-error chance recombinations. What force, then, brought them together and unified them in such a balanced form?

The only recombination machine that the naturalist can point to is a star, but here, the elements of the atom are already formed. Matter, energy, and space are all old by the time a star has been formed. Where did they come from? How was the electron formed in only one configuration? We know that the protons and neutrons are made up of quarks, but how did these quarks get shuffled into creating the hadrons?

The string theory now predicts that quarks are made of vibrating membranes. But where did these vibrating membranes come from, and what made them vibrate at their individual specific pitches to create the desired and individually distinct primal essence of the components of matter? How could the order of the quarks, as precise and unvarying as they are, found universally in the composition of each of the elemental particles of the atom, become so without directed or controlled energy? How could an undirected, chaotic explosion result in the ordered universal laws, which we observe through even the subatomic particles with their quantum fluctuations?

Now these are the easy questions. Here comes the really hard one: How is a spatial dimension created from an explosion? Some have now come to believe that dimensions are really expanded membranes. We can deduce, through mathematical equations, that there are ten spatial dimensions, but these do not tell us how they came to be. Why and how did it all begin at the real beginning, with only a singular design for the fundamental components of the atom and with at least a visible three-dimensional spatial component that allows for human life to exist?

The naturalist claims that the primordial Genesis Singularity exploded, thrusting all the matter and energy of our entire universe outward. Something happened that caused it to explode. They do not know what. They only know that it did. Some theorize that it was the interaction between matter and antimatter. But if there was an equal amount of matter to antimatter, as predicted by mathematics, why did it not completely annihilate the universe? We know of no way to transform radiant energy into anything other than pair production, where a particle of matter and a particle of antimatter are created so it is impossible for one form to be created in greater numbers than the other.

> If the universe in the first few minutes was really composed of precisely equal numbers of particles and antiparticles, they would all have annihilated as the temperature dropped below 1,000 million degrees, and nothing would be left but radiation. There is very good evidence against this possibility—we are

here! There must have been some excess of electrons over positrons, of protons over antiprotons, and of neutrons over antineutrons, in order that there would be something left over after the annihilation of particles and antiparticles to furnish the matter of the present universe (Weinberg 87).

The problem is that we have no material way to explain how that could happen. Mathematics predicts an exact number of particles and antiparticles. It could be explained if somehow there was a segregation of the matter and the antimatter, but we see no evidence of this pooling of antimatter anywhere in the universe. Neither do we have some valid mechanism that could segregate them. It is not just that matter survived, but that it survived in such a way that the particles were in the correct proportions to be able to form atoms, molecules, and life. There are no naturalistic explanations for this miracle that quite obviously shows a previsioned choice by a master designer.

The Missing Evolutionary Intermediates of the Atom

Somehow, all the subatomic quarks condensed from this intense radiant energy and the structure of the atom miraculously evolved with a single structural design. All atoms have a nucleus and concentric energy rings (shells) around them that, without a single exception, universally stipulate exactly how many electrons can be contained within them. Scientists call it the *Pauli exclusion principle*.

The design of the atom is universal. The number and type of its components determine the type of element created. But we do not have a variety of designs. One was chosen, but it can hardly be believed that is was chosen randomly. And it was chosen from the very first microsecond of the Big Bang. That is quite a hard pill to swallow if you are preaching random origins. I just don't have that kind of irrational faith.

The idea that the formation of the specific quarks and their precise and necessarily unique combinations that form the protons, neutrons, and electrons with the exact mass and electric charges necessary for them to function as an atom and simultaneously with the exact interplay of the four forces of nature that could allow such a structure to be stable, and furthermore, that it all miraculously congealed by sheer random luck is an irrational superstition of the first order. It is nothing more than an irrational speculation based solely on a priori metaphysical faith.

When one considers all the possible combination potentials that could have arisen and that could not have formed a stable atom in order for matter to exist as it does in our universe, it boggles the mind. How can anyone with an ounce of reason think this was a random coincidence? The sheer genius of its complicated and intricately balanced design cannot be logically considered as anything less than empirical evidence of a master designer.

Fortuitously, this incredible design balances the electrical charges within and allows for the recombination of these elements with other elements. All this must be counterbalanced by the fact that the atom, while combining with other atoms, must still remain somewhat stable and distinct; otherwise, the connection into a molecule could not work in producing different compounds.

The positively charged protons within the nucleus are tightly packed alongside one another. These positive charges are repelling against one another with such force that when broken, the release of energy is beyond what Darwin could have ever imagined in his wildest dreams. It is the terrible power of nuclear bombs. And yet they are held tightly packed in an infinitesimally small point by a precisely calibrated force called the strong nuclear force. That force has to be precisely as strong as it is for an atom to be able to function as it does in our universe. There is no wiggle factor. Just a fraction less or a fraction more, and our universe could not exist. How did random ordering come upon this exact magnitude in the very first try?

The electrons on the outside of this atom are spinning about or zigzagging within their corresponding shells with predetermined numbers allotted to them. These are negatively charged, and yet, while like forces repel, they remain in close proximity within the confines of their shells. But these shells within one atom are ruled by a precise exclusion factor that does not vary with any atom in the universe. The first shell around the nucleus can contain only two electrons. The second shell around the nucleus can contain only eight electrons, and the third shell can contain only 18. The fourth shell around the nucleus can contain only 32 electrons, and the fifth shell can contain only 50. The sixth shell can contain 72 electrons. Each interior shell must be filled before the next level is reached. Every atom in the entire universe abides by this singular design. It is universal and has been since the very first atom congealed. There is not a single atom in the entire universe that has ever been or shall ever be any different. How does random ordering accomplish this magnificent feat?

The last shell of any given atom that contains electrons is called the valence shell. Helium has only two electrons; hence, these are found in the first shell, called the "S" shell. The second shell is called the "P" shell and can contain up to eight electrons. But each of these shells contains subshells within them. There is always a pressure to fill the last valence shell in order for the atom to be more stable. So the union of two atoms that share electrons in order to fill this last valence shell is called a covalent bond. The genius design is unbelievable. It is the connections made in the valence shell with other atoms that allows for molecules to be formed. It is also what allows for compounds to be formed with different atomic elements. Miraculously, according to evolutionists, all of this developed through mere serendipity, without any prevision and external guidance, from the very first microsecond of the Big Bang. The materialist blindly believes that all of this evolved by just pure, blind chance!

But if evolution is true, where are the transpositions of electrons with variable charges? Where are the atoms that randomly created a different set of shells? Where is the smoking gun of evolution that surely would make up a preponderance of the material in our universe? It is not there! That means it happened that way on the very first try. That, my friends, can be considered nothing less than empirical evidence, not of a randomly ordered evolutionary process, but of an intelligently designed atom.

The idea that chance could find the exact formula to construct an atom in the very first try is so illogical that the element of faith outweighs the element of science. Here lies a deep problem for evolutionists—they have no evidence of the many transpositions that evolution should have left behind in order for us to follow the evolutionary path to our present state. Even if they had an infinite number of years to accomplish this miracle of the first order, it is more than doubtful that the order and symmetry observed in nature could have spontaneously risen through accidental means. It is a magnificent leap of unsubstantiated faith in an irrational speculation. I just don't have that much faith!

The universe was formed in such a way that all the parameters necessary in order for life to form were finely tuned, even from the very beginning of the Big Bang. The initial plasma state of all matter in a concentrated area was cooled as the universe expanded in a precisely chosen rate. This expansion rate was critical in the process of allowing matter to condense out of the plasma at just the right temperature and rate so that each of the particular particles were made at the exact proportions necessary to form all the basic elements of the atom.

But the real kicker is that this did not just happen in one section of the universe, but rather throughout its entire breadth and width, in total and complete compliance from the very first microsecond. This nearly perfect, even temperature throughout the entire universe as it expanded allowed for matter to be just about homogeneous all over. Not only did the fermions coalesce, but the four forces of nature also split apart from a singular force in a specifically balanced interplay between them to allow matter to exist as we know it. As we will see, this matter is filled with precise symmetries in such delicate balances that it is unimaginable that purely random ordering could have accomplished such a miraculous feat. Random ordering cannot rationally explain such precise and invariable harmony and supersymmetry in the design of all matter and in the universal laws of physics.

CHAPTER 6

● ● ●

THE FINE-TUNED UNIVERSE

The remarkable fact is that the values of these numbers seem to have been very finely adjusted to make possible the development of life. . . . This means that the initial rate of the universe must have been very carefully chosen indeed if the hot big bang model was correct right back to the beginning of time. It would be very difficult to explain why the universe should have begun in just this way, except as the act of a God who intended to create beings like us.

—Stephen Hawking

We must regard our existence as a miracle of such improbability that it is scarcely credible.

—Paul Davies

The evidence of an intelligent designer comes to us not only in the very act of the creation of matter and the space-time continuum, but also in the control over the ongoing process and in the aftermath of the Big Bang. If this expulsion of matter out of the white hole had taken place without divine direction (directed energy), and the resulting universe had consequently been governed only by random processes, then the result would have been a universe with a lumpy texture leading into an infinite number of black holes rather than galaxies.

Our rational observations clearly show that the universe is not unrestrained chaos. There is symmetry to the universe and to the laws that govern it. Were it not so, science would not exist. Our ability to perform experiments and conclude certain empirical data is due to the universal laws that govern our entire universe. If randomness were the matrix of reality, even our simplest acts would be inconceivable adventures since random variations would prevent us from using past experiences to predict anything about future outcomes.

We rely completely on the stability and universality of all physical laws in the universe. The laws that are true today were true yesterday and will be true tomorrow; that is, laws are absolute in nature. There is no room for relativism in chemistry, or the simplest mixture may randomly kill us. The very word *law* implies a fixed entity and not an evolving truth.

That does not mean that the universe is static; after all, it is always changing and dynamic. But the laws that govern those changes are ever unchanging, even if we have not yet figured them out. The laws are not dependent on our human perceptions. Science cannot function in a Kantian world. The idea that the observer by the act of observation gives reality to the thing observed would mean that there could be no absolute, empirical data with universal laws. Otherwise, an experiment done on one side of the world would have no bearing on the very same experiment being repeated in another part of the world and viewed by a different observer.

Newton's law of universal gravitation is the same on Earth as it is in all the stars in the constellation of Orion. The second law of thermodynamics is the same on Earth as it is in all the stars of the constellation of Draco. The law of conservation is universal in nature and does not vary in any other sections of our universe. Were it not so, we could not do even the simplest chemical problems. How does random ordering produce fixed and universally absolute laws? It is irrational to conclude that random processes birthed in chaos could manufacture universal laws. That is simply an oxymoron.

We see the evidence even in the subatomic particles of the supposedly ever-fluctuating quantum world. Here in the microworld, there are also certain absolutes that are fixed and immutable. For example, as revealed by George Uhlenbeck and Samuel Goudsmit, every electron in the entire universe, always and forever (at least as long as our universe exists), spin at one fixed, absolute, and never-changing rate. Not only does that rate never decay, but it is also exactly the same rate anywhere in our universe. It is a universal, intrinsic property of the electron. Is it rational to conclude that chaotic, random processes could produce fixed, never-changing spin rates universally?

The first law of thermodynamics states that the total energy of an isolated system cannot change; it is conserved over time, although it could change from one form into another. For example, the nuclear energy found in atoms can be converted to kinetic energy in an atomic blast. Yet the total amount of energy in a system is always preserved. In other words, energy/matter cannot be either created or destroyed. Yet the entire universe went from nonexistent to existent at a given point in time. How can that happen in a naturalistic universe whose very first law dictates that nothing can be either created or truly destroyed in our material reality? How can random chaotic functions produce the universal law of conservation?

The Enigmatic Proton

Then there is the case of the enigmatic proton. I love the proton. It is a positive little guy and stubborn like me. It is the backbone of all matter in the entire universe. We know the quark makeup of the proton. It ought to be possible for protons to decay into photons and positrons. Quantum physics tells us that anything possible is compulsory probable, but the enigmatic proton apparently does not know about quantum physics, and if it does, it apparently doesn't give a hoot. The proton does not decay. It refuses to break up into photons and positrons. That is quite fortunate for us, because that is what makes the atom stable and allows for our universe to exist as it does. The proton is truly the backbone of all matter. It is, however, a conundrum for naturalists who insist that there is no intelligent design to our universe.

> There is at least one very unusual feature of the Laws of Physics that seems very finely tuned with no anthropic explanation in sight. It has to do with the proton, but let's first review the properties of its almost identical twin, the neutron. The neutron is an example of an unstable particle. Neutrons, not bound inside a nucleus, will last only about twelve minutes before disappearing. Of course the neutron has mass, or equivalently energy, which cannot just disappear. Energy is a quantity that physicists say is conserved. That means the total amount of it can never change. Electric charge is another exactly conserved quantity. When the neutron disappears, something with the same total energy and charge must replace it. In fact, the neutron decays into a proton, an electron, and an antineutrino. The initial and final energy and electric charge are the same.
>
> Why does the neutron decay? If it didn't the real question would be, why doesn't it decay? As Murray Gell-Mann once quoted T. H. White, "Everything which is not forbidden is compulsory." Gell-Mann was

expressing a fact about quantum mechanics: quantum fluctuations—the quantum jitters—will eventually make everything happen unless some special law of nature forbids it.

What about protons—can they decay, and if so, what do they become? One simple possibility is that the proton disintegrates into a photon and a positron. The photon has no charge, and the proton and positron have exactly the same charge. It ought to be possible for protons to disintegrate into photons and positrons. No principle of physics prevents it. Most physicists expect that given enough time, the proton will decay.

But if the proton can decay, it means that all atomic nuclei can disintegrate. We know that atomic nuclei of atoms like hydrogen are very stable. The lifetime of a proton must be many times the age of the universe.

There must be a reason why the proton lives so long. Can that reason be anthropic? Certainly our existence places limitations on the lifetime of the proton. It obviously cannot be too small. Let's suppose the proton lives one million years. Then I would not have to worry very much about my protons disappearing during my life. But since the universe is about ten billion years old, if the proton lived only a million years, they all would have disappeared long before I was born. So the anthropic requirement for the proton lifetime is a lot longer than a human lifetime. The proton must last at least fourteen billion years.

Anthropically, the lifetime of the proton may have to be a good deal longer than the age of the universe (Susskind 2006, 189–190).

Notice that Susskind insinuates that man is somehow responsible for dictating the life of the proton: "Certainly our existence places limitations on the lifetime of the proton" (Susskind 2006, 189). This is the anthropic principle at work, trying to make humans the designers and creators of the universe. But let us leave that aside for the moment. Assuming that the age of the universe is, as Susskind theorizes, ten billion years (others place it at around 15 billion years), then a million times the age of the universe is 10^{16} years. And yet the life of a proton is vastly greater than this unfathomable age.

> But we know that the proton lives vastly longer than 10^{16} years. In a tank of water with roughly 10^{33} protons, we would expect to see one proton decay each year if the lifetime were 10^{33} years. Physicists, hoping to witness a few protons decaying, have constructed huge underground chambers filled with water and photoelectric detectors. Sophisticated modern detectors can detect the light from just a single decay. But so far, no cigar; not a single proton has ever been seen to disintegrate. Evidently the lifetime of the proton is even longer than 10^{33} years, but the reason is unknown (Susskind 2006, 190–191).

A proton that does not decay within one million times the total age of the universe is obviously not impacted in any way by any anthropic need. There is no anthropic connection between people's minds and needs with the creation process that took place long before people existed. The only connection to people is that God designed it in order to allow us to inhabit it. It is not an anthropic connection, but a deocentric connection.

How is it that sheer chaotic forces could randomly create in the first attempt a proton that is for all intents and purposes the foundation of all elements and, to the best of our knowledge, eternally stable?

To believe that random ordering would be responsible for such fine-tuned results is tantamount to lunacy.

There are universal laws that govern our universe; hence, there must be a universal lawgiver. Laws do not self-design. The presence of a law is indicative of a mind that conceived it. Those of us who believe in an open universe understand that the creator uses these laws to accomplish His will, even when we do not yet know what these laws are.

The supernatural is simply the natural that we do not yet understand. We do not promote a mystical magical concept of the universe. We believe that the laws, built into it by the creator, govern all of nature. But we also understand that through other laws that we may have no inkling of understanding yet, He is able to interfere in what we perceive to be the normal processes of nature because of our finite understanding, much in the same way that we may, for example, interfere in the normal course of a river by damming it up. What arrogance humans show when they think that only what they know and are able to understand is natural.

The Uniform Isotropic Homogeneity of the Cosmos

Why does the universe appear to be so neatly arranged in every direction in uniform isotropic homogeneity? If it all came from a random explosion without an intelligent guiding hand in the entire process, why are there not multiple sets of randomly generated laws without any single set of universally applicable principles? How can pure chaos create ordered laws? Why are there so many fine-tuned parameters with which this explosion must have complied in order for it to be able to support life?

Therefore, the only objectively rational deduction we can make is that random processes could not have created our universe. Is it not obvious to the objective observer that there is a design to every aspect and element of this vast and wonderful universe we inhabit? Does not this design logically beg for the existence of the master designer? The design argument is stronger now than ever before. Those who mock the design argument as antiquated and anachronistic thinking do not realize that their evolutionary argument dates back to Lucretius and is, in fact, the real antiquated and anachronistic argument.

Hawking clearly understood the enormous improbability of all these fine-tuned parameters having occurred simply by blind chance. He readily admitted it, and in spite of the fact that he was not thrilled with the notion of a supreme creator, he, at least at the beginning of his career, faced the facts. Unlike a great many of his colleagues, during his early years, he faced the implications of his mathematical findings. For this intellectual honesty during his early years, I am deeply impressed.

Speaking of the extremely narrow parameter in which this universe had to form in order to create life, he states.

> *The remarkable fact is that the values of these numbers seem to have been very finely adjusted to make possible the development of life. . . . For example, if the electric charge of an electron had been slightly different, stars either would have been unable to burn hydrogen and helium or else they would not have exploded. . . . Nevertheless, it seems clear that there are relatively few ranges of values for the numbers that would allow the development of any form of intelligent life. . . . This means that the initial state of the universe would have to have had exactly the same temperature everywhere in order to account for the fact that the microwave background has the same temperature in every direction we look. The initial rate of expansion also would have had to be chosen very precisely for the rate of expansion still to be so close to the critical rate needed to avoid recollapse. This means that the initial rate of the universe must*

have been very carefully chosen indeed if the hot big bang model was correct right back to the beginning of time. It would be very difficult to explain why the universe should have begun in just this way, except as the act of a God who intended to create beings like us (emphasis added) *(Hawking 129–131).*

Earlier, Fred Hoyle in his stellar studies realized that there is a remarkable chain of coincidences in the process of the creation of matter, without which carbon-based life could not have existed. The cooking of elements in stars is not the only physical way to create elements, as we shall later explore. High-intensity currents in plasma can create solid matter. Regardless, these intrinsic properties of matter are precisely fine-tuned in a very limited or narrow parameter, which precisely allows the existence of such things as the design of the atom so molecules could form, the precise balance between the four forces of nature that allows the atomic processes in our universe to take place, the universal nature of the laws of physics, the nuclear processes that give us a stable sun that gives light and warmth to the planet, the production of carbon in such quantities necessary for the existence of carbon-based life, and an enormous variety of other key factors, without which life as we know it could not have developed in our universe.

Hoyle realized that this remarkable chain of coincidences—the unusual longevity of beryllium, the existence of a disadvantageous level in O^{16}—were necessary, and remarkably fine tuned, conditions between the strengths of the nuclear and electromagnetic interactions along with the relative masses of electrons and nucleons. . . . First suppose that Beryllium 8 . . . had turned out to be moderately stable, say bound by a million electron volts. What would *be the effect on astrophysics? There would be many more explosive stars and supernovae and stellar evolution might well come to an end at the helium burning stage because helium would be a rather unstable nuclear fuel (Barrow and Tipler 253).*

Stars are, in essence, cooking pots for higher elements. But the recipe for cooking these elements is not by any means of the imagination a sure shot, had the properties of certain elements not been exactly fine-tuned to make this happen. Carbon could not exist unless beryllium and neutrinos possessed the unique properties that allow for this to happen in a nuclear furnace. The idea that random ordering caused this unique phenomenon is statistically irrational.

There are multiple ways that things could go wrong with the nuclear cooking. If there were no weak interactions or if neutrinos were too heavy, protons could not turn into neutrons during the cooking. The cooking of carbon is sensitive to the details of the carbon nucleus. One of the great scientific events of the twentieth century occurred when the cosmologist Fred Hoyle was able to predict one of these nuclear details just from the fact that we are here. In the early 1950s Hoyle argued that there is a "bottleneck" in the cooking of elements in stars like the sun. There appeared to be no way for the cooking to proceed past atomic number [read: mass] 4—helium. Nuclear cooking usually goes forward one proton at a time to form a heavier element, but there is no stable nucleus with atomic number 5, so there is no easy way to get past helium.

There is one way out. Two helium nuclei can collide and stick together to form a nucleus with atomic number 8. *That nucleus would be the isotope beryllium 8. Later, another helium nucleus could collide with the beryllium and form a nucleus with atomic number 12—good old carbon 12, the stuff of organic chemistry. But there is a fly in this ointment.*

Beryllium 8 is a very unstable isotope. It decays [falls apart] so rapidly that there is not enough time for the third helium nucleus to collide before the beryllium disappears—unless an unlikely coincidence occurs. If by accident there were an excited state—a so-called resonance—of the carbon nucleus with exactly the right properties, the probability for the beryllium to capture a helium nucleus would be much higher than expected. The likelihood of such a coincidence is very small, but when Hoyle suggested that such a coincidence might solve the problem of cooking the heavy elements, experimental nuclear physicists went right to work. And BINGO, the excited state was discovered with exactly the properties that Hoyle guessed. Just a small increase or decrease in the energy of the excited carbon nucleus, and all the work of making galaxies and stars would have been in vain; but as it is, carbon atoms—and thus, life—can exist.

The properties of Hoyle's carbon resonance are sensitive to a number of constants of nature, including the all-important fine structure constant. Just a few percent change in its value, and there would have been no carbon and no life. This is what Hoyle meant when he said that "it looks as if a super-intellect has monkeyed with physics as well as with chemistry and biology" (Susskind 2006, 182–183).

What Hoyle is saying is that there is an intrinsic connection between physics, chemistry, and biology that allows life to exist in our universe. The subtle adjustments that are finely tuned to one another in order for the necessary processes of life to exist are so extreme that it implies a super-intellect designer. One of these magnificent, fine-tuned parameters is the fine structure constant, which is the strength of the electromagnetic interaction between elementary charged particles. This happens to be almost exactly the value of 1/137. That value is unwaveringly universal.

Yet evolutionists ignore this teleological reality and continue to doggedly hold to their dysteleological dogma. The fine structure constant that Susskind speaks of is also known as Sommerfeld's constant, which is the precisely tuned strength of the electromagnetic interaction between elementary charged particles. If that constant were too strong, there could be no possibility for atoms to attach to one another, because atoms would remain isolated by the grip of the electromagnetic force. If it were too weak, the atoms could not hold together. There is no wiggle room in the exactitude of the strength necessary to create our universe with the higher elements necessary in order for life to exist. The fine structure constant must be exactly what it is.

Susskind apparently does not agree with Hoyle that a super-intellect monkeyed with physics, chemistry, and biology in order to design our universe to inhabit life. Instead, he nonchalantly ascribes these "unlikely bunch of accidents" as just accidents. But, in spite of his enormous faith in random ordering, he admits that the cosmological constant is a horse of another color.

> Taken together, these coincidences do seem like an unlikely bunch of accidents; but accidents, after all, do happen. However, the smallness of the cosmological constant is another matter. To make the first 119 decimal places of the vacuum energy zero is most certainly no accident. But it was not just

that the cosmological constant was very small. Had it been any smaller than it is, had it continued to be zero to the current level of accuracy, one could have gone on believing that some unknown mathematical principle would make it exactly zero. The event that hit us like the proverbial ton of bricks was the fact that in the 120th place the answer was not zero. No missing mathematical magic is going to explain that (Susskind 2006, 185).

We will deal more with the unique nature of the cosmological constant or vacuum energy in the next chapter, but for now, the point is that even when Susskind cannot explain away the sheer uniqueness of this fine-tuned phenomenon in our universe with any "missing mathematical magic," he still holds doggedly to his naturalistic presupposition by blind faith in a dysteleological origin for our universe. He states that "it is most certainly no accident," but he cannot bring himself to admit that it is a designed feature.

The evidence for an intelligent designer is obvious to those who can look at the evidence from an objective point of view. Susskind nonchalantly states, "Taken together, these coincidences do seem like an unlikely bunch of accidents; but accidents, after all, do happen" (Sussking 2006, 185). But the sheer number of these coincidences is so extreme that it is tantamount to foolhardiness to ignore the improbability of it all being just accidents. Even the simple things, like the distance of our planet to the sun, the very nature of our own particular sun, and the fact that we have a single moon at precisely the right size and distance from us—all these things are indispensable for life as we know it to exist on planet Earth.

Our moon is exactly one-fourth the size of our planet. That is a huge planet-to-moon ratio, not found anywhere else in the universe thus far. But it is the reason our planet has such a stable rotation around the sun with very little wobble. Any drastic changes in our wobble would spell sudden climactic changes, which would make it impossible for life to survive. The perception that chaos could have engineered such a multifaceted feat of precisely engineered special circumstances without which life could not exist is quite a step of blind faith.

But human nature is a thing to behold. And brilliant minds are sometimes clouded from simple truths by personal bias. It is the empirical evidence that so clearly illustrates the fallen state of the human condition.

CHAPTER 7

THE RIDDLE OF OUR SMOOTH UNIVERSE

If the primeval material was churned about at random it would have been overwhelmingly more probable for it to have produced black holes than stars, because the holes, being so much more disordered, can be produced in a vastly greater number of ways. For every star that formed, countless billions of more easily achieved black holes should have accompanied it."

—Paul Davies

The riddle that confronts evolutionists is not only to explain the fact that the universe is there, but that it was made and maintained in a universally uniform and smooth fashion from the very onset. The fact that our universe in every direction we look is basically smooth directly implies that a force was predicated upon it to maintain it thus from the very beginning. Otherwise, a random genesis governed only by chaotic forces would have resulted in the development of a universe that was lumpy in texture. Randomness cannot produce evenly distributed material and energy. Because of its irregular rate of distribution, it can only create an irregularly textured universe. An irregularly textured universe would have then led not to the creation of galaxies and stars, but rather to an overwhelmingly greater number of black holes. Our universe would have been filled with billions of black holes and very few, if any, galaxies. How can we explain our universe without a designer?

If the forces acting upon the expanding matter during the Big Bang were purely chaotic and void of any external interference, as posited by the naturalist paradigm doctrine of a closed system, then the subsequent aggregation of matter would have condensed only in disjointed and irregular fashion. A randomly ordered universe would have almost entirely turned directly into black holes. The condensation of matter would have simply bypassed altogether the stage of star formation.

Yet the universe could not have been absolutely smooth. There had to be just enough irregularity for matter to be able to accrue into the stars and galaxies before becoming black holes. The explosion had to thread the needle within an extremely narrow parameter out of an infinite number of possibilities in order for our universe to become exactly as it is. Too much disorder, and all you get are black holes. Perfect homogeneity, and there could be no aggregation of matter. All points would be equidistant and pulling at each other with equal gravitational strength—no galaxies, no stars, and no planets could form. In fact, not even atoms could form.

In addition to this precisely tuned texture, our universe had to pick a precise rate of expansion to thread that cosmic needle and allow our universe to form as it is. If it expanded too rapidly, it would not allow it to coagulate matter into stars and galaxies. If it expanded too slowly, it would all congeal again and collapse into another singularity—another giant black hole the size of the Genesis Singularity. This precise and exact expansion rate is what is referred to as the finely tuned cosmological constant, or vacuum energy. The term *vacuum energy* is associated with the speculation that dark energy in the vacuum of space produces a repulsion energy that causes the expansion of the universe.

The exasperating problem for the naturalist is that our universe is basically smooth with just the right amount of irregularity to allow star and galaxy formation—a very remarkable and fined-tuned precision that can hardly be considered the product of blind chance. It is because the universe is almost completely smooth that galaxies can exist. Had the texture of the expanding universe been lumpy, as would be demanded by a randomly guided genesis, then it would have necessarily resulted in a predominance of black holes and very few galaxies, if any. How does an explosion of this magnitude result in a smooth universe without directed energy to guide it so? How does random chance pick just the right expansion rate to allow our universe to form so it can form a universe that can inhabit life? It is inconceivable that a random explosion could proceed in such carefully choreographed homogeneity, not just from the moment of the explosion of the Genesis Singularity, but also all the way through to our present time.

The Chaotic Cosmology Theory vs. the Orderly Singularity School

At the beginning of the modern era, two views vied for supremacy in the evolutionary scientific community. First was the chaotic cosmology theory that posited that the initial conditions of the development of the universe were chaotic in nature. This, of course, was favored by those with a naturalistic presupposition since they saw the world as chance/serendipity created by the blind and undirected forces of atoms in collision. Some speculated that the apparent order in our immediate world might be an island in the ocean of chaos—so much for the uniformitarian dogma that was a foundational premise of Darwinian evolutionary thought.

The second view was the orderly singularity school, which posited the antithesis of the former in that the nature of the singularity was orderly; some of the noted proponents of this view at that time were British mathematician Roger Penrose and Stephen Hawking. The orderly singularity theory was, indeed, an evolutionary theory, nonetheless. Both theories equally maintained one of the central doctrines of the evolutionary religion's catechism—the doctrine of a closed system. The dogma of a closed system is a metaphysical presupposition promoted by evolutionists who simply state that no exterior interference in the processes that go on within our universe can exist—or no divine intervention. The universe is considered a closed unit that cannot be impacted from without. That assumption is not a scientifically devised or empirically proved fact. It is an a priori metaphysical choice, which at the very onset claims that no divine agency exists.

Our Judeo-Christian cosmological model believes that the evidence overwhelmingly points to an intelligent designer. The idea that the human will could act inside a closed system does not interfere with the scientific process that universally dictates that all within that system obey the laws of physics. If such intelligence exists, then that intelligence can equally act within the closed system as He wills, even though His existence is not limited to that closed system. He is not a cheap conjurer who ignores the very laws of nature that He created. If anything, He has knowledge of laws we cannot yet imagine. Although the proponents of the orderly singularity theory understand that our universe could not have developed without such ordered discipline, they cannot account for what causes such ordered discipline in a materialistic worldview that depends solely on random ordering.

This materialistic metaphysical position that has become the paradigm within the halls of academia is the result of the Enlightenment's fundamental rejection of any spiritual reality and the resulting bifurcation of the unity of truth, which was referred to in Machine or Man. In their divided field of knowledge, they claim that science is a realm in which absolute truths can be known, but theology is a realm in which no absolutes can be known. According to their Enlightenment ideology, the physical sciences are in the realm of objective data; on the other hand, theology is simply in a relativistic subjective category that can have no real bearing on the realm of science or space-time/historical reality.

The idea that God created the universe, as recorded in the biblical Genesis account, was therefore automatically dismissed as mythological garbage and incongruent with the scientific field of inquiry. The concept of the gradualist evolutionary development of all that exists (including life) within our universe through random accumulative changes now became the paradigm of the academic and scientific institutions. Historians mark the moment of this bifurcation of truth when the humanities were divided from the physical sciences in our universities.

In direct contrast, our Judeo-Christian cosmological model believes in the primordial and fundamental doctrine of the unity of truth. Because all truth emanates from God, there can exist no artificial division of truth. What God has revealed as true, whether it is theological, historical, scientific, sociological, or relating to any other aspect of reality, is absolutely true. The lordship of God over the entire universe encompasses all reality—the physical as well as the spiritual. Hence, there is no division between the physical world and the spiritual world; they are both part of the same reality.

Most in our postmodernist culture would view this position as untenable and incompatible with science. That is the reason for the writing of this series. I propose to the reader that under the scrutiny of the scientific process, our Judeo-Christian cosmological model will show that it is not only more rational, but it has an enormously greater preponderance of scientific empirical evidence to support it.

The chaotic cosmological model was in favor before the universal acceptance of the second law of thermodynamics in the 1930s, after which it was shown that the entirety of the universe has the same time direction in which the overall entropy always increases, regardless of how far one extrapolates into the future. That is, it states unequivocally that everything in the universe goes from order (low entropy) to disorder (higher entropy) as time moves forward. In plain language, all things run down and break down in time. Therefore, the idea that disorder (chaos) could evolve into order became a bit more difficult to swallow. The chaotic cosmological model began to sputter and fade out.

In other words, the absolute universality of the second law of thermodynamics then forces upon us the obvious conclusion that if the world is increasing in entropy (disorder), then it must have had a beginning with much less entropy (more order) than exists at the present. And if in the present we observe a universe that seems to be almost completely smooth and ordered (homogeneous and isotropic), then the initial explosion of the singularity must have been extremely smooth and ordered from the first fraction of a second during the Big Bang—so smooth and ordered that it could hardly be believed that such order could come from the gradualist random ordering process that is at the heart of the evolutionary paradigm.

In complete antithesis, our Judeo-Christian cosmological model simply states that the very elements that began the Big Bang singularity not only came from the radiant energy of God, but the entire process was shaped and designed by His willful and active role during the entire process. It must have taken place with some exterior directed force acting upon the explosion of the Genesis Singularity to maintain it in such order during the most violent explosion ever in the history of the universe. The task for the naturalist became quite difficult, for they had to come up with a materialistic way to explain how this order could have arisen and been maintained without the directive force of a master designer.

The intrinsic properties of an explosion are, by nature, chaotic; that is, in all known explosions, there is by definition an enormous leap in entropy (disorder). How is it that in the most powerful explosion of all eternity, there was a pervasive and overriding ordering process throughout the entire event that allowed our universe to expand at an absolutely critical rate and end up in a smooth universe? That scientists could believe that an explosion of that magnitude could be controlled by random forces to such quiescence is a leap of faith of astronomical proportions, to say the least.

Paul Davis reluctantly admits:

> *It is here that we run into one of the greatest mysteries of cosmology. As explained observations show that the universe is highly symmetrical and uniform, both in the way that galaxies are distributed through space and in the pattern of expansion motion. If the universe that erupted from Jiffyland consisted of myriads of causally independent regions, why should they all have cooperated to form a smooth and orderly motion? If the universe began at random, it would have started out by expanding in a highly turbulent and chaotic motion, and only an infinitesimal fraction would appear in the smooth, uniform, isotropic motion that we actually observe. Out of all the remarkably chaotic motions with which the universe could have emerged from the Big Bang, why has it chosen such a disciplined and specialized pattern of expansion? (Davies 156).*

Davies inadvertently grants the universe the ability to choose, since by all measures, the improbability of this happening implies an obvious, willful choice. He subconsciously states the obvious inconsistency with his naturalistic position. Many possible models were proposed, but none could explain such a smooth expansion with enough minute lumps to cause the creation of stars and galaxies. The balance to achieve this is incomprehensible without the express form of some guiding force or exterior control.

Of course, this would be direct evidence of the possibility of the existence of an intelligent designer and is therefore metaphysically unacceptable to evolutionists. Yet they readily admit that this ordered arrangement of the universe is a practical impossibility if guided only by pure chance. The staggering magnitude of the improbability is, from an objective point of view, completely insurmountable. These scientists, if they count themselves honest, must mathematically accede the obvious reality, even against their emotional bias. Davies reluctantly admits this.

The likelihood of the present arrangement arising by chance appears to be negligible (Davies 157).

It appears to be negligible? That is a tad bit of an understatement. It would be like throwing a stick of dynamite into a pile of firewood and expecting to see, after the smoke clears, a nicely painted two-story house with a white picket fence, a two-car garage, all the appliances (dishwashers, washing machines, dryers, toasters, blenders, air conditioners, etc.), and the furniture neatly placed within the decorated rooms full of hanging paintings, mirrors, and flower in vases. Without some incredibly powerful force directing energy into the explosion and maintaining it in an ordered, smooth fashion, chaos would have ensued, and life could not have existed in such a universe. No evolutionist can deny that our universe is a highly ordered hierarchal system governed by absolute universal laws of physics. Ernest J. Sternglass declared when he began to conceive his theory of the primordial atom in the Big Bang, reworking Lemaître's original idea:

> The Universe appears to be a highly ordered hierarchical system *composed of rotating systems of increasing size, as first envisioned by Immanuel Kant two hundred years ago, rather than a random collection of galaxies* (emphasis added) *(Sternglass 81).*

Davies further accedes:

> *It seems, therefore, that the cosmic background radiation is testimony to the fact that the universe was born in disciplined—Quiescence, right back as far as one ten - million - billion - billion - billion - billionth of a second—which is quite a conclusion!*
>
> *If the above reasoning is correct, and some cosmologists are skeptical, it returns us to the paradox of why the universe began so smoothly. It is here that the anthropic principle may be of help (Davies 160).*

Well, I quite agree that the anthropic principle might be a help to Davies here, not in actually resolving the dilemma but in assuaging his angst over the obvious. The precise consistency of the universe beginning at the Genesis Singularity and all the way through its outward expansion is essential in determining the final product formed.

Now this process did not gradually become ordered. Gradual random processes form the fundamental axiom of the evolutionary paradigm to explain order coming from disorder. But this was nothing like that. It was born from the onset in a specifically predesigned order, as Davis claimed: "The universe was born in disciplined – Quiescence, right back as far as one ten – million – billion – billion – billion – billionth of a second – which is quite a conclusion!" (Davies 160).

There is no gradual, random process anywhere in the mathematics of the Big Bang that can support the evolutionary paradigm. Quite the contrary, it matches perfectly with our Judeo-Christian cosmological model. It not only had to begin with a homogeneously spread temperature, but it had to be exactly the right temperature coupled with exactly the right expansion rate. If it were too hot, the energy of expansion would have been too great. The universe would have exploded outward with such force that no atoms could have ever been formed. The universe would have been a thin soup of indistinguishable subatomic particles flying ever more distant from one another.

It is as if the material within it was somehow forced to become almost completely homogenous, all through the developmental phases of the explosion. Had that not happened, the result would not have led to the formation of galaxies and stars but rather to black holes. Instead, what we observe in our universe is that the formation of galaxies preceded the formation of black holes. These are then formed either inside the galaxies as the age of the galaxy begins to show the effects of the second law of thermodynamics and the ordered begins to become disordered, or as the collapse of large stars that run out of fuel. The black holes are, in effect, the death of the stars and galaxies.

We have come now to understand that many galaxies indeed have black holes within them. But what this means is that the formation of these galaxies would have been bypassed, and the black holes would have been created before the more ordered galaxies. As it is now, the black holes are, in fact, formed as the central cores of the galaxies begin to implode or as the stars die. It is the natural terminus of the increase in entropy universally determined by the second law of thermodynamics.

Hence, the galaxies are formed first, and the development of a black hole is then the logical sequence of events, following the course expected through the second law of thermodynamics. How could these galaxies form in such an ordered fashion so as to not end up as black holes from the very beginning? And how could they further develop into clusters and superclusters through random processes without initially having formed into super-giant black holes? The slightly fractal nature of our universe seems to be designed within such a narrow parameter of possibilities that it is impossible to concede that random chance happenstance could account for this structure. If the underlying presupposition of the Darwinian model is correct, then there should not be any galaxies out there. And more importantly, if by some miraculous chance they were created, they should be randomly situated and not in the ordered clusters and superclusters that we observe today. In fact, there should be very few stars at all. "There shouldn't be galaxies out there at all, and even if there are galaxies, they shouldn't be grouped together the way they are" (Trefil 3).

Undirected random processes cannot account for the formation of our magnificently ordered universe. Ironically, Davies understands that and agrees with it, yet he cannot seem to bring his mind to accept the obvious inference.

> *However in general grounds it is clear that if the material were very lumpy then black hole formation is more likely to occur than if the matter were smooth and evenly distributed.* It seems safe to suppose that a universe that began in a very inhomogeneous condition would emerge from the Big Bang populated not by stars but by black holes (emphasis added) *(Davies 165).*

He continues:

The second law states that the total disorder always increases with time. The black hole obeys a law, which says that it always increases with time, so one suspects that the size of the hole is a measure of its degree of disorder. This guess was confirmed when the connection between the temperature of a black hole, as computed by Hawking, and its mass was examined: the black hole turns out to obey the same relation between disorder and temperature as does a gas, if the area of the hole is used as a measure of disorder. The area is in turn related to the mass of the hole so we have at hand a means of comparing the degree of disorder of a given mass of material with the equivalent disorder that would be achieved should that material be discarded down a black hole. In the case of one solar mass of matter the black hole disorder works out at several billion, billion times that of the actual sun, a result that carries an ominous implication: All else being equal , there is an overwhelmingly greater likelihood of the material of the sun being inside a black hole than in a star. *The crucial qualification here is all else being equal. Evidently, all else is not equal in our universe, or there would be no sun, or any other stars.* If the primeval material was churned about at random, it would have been overwhelmingly more probable for it to have produced black holes than stars, because the holes being so much more disordered can be produced in a vastly greater number of ways. For every star that formed, countless billions of more easily achieved black holes should have accompanied it. . . . *The relationship is therefore an escalating one, for when the numbers become large a small amount of additional disorder represents a highly greater probability.* In the case of the sun, whose disorder is only one hundred billion billionth of the equivalent black hole, the chances against the sun, rather than the black hole, emerging from a purely random process will be roughly one followed by the same number of zeros! That is one followed by one hundred billion, billion zeros, which is pretty improbable by any standards.

If the same argument is applied to the entire universe, the odds piling up against a starry cosmos become mind boggling: one followed by a thousand billion, billion, billion zeros at least (emphasis added) *(Davies 168–169).*

When dealing with such large numbers, our minds are incapable of understanding in any concrete way the unbelievable size that it represents. In order for us to begin to capture the true significance of such large numbers, we must equate them with something that we can compare.

We know that a minute contains sixty seconds. One hour contains sixty minutes, so one hour contains 3,600 seconds. One day has 24 hours, so one day contains 86,400 seconds. One year has 365.25 days, so one year contains 31,557,600 seconds. If we were to compute the number of seconds that have elapsed in the supposed 15 billion years that evolutionists claim our universe has existed, we would end up with a 1 with 18 zeroes after it (10^{18}). That is, there are 1,000,000,000,000,000,000 seconds in our supposed evolutionary history. Yet this number is dwarfed by the improbability of our universe having been created through chaos; that is, "one followed by a thousand billion, billion, billion zeros at least" (Davies 169).

Imagine our smallest interval of time, the amazing jiffy, which is the measure of the time an atom vibrates one time. Now this atom will oscillate once in a million, billion, and billionth of a second. Imagine how many jiffies there will be in one hour, in one day, in one week, in one month, in one year, in one century, in one millennium, in one million years, in one billion years, in 15 billion years, the supposed age of our universe. If you could count the number of jiffies in 15 billion years, the number is a 1 with 40 zeros (10^{40}).

That is, there are 10,000,000,000,000,000,000,000,000,000,000, 000,000,000 jiffies that have transpired in the supposed 15 billion years of the evolutionary history of our universe. Compare that enormous number to the number calculated by Davies that has "a thousand billion, billion, billion zeros at least." All the books in our world could not contain the number of zeroes in that number.

If we add the number of atoms in every object within this universe, the number is a staggering 1 with 80 zeros (10^{80}). Yet this pales compared to the odds for our universe having evolved without directed energy. Can you understand why a thousand billion, billion, billion zeros is an impossible odd? This is the statistical hurdle that evolutionists must leap by blind faith. We must rationally conclude that evolutionists are truly people of deep blind faith. I wish I could have that kind of faith, but I don't.

The crucial qualification missing in Davies's model is an intelligent designer. There is no other way to explain this crucial choice made at the very moment of creation in a naturalistic system. The statistical hurdle is insurmountable and completely illogical for any practical consideration. This probability is so minutely miniscule compared to the immense other probabilities that would have created a chaotic universe filled with black holes that it is really amazing how a rational scientist, knowing full well the immensity of this improbability, could actually choose to believe that it just happened by accident. Is this not the very definition of an irrational bias and the very antithesis of what the scientific process is supposed to be?

But naturalists are undeterred by such odds. Their atheistic metaphysical presuppositions trump all empirical data. Against all reason, they are now trying to sell the idea that black holes are responsible for the creation of galaxies. This, of course, would render the second law of thermodynamics void, for it would predict that entropy has decreased in the universe. But we will deal with this absurd proposition later.

The Homogeneous Expansion of the Genesis Singularity and the Horizon Problem

Most modern cosmologists hold to the idea that the force of gravity was largely responsible for molding the universe. This slow, gradual process of the accretion of matter due to gravity of course requires very long spans of time. Our Judeo-Christian cosmological model, however, disagrees. Our position is that the much more powerful electromagnetic force was the prime molder of our universe. But we will address that later.

The Standard Model of the Big Bang stipulates that during the first 300,000 years of the universe, the Genesis Singularity had exploded and expanded to a volume of 300,000 light years from the initial location of the singularity. According to scientists' calculations, the matter within this volume was in a plasma state, and the photons were encapsulated within the dense soup of plasma, which was so dense that light could not escape. In other words, the universe was still opaque. Darkness was over the face of the deep.

The expansion has two critical effects. First, it lowers the temperature and allows the high-energy particles zipping about in frenzy and smacking into each other to slow down and eventually coalesce. Second, it reduces the density as particles move away from one another and avoid smacking into each other more. This is particularly important so that matter created from that original plasma energy might have a chance to avoid striking its antimatter and annihilating. That precise expansion ratio is critical to both these processes. That ratio of expansion had to be carefully chosen right from the beginning in order for our universe to form as it did. Had the material remained too dense, the chances of matter striking antimatter would have been greater and would have consequently annihilated all particles. Our universe would be completely empty of any matter.

According to scientists' model from this critical period during which the volume of the universe reached a distance of 300,000 light years from its point of origin, which I call the Alpha Point, density was reduced enough in order for light to escape—and there was light. Yet all during that process, some force must have been acting on the plasma soup to maintain its homogeneity. One look at our sun, which is a ball of spinning plasma not unlike the expanding Big Bang, and you can see magnetic forces that twist and turn loops, creating a caldron of churning matter that is anything but homogeneous as occasional solar storms shoot out huge coronal mass ejections. The conundrum for the evolutionist is that today, some 15 billion years after the Big Bang, the universe remains basically (although not entirely) homogeneous throughout its entire length and breadth. That is physically irreconcilable with the randomly generated naturalistic model.

In other words, if we look due west to the edge of our universe, we find that the background cosmic radiation is almost exactly the same as if we were to look due east at the opposite edge of the universe. That does not mean that the star field will look identical in the different hemispheres. It means that the energy density of the universe is, to an exacting degree, almost completely homogeneous throughout every hemisphere at which we can look.

When one considers that the volume of the universe has expanded to 10 to the 88th power from that critical point 300, 000 years from the Big Bang, that uniformity should not exist if the universe is ruled by chaos and randomness. Not only is it inconceivable to have started in homogeneity, it is even more inconceivable that having expanded to 10^{88} of its size, it could still be held in such uniformity, if the universe is run only through random chaotic processes.

That problem for those who promote the idea that our universe is guided solely by random processes did not begin 300,000 years after the Big Bang. Before that critical point, when light was trapped within matter in a broiling plasma state, before photons were decoupled from matter and light became visible, uniformity could not be expected, either. One look at the hot ball of plasma we call our sun shows us that there are constant swirls and currents and even solar earthquakes created by the churning plasma. How could a ball of plasma with billions of times more energy in the form of electromagnetic currents and heat than our sun be controlled and tamed into a homogeneous distribution of energy by undirected random processes? According to the Heisenberg Uncertainty Principle of quantum mechanics, the universe could not have exactly the same properties everywhere when it precisely predicts that we cannot even know from one point to the other what state that property is in at any given moment in time. What finite force could accomplish all these things in a naturalistic system?

How can random finite forces maintain this homogeneity from horizon to horizon in the expanding universe? The standard physical theories, as presently understood, tell us that information cannot travel at speeds greater than the speed of light. If we look at a galaxy "A" at the western edge of our universe, some 15 billion light years away from the point of the singularity, and at galaxy "B" some 15 billion light years in the opposite direction, there are 30 billion light years between the two opposite ends.

In the 15 billion years of time that has passed, there is not enough time for the two opposing galaxies at either end of the universe to have any communication between them. No particle or force can travel 30 billion light years in 15 billion years. How, then, could they remain homogeneous when there is no causal connection between the two extreme ends? No particle or boson can travel faster than the speed of light. The magnitude of this problem is enormous for the Standard Model. Cosmologists have dubbed this glaring inconsistency the "horizon problem." It is a paradox that is unexplainable unless there is a God who holds all things together that spans beyond our finite universe, as our Judeo-Christian cosmological model recognizes.

There is no present physical law that can explain how our universe continues to expand in homogeneous fashion without some unknown causal connection that maintains it so, which can span the entire universe. It is as if something outside of our visible space-time continuum is acting upon our universe. Hmm! I wonder what that could be?

The fact that our universe developed and continues in this ordered fashion is the empirical evidence of a dynamic creator as opposed to a static creator (deism). It is the scientific evidence of a universe that is not anthropocentric, but deocentric. It is the evidence that our universe is not ruled by randomness but by a willful design by an immense power that can span its influence over every inch of our entire universe simultaneously.

The Four Fundamental Questions

There are several considerations we must make to understand the true nature of our universe:

1. Is the universe finite or infinite? We have already addressed the fact that the universe had a beginning. But some say that although it had a beginning in time, it was born infinite in spatial dimensions. We will address this.

2. What is the shape of our universe? As we will see, the shape of our universe determines whether it can be infinite or not.

3. Can the large-scale design observed in our present universe be formed through purely random processes? It is here that the Rapid Inflationary Model provides a materialistic mechanism to explain the isotropic and homogeneous texture of our present universe.

4. But is our present universe really completely isotropic and homogeneous? Can the Rapid Inflationary Model explain the findings of COBE (Cosmic Background Explorer), WMAP, and the Planck satellites that mapped the microwave background radiation in our entire universe?

Our Judeo-Christian cosmological model differs drastically from the naturalist model. Let us compare the claims and the empirical data that are used to support both models.

CHAPTER 8

• • •

THE CENTER OF THE UNIVERSE AND OUR JUDEO-CHRISTIAN COSMOLOGICAL MODEL

A t this point, there are three competing ideas regarding the shape of our universe. The most widely held idea is the spherical model, like the surface of a balloon. In this model, space is curved. Most naturalists have traditionally accepted the spherical model because their preference is to believe that the universe has no center. The second one is that the universe is shaped like a saddle. Here, space is also curved.

Then there is the proposition that the universe is flat. In this model, space is flat; that is, the large-scale structure of the universe is Euclidian in structure.

Diagram 8.1: The three possible shapes to our universe as proposed by modern cosmologists: 1. spherical, 2. saddle-shaped, 3. flat

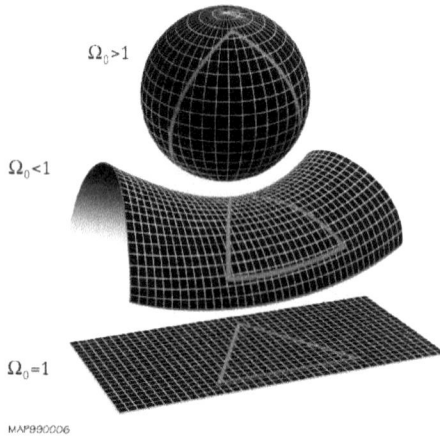

$\Omega_0 > 1$

$\Omega_0 < 1$

$\Omega_0 = 1$

MAP990006

To illustrate the flat model, imagine a large flat sheet of rubber (representing our space-time universe) containing within it the same painted dots that represent the matter within our universe. If we stretch this rubber sheet outward, the painted dots move away from one another at different rates. The painted dots nearest to the edge move faster than the dots near the center of the sheet. If we were to choose a dot that is halfway from the center of the sheet as our observation post, we would see all dots beyond our dot (toward the outer edge of the sheet) as moving away from us. The starlight from these dots would be red-shifted. If we were to look in the opposite direction, toward the center of the sheet, we would see those dots as moving toward us. Their starlight would be blue-shifted.

The percentage of the number of stars that are blue-shifted to those that are red-shifted would depend on the position within that sheet from which we take the observations. The closer we are to the outer edge of the sheet, the more blue-shifted stars we would observe. We, being on the outside rim, would see most stars coming toward us. The closer we are to the center of the sheet, the more red-shifted stars we would observe. Only if our observation point is exactly in the middle of the flat sheet could we see all stars as being red-shifted. Now, that does not mean that the shape of the universe is two-dimensionally flat. It could be a sphere, but it would then be a solid sphere that expands outwardly in all directions. What we mean by the universe being flat is that space is not curved; it is Euclidian.

Since the evidence of the red-shifted stars is almost 100 percent from our perspective on Earth, then if space is flat, Earth must be at the center of the universe. Naturalistic theorists would tell you that there is no center from which the universe expands, because they have subjectively chosen to begin the calculations of the Big Bang with the presupposition that the spatial dimensions of the universe were from the beginning infinite and boundless. But this is merely a subjective bias, a speculation not founded on any mathematical or scientific necessity. In fact as we will see later, it is contrary to the science we know regarding our matter (energy)/space-time continuum.

It is because the composition of the universe seems uniform in every direction we look from our planet and all stars seemed red-shifted from Earth's point of view that only two possibilities exist for explaining this phenomenon. Either we are propitiously located at the center of the universe, or the universe is infinite in size and, therefore, looks the same in all directions. Naturalists promoting the Copernican principle prefer that the universe be infinite. Evolutionists always prefer infinity to finiteness. It gives them a much larger span of time to accomplish the small, gradual changes necessary for random ordering to seem more plausible by the sheer number of chances to evolve into a more complex fashion.

The idea that Earth would be at the center of the universe contradicts their cherished Copernican principle, which is the antithesis of our Judeo-Christian cosmological model. Therefore, they insist that it must have been infinite at its inception since the equations would render a universe that appears as it does but, most importantly for them, without Earth in any significant position in the universe.

The idea that Earth is at the center of the universe, held by our Judeo-Christian Cosmological Model is a repugnant position to those who hold to an atheistic presupposition. But in reality, their choice to begin with an infinite universe is nothing more than a mathematical fix to promote their ideology, since being in the center of the universe would reinforce the opposite idea that our planet was specially designed. There is no mathematical necessity to begin with an infinite universe. There is no empirical evidence that forces them to begin with an infinite universe. There is only a metaphysical preference to begin with an infinite universe. In other words, because they are not willing to consider the Genesis account, which places Earth at the center of the universe, their mathematical conclusions are tainted by their ideological bias.

From the point of view of existentialists or naturalists, a universe that gives Earth no specific importance in its placement within it is quite appealing due to their metaphysical presupposition that excludes God. This is the foundational metaphysical proposition of their Copernican principle. Since the choice of a finite and bounded universe necessarily implies a central position, it is categorically ignored as a possibility. It is precisely this natural bias against the idea that the universe is designed with purpose related to humanity that causes them to choose one possibility over another. This overt antipathy illustrates the obvious subjective metaphysical bias of their intent to discount anything that would smack as a special feature to our home planet and which would logically infer a design by a creator. There is no mathematical reason for choosing one model over another. The choice is a metaphysical one.

For naturalists to be able to exclude God from any consideration, they must exclude any importance or purpose to the existence of our planet and the uniqueness of life we find therein. To admit to a special creation of our planet as the center of a finite universe would be to admit the possibility of a personal creator who has always existed before and outside the universe.

As we shall now see, our Judeo-Christian cosmological model, which begins the calculations with a finite and bounded universe, is a more rational scientific proposition that matches the observable data more completely. The same calculations, using the laws of general relativity, would also explain why the universe looks as it does, given that our planet would be close to the center of its inception.

Is Our Universe Finite or Infinite?

The cosmological principle is the evolutionary assumption that space is infinite or unbounded and exactly homogeneous. The Copernican principle is the idea that Earth has no special place of significance in the universe. These two metaphysically biased cosmological assumptions form part of the catechism of the Darwinist evolutionary religion. It strikes me as ironic how many principles the unprincipled relativists are prone to invent for their Darwinian catechism.

Fundamentally, there are only two choices for the shape of space: it is either flat or curved. The answer to that will allow us to understand the overall shape of our universe. However, cosmologists tell us that there are three basic possible shapes that our universe could have taken as it expanded from the Big Bang. Some are possible if space is curved, and others are possible if space is flat. Each of these can also be either finite in extent (bounded) or infinite in extent (unbounded). The third consideration is the position of Earth in relationship to the universe. Each combination of these alternatives can be tested in order to determine which one is the actual and physical representation of the shape of our universe.

What does it mean that space is flat or curved? If space is curved, then measurements between faraway galaxies would show that light between them is curved. If the universe is spherical, as many of the proponents of a spherically shaped or saddle-shaped universe declare, then the measurements between three separate galaxies located far from each other would show curved lines, either convex (sphere) or both convex and concave (saddle).

In a flat universe, the lines triangulating three galaxies located far from each other would be straight. That is called a Euclidian universe, after the famous Alexandrian Greek mathematician. Most of the Western population has used Euclid's book of geometry, even well into our last century.

But also crucial to understanding the observational data, we must consider the position of Earth (our observation point) in that universe and whether or not the universe is bounded and finite or unbounded and infinite. Thus we will now consider five possible combinations of the three possible shapes of our universe and analyze their alignments with the current empirical data and our physical observations.

Option 1: A Spherical and Finite Universe

Some have proposed that the shape of our universe is as the skin of a sphere (such as the skin of a balloon). The central part of the sphere, or core (the air in the balloon), is not included as part of the universe in their model.

To illustrate the spherical model, imagine painting dots on the surface of a perfectly spherical balloon representing the matter within the universe. As we blow up the balloon, the surface increases in size, and the dots move away from one another at a symmetrical rate. This allows the naturalist to claim that at any point on that balloon, looking in all directions, the universe is expanding away and the starlight would be seen as red-shifted. Because that is precisely what we observe from our point of view on Earth, it has become the most popularly accepted choice by evolutionists.

This model does not have a center and, therefore, agrees with their cherished Copernican principle, which gives Earth no special place of significance in the universe—a very endearing proposition to deophobes.

In addition, due to its spherical shape, it has a very unusual property. One could travel infinitely and never reach the end. If the universe is spherical, then space is curved. If our universe is spherical (like the skin of a balloon) and light from the galaxies on the other side of the sphere is being bent around the circle to reach us, then we would always see the same number of stars and galaxies, no matter which way we looked. That shape could explain why our universe looks the same in every direction. The term used for this phenomenon is *isotropic.*

The key evidence used to show that our universe is isotropic is the microwave background radiation, which seems to be the exact same temperature in every direction we can measure. This is thought to be the afterglow of the Big Bang, and therefore it implies a universe that has expanded uniformly and isotropically. The spherical model then matches well with this.

The most important feature of this curved space model is that as the sphere is expanding, all points seem to be moving away from all other points. In other words, no matter what point of reference one chooses to measure from within the universe, all stars would look red-shifted as it expanded in size. This also matches our observations, for practically all the stars are seen as red-shifted, or moving away from us.

In addition, the spherical universe then has no real center. As the skin of the sphere expands, it no longer has any linear connection to the Alpha Point at the center of the sphere. This view then aligns quite nicely with their Copernican principle that assigns no place of importance to Earth within the structure of the universe.

Another great appeal of this spherical universe for the evolutionist is that the sphere makes the universe seem infinite, because one can travel in any direction around the sphere for infinity. However, that also means that you could theoretically end up where you started— quite an unusual proposition.

In this model, space is curved. In other words, lines triangulating three different galaxies located far from each other would not be straight but curved as it follows the spherical shape of space in the universe. Space would be convex.

Now they claim that it is unbounded because it can be traveled for infinity, but that is not exactly true. Their reasoning is faulty. Anything that is said to have a shape by definition has a boundary that creates that shape. It is simple elementary logic. A careful examination of the following diagram will illustrate that it is only unbounded in two directions of our three-dimensional space-time continuum.

In a spherical universe, space is curved, and a cross section would appear as a donut, but it is, in fact, a complete sphere. Both Y and Z vectors are unbounded, but the X vector is bounded in both ends. Hence, it is not completely unbounded and cannot be truly infinite in all vectors of our three visible dimensions. The spherical model is thus a finite universe.

Diagram 8.2: Cross-Section of Spherical Universe

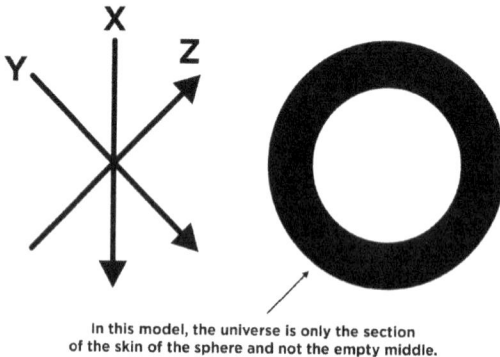

In this model, the universe is only the section of the skin of the sphere and not the empty middle.

Since the spherical model is bounded, then by definition it also cannot be considered truly infinite. Space does not extend into infinity in all three vectors of the visible space-time continuum if it has the bounds of the skin of a balloon. Any shape that has a boundary cannot be considered infinite. It may be infinite in any of the directions not bounded, but where bounds exist, infinity does not.

In fact, if it were truly infinite, it could not even be a solid sphere. In a solid sphere, by definition the outside of the sphere is a boundary. Any boundary precludes infinity. A truly infinite universe cannot have any shape. It must have no end in all directions.

Option 2: A Finite Saddle-Shaped Universe

There is a second potential shape to our universe that has been considered by cosmologists, where space would also be curved and turned on itself as in the shape of a saddle. From an observation point at the center of the saddle, space would be concave looking in one direction. But from that same observation point, looking at an angle of 90 degrees from the previous direction, space would be convex (see Diagram 8.1).

If our universe is shaped like a saddle, then it cannot extend unto infinity in all areas. It is limited by its boundaries. As one observes in any direction, light from the galaxies on all the other sides of the saddle is being bent around the skies and would appear isotropic. This matches our observable data.

In a saddle-shaped universe, lines triangulating three galaxies located far from each other would be curved in either convex or concave curves, depending on which direction one observes and their position in the shape of the saddle. Space would be curved. However, once again, this model has bounds because it has a shape and cannot therefore truly be considered an infinite universe, because anything with shape has bounds.

Option 3: A Flat and Unbounded or Infinite Universe

There are two options in this category:

Option 3A: Flat and unbounded universe without a beginning and without Earth at the center

If our universe were flat and infinite but not completely homogeneous, then no matter which way you looked, the universe would still appear the same for infinity. That is, the universe would seem isotropic in all directions, because no matter which way we turned, we would be looking toward infinity. Looking in any direction toward infinity, we would see a fairly equal distribution of stars and galaxies, even though it would not be an exactly homogeneous distribution.

In an infinite universe, there is no center; hence, the location of Earth would be inconsequential. This aligns well with the evolutionary Copernican principle. If there were no beginning, then one would expect, in a randomly ordered reality, for the stars to be moving about in random order. Viewing outward from Earth, we would expect to see an equal number of red-shifted stars and blue-shifted stars. This was the expectation of evolutionary cosmologists when they held to the Steady State theory prior to Einstein's and Hubble's contributions.

Einstein's equations, however, showed that space, time, and energy/matter had a beginning. Hubble showed that the universe is not static, but expanding outward. Practically, all stars are red-shifted and moving away from us. This proposition, therefore, has already been disqualified.

This obviously does not match our observations. But evolutionists have proposed another option to a flat and unbounded universe in which all galaxies are exactly homogeneously distributed throughout the universe. This qualification allows all observations from any given galaxy to appear as though all other galaxies are red-shifted, from their perspective.

Option 3B: A flat infinite or unbounded universe with Earth not at the center of the universe and with a completely homogeneous texture (the Cosmological principle)

In an infinite and unbounded universe, there is no center. Hence, the position of Earth is inconsequential. If every galaxy in the universe were exactly equidistant from one another (in other words, if the universe would be perfectly homogeneous and unbounded), then an expanding universe could appear exactly isotropic. All points would appear red-shifted, no matter what galaxy from which one chose to view the universe.

This is the Cosmological principle, which is fundamentally the necessary foundation from which the Copernican principle can be established. It allows for an explanation of the red-shift observations from our planet without having to start with Earth at its center, since in an infinite universe there is no center.

Here are the three key qualifiers:

1. Each galaxy must be equidistant in order for our universe to be perfectly homogeneous.
2. The universe must be boundless (that is, infinite in size) and without a center in order for the Copernican principle to be plausible.
3. The universe must be expanding.

If the universe is bounded, then by necessity it has a center, and the expansion of the universe would not result in this fashion. Therefore, this unbounded model is highly preferred by evolutionists motivated to propagate their cosmological principle and the Copernican principle, which gives Earth and our solar system no special place in the universe.

Let us compare four galaxies in order to illustrate this model.

We will name these galaxies, A, B, C, and D. In accordance with the three prerequisites of this model, each of these galaxies, expanding among each other, is equidistant to each other (exactly homogeneous), and the universe is unbounded (infinite). The length of the arrow in the illustration symbolizes the degree of red shift seen from the position chosen to view outwardly. The longer the arrow, the more red-shifted is the light being measured by the viewpoint galaxy.

Choosing galaxy B as our first point of observation, we would see that galaxies A and C are equidistant, and their red shifts are comparable, but galaxy D is further away, and its red shift is greater. Galaxy E is even further away, and its light is therefore red-shifted the most.

Choosing galaxy C as our point of observation, galaxies B and D would have equally sized red shifts, but galaxies A and E would have a longer red shift since they are further away.

Choosing galaxy D as our point of observation, we would see galaxies E and C as red-shifted the least, and galaxy B would have a greater red shift. Galaxy A would have the greatest red shift since it is the furthest away.

Diagram 8.3: A flat, infinite universe with Earth not at the center of the universe and with a completely homogeneous texture

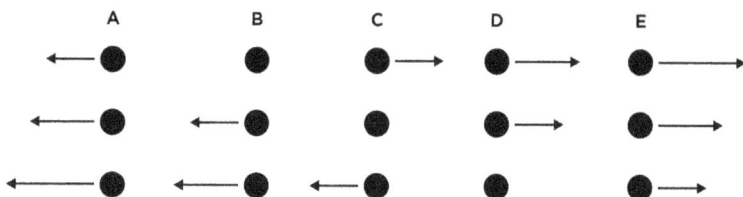

The homogeneous distribution of the galaxies in our universe could then theoretically provide us with an explanation of the almost exclusively observed red-shifted light that covers our entire sky. This is the preferred view by Steven Weinberg, winner of the 1979 Nobel Prize in Physics, who very much likes the idea that our solar system and Earth have no special place of significance in the universe.

> It appeared that aside from a few close neighbors like the Andromeda Nebula, the other galaxies are generally rushing away from our own. Of course this does not mean that our galaxy has a special central position. Rather, it appears that the universe is undergoing some sort of explosion in which every galaxy is rushing away from every other galaxy.

The interpretation became generally accepted after 1929, when Hubble announced that he had discovered that the red shifts of galaxies increase roughly in proportion to their distance from us. . . . This hypothesis is so natural (at least since Copernicus) that it has been called the Cosmological Principle by the English astrophysicist Edward Arthur Milne.

As applied to galaxies themselves, the Cosmological Principle requires that an observer in a typical galaxy should see all the other galaxies moving with the same pattern of velocities, whatever typical galaxy the observer happens to be riding in (Weinberg 21–22).

Of course, the entire scheme of this proposition is dependent on three major assumptions:

1. The first is the premise that there is an absolute homogeneity throughout the entire cosmos. All points of reference must be equidistant. If so, then viewing from any galaxy, all galaxies would be seen as also receding from them. Here, it is important to distinguish between isotropic and homogeneous. When we look in every direction from our point of view on Earth, there seems to be an equal number of galaxies distributed in every direction—it appears isotropic. Homogeneous means that these galaxies are exactly equidistant from one another, so the texture of our baryonic matter within the universe is almost completely smooth.

As can be seen, the velocities obey the Hubble law: the velocity of any galaxy as seen by any other is proportional to the distance between them. This is the only pattern of velocities consistent with the principle of homogeneity (Weinberg 22).

Clarification is necessary here. Hubble's law does not say that all galaxies are homogeneously distributed. It says that their velocity moving away from us varies proportionally to their distance from us. The farther they are, the faster they are moving and receding from us. The speed is proportional to the distance.

In the next few sections, we will examine the results of the three satellites sent to map the cosmic microwave background radiation (CMBR) of the universe to see if the baryonic matter and the CMBR are exactly homogeneous. But for now, let us consider the other two assumptions.

2. The second assumption is that the Big Bang singularity became instantly infinite. The truth is that evolutionists can claim that space is unbounded until the cows come home. There is no way to have a beginning singularity with an unbounded universe. The Genesis Singularity of the Big Bang, by definition, must be bounded in order to be a singularity. Without a bounded singularity, there can be no singularity. To claim that a singularity with definitive boundaries expanded to infinity in a finite amount of time is an irrational speculation that has no credible mechanical process of achieving it.

No spatial dimension can achieve infinite status from a finite start with a bounded singularity, since it would have to become instantaneously infinite, which is a physical absurdity. Not only is there no mechanism to begin with a bounded universe that can change it to an unbounded universe, there is also no empirical data that suggest that our universe is unbounded to begin with.

In order for a flat universe to become infinite in space, it must extend in all directions for infinity. That is an impossible physical process that cannot be rationally supported as culminating in a finite amount of time since the Big Bang. In a spherical universe, like the skin of a balloon, there does not have to be a center. But in a flat universe, there has to be a center at the Alpha Point, where the Big Bang took place. This model is impossible to reconcile with a beginning. Anything that begins bounded cannot become unbounded. It is an irrational leap of faith that cannot be supported by any empirical mechanical process. Quite simply, if there is an Alpha Point, there is a center. This assumption is an unsupported wishful speculation.

3. The third assumption is that the universe is dynamic. It is expanding. This assumption has been verified, and, in fact, new evidence seems to point out that it is expanding at a much faster rate than previously documented. A team of researchers led by Adam Reiss (2011 Nobel Prize in Physics) and using data from the Hubble Space Telescope has calculated that the farthest edges of our universe may be expanding as much as 9 percent faster than previously believed. Using the telescope, researchers tracked the movement of 2,400 stars and 300 supernovas. Their measurements gave the Hubble constant a new expansion rate of 73.2 kilometers per second per megaparsec. According to their calculations, that would mean that in 9.8 million years, the universe would double in size.

But they are failing to recognize the time dilation phenomenon experienced at the edge of our universe as space is being stretched. If the Judeo-Christian cosmological model is correct, then the doubling of the size of our universe will take place long before just one thousand years. The increase in age and entropy will also cause an increase in the number of quasars we will observe as well as an increase in their distance from us. We therefore predict that in the coming few decades, the percentage of the expansion rate of our universe will dramatically increase. We also predict that in not too long, the observation of quasars and supernova will begin to cross beyond the 15-billion light year horizon limit of our present observations. In fact, we predict that when that doubling is reached, space-time will begin to crumble backward as the fabric will be torn or ripped apart by the enormous entropy of the death of galaxies at that point. Our universe will then reach the Omega Point as that rip travels backward toward the Alpha Point. Time will tell if the Judeo-

Christian model is correct. Let us now discuss the fourth option to the shape of our universe.

Option 4: A Flat and Finite Universe

There are two options in this category.

Option 4A: A flat and finite or bounded universe without Earth at the center and homogeneous.

What if the universe were flat, finite, or bounded? A flat and bounded universe by necessity has a center point—an Alpha Point. If our universe were flat, finite, or bounded, and if our home planet were not near the center, then, as we looked out into all directions of our universe, we would see more galaxies on one side, the side farthest to the edge of the universe. In other words, one side would have more galaxies than the other, and our universe would not be isotropic. The universe, however, appears isotropic in whatever direction we look from Earth, so that does not match with reality.

Even if the universe were exactly homogeneous and all galaxies were moving apart equidistantly according to their distance from the Alpha Point, the fact that we are not at the Alpha Point in this model means we will see one side with more stars than the opposite side, and we will also see some blue-shifted stars.

Diagram 8.4: A flat finite universe with Earth not at the center of the universe and homogeneous

The diamond represents the center of the universe while the circle represents our earth.

If Earth is not at the exact center of this flat, finite, and bounded universe, then we would see the light of the stars on the opposite side of the Alpha Point as more red-shifted than the side we are inhabiting.

The side our planet happens to reside in would then have many blue-shifted stars, which would point to the center of the universe.

Since the observational data show us that almost all stars are red-shifted, this model with Earth not in the center also does not match reality. Since we now know that our universe had a beginning, then in a flat, bounded universe, all things would be expanding from that central point, the Alpha Point. Even though this Alpha Point is contested by evolutionists who wish to imagine a universe that has no center in order to escape giving Earth any special significance to its location, they have absolutely no empirical evidence to disprove this possibility. A bounded universe cannot escape having a center, or Alpha Point.

If this flat, finite universe has a center, then, depending on where we are in relation to that center, we would see some stars coming toward us (blue-shifted) and some stars moving away from us (red-shifted). Therefore, depending on Earth's position relative to the Alpha Point of the Big Bang, we would see that some stars are moving toward us, and others are moving away from us. In other words, the starlight of the stars moving toward us would be blue-shifted, and the starlight of the stars moving away from us would be red-shifted.

Once again, the observational data do not match this. From Earth's perspective, almost all stars around us are moving away from us. Our observation from Earth is that starlight is red-shifted in every direction. This model, with Earth not in the center, does not match our observational data.

Option 4B: A flat, finite, or bounded universe with Earth at the center and largely isotropic but not entirely homogeneous.

A flat, finite universe must have a center, the Alpha Point for the Genesis Singularity. The empirical evidence shows us that almost all starlight as viewed from Earth is red-shifted. The number of blue-shifted starlights is so miniscule that it would be impossible to believe that we are in a flat universe unless we were at the center of the universe. If Earth is near the center of the universe, all starlight would appear red-shifted. This matches with observable data.

If we lived in a flat, finite, and bounded universe and our planet were located in the center of the universe, then we would see generally isotropic skies everywhere we looked into the cosmic skies. This also matches the observable data in our cosmic skies. The universe in this model would not need the galaxies to be exactly homogeneous.

This, however, would contradict the cosmological principle that states that our universe is infinite or unbounded and exactly homogeneous. It would further contradict the Copernican principle that Earth has no special significance in its location in the universe. A flat, bounded, and finite universe with Earth at its center would then be an outright refutation of the standard evolutionary cosmological model.

On the other hand, it would match perfectly with our Judeo-Christian cosmological model that states God created our universe a finite time ago with the express purpose of providing a universe for humankind to inhabit. This means that the purpose for the entire universe is tied to God's purpose for humankind. This further means that Earth does have a special significance in more ways than just its location in the universe.

Furthermore, our Judeo-Christian cosmological model claims that this universe is both finite in space and time. It has a beginning and an end. It has an Alpha Point and an Omega Point. Therefore, it cannot be unbounded.

Diagram 8.5: A flat, bounded and finite universe with the Earth at the center of the universe. This is the position of our Judeo-Christian cosmological Model.

Our Judeo-Christian cosmological model specifically notes that God created Earth for humans to inhabit, and for this reason, He created the universe around it. That implies a central location in the universe. When considering all the possible places Earth could have formed in a randomly generated universe, the idea that Earth would be propitiously found at the very center by blind chance is a statistical improbability of such magnitude that it would be, for all practical purposes, a miracle of the first order. It is, in effect, irrational to believe that random ordering could have accidentally placed Earth at the exact center of our universe. It is for this reason that evolutionists have preferred the models that are unbounded and infinite in order to prop up their Copernican principle, which aligns better with their atheistic presupposition.

If random evolutionary forces created the universe, the chance of Earth ending up at any position other than the center is almost certain. In other words, the likelihood of our planet ending up at a point where there would be a preponderance of mixed red-shifted stars together with blue-shifted stars would be an almost absolute certainty. But that is not how our universe appears. Almost all stars from our observation point on Earth are red-shifted, meaning they are all moving away from us in every direction around us.

Hence, the only way evolutionists can explain away this phenomenon is to say that the universe has no bounds and is exactly homogeneously distributed. Therefore, viewing outward from every other galaxy in space would look like all galaxies are also red-shifted. In effect, we now have three questions to answer before making the right choice:

1. Is space curved or flat?
2. Is space infinite or unbounded?
3. Is the baryonic matter in the universe universally homogeneously distributed?

1. Is Space Curved or Flat?

So we come to the point that the first defining question we need to answer in order to know which model matches reality is whether space is curved or flat. The empirical evidence has now shown that space is not curved. The WMAP satellite information gave us conclusive data that show our universe is, in fact, flat or Euclidian.

> What did WMAP find? It found that Euclid was right!
> Space is flat (Susskind 2006, 160).

What WMAP discovered is that the triangulation of three different faraway galaxies produced a flat Euclidian triangle. Space is flat. Therefore, the sphere and saddle-shaped models are incongruent with the empirical data garnered by all three of the mapping satellites sent in the last two decades. That disqualifies both Option 1A, Spherical Universe, and Option 2A, Saddle-Shaped Universe, which also requires space to be curved.

In order to choose between the remaining models, we must now consider if the universe is unbounded and homogeneous.

2. Is the Universe Unbounded?

The general theory of relativity has forced evolutionists to accept that our universe had a beginning, but they are still attempting to whitewash this beginning with infinities that would help them desperately avoid a creator. Therefore, in spite of the fact that they cannot deny that time, space, and matter had a beginning and that the entire material universe began at a single and bounded point in space, they have chosen to believe that space was born infinite and unbounded. From a rational perspective, naturalists have chosen to accept a view that is absurdly irrational in order to bolster their deophobic theological bias.

> It is natural to ask how large the universe was at
> very early times. Unfortunately we do not know,

and we are not even sure that this question has any meaning. As indicated in Chapter II, the universe may well be infinite now, in which case it was also infinite at the time of the first frame, and will always be infinite. On the other hand, it is possible that the universe now has a finite circumference, sometimes estimated to be about 125 thousand million light years. (The circumference is the distance one must travel in a straight line before finding oneself back where one started.) (Weinberg 105).

Weinberg purposefully chooses out of the many only two possibilities that align with the evolutionary cosmological principle and the Copernican principle, without even admitting that there are other possibilities. His first choice is that the universe began infinite in size. His second choice is the spherical model (the skin of the balloon), which also, although being finite in extent, supports the evolutionary Copernican principle that gives Earth no special significance in its position in our universe because it has no center. The spherical model, however, requires space to be curved, and we have already documented that it is flat. Therefore, this choice is inconsistent with reality and is now roundly disqualified.

Regarding his first choice, it is an oxymoron to state that the Big Bang had a singularity, which, by definition, requires a tiny bounded space of finite size that contained all the matter/energy in our universe crunched into a point of radiant energy of infinite density, and then magically claim that it simultaneously began unbounded.

The Genesis Singularity, by definition, is a finite point in place at the very beginning of time that contained all the finite matter in our universe, crunched into the form of radiant energy with an infinite density. That point in time is the Alpha Point—the beginning of space-time. There can be no singularity without that space being bounded. There can be no singularity without bounds. An unbounded singularity is not a singularity. It is the restriction of the bounds that causes it to be crunched into a singularity. If space-time began with bounds, it cannot become unbounded.

The idea that space could begin in a bounded singularity and then instantaneously turn into an infinite state, without an infinite amount of time to become so, cannot be labeled as anything less than ridiculous and desperate speculation that borders on insanity. How could this finite point in space become infinite without infinite time to accomplish it?

One could rationally say that our universe will expand for an infinite amount of time, but at any point on that inflationary road, when the universe is measured, it is still finite and bounded if it had a beginning. A finite beginning will never reach infinity because it needs either an infinite amount of time to reach it or an infinite speed. Neither of these two options is scientifically possible. There is no scientific mechanism known to accomplish either of these two choices.

Beyond that, the evolutionist slyly ignores that space and time are intertwined and that it would be impossible to do something to one that will not simultaneously alter the other. In order for space to become instantaneously infinite, there must be no causal connection between space, matter, and time.

Just think about the ridiculousness of their proposition for a minute: They are essentially claiming that nothing could become instantaneously something infinite. This is an absurd speculation that borders on insanity. Space cannot simply become instantly infinite without affecting time. They are not mutually exclusive aspects of our continuum in this universe. They are causally connected. Reason, however, demands that if our universe had a beginning in space and time, then it has an end and is finite. To think otherwise is absolutely irrational.

In describing the first frame, the first moment of the Big Bang, Weinberg states:

> Since the temperature of the universe falls in inverse proportion to its size, the circumference of the universe at the time of the first frame was less than at present by the ratio of the temperature then (10^{11} °K) to the present temperature (3 °K); this gives a first-frame circumference of about 4 light years. None of the details of the story of cosmic evolution in the first few minutes will depend on whether the circumference of the universe was infinite or only a few light years (Weinberg 106).

Am I missing something here? Which is it? Does a "circumference of about 4 light years" bind the universe in an event horizon, or is it infinite and unbounded? If the temperature of the early universe falls in inverse proportion to its size, meaning it cools as it expands, how then can Weinberg claim that the size of the universe does not impact the cosmic developments? The size is absolutely foundationally important because the expansion rate determines the cooling rate and the cooking time of the material at the correct temperatures necessary to end up with the correct proportions of subatomic particles necessary for life to exist. The idea that the circumference of the universe is immaterial to the events that took place is silly rubbish that can only be defended by the notion that the matter expanding would be disconnected to the infinite spatial dimension he is proposing just magically appeared. This, of course, goes contrary to Einstein's understanding of the role of space-time in its relation to matter and gravity.

The evolutionists want to have their cake and eat it, too. They say that space is infinite and unbounded, but then they say that there was a rapid inflationary period in which the universe grew to almost its present size in a fraction of a second. Well, which is it? If the spatial dimensions of our universe are infinite, then it could not be expanding since one cannot expand infinity. Anything that is expanding is finite. If our universe expanded, then it had bounds and was finite in size.

It is therefore irrational to logically claim that at the same time our universe is expanding it could be infinite in extent. One could say that space within an infinite universe can be undulating and dynamic, but it would not be monolithically expanding as we observe it. In fact, it is accelerating as it expands, which means logically that its bounds are stretching. One could even say that matter is expanding within an infinite spatial dimension, but to do so would mean that matter is not anchored to space-time. Everything we know about the interconnection of matter and space-time would be null and void.

The undeniable fact that our universe had a beginning cannot be reconciled with an unbounded universe evolving from it. We will explore that more deeply in the following section.

The Relationship between Space, Time, and Matter

Einstein's equations have forced the evolutionary scientific community to acknowledge that our universe had a beginning. The mathematical calculations bring us to a point at which the entire universe was bound inside a singularity, such as the singularities that exist today within black holes, except that this singularity contained everything in our universe and would have been many orders of magnitude more powerful than any singularity that now exists or ever will exist in our universe.

It would be irrational to claim that a singularity, bounded by a single finite-sized space-time coordinate point, could contain an infinite amount of matter in a finite, limited area. That is simple elementary logic. So we know that the matter in our universe is finite. We also know that it began within a single point of space-time. Hence, time, space, and matter cannot be infinite, because each of them began at a finite point. Yet the evolutionary cosmologists continue to claim that space was born infinite from the Big Bang.

They disregard the intimate causal relationship between space, matter/energy, and time. Not one of these three things could exist in our present universe without the others. You cannot have space without time. If you do not have time, there is no *when* that the space existed. You cannot have matter/energy without space. If there is no space, there is no *where* for matter/energy to exist. Conversely, if there is no energy/matter, you cannot have space, because space cannot exist without the intrinsic zero-point energy and the Higgs field with its non-zero energy component. Spatial dimensions are not a void or vacuum. They have structure, they are moldable, and they have an intrinsic energy.

Both matter and energy impact space. It is made denser by the presence of matter or energy.

> *The earth is going around the sun a little faster than it otherwise would if the sun were not hot, because the energy in the sun's heat adds a little to the source of its gravitation. At super-high temperatures the energies of particles in thermal equilibrium can become so large that the gravitational forces between them become as strong as the other forces. We can estimate that this state of affairs was reached at a temperature of about 100 million million million million million degrees (10^{32} ^{0}K) (Weinberg 145–146).*

At this high temperature, somewhere in the first nanoseconds of the Genesis Singularity, all the four forces of nature were unified. So we see that both matter and energy increase gravity, which makes space denser. The effect created by the warping of space into a denser form is what is perceived as an attractive force. Moreover, time is an intrinsic component of space, and its rate is dependent on the density of space. There can be no separation of time, energy/matter, and space. They are inextricably intertwined in a continuum.

Russell Humphreys, in his enlightening book *Starlight and Time*, explains the naturalist bias for this arbitrary assumption of a universe beginning infinite and boundless.

> Why do Big-Bang cosmologists use as their starting point the assumption (which seems quite contrary to common sense) that the universe has no boundary? Is there some good scientific reason, or is it perhaps demanded or even suggested by well-established, experimentally backed theory, like general relativity?
>
> The answer is no. It is an arbitrary assumption, called the "cosmological principle," or more recently the "Copernican principle." This assumes that (whether the universe is finite—like that of the ant on the balloon—or infinite) there is no edge and no center. On a large enough scale, matter is evenly distributed around us. Therefore, it is asked, if there were an edge, then why don't we see more galaxies on one side of us than the other?
>
> This would be easy to explain if we were in a special place close to the center. Such a "special arrangement" is exceedingly improbable on a chance basis. It therefore strongly smacks of purpose, and is thus unpalatable to most theorists today, who prefer to believe in a universe ruled by randomness. So it is simply assumed that there is no center, and no boundary. . . . However if the universe is bounded, then there would be a center of mass and a net gravitational force, and we could begin to consider the time-distorting effects of gravity on a massive scale (Humphreys 18–19).

Humphreys was awarded his doctorate in physics from Louisiana State University and has worked in the fields of nuclear physics, geophysics, and pulsed nuclear physics and on the Particle Beam Fusion project. His book *Starlight and Time* offers an alternative to the cosmological principle and the Copernican principle (the naturalistic assumptions that the universe is infinite and unbounded at its inception and that there is no special significance to the position of Earth) and is every bit as valid and conforms more accurately to the scientific evidence.

I do believe Humphrey goes very easily on the language when he says that their idea of beginning with an instantly infinite universe "seems quite contrary to common sense." Think carefully about the ridiculousness of this proposition. To begin with, how can anyone assert with credibility that our universe was born from nothingness? Did it just magically appear? How does nothing become instantly something in a materialistic reality? What materialistic mechanism can cause nothing to magically become everything? It is crazy to even think that nothing could become something instantly, but to further state that this something became instantly infinite is literally light years beyond hutzpah.

The real truth is that nothing outside of God is boundless. All things finite are bounded. Without God as the starting point, there is no naturalistic explanation that can explain the origin of the Big Bang Singularity.

Throughout our history, humankind has always raged against God. Time after time, in every generation and in every continent, people have denied the infinite God in order to prop up their own minds as the arbiters of truth and morals. In doing so, they worship the creation rather than the creator. Postmodernists may not fabricate gods of clay, wood, and stone to put on their mantles, but they worship idols nonetheless. They make the mind their god.

They purposefully insist that the creation is infinite in order to deny the infinite creator. In their quest for absolute moral autonomy, they deny the obvious design and symmetry of our universe and seek to erase the reality of our finiteness. A finite universe is incompatible with naturalism. No naturalistic explanation could ever explain its origin. Because of their underlying antipathy toward a creator, they have a deep-seated felt need to make the universe infinite, and in so doing, they attempt to avoid the pesky problem of a beginning, which cannot be avoided in a finite universe.

This choice to make the universe infinite by itself is a strictly faith-based subjective assumption, which is far more illogical than the proposition of our Judeo-Christian cosmological model that the energy of God (the voice of God, since matter is interchangeable with energy) formed the elements of the creation of our universe at a given point in time with a bounded and finite size. This proposition, from a purely scientific perspective, is less speculative and fits neatly with the observable and empirical data we have, as well with the laws of physics we know.

Frankly, this idea that the universe was born infinite seems to me not only an illogical and irrational improbability, but it just simply requires a great deal more faith to believe it. Our Judeo-Christian cosmological model that accepts the literal narrative of the biblical Genesis in which the universe, finite and bounded, was created by an infinite being with intelligence, fits perfectly with the empirical data that space, matter/energy, and time are interrelated and inseparable elements of His creation, which had a beginning point.

If matter and time are not infinite and space is, then there can be no causal connection between a static infinite space and the dynamic finite matter within it. If space were infinite, then it would be static. One cannot expand beyond infinity. How, then, is it that our universe is expanding?

That makes no sense and, in fact, is contradicted by the general theory of relativity, which plainly teaches us that the interrelationship between space, time, and the gravity of matter is inseparable. All physical evidence and the laws of physics lead us to the conclusion that there is an inseparable causal relationship between space-time and matter. What is done to one impacts all the others. How, then, can the universe become infinite and unbounded instantly without impacting matter and energy? That is simply an absurd speculation, not hard science.

Our Judeo-Christian cosmological model states that our universe is finite and bounded and could not have become unbounded in a finite amount of time. Because it is finite and bounded, space-time and matter are able to expand outward. It is more likely that the very fabric of a finite-sized and bounded space-time is being stretched and spun at an accelerating rate, because God set the universe spinning at the very beginning. Therefore, all the matter within its visible and invisible space-time dimensions is subsequently speeding outward in centrifugal force and stretching. Our model does not disassociate space, time, and matter. We maintain that there is an unbreakable causal connection between them. Any physical impact on one affects the others.

It is irrational to assert that the expanding galaxies are expanding within the already infinite universe and at the same time maintain that any causal connection exists between the galaxies and the matrix of space itself. Matter, then, would be free-floating and not anchored in space. That would mean that there is no interconnection between the spatial dimensions and matter. The universe would be in a dichotomy where matter is dynamic and space is static. This, to any thinking person, is an illogical absurdity and would, in fact, be stating that Einstein's theory, which explains the force of gravity as the warping of space-time, is wrong.

In contrast, the observational data and the mathematical formulas of physics tell us that matter (mass) produces gravity, which impacts the density of space-time. They are intertwined and inseparable. In fact, it is the density of the spatial dimensions that determines the rate of the passage of time relative to whatever point of observation we choose throughout the universe. If we choose an observation point where space is dense, time moves more slowly than another point of reference where space is thinned out or stretched.

When one acknowledges the observational findings of the last three satellite probes to map our universe and the analysis of the red-shift data regarding galaxy groupings, it defies rational thought to hold to the perspective that the universe was created through purely random events. Our universe is filled with elegance and supersymmetry that point not to a randomly generated structure but to a master designer. He placed us in the center of His universe to manifest His design and consequently make known the reality of His existence to all who would seek His face. From a scientific perspective, this is much more reasonable, and it is what Humphreys suggests.

> *However, what if we begin our calculations with the opposite assumption, equally scientifically valid, namely that matter in the universe has a center and an edge (is bounded)? This makes more sense and is also scripturally far more appropriate. When we feed in this, plus the same observations, into general relativity, quite a different cosmology falls out.*
>
> *I call it a "White Hole" cosmology for reasons to become clearer shortly—and it just happens to solve the problem of the starlight travel-time rather neatly (Humphreys 21).*

Humphreys explains our universe and the appearance of immense ages in faraway galaxies as the result of gravitational time dilation, from the natural process of the inception of our universe. His theory that a white hole is the mechanism, which shot out the matter and space-time contained within the Genesis Singularity to seed our universe, is quite compelling. His calculations begin with the more sound assumption that our universe is bounded and finite.

The almost infinite curvature of space-time, which would have resulted around the rim of the critical circumference of a white hole from an inception of a finite and bounded universe, would, in fact, cause massive time dilation near its event horizon (critical circumference). That would adequately explain why time on Earth near the center of the universe and exiting last through the initial critical circumference of the primordial Genesis Singularity would have moved only several thousand years. On the other hand, matter on the outer fringes of the universe would have moved so much faster in order to have the necessary escape velocity to cross the primordial critical circumference, (PCC) and by its extreme stretching, it would have aged several billion years more than matter exiting last through the progressively weakening PCC.

But we will deal with the age of the universe later; for now, we must return to the models of the shape of our universe. We have shown from a rational perspective that the idea of a beginning of space-time and matter/energy is irreconcilable with an unbounded and infinite universe. The last question now remains: Is the universe homogeneous?

3. Is the Design of Our Universe Homogeneous?

Option 3B: Do the empirical data support the cosmological principle —that is, a flat infinite or unbounded universe with Earth not at the center of the universe and with a completely homogeneous texture?

We have already answered the first two questions:

1. Space is flat.
2. Space cannot be infinite and unbounded if it had a beginning.

To answer the third question, we must ask two other questions:

1. Is the baryonic matter in our universe evenly distributed in large-scale structures?
2. Is the CMBR completely isotropic? Is the ambient energy of the afterglow of the Big Bang exactly isotropic?

These two requirements are the necessary ingredients that can substantiate the standard cosmological evolutionary model. The empirical data regarding Baryonic matter and its large-scale distribution in our universe tells us that it is not exactly homogeneous.

Our Galaxies Are Arranged in Seven Concentric Circles around Earth

Red-shift studies indicate that our universe seems to have been previsioned and designed with a coded message in its structure. The disappointment for naturalists who insist the universe was created by random forces does not end with a beginning that is incompatible with an eternal and infinite universe. The galaxies are also not homogeneously distributed, as their cosmological principle requires. The galaxies in our universe seem to be concentrated within seven concentric spheres that surround the Earth-sun system. The observable evidence clearly shows that our Earth is, in fact, in a very specially designated region of the universe at the very center of these seven concentric bands of galaxies.

> *In 1970, William J. Tifft, astronomer at Steward Observatory at the University of Arizona examined the redshifts of various galaxies and found that they were all distributed at specific spherical distances from Earth, namely in multiples of 72 km/sec, and a smaller grouping of 36 km/sec.*
>
> *To picture this in your mind's eye, it is like bands of galaxies, with each band separated from the other in evenly spaced and proportional rings. Tifft's findings were quite shocking to the field of astronomy, since not only were the more obscure sources such as gamma-rays and quasars showing Earth in the center of the universe, but now the common galaxy, which was far more numerous and readily observable, was*

showing precisely the same centrality of the Earth. . . .
Sky and Telescope, which is by no means a geocentric
periodical, says of Tifft's reports: "Quantized redshifts
just don't fit into this view of the cosmos [the Big Bang
View], for they imply concentric shells of galaxies
expanding away from a central point, Earth (Sungenis
and Bennett 2014a, 390–391).

The Copernican principle, which adamantly states that Earth has no special significance in the universe, is nothing more than an artificially fabricated extrapolation of the heliocentric discovery that Earth is not the center of our solar system—the sun is. But the solar system is the center of our universe. The empirical evidence documented by red-shift studies show that our galaxies are found in seven quantum bands that form concentric shells around Earth. This observable data is at odds with the mainstream naturalistic bias that our universe was generated by random causes in a homogeneous distribution of galaxies and that Earth has no special significance in its placement in the universe.

The cosmological principle that demands a homogeneously smooth universe has been shown to be in opposition to the empirical data collected regarding the location and distribution of galaxies in our universe. Their unbounded and homogeneous model has been disqualified by empirical reality. Although homogeneity within the bands seems to be maintained, there is a break of empty space between each band where no galaxies are found. That makes the design of the universe even more complex. It cannot be relegated to random distribution patterns even with a rapid inflationary period, which cannot account for the empty spaces between the seven bands of galaxies. This empirical data completely refute the rapid inflationary period, which was the hail-Mary pass of evolutionists to explain the seeming homogeneous and isotropic nature of our universe.

These studies have been done and redone; they have been in the record since the 1970s. But the scientific community seems to have no interest in addressing them. They have simply ignored the pink elephant in the room, and with good reason. Baryonic matter is not distributed homogeneously throughout the universe. The Rapid Inflationary Model cannot account for this symmetrical shell superstructure of our universe. The evidence debunks all evolutionary mechanisms in cosmology.

Not only does this large-scale design of our galactic bands show that randomness could not have birthed our universe, but it also shows that Earth is at the very center of the universe. It does have special significance and therefore also disqualifies the Copernican principle and the cosmological principle and their dysteleological metaphysical doctrines.

I would like to emphasize here the significance of W. G. Tifft and W. J. Cocke's discovery, which, in my opinion, is much deeper than most realize (Tifft and Cocke). I suspect that these seven concentric bands are, in fact, seven different, albeit visible, time dimensions. As each of these concentric bands stretches further from the Alpha Point, space is stretched, and time speeds up in direct relation to the density of space in each region. Hence, the rate of time in each of these bands is a function of the density of space, which increases in rate in relation to their distance from the Alpha Point—Earth. That is, the furthest bands from the Alpha Point where space is stretched the most the rate of time is increased proportionally.

The handiwork of God explicitly evidences His design so humans can know the reality of His majestic power. But we must understand that this is not some obscure discovery not ratified by other scientists. These findings have been confirmed by multiple studies, and yet evolutionists continue to act as if the king is clothed.

> *D. Koo and R. Krone, two University of Chicago scientists, did the same redshift analysis on galaxies. Their*

results were identical to Napier's and Guthrie's and even made it to the New York Times. They conclude, "the clusters of galaxies, each containing hundreds of millions of stars, seemed to be concentrated in evenly spaced layers" [i.e., concentric spheres around the Earth]. Incidentally, for those who see symbolic significance in numbers, the number of "evenly spaced layers" discovered by each team of astronomers is seven. There are seven evenly-spaced layers in the north direction, and seven evenly-spaced layers to the south. Koo admits that astronomers are very disturbed at this spacing, obviously because it gives evidence of intelligent design and geocentrism (Sungenis and Bennett 2014a, 393).

The evidence of these seven concentric shells or bands of galaxies disqualifies the Option 3B cosmological principle and supports our Option 4B Judeo-Christian cosmological model of a finite or bounded universe created by the one true God of the Holy Scriptures, as purposed for the habitation of humans.

The number 7 is numerically associated with God the Father and the Genesis record of the creation of our universe in seven days. But its symbolic significance is much deeper and shows God's providence in an amazing way. As we shall soon see, our universe has doubled in size six times since the moment light was separated from darkness. The Judeo-Christian model predicts that when the seventh band of galaxies doubles for the seventh time, God's Omega Point will be reached. When the seventh seven is reached, the Time of Jubilee shall come, the Time of Timelessness.

The centrality of our solar system in the universe is verified by many other empirical findings, about which most of us have been kept in the dark. Did you know that all the stars in the universe are saluting us?

The Barr Effect: The Axis of Binary Stars Point toward Our Solar System

God has left no stone unturned in urging us to see His masterful design of the cosmos. There is this peculiar phenomenon that evolutionists have no way of explaining from their dysteleological worldview. They call it the Barr Effect. It turns out that most binary stars are spinning around each other with the axis of their systems pointing toward Earth. The preponderance of twinned stars aligned toward us is so great that it is obvious that the few stars that are not pointed toward our solar systems may be those that have been disturbed by collisions since their time of creation.

How does a randomly generated universe accomplish such a feat? When one considers that a little more than 50 percent of the stars in our universe are binary, tertiary, and even quaternary, whose axes point toward us. That means that our particular solar system is central to the entire universe.

In astrophysical terms this means that the orbital axis of binaries are situated with respect to the Earth. Since binary stars are seen over the 360 degrees of visual space, this means that the axis of each binary system is pointing toward the Earth as if the Earth were the center of a giant merry-go-round and the axes were arrows. Without admitting to any possibility that the binaries show Earth is the center of the universe, astronomers instead prefer to attach innocuous names to such phenomena, this particular one being the "Barr effect," after the astronomer J. M. Barr. Barr's original study found that of the 30 spectroscopic binaries he analyzed, 26 had longitudes of periastron between 0 and 180 degrees, which means that they were oriented toward Earth as their center (Sungenis and Bennett 2014a, 404–405).

It is obvious that reason, logic, and empirical data have nothing to do with the naturalistic predisposition of those who attempt to blot out the sun with their thumbs. But the naturalist stubbornly holds on to their biased belief that our universe is infinite in size and that the Earth-sun ecliptic is not at the center of a randomly generated universe. But the empirical evidence says otherwise. How could a randomly generated universe be so arranged as to be saluting the Earth?

Our Judeo-Christian cosmological model, which stipulates that our universe is finite in size and content, is vindicated by the empirical data. This is the model that best matches our observable reality and the empirical data recorded by the mapping satellites. The space-time continuum is finite. The matter/energy within it is finite. The fact that almost all the stars in our universe are, from our point of view on Earth, red-shifted is much better explained scientifically by the Earth's central position in an expanding but finite universe.

It is elementary. If the universe is finite and bounded, then we would, by necessity, have to be close to its point of origin (Alpha Point) in order for it to appear as the center of the seven concentric bands of galaxies that surround us and for it to seem generally isotropic in every way that we view outward from our planet. When we view outward, we are then, in essence, viewing in the direction in which the Genesis Singularity exploded outward from the Alpha Point, and all stars are then moving away from us. The Earth is located in a very special and significant place, which matches exactly the Genesis account of creation.

The fact is that evidence that supports our central position in the universe abounds, but it is simply ignored as a red-headed stepchild by those who wish to see the king still dressed. There is yet more to consider.

The Evidence of the Observed Gamma Ray Distribution Pattern

The naturalistic presupposition that our universe is infinite and unbounded is refuted by other observational data. If our universe is infinite, why is every inch of our sky not lit up with gamma ray bursts or their afterglow radiation? If we have an infinite number of stars in an infinite universe, then every inch of the Earth would have been bombarded by long gamma-ray bursts coming from an infinite number of supernovae that would have long ago sterilized the Earth. Our night sky would be completely white with starlight that has streamed from infinity. This is Olbers's Paradox, which we will address in more detail in Chapter 11. For now, the point is that instead of an infinite number of gamma-ray bursts, what we observe is that these long gamma-ray bursts are finite in number, and, just as important, the vast majority of them are coming from the distant galaxies at the far edge of our universe.

What are gamma-ray bursts, and where do they come from? When the first gamma-ray burst was discovered, scientists believed this powerful beam must have come from our galaxy due to its tremendous intensity. Since then, the study of the spectra of their afterglows has shown conclusively that they actually come from the far reaches of our cosmos. The correlation between these long gamma-ray bursts and extremely distant galaxies is now widely accepted.

The vital clue is that the vast majority of these powerful gamma-ray bursts are found emanating within a specified region from 12 billion light years away to the outer edge of our universe. In fact, although their luminosity is certainly strong enough for us to see them, nothing is seen beyond 15 billion light years distance. There are no gamma-ray bursts from beyond 15 billion light years. This is observational data that our universe has an outer boundary, albeit expanding outward at an increasing but finite rate.

When enormous stars die in a supernova, they can radiate intense beams of concentrated gamma rays, which can be detected here on Earth. Fortunately for us, the galaxies closest to us do not exhibit the same preponderance of dying stars. These bursts could fry our planet in seconds. It is in the oldest areas of our universe, at the outer edge, that we find the greater preponderance of these unbelievably bright bursts that are equivalent to the total energy put out by the lifetime of our sun.

Curiously, it seems that these deadly gamma-ray bursts display a bimodality, whereby one group has an average duration of three seconds and the other an average duration of 30 seconds. It is as if they come in discreet quantum bursts, which may be a clue to the time dilation difference of their specific regions. The afterglow of the intense energy beam can also be detected several hours after as a result of the gamma ray's impact on the dust in space through its line of trajectory. This interaction with space dust causes other detectable electromagnetic signals that allow us time to trace their paths through the universe.

But the main point to be observed is that there is an outer boundary at the edge of our universe after which nothing is observed. Yet the intensity of these stupendous bursts of energy is so powerful that if they existed many billions of light years beyond this present outer boundary, we could still observe them. If they were there, we would be able to see them. But there are none beyond this event horizon of our universe. Our universe is not unbounded.

The observable data simply imply that they aren't visible because our universe has an outer boundary. The Judeo-Christian model states that our universe is finite and bounded and that our space-time continuum ends at this edge. If our model is correct, then we can predict that as our universe continues to expand, this edge will move further away from the Alpha Point. Hence, in the future, we should be able to see beyond the 15 billion light years distance of the present edge of our universe. Our Judeo-Christian cosmological model predicts that in the future, we will be able to see gamma-ray bursts emanating from beyond our present 15 billion light years limit.

Most of us have heard the term *time dilation*. We have learned from Einstein that in the same interval, two observers can experience different rates of time. But where there is time dilation, there is also time constriction if we look at it from the opposite viewpoint. If observer A were located in an area of higher space density (toward the center of the universe) than observer B (toward the outer horizon of the universe), then time would be slowed down for observer A by the crunching of space-time. On the other hand, the stretching of space-time would speed up time for observer B toward the outer horizon.

From the viewpoint of observer A, the time of observer B would be dilated. From the viewpoint of observer B, the time in the area of observer A would be constricted. That means that observer A, looking at a gamma-ray burst coming from the area of observer B where space-time is stretched and hence time is sped up, would see an event that took a much longer time to unfold out there where time runs faster. But observer A would see this crunched into a much slower time frame toward the center of the universe.

Hence, the extreme luminosity of these gamma-ray bursts may also be partly explained by the fact that the outer edges of our universe are in a region where space is in fast-forward time. Here, at the edge of our expanding universe, space is being stretched as it moves outward at an accelerating pace. When we stretch space, we speed up time; so those three seconds or 30 seconds in which we see the varied gamma-ray bursts from our viewpoint (where time is running more slowly) are, in fact, a condensed or constricted view of an area of space-time where time is dilated and running faster.

In essence, when we look toward the edge of our universe, although it took that light 15 billion years to get here, we are not really looking into the past; we are, in fact, looking into the future. We are seeing enormous spans of time in a condensed or constricted time frame within our region of space-time; thus the power of their luminosity is so great. We are looking at a stream of electromagnetic energy that has been flowing in that region for a much longer period of time than our observational frame of reference on the Earth, which has reduced that time to either a three-second or 30-second interval, depending on its distance from us. (For a more complete treatment of time dilation and constriction, see the section, "The Stretching of Space-Time" in Chapter 14.)

The second observation is that these long gamma-ray bursts emanating from exploding massive stars are all indications that, in fact, stars and galaxies are dying at an increased rate toward the edge of our universe. Those galaxies and stars are reaching the end of their lifetime and following the prescribed and absolutely universal laws of entropy. That is not only evidence that the rate of time in these outlying areas of the edge of our universe is responsible for the decay of these stars and galaxies, but also that our universe is, in fact, finite.

Therefore, the third observation we can make is that our universe will eventually decay and die. It would be irrational to assume that this multifaceted display of greater entropy does not mark a physical outer potential bound for the life and extent of our universe. The speculation that stars, galaxies, and space exist into infinity stands against all the direct observable data. There exists no evidence that contradicts our Judeo-Christian cosmological model that our universe is finite and bounded. In fact, what the observational data indicate is that random ordering could not have created the large-scale symmetrical design of our universe, as we shall expound later.

But I would like to take a slight detour here to comment on the false evolutionary notion that the outer edge of our universe displays the early stages of our existence and, furthermore, to counter their dysteleological obsession.

The Dysteleological Obsession and the Copernican Principle Delusion

The proponents of the Copernican principle have a natural affinity for making black holes important to the evolution of the universe and galaxies because they are the largest producers of entropy or disorder. Therefore, naturalists would love very much if black holes were the creators of our universes and galaxies, since that correlates with their basic metaphysical doctrine that order can be created by disorder and randomness. That is their dysteleological obsession.

But they have the cart before the horse. Supermassive black holes did not create our galaxies. Our galaxies are dying and becoming supermassive black holes. The second law of thermodynamics is not reversed by black holes. Quite the contrary, black holes are the final product of the second law of thermodynamics—no greater entropy can be found in the universe. The black hole is the omega point of the second law of thermodynamics, not the Alpha Point of our galaxies.

Now black holes are not always black. Some supermassive black holes become quasars, the brightest objects in the universe. When a supermassive black hole becomes so large that its gravitational pull attracts galactic matter near it with such force that it reaches almost the speed of light before entering the critical circumference, it becomes a quasar.

That enormous friction and heat produced by galactic matter traveling near light speed becomes too much for the black hole to swallow, and this extra energy simply explodes outward in intense electromagnetic bursts. In other words, such an overwhelming amount of energy from the material being dragged into the critical circumference cannot be completely absorbed by the glutted black hole, and therefore, the excess creates jets of electromagnetic radiation that shoots outwards from the poles.

These enormous beams of electromagnetic radiation are so powerful that their emission of energy is greater than anything else in our universe that we have observed to date. It is, in fact, the telltale symptom of very old black holes, which have become gargantuan. Quasars are the last gasps of dying galaxies. But that does not mean that quasars only exist in the outer fringes of our universe. Galaxy collisions can accelerate the formation of such enormously supermassive black holes. Some are found closer, but the great preponderance of them are found in the oldest regions of our universe.

Quasars Mark the Death of Galaxies, Not the Birth of Stars

A 2007 study by S. V. Pilipenko of the Moscow Institute of Physics documents that the preponderance of quasars increases with their distance from us. In other words, the areas of our universe where time has been sped up the most contain the largest number of quasars. Quasars are simply very old, giant, feeding black holes that are only visible to us because of their enormous emission of energy in the form of these powerful electromagnetic jets.

Because no light can escape a black hole, we normally cannot see where a black hole exists. But when black holes are feeding, the accretion disc of material being ingested as it spins wildly around them lights up from the enormous turbulence and friction created, which allows us to see them. Not only does the black hole cause the matter around it to spin at near light speed, but the very space-time fabric is also being dragged and massively contorted. Scientists call this "frame dragging." Black holes literally swallow space-time as well as matter and energy. It is the eater of worlds.

The material surrounding these galactic monsters along with the severely contorted space-time are relentlessly sucked toward the black hole and spun around it at unimaginable velocity by the massive gravitational pull of the powerful singularity deep in the core of these giant black holes. All matter turns to plasma, and the accretions disc becomes visible. The angular momentum of the spinning material and the contorted space-time frame being dragged around it develop a centrifugal force at these enormous speeds that tries to counter the gravitational attraction. This process slows down the rate at which the black hole can feed, so the spinning disc around it seems to be resisting its eventual and inevitable demise. The spinning plasma creates enormous amounts of electromagnetic energy. When enough material is built up so the feeding black hole cannot eat it fast enough, the accretion disc becomes unstable and shoots outward from both poles of the black hole in giant bursts of energy that travel for millions of light years.

The term *quasar* stands for *quasi-stellar radio sources*. The enormous luminosity of quasars is hard to imagine. The enigma presented to scientists is that these bright lights from quasars are coming from the very edge of our universe and ought not to be as bright at that great distance. These powerful lights may be as much as 100 times the luminosity of all the stars in the Milky Way combined.

The enormous power of these jets of energy cannot be understated. These powerful beams of excess electromagnetic energy are literally shot outward from the poles of these spinning, supermassive black holes as they violently feed on the galactic material sucked in from the outskirts of their event horizon. Reason dictates that they have become so powerful simply because of their long ages. The longer a black hole feeds, the larger it will become as it continues to swallow the galaxy it inhabits as a cancer from within. Some of these may even have been created by galaxy collisions that merged two or even more central black holes that typically form in the center of galaxies as they age.

The mass inside their singularities may be as much as billions of times the mass of our sun. They have, indeed, swallowed many worlds. The brightest quasar in our night sky is 3C273 in the constellation Virgo. It is, in fact, one of the closest to us, which makes it that much brighter. This quasar is so bright that if we were to place it 33 light years from the Earth, it would still shine as bright as the sun shines to us only 93 million miles away. Its luminosity is 4×10^{12} times stronger than our sun's. Mind you, it only takes eight and three-fourths minutes for light from our sun to reach us. Our nearest star, Alpha Centauri, is four and one-fourth light years away. This quasar would be some eight times the distance of Alpha Centauri from us, and it would still shine as bright as our sun. The luminosity of quasars is simply mind-blowing.

When we say that some black holes are not black, we are not saying that matter, which enters into the event horizon, can ever come out, except for the tiny amount of energy known as Hawking radiation that we spoke of earlier. Small black holes actually radiate more Hawking energy than larger black holes, whose gravity is stronger and their entropy within is greater. But these amounts are so miniscule as to be practically negligible.

For all practical purposes, nothing can escape the power of the singularity once it crosses the event horizon or critical circumference—not even light. In other words, the energy spewed in giant electromagnetic beams from these supermassive black holes is derived from the mass of the outlying accretion matter and the electromagnetic energy produced by the wildly turbulent plasma in the swirling accretion disc. The speed of the matter falling toward the critical circumference is so great that it overloads the ability of the black hole to swallow the matter, and the excess energy is regurgitated outward in powerful beams of electromagnetic energy before entering the event horizon.

Their brightness from our perspective on Earth is due to two reasons. First, they have grown to an enormous size due to their long ages. That means that the critical circumference has moved outward from the singularity and is capable of bringing in more of the material in its galaxy.

The second reason these quasars seem so enormously bright to us is because we are seeing a large time frame from that distant region compressed into our much slower time frame. Space-time in the farthest regions of our universe is being stretched outward at an enormous rate, thus speeding up time in that region. So from our viewpoint in time, where space-time is much denser, we see the great speed of these galaxies because they are moving in fast-forward time compared to us.

Large components of time in these areas where space is so stretched are being funneled into our much slower time frame, and therefore, the luminosity is much greater than expected at that distance. For this reason, it seems to us that galaxies at the end of our universe are rotating at speeds beyond the normal bounds of physics. We need not add this magical invisible dark matter. Nor do we need to discard our equations that clearly limit the speed of light to understand what is happening. We simply need to adjust the rate of time according to the stretching of that region of space.

The evolutionist bucks the idea that adjusting the relative time frames fixes the problem, because it would mean that our area of space has not undergone the enormous spans of time necessary for the slow and gradual evolutionary process of gravity-induced aggregation to create our galaxies in our neighborhood of the universe. They want the edge of our universe to be the past and not the future. They want these black holes to be the creators of galaxies and not the cemeteries of galaxies. They want our time frame here to be in the billions of years in order for evolution to have the chance they need. They think that the outskirts of our universe reflect the early universe. It may take 15 billion years for that light to travel to Earth, but that region of space is so stretched that space-time is moving proportionally faster in relation to their distance from us. Therefore, they are, in fact, looking into the future.

The fact that the preponderance of these quasars is directly proportional to their distance from us evidences the magnitude of the stretching of space-time and the subsequent greater age in that region. It is proof that they are not the markers of the beginning of galaxies, but rather the markers of their death.

In a study submitted in April 2007 titled "The Space Distribution of Quasars," Pilipenko writes:

> It is for interest of models of the origin of quasars to establish whether their degree of clustering evolves with time. To investigate this issue Porciani et al subdivided their catalog into three redshift bins and calculated their correlation factor in each of these bins. They argued that r_0 increases by a factor of 1.4 as z increases from 1 to 1.9 with a significance level of

3.6 sigma. Shen et al analyzed the quasars correlation function at $z > 2.9$ and found the correlation length r_0 to increase substantially with redshift $r_0 = 17\ h^{-1}$ Mpc and $r_0 = 24\ h^{-1}$ Mpc at $2.9 < z < 3.5$ and $z > 3.5$ respectively (Pilipenko).

In other words, the preponderance of quasars increases tremendously with the distance from us. That is, those regions of our universe that are beyond 12 billion light years from us contain the largest number of quasars. That means that these regions of our universe where time has been sped up the greatest (because space-time has been stretched the farthest) and have therefore aged the most are the ones most commonly exhibiting this advanced state of the natural decay of the galaxies. These faraway galaxies are more rapidly being consumed by their now gigantic central black holes that have been feeding and growing to enormous sizes for a very long time. The unfounded speculation that black holes played any part in the formation of our galaxies is here countermanded, as we see that black holes are, in fact, the measure of the devolution of galaxies as they age. Again, the data contradict the randomness origins doctrine of the Copernican principle and expose their delusional obsession with dysteleological origins.

Studies of the mass of supermassive black holes and the orbital speed of the stars in the outer fringes of their constituent galaxies are deceptively used as propaganda by evolutionists for their speculation that black holes have a primary role in the development of that galaxy. Quite the contrary, what it implies is that in the areas of our universe toward the outer edges, time has sped up by the stretching of space. Hence, the stars are moving in fast-forward time, and, from our perspective, are moving much faster than the slower time frame in our sphere of the universe where the space-time fabric is denser. Therefore, the black holes have increased in size due to the increase in entropy of the total system through the passage of enormous spans of time.

Simply put, the more time that has elapsed, the greater the size of the black holes. Hence, we can extrapolate backward and declare that at the beginning of time, there were much smaller black holes and perhaps even fewer of them, maybe even none. If we go back far enough, there were no black holes. They could not therefore be the responsible agent for the formation of infant galaxies. But there is yet another very important fact about these quasars that evidences our central place in the universe.

Quasars Are Arranged in 57 Concentric Circles with the Earth at the Center

The quasars refute the dysteleological presupposition of evolutionists, but the arrangement of these quasars as previously documented has been corroborated by a study done by astrophysicist Yatendra P. Varshni, which also points to the centrality of the Earth in the universe, thus pounding another cosmic-sized nail in the coffin of the Copernican principle.

> *Astrophysicist Yatrenda P. Varshni did extensive work on the spectra of quasars. In 1975 he catalogued 384 quasars between redshift of 0.2 and 3.53 and, amazingly, found that they were formed in 57 separate groupings of concentric spheres around the Earth. He made the following startling conclusion:*
>
> *[T]he quasars in the 57 groups . . . are arranged on 57 spherical shells with the Earth as the center. . . . The cosmological interpretation of the redshift in the spectra of quasars leads to yet another paradoxical result: namely that the Earth is at the center of the universe.*
>
> *Varshni first based his calculations on the spectra of the quasars and then did a second test on their actual redshifts. Both tests produced the same results.*

> *Varshni concludes that if his analysis is correct for quasars, then . . .*
>
> *The Earth is indeed the center of the universe. The arrangement of quasars on certain spherical shells is only with respect to the Earth. These shells would disappear if viewed from another galaxy or quasar. This means that the cosmological principle will have to go. Also it implies that a coordinate system fixed to the Earth will be a preferred frame of reference in the Universe (Sungenis and Bennett 2014a, 379).*
>
> *Varshni calls his unexpected findings as paradoxical because he is firmly planted in the ideology of the Copernican principle (cosmological principle). Nevertheless, he attempts to be honest with the empirical data. To his credit, he is honest enough to admit that the existence of these concentric shells of quasars that places the Earth at the very center of these concentric shells simply devastates the foundational ideology of the Copernican principle and the cosmological principle, which blindly insists that the Earth has no special significance in its position in our universe and that our universe is infinite and completely homogeneous. As he admitted, "This means that the cosmological principle will have to go. Also it implies that a coordinate system fixed to the Earth will be a preferred frame of reference in the Universe" (Sungenis and Bennett 2014a, 379).*

This finding, of course, came before the discovery of the quadrupoles and octopoles in the cosmic microwave background radiation, which pounds the last cosmic nail in the coffin of the Copernican principle. Unfortunately, I would advise not to hold your breath expecting the naturalists to concede the obvious. More often than not, the underlying metaphysical faith of evolutionists trumps any obvious data that contradict it. The data are simply ignored as they continue their propaganda machine to rationalize a universe not created by a God who would hold them morally accountable. It is the natural behavior that has become typical for the evolutionary priesthood. It is the classical symptom of those with a chronic, malignant case of deicide.

Therefore, the idea that random chaotic events randomly brought forth our universe cannot be correlated with the data observed and documented by science. The empirical fact documented by Varshni that these quasars are found in 57 concentric circles extending outward with the Earth at its center completely refutes the Copernican principle.

Moreover, black holes do not begin or control the evolution of galaxies or the creation of universes. The preposterous speculation that the intense beams of erupting supermassive black holes are responsible for the birth of stars is a speculative leap of faith, which contradicts known physical laws. Their speculation is that these beams can trigger the formation of stars in dust clouds of outlying galaxies.

These intensely powerful bursts of electromagnetic energy, which includes the deadly gamma rays, contains so much power and energy that their luminosity is more than 100 times greater than all the stars in our galaxy put together. These would not in any way increase the magnitude of the force of gravity between the particles in any cosmic dust cloud in order to condense the cloud into a star. To believe so is nothing more than fanciful, wishful thinking to prop up a failed gravitational model. The opposite is the case. Such a powerful beam of energy would incinerate everything in its path and cause a violent heating of the cosmic dust cloud that would violently expand outward rather than condensing into a star. Only if the electromagnetic beams were in the form of Birkeland currents could this process be constructive rather than destructive.

Some of these supermassive black holes are close enough to us that we can better observe the impact of their powerful electromagnetic beams. An example of this expanding effect can be observed in NGC 4258, also known as Messier106. This black hole, located in the northern constellation of Canes Venatici, is somewhere around 21 million light years from our solar system. Because hydrogen molecules emit light when they become ionized or turned into plasma, we are able to observe radial arms that extend outward through the galaxy from its interior black hole. High-energy jets emanating from the feeding supermassive black hole (estimated to be 40 million solar masses in size) shoot out into the galaxy, piercing the disk and the surrounding halo of its dust cloud.

The impact of these massive beams of high-energy radiation on the dust cloud is clearly observed. The dust cloud expands outward with violence and creates a sort of expanding cocoon around the beams of intense radiation. We can see this interaction as the shock waves heat the clouds, push them outward, and release radiation in optical wavelengths.

Furthermore, because of the wobble of the supermassive black hole, the beam actually travels through the galaxy. For this reason, we can also see the areas where it once impacted the galaxy as diffused areas of the cloud. Quite contrary to their speculation, we do not see an increase in the power of gravity to overcome the expansion and diffusion created by these powerful beams to form stars. The more apt description is that these giant death rays are incinerating their galaxy as the wobbling black hole causes the electromagnetic beams to stream out like a runaway firehouse of death. They are, indeed, the eaters of worlds.

Naturalists' bias toward a model that would make black holes the creators of stars and galaxies is nothing more than the reflection of their imagination that our universe was created by randomness. Their deep-seated, gut-felt need to make randomness the creator of this universe is here made evident to any casual observer. It is their

NGC 4258 (M 106) with its high-energy jet belching from the supermassive black hole in the center of the galaxy. These spiraling arms are shown in red by the ionized light emitted by hydrogen atoms.

dysteleological obsession and delusion that biases their speculations. Gravity simply does not have the power to condense a superheated cloud of cosmic dust into a star.

It is the electromagnetic energy, not gravity, that most likely forms stars. If those electromagnetic beams belching from black holes form Birkeland currents in the ionized gases, then it is possible that solid matter could be formed from the plasma. But black holes are not needed to create plasma. Plasma existed from the first moments of the Big Bang. And a spinning ball of plasma is able to generate its own intense currents. Therefore, black holes are simply not necessary as the creator of stars and galaxies in the way the evolutionists so desperately hope.

How, then, did our universe come into existence? Evolutionists have bet the farm on the Rapid Inflationary Model. Critical to their theory is the second question we are asking regarding the evidence for the Cosmological principle: Is the CMBR completely isotropic? In other words, is the ambient energy of the afterglow of the Big Bang exactly isotropic throughout the entire universe?

The Ambient Energy (CMBR) of the Afterglow of the Big Bang Is Not Universally Isotropic

You will remember that the great problem for the naturalist has been to explain the almost homogeneous and isotropic nature of our entire universe. Little did they imagine that the galaxies would only be homogeneous within each of the seven concentric bands that surround our solar system. There is no homogeneity between the seven bands. But there is a level of isotropy and homogeneity within those bands that is undeniable and difficult to reconcile with a randomly generated universe. Their mathematical fix was to come up with a theory that claimed the universe magically expanded at such a rapid rate for a very short period of time. In this way, all things remained completely homogeneous and universally isotropic throughout that brief elastic process.

Evolutionists hold on to their rapid inflationary model in order to explain the almost miraculous nature of the generally homogeneous and isotropic properties of most of the entire universe. Because random ordering could not have achieved such universal isotropy and homogeneity as we have observed in the microwave background radiation, evolutionists had to come up with some rational way to achieve this from a naturalistic cause. By declaring that our universe has expanded even beyond the speed of light at an early period, they could therefore justify the isotropic and homogeneous nature we have observed by measuring the cosmic background microwave radiation. Before learning of the new information from mapping satellites, researchers expected the CMBR to be almost exactly the same everywhere we looked in our universe.

Such homogeneity would be in complete contradiction to a randomly ordered universe, and so their Rapid Inflationary Model allowed them a mathematical fix to explain our present smooth appearance in the universe. By inserting a rapid inflationary period at the beginning of the Big Bang that moved outward at speeds beyond the speed of light, they can magically provide a materialistic explanation for this almost miraculously smooth universe. In this way, they could artificially create a universe that remains isotropic and homogeneous in its structure and reach near our present expansion without chaos disturbing the homogeneous and isotropic fabric, as would be expected through the increase in entropy predicted by the second law of thermodynamics if the process would have not been almost instantaneous.

Naturalists, wishing to explain the generally isotropic and homogeneous appearance of our universe, thus developed in the 1980s the Rapid Inflationary Model. Alan Guth, at the Massachusetts Institute of Technology, proposed that the early universe went through an inflationary stage, a rapidly expanding stage, at a highly increased rate that superseded the speed of light.

In addition to Guth's model, Andrei Linde, in 1983, presented a variation called the Chaotic Inflationary Model. Of course, these models were developed in order to attempt to explain the paradox that a smooth and seemingly homogeneous creation has occurred from a presumably chaotic explosion, within a randomly guided naturalistic system.

Let us first examine the basic outline of this inflationary theory. The theory posits that our universe expanded at a speed greater than the speed of light into an enormous size between 10^{-36} seconds after the Big Bang to sometime between 10^{-33} and 10^{-32} seconds. Within these tiny fractions of time, something (no one knows what that something is) caused the universe to expand almost to our present size. Then, quite miraculously, that unknown something caused it to stop this rapid inflation, and then our universe began to once again expand but at the much reduced present rate.

The newest speculation is that dark energy, which they claim had been basically ineffective while the force of gravity overpowered it in the more compact, earlier universe, began to counter the force of gravity now that it had expanded far enough. In other words, the expansion of the universe reached a critical point when dark energy overpowered gravity. We will address this speculative dark energy later, but for now, let us address this idea of a hyper-light-speed expansion.

You might say, "Wait a minute, didn't Einstein say that nothing can move faster than the speed of light?" The answer would be yes, but it was the universe expanding many times faster than the speed of light, not the matter within it. It is not quite clear just what kind of random mechanism could engineer the initiation of such a feat that would so fortuitously begin this inflationary period and end it at just the right time so the consistency of our universe could then produce galaxies and stars rather than black holes.

Through the use of Gaussian distribution equations, cosmologists can show that the distribution of this plasma ball could retain this homogeneity to a very large size through this magical hyper-light-speed expansion. Gaussian distribution equations are used to solve average distributions and heat diffusion patterns. These equations form the lynchpin of the Rapid Inflationary Model. They provide a naturalistic mechanism to explain the isotropic distribution of the cosmic microwave background radiation that scientists have documented as the leftover or afterglow of the Big Bang. But for this rapid inflationary process to be supported by the Gaussian equations, the entire universe must be both almost exactly homogeneous and almost exactly isotropic.

So scientists set out to ratify this necessary assumption to make their standard cosmological model an airtight theory. In order to more adequately map the distribution of this cosmic microwave background radiation, scientists have sent out three satellite probes in the last few decades. Each succeeding satellite improved on the previous with more sensitive equipment. The CMBR of our entire universe has been now repeatedly mapped with ever more

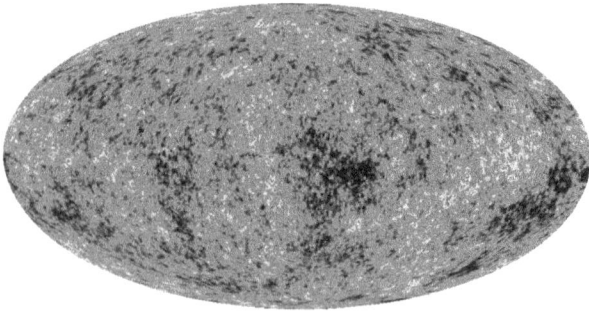

Cosmic background radiation map of our universe

increasing resolution. Most of us have heard about the magnificent accomplishment of mapping the afterglow energy of the Big Bang that brought forth our universe. Pictures of this map were circulated in many periodicals.

However, the big surprise that is not so well promoted is that the sophisticated radiometer in the mapping satellites detected several areas where the cosmic background radiation was not as homogeneous and isotropic as the rest of the universe. That is, the more detailed mapping of the CMBR showed that the universe is not completely isotropic and homogeneous, as required in order for the Rapid Inflationary Model and the standard cosmological evolutionary model to be true.

The evidence clearly shows that our universe has distinct and mysteriously located anisotropic areas that formed quadrupole planes and octopole planes, which could not have been made by the mathematical fix concocted through the cherished, magical Rapid Inflationary Model. Moreover, to evolutionists' great dismay, these symmetrically located areas showed an alignment to the sun-Earth ecliptic (the plane of the Earth's rotation around the sun). That was a torpedo in the broadside they were not expecting and little appreciated. Their disdain for this alignment with the sun-Earth ecliptic earned it the auspicious title, the axis of evil.

The Axis of Evil: The Alignment of the Dipole to the Earth and Sun's Ecliptic

So disturbed were the evolutionists when they analyzed the information gathered by their first mapping mission launched in 1992 (COBE) that they thought it had to be a mistake and launched a second and even more powerful satellite, the WMAP, in 2001. But when that returned with even clearer results, confirming that the dipole was, in fact, aligned with the Earth-sun ecliptic and the quadrupoles were aligned with our equinoxes, they still did not want to believe it and sent another more powerful probe, hoping to solve their enigma.

Through the European Space Agency, a third probe called Planck was sent out in 2009 to redo the measurements. Unfortunately, to their great consternation, these symmetrically located anomalies of anisotropy were refined even further and again verified by the Planck satellite. In my world, three strikes means you are out. What is disturbing to me is that that none of this information has been made easily available to the public. One cannot find it on the NASA website. It was not covered by the science periodicals that are printed for the public. One must dig through the technical periodicals to find this information.

The grave problem for evolutionists is now threefold:

1. How does random ordering through Gaussian equations explain these peculiar areas of anisotropy? The Gaussian equations used to prop up the Rapid Inflationary Model cannot form these areas of anisotropy.

2. How can random ordering create such areas of symmetry in these propitiously placed pockets of anisotropy, which were mapped by all three satellites? Random ordering cannot create symmetrical patterns of this size and magnitude. This is evidence that the creation of our universe has a previsioned and intentional design.

3. How can these areas of anisotropy be aligned with our solar system when their cherished Copernican principle tells us

that the Earth has no special place of significance in the universe? The alignment to our solar system is empirical evidence that whoever designed this universe meant to draw attention to our home planet.

Holy cow! The evidence was not one torpedo, but three in one. I call that poetic justice. These dipoles, quadrupoles, and octopoles that were formed are completely incompatible with the Rapid Inflationary Model, the cosmological principle, and the Copernican principle. The standard cosmological model of the evolutionists received three mortal bullets in the heart, and the general public knows nothing about it.

For all intents and purposes, the explanatory power of the Rapid Inflationary Model is now useless, because it could not have created these areas of anisotropy. In order for this model to become a real potential answer, every square inch of space would have to be exactly isotropic. It is now known that it is not. But even more alarming is that the areas of anisotropy are not randomly located. They form a symmetrical pattern that places the Earth-sun ecliptic at the very center. Hmmm, I wonder how that happened.

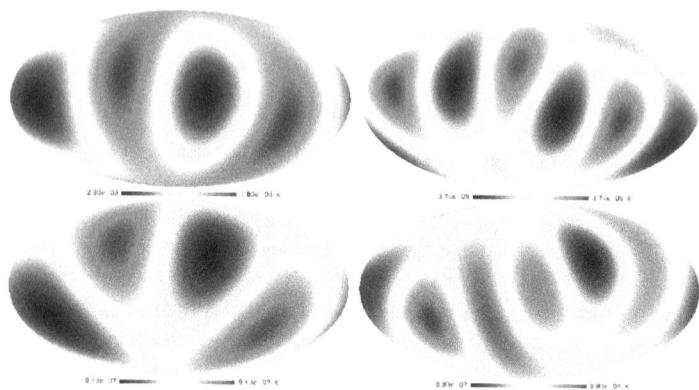

The symmetrically aligned nisotropic dipole, quadrupole, sextapole, and octapole areas of the CMBR discovered by the mapping satellites

Here is what we discovered: These symmetrically located areas of anisotropy are, in fact, quite aligned with each other. But not only are they symmetrically aligned, lo and behold, they are also aligned orthogonally to the Earth-sun ecliptic. And if that does not knock you off your seat, even more amazing is that they are aligned to the equinoxes and solstices that mark our quadrants in the trajectory of our revolution around the sun. That they would be aligned with our equinoxes is a momentous discovery that shows the Copernican principle is flat out dead wrong.

I believe that this purposeful alignment with our equinoxes is a message from the creator, which He inscribed during that moment of creation and prepared for us today when He knew we would gain the knowhow to observe it. That alignment with the equinoxes is an indication of the cosmic battle between good and evil that is symbolized by the battle between light and darkness.

The equinoxes mark the times of our year when the daylight and nighttime are equal. The summer solstice is the longest day and shortest night. The winter solstice is the longest night and shortest day. The vernal or spring equinox is the time of the year when daylight begins to grow longer than darkness. It does so until the summer solstice when the day is the longest and the night is the shortest in the year. It is the spiritual symbol of light gaining ascendancy over darkness. It marks the time of Passover and the death of Christ.

The autumnal equinox marks the time of the year when, from that day forward, darkness gains ascendancy over light. It marks the time of Yom Kippur, the day that will mark the atonement of the Earth by the returning Messiah. It also symbolically represents the most important day for the occult, and for this reason, it is their high holiday when Halloween and many other occult traditions such as the Samhain holiday are observed. Darkness from that day forward grows longer until the winter solstice, when nighttime is the longest and daylight is the shortest in the yearly path we revolve around the sun.

The four points of the equinoxes symbolically represent the cosmic battle between good and evil in our world. That knowledge was held in common by all ancient cultures. And God, in His marvelous design of our universe, provided for those who seek His face the evidence of His prevision and providence within the very structure of our universe. How utterly magnificent is His providence!

This alignment with our equinoxes can only mean that they were designed thus in order to bring special attention to the Earth and our solar system at the center of the universe. Such a combination of precise alignments between the large-scale structure of the universe and our plane of revolution around the sun cannot be rationalized with a random genesis to our universe. And equally important, it cannot be explained by the mechanism of the Rapid Inflationary Model.

Actually, it disqualifies the Rapid Inflationary Model as the mechanism for the overall homogeneous and isotropic nature of our universe. No Gaussian equations can account for the specific structures encountered by our satellite probes that measured the background microwave radiation of our universe. The naturalists are back at square one, even though they act as if the king is still dressed.

Trying to put an evolutionary spin on these perfectly aligned quadrupoles, evolutionary cosmologists, instead of admitting that the Gaussian equation could not have formed our universe through their Rapid Inflationary Model, are now intimating that these are simply anisotropic areas that may give us clues to galaxy formation. But hardly any mention is made in any public periodicals of their symmetrical alignments with the ecliptic of the Earth, the sun, and the equinoxes that divide our year into four seasons. Hardly any popular magazines or newspapers published this disturbing information for evolutionary cosmologists.

And certainly no attention has been given to the fact that the Rapid Inflationary Model is incapable of explaining how these areas of anisotropy can be created in such alignment and in such symmetrical alignments with our Earth through Gaussian equations. In essence, the rapid inflationary period should now be written "RIP" for "rest in peace." For all intents and purposes, the observable data rule out a rapid inflationary period as a legitimate theory of how our universe developed the large-scale structure it exhibits.

So disturbing are these alignments of the CMBR with our solar system to evolutionists that the dipole was dubbed the axis of evil by Kate Land and João Magueijo, who viewed it as an evil thing to oppose the Copernican principle they so cherish. That evidence completely contradicts the predictions of their standard cosmological model and their Rapid Inflationary Model. That the dipole of the CMBR of the entire universe would be pointed to our solar system would be analogous to claiming that our Milky Way galaxy and all it contains would be aligned to a single molecule sitting on top of my little finger.

On the other hand, our Judeo-Christian cosmological model fits like a glove with the observable evidence. The universe bears the fingerprint of a master architect who created such large-scale structures in order to point toward the Earth and manifest His prevision and intended purpose for His creation. But the most magnificent aspect of this discovery is that it also provides evidence of the creation song of God.

The Creation Song of God

There is yet one more discovery regarding these areas of anisotropy that shows us our universe is the creation song of God. It seems that there is a musical harmony to their relationship that cannot be explained by random ordering.

> *Like an orchestral movement, the WMAP results can be analysed as a blend of patterns of different spatial frequencies. When Magueijo and Land looked at the hot and cold spots this way, they noticed a striking similarity between the individual patterns. Rather*

than being spattered randomly across the sky, the spots in each pattern seemed to line up along the same direction (Sungenis and Bennett 2014a, 329).

Random ordering cannot produce the harmonics of the CMBR and the resulting specified alignments to the Earth. It is nothing less than an obviously intended tapestry that bears the fingerprint of God for those with eyes to see. And so we see that not only were these patterns not randomly situated, but they also were found to be pointing toward the Earth-sun ecliptic as a central hub.

And to put the icing on the cake, the patterns were found to be in a harmonic state, as if the universe were created as a song of God. The evidence affirms our Judeo-Christian cosmological model that claims our universe was created by the voice of God—His love song—and disqualifies the naturalistic dysteleological model completely.

The harmonic multipoles of the CMB are analogous to the harmonics of musical vibrations. When a string on a violin is plucked it vibrates very fast. In turn, the air molecules vibrate and sound waves travel to our ear. But the note made by the violin makes the string vibrate in a very complex manner. First, is the basic or fundamental note, but many other notes appear that, when all the notes are combined, makes the sound that is unique to a violin, as opposed to a cello. . . . In a similar way, the CMB monopole is the fundamental note, but can then be divided into higher harmonics, such as dipole, quadrupole and octupole. Whereas the various harmonics of musical tones will create a different note, the CMB harmonics will create different orientations or directions for the microwaves. The astounding fact for the CMB harmonics is that all of them point to the ecliptic and equator of the Earth (Sungenis and Bennett 2014a, 323).

The Death of the Rapid Inflationary Theory, the Cosmological Principle, and the Copernican Principle

So the problem for the evolutionists begins with the underlying premise that random ordering has been the architect of our universe. The detection of the CMBR being almost completely isotropic sent them flying for an explanation that could account for it without relying on divine intervention. Thus came the Rapid Inflationary Model, but then they discovered that the CMBR is not completely isotropic since it has distinct areas of anisotropy that form a peculiar pattern that is not only symmetrical but connected to our solar system. This empirical data completely disqualify their inflationary proposition, but as I see it, the problem for their theory stems further back. It stems from their hyper-light-speed inflationary hypothesis

Einstein's theory says unequivocally that nothing moves faster than the speed of light. Nothing in our entire universe can run faster than the speedy photon. Evolutionists, however, claim that matter did not inflate beyond the speed of light, but rather it was the matrix of space-time in the universe that contained it, which inflated it at hyper-light-speed. Let's think about that for a moment.

Picture a deflated balloon with seven blue dots along its length and seven other red dots along its width forming a cross. If the dots are connected to the skin of the balloon, then as the balloon is inflated, the dots move further apart at a parallel rate to which the balloon is being inflated. The red and blue dots grow in distance to each other and also grow in size. That is, matter is becoming less dense as space is being stretched. So the red and blue dots not only increase in size but also in distance to one another. As the universe expands, the dots are inflating symmetrically; hence, they remain isotropic and homogeneous. That part of the theory does seem to have some explanatory power.

However, you will also notice that as the balloon inflates, the dots are moving. If, instead, those dots were somehow floating inside the balloon and not connected to the fabric of the balloon, then we could inflate the balloon without impacting the motion of the dots. But in our universe, matter is connected to its space-time suprastructure that envelops it and reacts to it. It cannot exist independent of the space-time suprastructure. It is, in fact, anchored to it. It is like the dots on the skin of the balloon anchored in space-time.

Any inflation of space-time will therefore also inflate the matter it contains accordingly and cause it to move away from one another. It is therefore impossible for space-time to inflate without inflating the matter within it, because space-time and matter are connected in a continuum. If it ever took place at all, the hyperinflation of the universe could not have taken place at speeds that would make matter inside it travel beyond the speed of light. So this rather simple example shows us that the universe could not have hyperinflated as claimed without causing the matter inside it to go beyond the speed limit of reality.

Now this model is, of course, the spherical model that we spoke of earlier where space is curved. We now know that space is flat. Hence, the inflation from a starting point is more like a solid sphere that moves outward by the angular momentum of its spin along x and y vectors. There is a center, the Alpha Point and there is a stretching that varies in intensity as one moves outward from the center to the edge of the universe on either the x or y vector. Any stretching that begins to reach the speed of light would automatically cause the baryonic matter within it to increase in mass. This means it could never go beyond the speed of light and would then cause the universe to slow down as the increase in mass creates an increase in gravity and counters the outward angular momentum. Hence, this idea of a hyperinflationary period is physically impossible.

In order for space to inflate faster than the speed of light without allowing matter to inflate faster than the speed of light, space would have to be disconnected to the baryonic matter it encapsulates. This disconnect flies in the face of all we know about the interconnections that produce our space-time continuum.

But this is not the only problem with this mathematical fix. Any change made to the density of space would not only have a direct impact on the density of matter but also on the rate of time within it. An event of hyperinflation of space-time would mean that the rate of time would have equally been hyperinflated, causing extreme aging. But the inflation of the universe would have to be accomplished in a uniform fashion throughout the entire universe in order for it to remain smooth. That seems to contradict our present observations, which show that our universe is expanding at different rates, depending on their distance to us. Those farthest away are expanding fastest, and even more importantly, that expansion is, in fact, increasing in velocity. Our universe is not expanding at a steady, even pace. It is an accelerating pace. How then could it have hyperinflated, slowed down, and then begun to accelerate again?

The rapid inflationary idea is basically that the universe underwent a magical expansion of enormous proportions by an unknown mechanism to an unknown hyperdrive speed (faster than the speed of light) that would, by another unknown mechanism, put on the brakes at just the right moment and then stop the hyperdrive expansion of the entire universe, all of this without changing the isotropic and homogeneous consistency of the material within the universe. But that is not all. Once the hyperdrive stops, the universe must once again, by another unknown mechanism, begin to expand outwardly with an exact acceleration rate so that stars and galaxies can form and so it can match our present observations.

What magical wand could make the universe expand faster than the speed of light and then magically slow down? It is a bit of a mystery. There is no known mechanism that could cause magical "space brakes." But the biggest trick is yet to come. Now, after having slowed down from that supralightspeed inflationary period, something has to cause it to start inflating all over again after having applied the brakes. It seems like the universe acts like my grandmother used to drive. She would alternate between the accelerator and the brakes, making my head bop back and forth.

"What?" said Foolwinkle, scratching his head, "I don't get it."

"What's not to get?" said Rocksy, shrugging his shoulders.

Foolwinkle wagged his finger at Rocksy and said, "You said we don't believe in miracles."

"This is not a miracle," complained Rocksy, "Can't you see we are using scientific language to explain it?"

"If it's not a miracle, then what materialistic mechanism could cause this miraculous expansion that, against all our science, progressed at speeds beyond that of light?"

"Well, that's easy. It could have been anything. Why, it could have been dark energy." Said Rocksy with a Cheshire grin, looking up at Foolwinkle.

Foolwinkle stood, rubbing his chin and tapping his foot as he thought for a moment. And then he said, "That makes no sense either. You said that dark energy was incapable of working when the universe was tightly packed and gravity was at its strongest. You said that only after the universe had spread out thinly could dark energy begin to have discernible effects. It could not be dark energy."

"Well, I suppose you are right, but that does not matter," said Rocksy, getting visibly agitated by Foolwinkle's foolish questions.

"What do you mean it doesn't matter?" responded Foolwinkle angrily. "You are telling me that something made the entire universe expand faster than our laws of physics allow anything to travel, but you don't know what power was capable of doing it? That is a miracle. You don't even know what caused it to stop."

"My dear Foolwinkle," responded Rocksy with a slow, condescending tone. "Those are only minor points, you see? What is important is that we can use this to explain how the universe looks the way it does without, you know, using the G word."

But, all of this is rather a moot point. The rapid inflationary period has been shown to be incompatible with the present structure of our universe. As we have previously noted, the evidence garnered by the three space probes sent to map the microwave background radiation revealed areas of anisotropy that are propitiously placed in alignment not only with the sun-Earth ecliptic but with our equinoxes, and these cannot be explained in any way by the Rapid Inflationary Model, which cannot produce such areas of anisotropy. For all intents and purposes, it is dead.

Moreover, the alignment of these areas of anisotropy with our solar system points to the uniqueness of the Earth and cannot be explained by any other materialistic and randomly generated mechanism. It points to a purposefully designed anomaly that serves God's intent of leaving His fingerprint for those who are willing to see the truth. The Rapid Inflationary Model cannot create symmetrical areas of anisotropy that are placed in specified positions that form alignments with our sun-Earth ecliptic and our equinoxes. Such specified information is indicative of a mind and not random chaotic processes.

The following excerpt from *Advances in Astronomy*, Volume 2010 documents the truth about these extraordinary alignments and their incompatibility with the Rapid Inflationary Model. It readily admits that these fantastic alignments are "inconsistent with Gaussian random, statistically isotropic skies." Furthermore, they are completely baffled at the correlation to our solar system: "Particularly puzzling are the alignments with solar system features. CMB anisotropy should clearly not be correlated with our local habitat" (Copi 2010).

B. Planarity and Alignments

Tegmark et al. and de Oliveira-Costa et al. first argued that the octopole is planar and that the quadrupole and octopole planes are aligned. In the work of Schwarz et al., followed up by Copi et al., we investigated the quadrupole-octopole shape and orientation using the multipole vectors. The quadrupole is fully described by two multipole vectors, which define a plane. This plane can be described by the "oriented area" vector:

$$\sim w\,(\dot{}\,;i\,,j) \equiv v\hat{}\,(\dot{}\,,i) \times v\hat{}\,(\dot{}\,,j)$$

(Note that the oriented area vector does not fully characterize the quadrupole, as pairs of quadrupole multipole vectors related by a rotation about the oriented area vector lead to the same oriented area vector.) The octopole is defined by three multipole vectors which determine (but again are not fully determined by) three area vectors. Hence there are a total of four planes determined by the quadrupole and octopole.

In the work of Copi et al (2010), we found that:

(i) the four area vectors of the quadrupole and octopole are mutually close (i.e., the quadrupole and octopole planes are aligned) at the 99.6% C.L.;

(ii) the quadrupole and octopole planes are orthogonal to the ecliptic at the 95.9% C.L.; this alignment was at 98.5% C.L. in our analysis of the WMAP 1 year maps. The reduction of alignment was due to WMAP's adaption of a new radiometer gain model for the 3 year data analysis, that took seasonal variations of the receiver box temperature into account—a systematic that is indeed correlated with the ecliptic plane. We regard that as clear evidence that multipole vectors are a sensitive probe of alignments;

(iii) the normals to these four planes are aligned with the direction of the cosmological dipole (and with the equinoxes) at a level inconsistent with Gaussian random, statistically isotropic skies at 99.7% C.L.;

(iv) the ecliptic threads between a hot and a cold spot of the combined quadrupole and octopole map, following a node line across about 1/3 of the sky and separating the three strong extrema from the three weak extrema of the map; this is unlikely at about the 95% C.L.

These numbers refer to the WMAP ILC map from three years of data; other maps give similar results. Moreover, correction for the kinematic quadrupole—slight modification of the quadrupole due to our motion through the CMB rest frame—must be made and increases significance of the alignments. . . .

While not all of these alignments are statistically independent, their combined statistical significance is certainly greater than their individual significances; for example, given their mutual alignments, the conditional probability of the four normals lying so close to the ecliptic is less than 2%; the combined probability of the four normals being both so aligned with each other and so close to the ecliptic is less than 0.4% × 2% = 0.008%. These are therefore clearly surprising, highly statistically significant anomalies—unexpected in the standard inflationary theory and the accepted cosmological model.

Particularly puzzling are the alignments with solar system features. CMB anisotropy should clearly not be correlated with our local habitat. While the observed correlations seem to hint that there is contamination by a foreground or perhaps by the scanning strategy of the telescope, closer inspection reveals that there is no obvious way to explain the observed correlations. Moreover, if their explanation is that they are a foreground, then that will likely exacerbate other anomalies that we will discuss in section 4.2.

Our studies . . . indicate that the observed alignments are with the ecliptic plane, with the equinox, or with the CMB dipole, and not with the Galactic plane: the alignments of the quadrupole and octopole planes with the equinox/ecliptic/dipole directions are much more significant than those for the Galactic plane. Moreover, it is remarkably curious that it is precisely the ecliptic alignment that has been found on somewhat smaller scales using the power spectrum analyses of statistical isotropy (Copi).

Let me isolate from the previous technical quote some of the key points for the reader:

1. There are a total of four planes determined by the quadrupole and octopole.

2. The quadrupole and octopole planes are orthogonal (perpendicular) to the sun-Earth ecliptic.

3. The sun-Earth ecliptic threads between a hot spot and a cold spot of the combined quadrupole and octopole map, following a node line across about one-third of the sky and six separating the three strong extrema from the three weak extrema of the map.

4. The maps from the other mapping satellites give similar results.

5. The study indicates that the observed alignments are with the ecliptic plane of our sun–Earth system and with the equinoxes or with the CMB dipole and not with the Galactic plane. In other words, the alignments are not with the Milky Way plane but with our solar system specifically.

6. Their perplexing conclusion: "Particularly puzzling are the alignments with solar system features. CMB anisotropy should clearly not be correlated with our local habitat" (Copi).

The data are irreconcilable with Gaussian equations used to legitimize the Rapid Inflationary Model: "These four planes are aligned with the direction of the cosmological dipole (and with the equinoxes) at a level inconsistent with Gaussian random, statistically isotropic skies" (Copi).

7. And what is their rather weak conclusion regarding the standard inflationary theory? "These are therefore clearly surprising, highly statistically significant anomalies—unexpected in the standard inflationary theory and the accepted cosmological model" (Copi).

That is the understatement of the century. Here is a bit of advice. If the data do not correlate with the present evolutionary paradigms of the Rapid Inflationary Model, the cosmological principle, and the Copernican principle, don't ignore, minimize, manipulate, hide, or destroy the data. Change your presupposition to match the data. That, my friends, is called scientific integrity.

Here is the bottom line: Our sun-Earth ecliptic is central to the entire structure of our universe. The evolutionary model has been lanced through the heart by indisputable empirical data. Our Judeo-Christian cosmological model has been vindicated yet again by true science. And yet this information is being suppressed by those whose deophopic predilection blinds them to the truth.

The indisputable truth is that the anisotropy analysis of the correlation of the harmonics between the monopole, dipole, quadrupoles, and octopoles are so aligned with each other and the Earth-sun ecliptic that it rules out any consideration of an origin by some local effect. The cause must be universal in scope. It can only be explained by some action that took place at the moment of the initial explosion of the Genesis Singularity at the Alpha Point. Therefore, their anemic attempt to link these with galaxy formation is nothing more than a not-so-well-designed attempt at misdirection—nothing more than Orwellian doublespeak.

The problem for the naturalist is not just that the evidence clearly stipulates that the Earth-sun ecliptic and the equinoxes are aligned with these anomalies, but that it proves that the rapid expansion model that was created to explain the generally isotropic and homogeneous qualities of our universe cannot form these multipoles. Thus their reluctant statement, "These are therefore clearly surprising, highly statistically significant anomalies—unexpected in the standard inflationary theory and the accepted cosmological model" (Copi).

Their wording is laughable. That it is "unexpected" is the understatement of the century. The more proper statement is that their Rapid Inflationary Model and their standard cosmology based on their a priori metaphysical doctrine of the Copernican principle and their cosmological principle that insists on universal homogeneity is irreconcilable with the empirical evidence garnered by all three mapping satellites. The standard evolutionary cosmological model is dead in the water—period.

The Rapid Inflationary Model cannot create these highly structured anomalies. It can only explain a completely isotropic environment that is universally isotropic and absolutely homogeneous. Therefore, their model of an infinite flat and homogeneous universe has been disqualified. In short, the death of the Rapid Inflationary Model and their standard cosmological model that insists in an infinite universe was sealed by the observational facts, even though they stubbornly still act as if the king has clothes.

This axis created by these areas of anisotropy, the cosmological dipole, is so abhorred by the deophobes that the naturalists have labeled it the axis of evil. They think it is evil because it contradicts their evolutionary religion. I find it quite amusing that a materialist could call anything evil, since in their worldview, what is just is. There is no good and no evil. Such is human nature that we cannot escape our humanity because we are hardwired by the breath of God in us to intuitively know that evil is real. Even in their rebellious angst, they betray the fact that they are not just an accident in the universe. They are not just organic machines void of any transcendental value. The creator who designed this incredibly sophisticated symmetry of our magnificent universe for those who care to know the truth has given them a spirit.

But there is yet a much deeper symbolic significance to the propitious alignment of these anisotropic anomalies for those who have eyes to see and ears to hear. In 2010, anticipating the results of the last satellite, the Planck satellite sent by the Europeans, Professor John P. Ralston, Department of Physics and Astronomy at the University of Kansas, stated in an abstract:

> *Consistent signals of anisotropy have been found in data on electromagnetic propagation, polarizations of QSOs and CMB temperature maps. The axis of Virgo is found again and again in signals breaking isotropy, from independent observables in independent energy regimes. There are no satisfactory explanations of these effects in conventional astrophysics. . . . To summarize, our studies find there is nothing supporting isotropy of the CMB, and everything about the data contradicting it. . . . The PLANCK observations of polarization data from the CMB are eagerly awaited. We can predict with reasonable certainty that correlations contradicting isotropy will be seen; spontaneous alignments of polarizations will occur along the axis of Virgo (Sungenis and Bennett 2014a, 388).*

In a nutshell, "To summarize, our studies find there is nothing supporting isotropy of the CMB, and everything about the data contradicting it." Now, I find Ralston's statement a bit more honest. I applaud him for his candor. As it turned out, the PLANCK observations, being the third satellite to map the CMBR in the entire universe, completely corroborated the data from the two previous probes in even greater detail.

Some of you may be surprised to learn that the trajectory of our planet through space began in Virgo and is heading toward Leo. For those who wish to see the documentation of the trajectory of our planet through space, refer to a helpful article entitled, "The pre-launch Planck Sky Model: a model of sky emission at submillimetre to centimetre wavelengths."* This is highly significant in the symbolism of these two constellations and the providence of God. Pardon me if I digress here for a moment, but the significance of Virgo must be understood in order for us to really understand the message God has left in His creation so humans can realize His authorship of the creation song of our magnificent universe.

The Significance of the Dipole Alignment and the Sun-Earth Ecliptic in Human History

Some of my friends have suggested that I not deal with the subject of the zodiac in this book and that it should be kept separate from the scientific data. I have decided not to heed that advice because I believe strongly in the Judeo-Christian doctrine of the unity of truth. When we think in those terms, we are, in fact, abrogating to the evolutionists the platonic division they have invented between the metaphysical and the physical. There is no such division in reality. All truth is unified, and all truth emanates from God, who is sovereign over all reality.

But, we must begin by saying that the fortune-telling astrology of the occult is, in fact, a corruption and distortion of the real purpose God had for the previsioned placement of the constellations. What I will relate now has absolutely nothing to do with modern-day astrology. It must be noted that initially the constellations of the zodiac depicted the ancient story of the cosmic struggle between

---

* Delabrouille, J., M. Betoule, J.-B. Melin, M.-A. Miville-Deschenes, J. Gonzalez-Nuevo, M. Le Jeune, G. Castex. et. al. 2012. "The pre-launch Planck Sky Model: a model of sky emission at submillimetre to centimetre wavelengths." *Astronomy & Astrophysics*. 5. https://arxiv.org/pdf/1207.3675.pdf.

good and evil. It contained a prophetic message of the struggle between Satan and God that continues even now.

Hebrew tradition claims that Seth and Enoch long ago cataloged the specific groupings of these constellations as they received divine inspiration. Today, the purpose and significance of the constellations have been corrupted and profaned by the occult and used for personal divination. It is that occult narrative that our culture has accepted as the meaning of the stars that hides the true and original meaning ascribed by Enoch and Seth. For a deeper study on this issue, see "The Witness in the Sky" in Chapter 15.

There are 12 constellations that, from our view on Earth, are found within the ecliptic of the sun. That is, the sun seems to go through these 12 specific constellations as we view them from Earth. It is that ecliptic that is aligned with these symmetrically placed areas of anisotropy throughout the four quadrants of our universe. Therefore, in my humble opinion, I think God is drawing attention to the message that He has given humankind regarding the ancient cosmic battle between good and evil that is contained in the 12 signs of the zodiac that follow the ecliptic. This is corroborated by the fact that they are also aligned with the equinoxes.

The constellation that rises on the eastern horizon during the spring equinox is called the ruling house for that epoch. Because the Earth has a slight wobble, the constellation that marks the rising of the sun as we look east on the spring equinox changes approximately every 2,160 years. This movement from one constellation to the other is called the precession of the equinoxes. The specific constellation that marks the horizon due east when the sun rises in the spring equinox at this time is Pisces, which is considered the ruling constellation for our epoch. Each of these constellations tells of God's plan for the ages. They show the providence of God in the affairs of humans, for He knows the end before the beginning.

In our present age, we are nearing the time when Pisces, which began rising on the spring equinox during the time Christ was born, shall give way to the constellation of Aquarius. During our epoch, while Pisces rises every year on the spring equinox, Virgo rises on the autumn equinox, for it was in this epoch of Pisces that the virgin gave birth to the seed of Eve who would wound Satan in the head. So the prophecy of the coming of the deliverer from the seed of Eve in Genesis 3:15 was fulfilled in this epoch.

The constellation of Pisces is composed of two fish tied by their tails by a band to the neck of Cetus (Leviathan). They represent the Gentiles and the Jews, which together comprise the church. The New Testament describes the Gentiles as wild branches grafted to the natural tree of the olive tree of Israel.

> *But if some of the branches were broken off, and you, although a wild olive shoot, were grafted in among the others and now share in the nourishing root of the olive tree, ... For if you were cut off from what is by nature a wild olive tree, and grafted, contrary to nature, into a cultivated olive tree, how much more will these, the natural branches, be grafted back into their own olive tree (Romans 11:17, 24).*

This was predicted by Moses, and the symbolism is found within the very instructions given to the people during the celebration of the Feast of Succoth, which represents God's protection over His people during the time of the wandering in the desert of Sinai. It also symbolizes the protection that shall be given to them again in the wilderness when Elijah prepares the way for the Messiah. These are the instructions given in the book of Nehemiah as the Jews returned from Babylon and began to build the second temple. In that day, Israel, the natural olive tree, shall accept the wayward Gentiles that come to worship God. Jerusalem shall be known as the City of Truth.

> *That they should proclaim it and publish it in all their towns and in Jerusalem, "Go out to the hills*

and bring branches of olive, wild olive, myrtle, palm,
and other leafy trees to make booths, as it is written"
(Nehemiah 8:15).

The very same thing experienced by Nehemiah shall happen again when the Lord returns to free Israel from the grip of the antichrist. In that day, Ezekiel's temple shall be built. But first, seven years of great tribulation shall come upon humankind. In the middle of those seven years, the antichrist will attack Jerusalem. Hear the words of Elijah, my brothers, for he shall tell you of the hiding place beneath the eagle's wings in the wilderness. There you shall be in the ark of the second earth. Hear the words of Yeshuah—if you obey, you shall be saved. If you disbelieve, you shall not survive the fangs of the serpent that comes from the north:

> *"So when you see the abomination of desolation spoken*
> *of by the prophet Daniel, standing in the holy place*
> *(let the reader understand), then let those who are in*
> *Judea flee to the mountains. Let the one who is on the*
> *housetop not go down to take what is in his house,*
> *and let the one who is in the field not turn back to*
> *take his cloak. And alas for women who are pregnant*
> *and for those who are nursing infants in those days!*
> *Pray that your flight may not be in winter or on a*
> *Sabbath. For then there will be great tribulation, such*
> *as has not been from the beginning of the world until*
> *now, no, and never will be. And if those days had not*
> *been cut short, no human being would be saved. But*
> *for the sake of the elect those days will be cut short.*
> *Then if anyone says to you, 'Look, here is the Christ!'*
> *or 'There he is!' do not believe it. For false christs and*
> *false prophets will arise and perform great signs and*
> *wonders, so as to lead astray, if possible, even the elect.*
> *See, I have told you beforehand. So, if they say to you,*

'Look, he is in the wilderness,' do not go out. If they say, 'Look, he is in the inner rooms,' do not believe it. For as the lightning comes from the east and shines as far as the west, so will be the coming of the Son of Man. Wherever the corpse is, there the vultures will gather.

"Immediately after the tribulation of those days the sun will be darkened, and the moon will not give its light, and the stars will fall from heaven, and the powers of the heavens will be shaken. Then will appear in heaven the sign of the Son of Man, and then all the tribes of the earth will mourn, and they will see the Son of Man coming on the clouds of heaven with power and great glory. And he will send out his angels with a loud trumpet call, and they will gather his elect from the four winds, from one end of heaven to the other" (Matthew 24:15–31).

In that day, the two olive trees that, for the first half of those seven years, have been preparing the way shall be captured by the King of the North invading Jerusalem. The King of the North, helped by Azazel who has come with Shemihaza as the two towers that help Lucifer, shall capture Elijah and Enoch in Jerusalem. And in the very midst of that seven-year period, they will be killed before the watching world.

But three and a half days after their death, they will resurrect before the eyes of the entire watching world. And Israel, safe beneath the eagle's wings in the wilderness, shall also resurrect in 1,260 days, or three and a half years. At the end of God's judgment upon the wicked, Israel will come out from underneath the eagle's wings to cross the Jordan River and enter into the Promised Land with Yehoshua. In that day, Jericho shall crumble by the roar of the Lion of Judah, with an earthquake so great that every single city in the world shall crumble to the ground at the roar of the Lion, who is also the Lamb.

And a great sign appeared in heaven: a woman clothed with the sun, with the moon under her feet, and on her head a crown of twelve stars. She was pregnant and was crying out in birth pains and the agony of giving birth. And another sign appeared in heaven: behold, a great red dragon, with seven heads and ten horns, and on his heads seven diadems. His tail swept down a third of the stars of heaven and cast them to the earth. And the dragon stood before the woman who was about to give birth, so that when she bore her child he might devour it. She gave birth to a male child, one who is to rule all the nations with a rod of iron, but her child was caught up to God and to his throne, and the woman fled into the wilderness, where she has a place prepared by God, in which she is to be nourished for 1,260 days.

Now war arose in heaven, Michael and his angels fighting against the dragon. And the dragon and his angels fought back, but he was defeated, and there was no longer any place for them in heaven. And the great dragon was thrown down, that ancient serpent, who is called the devil and Satan, the deceiver of the whole world—he was thrown down to the earth, and his angels were thrown down with him. And I heard a loud voice in heaven, saying, "Now the salvation and the power and the kingdom of our God and the authority of his Christ have come, for the accuser of our brothers has been thrown down, who accuses them day and night before our God. And they have conquered him by the blood of the Lamb and by the word of their testimony, for they loved not their lives even unto death. Therefore, rejoice, O heavens and you who dwell in them! But woe to you, O earth and sea, for the devil has come down to you in great wrath, because he knows that his time is short!"

And when the dragon saw that he had been thrown down to the earth, he pursued the woman who had given birth to the male child. But the woman was given the two wings of the great eagle so that she might fly from the serpent into the wilderness, to the place where she is to be nourished for a time, and times, and half a time. The serpent poured water like a river out of his mouth after the woman, to sweep her away with a flood (Revelation 12:1–15).

Thou art that woman, Israel. Obey the words of Elijah who will tell you where to find the Great Succoth set for you. For you shall be the priests of God in the Third Earth ruling with Aquarius. Therefore, the next constellation that shall rise in the spring equinox when the precession of Pisces is complete is Aquarius. It will mark the end of the Second Earth and the beginning of the Third Earth. Aquarius is the water bearer, and it has only one fish, which is freely swimming in the water that it is being poured on the Earth. Water is the symbol of the Holy Spirit and marks the coming of the age of the Davidic Kingdom.

The singular fish, no longer tethered to Leviathan, represents the union of Jews and Gentiles in the Davidic Kingdom when we are freed from Leviathan and the Messiah reconstitutes the Earth. The scriptures tell us that during this time, Satan shall be chained in the Great Abyss. He will no longer hamper the children of God, symbolized by the fish being no longer tethered to Leviathan, but instead freely swimming in the water.

When Aquarius rises on the spring equinox, Leo shall rise in the autumn equinox. Hence, both Virgo and Leo have a specific unfolding between the Age of Pisces and the Age of Aquarius. Leo is the Lion of Judah who shall reign in the Davidic Kingdom of Aquarius, foretold long ago by Israel on his deathbed.

He crouches, he lies down as a lion,
And as a lion, who dares rouse him up?
"The scepter shall not depart from Judah,
Nor the ruler's staff from between his feet,
Until Shiloh comes,
And to him shall be the obedience of the peoples"
(Genesis 49:9–10 NASB).

But what holds true for the future in the witness in the sky also held true for the past. Prior to Pisces, Aries began rising in the spring equinox when Abraham was born. The ram caught in the thicket provided the substitutionary sacrifice for Isaac, and thus it was through the seed of Abraham, Isaac, and Jacob that the Messiah was to be born and die on the Passover day during the spring equinox. The importance of the shofar (ram's horn) announcing the New Year, Rosh Hashanah, is made evident to those with eyes to see and ears to hear.

The significance of the trajectory of the Earth in the cosmos from Virgo toward Leo is, in fact, the story of the Judeo-Christian message that spans human history from Adam and Eve to the coming of the Lion of Judah and the consummation of the ages.

The Significance of Virgo

The tradition of the ancients gave Virgo first place in the astrological charts as the first to rise on the spring equinox. Much later, when the occult corrupted the Zodiac's message, this was changed, putting Capricorn in its place.

> *Those who claim very high antiquity for the zodiacal signs assert that the idea of these titles originated when the sun was in Virgo at the spring equinox, the time of the Egyptian harvest. This, however, carries them back nearly 15,000 years, while Aratos said that Leo first marked the harvest month (Allen 461).*

From the very beginning, the story of Virgo and Leo were understood by ancient people. Today, all of this knowledge has been corrupted by the occult, and modern man knows nothing of the true ancient wisdom. What does Virgo represent? Here, we find a beautiful young virgin, whereby all the mythologies concur that she is both pure and virgin. In her left hand, she holds an ear of wheat called the "spica," and on this spica are the individual seeds of the wheat. The brightest star in this constellation is found here, and in the Hebrew it is called Zerah, which means seed.

But, how does a virgin bring forth seed? In her right hand, she holds a palm branch. Here is the second brightest star of this constellation. In the Hebrew it is called Tsemech, which means branch. Who is this branch?

If you remember the prophecy of Genesis 3:15, God claimed from the very beginning at the garden in Eden that from the seed of the woman would come the one, who will be called the branch by later prophets. He is the one who would defeat the serpent who beguiled Eve. It is, therefore, no coincidence that the Jewish prophets called the Messiah the branch. The symbol of Virgo holds that branch in her right hand, symbolizing the fact that He would come through her genetic branch.

Isaiah prophesied that from the loins of Jesse, the father of David, the Messiah (that branch) would come.

> There shall come forth a shoot from the stump of Jesse, and a branch from his roots shall bear fruit. And the Spirit of the LORD shall rest upon him, . . . but with righteousness he shall judge the poor, and decide with equity for the meek of the earth; and he shall strike the earth with the rod of his mouth, and with the breath of his lips he shall kill the wicked (Isaiah 11:1–2, 4).
>
> In that day the branch of the LORD shall be beautiful and glorious, and the fruit of the land shall be the pride and honor of the survivors of Israel (Isaiah 4:2).

"Behold, the days are coming," declares the LORD,
*"when I will raise up for David a righteous Branch,
and he shall reign as king and deal wisely, and shall
execute justice and righteousness in the land. In his
days Judah will be saved, and Israel will dwell securely.
And this is the name by which he will be called: 'The*
LORD *is our righteousness'" (Jeremiah 23:5–6).*

It is obvious that this righteous branch, which is prophesied to
be a descendant of David, will "in that day" become King of Israel,
bringing justice and righteousness back to the world. But most
importantly, He will be "the Lord our righteousness" in the flesh.
The phrase "in that day" always refers to "the day of the Lord." These
are used interchangeably throughout scripture and refer to the end
times when the Lord returns to claim the Earth at the precise time
that the antichrist takes over the entire world exactly in the middle of
that seven-year tribulation period.

The name of the Messiah as He returns to defeat the antichrist is
then clearly stated as God. It leaves no doubt that this man, born of the
seed of a human being, David, would be the Lord our righteousness.
But why is Virgo symbolized as a virgin when she is the mother of all
humankind? Because through her would come a virgin who would
conceive the branch.

Therefore the Lord Himself will give you a sign. Behold,
the virgin shall conceive and bear a son, and shall call
his name Immanuel (emphasis added) *(Isaiah 7:14).*

Immanuel means literally "God with us." Therefore, this seed of
Eve will be both human and God. Let me at this point once again
reiterate that we do not believe that any human could ever become
God. However, we do believe that if God chose to, He could take on
human form—nothing is too great for Him.

The symbol of Israel has, for this reason, always been the palm branch. And it is for this reason, on the day that Christ entered the City of David in the event called the triumphal entry, that palm branches were placed on the way, lining the path through the Eastern Gate, or the Shushan Gate, through which he entered into Jerusalem. Christians celebrate this day as Palm Sunday in the Easter holiday (Passover), which coincides with the spring equinox and the time of the firstfruit of the harvest.

The constellation of Virgo then gave people at this very early date three important images imbued with prophetic meaning:

1. **A virgin:** Somehow, the birth of this Messiah that would defeat the great deceiver would come through a virgin.

2. **The spica (contains the seed):** This is obviously tied to two significant prophetic symbols. First, she is the one from whom all of humanity will emanate; therefore she bears the seed of humanity. Second, it is through her seed that the Messiah will come. The Genesis 3:15 prophecy foretells the humanity of the Messiah, and this seed from the virgin is the hope of humanity to defeat the ancient foe, Satan. This is specifically important because Christ is only individual in the world who can actually trace his ancestral branch all the way back to Eve, through David, Israel, Abraham, Noah, Enoch, and Adam. His genealogy, as preserved in the New Testament, is therefore the only surviving genealogy in the world that can claim this genealogy and trace it to the time of the second diaspora. Any other claim is undocumented by any ancient manuscripts.

3. **The branch:** The Jewish prophets later recognized the branch—the Messiah, which is of her seed—as the branch associated with the Feast of Booths, or Succoth, a symbol of protection and provision. Thus, the Messiah would be the provider and the protector of humanity in its darkest moments when the great opposer will seek to destroy all

the children of God. The New Testament tells us that in the middle of the seven-year tribulation period, the enemy of humanity will attack Jerusalem, seeking to kill every Jew on Earth. God shall provide a place of hiding for those who obey the warnings of Elijah, who shall come to prepare the way. This is the great Succoth I spoke of earlier.

Virgo is then at once the first mother, whom we call Eve, as well as the nation of Israel and Mary, through whom Eve's seed entered into our space-time continuum as the Son of God.

Now, it is also important to understand that every culture in the world associates their later corrupted female deities, who depict the common motif of the "great mother," as associated with Virgo. This is so because of the great prophetic message reflected in their cultural memory of the once universally held truth that was related by the Genesis 3:15 prophecy to all the descendants of Noah. From the very beginning, it was predicted that through the seed of the virgin (Virgo), the coming Messiah would defeat Satan.

For this reason, Satan has hated Eve and all who are associated with the funneling of the seed of Eve up until the Messiah was born, and this hatred and violence shall continue unabated until the Lion of Judah returns to defeat him in the Battle of Armageddon. Satan, therefore, felt it a priority to corrupt and deviate that message, which was clearly established in the stars from the very beginning. Unfortunately, we must acknowledge that he has quite successfully perverted it in order to keep its true meaning hidden from humanity. In this way, the worship of the mother goddesses began, and the meaning became corrupted into a fertility cult. Thus, we find in every culture the perversion of Virgo into a sex goddess.

Returning now to the cosmological dipole, which is now corroborated by the data to also be aligned with Virgo, we can begin to appreciate the forethought, prevision, and purpose for God's intentional design of these areas of anisotropy, which point to the very beginning of the history of humanity and the creation of the universe in Virgo and to the Lion of Judah who shall come to rule the universe from Jerusalem in Leo.

That this anomaly in the CMBR would be aligned, not only with the rotational axis of the Earth around the sun and the equinoxes of our year as well as the very first constellation that arose over the vernal equinox at the time of Adam and Eve, but also to the constellation of the Messiah, Leo, which cannot be labeled as anything else but miraculous and proof of an intelligent design for our universe from the first moment of the Alpha Point.

The alignment of the dipole to the axis of the Earth-sun ecliptic and also with Virgo is a magnificent symbol of God's providential will. It marks our beginning, the person from whom we came. But if we consider that our planet and sun are moving at 371 km/second toward the constellation of Leo, we can begin to see the end result of what Virgo prophesied in the stars at the beginning—the coming of the Lion of Judah. Virgo was the first sign, the Aleph. Leo is the last, the Tau.

The Significance of Leo

What is the significance of Leo? The 2nd sign after Virgo which signals the direction of our planet in history and in the witness in the sky is the regal king of the universe, Leo. It is the same lion which the angel spoke of to the Apostle John in Revelation when he cried over the scroll of seven seals. The angel said to him, "Stop weeping; behold, the Lion that is from the tribe of Judah, the Root of David, has overcome so as to open the book and its seven seals" (Revelation 5:5).

> Leo was occasionally represented in antiquity as standing over a snake, so it is fitting that this constellation is next to Hydra. Heavenly Leo, associated with the sun, defeats the chthonic snake, again representing the triumph of light over darkness. . . . The Romans called one of this constellations bright stars

Regulus, meaning, "Little King." It was considered a
ruler of the heavens: it was called King in Babylonia,
"the Mighty" in India, and "The Great" in Persia (Mc-
Donald 56–57).

The lion is at the top of the food chain, fearing no animal and having no predator. Its regal demeanor and awesome power have earned it its undisputed title, King of Beasts. With a single stroke of its front paw, it can rip through four inches of sinew and cut through the backbone of an ox. Its long ivory teeth and powerful jaw allows it to crush the skull of a horse or zebra with one bite, yet with all its raw power and size, it is as nimble and quick as a household cat.

There is nothing more disconcerting to a camper in the African Savannah than the roar of a lion in the night. The depth and tone of the roar strikes fear at the heart of the stoutest of men and can be heard for miles. This is the image of the returning avenger who will bring down the tyrannical and oppressive one-world government of Satan. This is the Lion of Judah in all His glory, pouncing from the heavens upon the evil forces of the antichrist, rending and tearing it apart.

When in that day the roar of the Lion of Judah is heard around the world, He will tear to pieces and utterly destroy the armies of Satan, bringing to naught all the works of humanity with an enormous earthquake that will bring down every building in every city of the entire world. He will reduce the Second Earth to rubble. It is with His deafening roar that the Second Earth will utterly be destroyed. It is then that He will usher in His long-awaited kingdom on the seventh day, the Day of the Lord, and from that kingdom He will rule the surviving believers.

The prophet Joel, writing around 2,700 years ago, saw this day and uttered this dark, ominous portent:

Proclaim this among the nations: Consecrate for war;
stir up the mighty men. Let all the men of war draw

near; let them come up. Beat your plowshares into swords, and your pruning hooks into spears; let the weak say, "I am a warrior."

Hasten and come, all you surrounding nations, and gather yourselves there. Bring down your warriors, O LORD. Let the nations stir themselves up and come up to the Valley of Jehoshaphat; for there I will sit to judge all the surrounding nations.

Put in the sickle, for the harvest is ripe. Go in, tread, for the winepress is full. The vats overflow, for their evil is great.

Multitudes, multitudes, in the valley of decision! For the day of the LORD is near in the valley of decision. The sun and the moon are darkened, and the stars withdraw their shining.

The LORD roars from Zion, and utters his voice from Jerusalem, and the heavens and the earth quake. But the LORD is a refuge to his people, a stronghold to the people of Israel.

"So you shall know that I am the LORD your God, who dwells in Zion, my holy mountain" (Joel 3:9–17).

When the Lion of Judah once again steps on Mount Zion and His roar reverberates throughout the entire universe, it will cause the heavens and earth to tremble. Then deliverance will arrive, and the enemy will be utterly destroyed. It is not by coincidence that the lion is often pictured as the symbol of royalty and, more specifically, God in many cultures.

It is significant that the constellation of Leo was present at every one of the major judgments of humanity. He was rising over the vernal equinox (rising on the spring equinox) at the judgment after the expulsion from the garden of Eden. At the judgment of the Great Flood, when the First Earth died, He was rising at the winter solstice. He will be present at the future judgment of the nations when the Second Earth dies, rising on the autumnal equinox. It is during Yom Kippur that the Lion of Judah shall come to bring deliverance to humankind. The real Lion has never been conquered and will, in fact, conquer the impostor lion when He returns to judge the world as Orion, and then He shall restore the Earth as Aquarius.

The lion symbol, therefore, speaks of fury and judgment resulting in the destruction of life as well as the institution of new life, and the names given to this constellation describe in detail the burning fury of this day of reckoning.

> *Here is the great Lion in all the majesty of His fierce wrath—Aryeh, He who rends; Al Sad, He who tears and lays waste; Pimentekeon, the Pourer-out of rage, the Tearer asunder; Leon, the vehemently coming, the leaping forth as a consuming fire (Seiss 134).*

The names of the stars of the constellation Leo continue in the same theme:

1. Rigel or Regulus: means the feet which crush.
2. Denebola: means the judge, the Lord who comes in haste.
3. Al Giebha: means the exalted, the exaltation.
4. Zosma: means the shining forth, the epiphany.
5. Minchir al Asad: means the punishing or tearing of Him who lays waste.
6. Deneb Al Eced: means the judge coming.
7. Al Defera: means the putting down of the enemy.

In short, the symbolism created by the structure of the universe is an alignment that points to the very first zodiac sign of Virgo, which marks the beginning in the garden of Eden and the very last, or 12th zodiac sign of Leo, which marks the return of the Messiah to our planet in order to destroy the forces of evil and restore the Earth.

In that day, Isaiah tells us that the lion will lay down with the lamb, and the profaned nature of predation shall be done away with.

Peace shall rule our planet as swords are turned into plowshares. I don't know about you, but understanding the message God has left for us in the very structure of the universe He previsioned and created by His voice is utterly mind-blowing. The overwhelming evidence shows that our universe is not the result of random ordering, to the great disappointment of those who ardently promote the Copernican principle. It has been designed with prevision by God to show His omnipotence, omniscience, and providence over the affairs of humanity.

Heliocentric vs. Geocentric Cosmology

There are several more deductions we can make from the data given to us by these three space probes that mapped the microwave background radiation of our universe. The alignment is not with the plane of Earth's equator, but with the plane of its orbit around the sun (the ecliptic). Thus, it is based on a heliocentric scheme and not just the Earth. This seems to indicate that the Earth's equator may not have been originally inclined at the 23 degree angle it now has. That is, the equator would probably have been aligned with the plane of its orbit around the sun when God created the First Earth.

Furthermore, it also seems to indicate that our orbit around the sun may once have been less elliptical, and that would have made our lunar and solar calendars line up perfectly. Thus, the universally accepted traditional lunar calendar used by all civilizations immediately after the Great Flood makes sense. The traditional Hebrew calendar, as well as that of all the ancient civilizations, was based on 12 months of 30 days that were in synchrony with the phases of the moon. Thus, the year was composed of 360 days under the more ancient lunar calendar. But our solar year is determined by our orbit around the sun and is presently 365¼ days long.

I suspect that prior to the Great Flood there was no difference between the lunar and solar year. The Earth's rotation around the sun was probably more circular, and thus, the lunar year and the solar year were both 360 days. It was the mechanism that brought forth the Great Flood; namely, the seven meteor impacts that lengthened our solar year and tilted the Earth to its present form by the power of their combined forces of impact (See Book 4 of this series, *Death of the First Earth*). Our rotation around the sun then increased in distance as the meteors struck right after the autumn equinox. This catastrophe thus extended our time around the sun by 5¼ days. The equinoxes, marked by our position around the rotation of the sun, are presently just a bit off their original alignments, and thus, the quadrupole evidences this small differentiation due to our now slightly elongated orbit around the sun. Our Second Earth is not as the First Earth was meant to be.

We can also deduce from this large-scale cosmic alignment of these areas of anisotropy that they were created during the formation of our universe and prior to the catastrophe that brought the Great Flood to end our First Earth. Hence, their alignments would have been even more remarkably exact when the Earth's orbit was as God had intended in Genesis.

In fact, I also believe that the moon was impacted by the same disaster, and its orbit actually sped up to be now on average 27.32166 days (27d 07h 43m 12s). The actual orbit of the moon around the Earth is called the sidereal month. But there is also what is called the synodic month. As the moon traverses its path, it creates from our perspective on Earth the different phases formed by our shadow on its surface. The synodic month is 29.53059 days (29d 12h 44m 03s). That is almost 2.21 days longer than the sidereal month. But it is the synodic month that allowed ancient man to calculate the lunar calendar, and therefore it is the synodic month that I believe used to be exactly 30 days in duration during the First Earth (see Chapter 19 for more information regarding the moon).

These large-scale cosmic alignments with our sun-Earth ecliptic are direct evidence of a heliocentric system and not a geocentric system, since the alignment is not with the plane of the equator of the Earth. Today, our 23-degree tilt makes our equator out of sync with the original alignment that it probably had with the ecliptic around the sun. There is a very important symbolic message in this shift of the equator. The sun is the symbol of God, and the Earth is the symbol of humankind. The original relationship with the Earth and the sun was created in complete harmony, as the Earth's equator was probably aligned with the ecliptic or our path around the sun.. That harmony has been broken by the fall of man. It is illustrated by the tilt of our equator away from the ecliptic and, therefore, our rejection of the spiritual symbolism of the actual course of our solar system through the cosmos toward Leo, the Messiah.

Our Judeo-Christian cosmological model states that the Earth is central to the universe in the overarching meaning of God's providential and previsioned design, which is symbolized as well by our physical location in relation to the entire universe. It was created before all the stars (including our sun), other planets, and the moon. But that centrality does not necessarily mean that the sun and planets revolve around the Earth. This is an illegitimate extrapolation from the narrative of Genesis. Our centrality in purpose and importance to God's plan of the ages does not require that the laws of physics be countermanded so the larger bodies (such as the sun) should revolve around the smaller bodies (such as the Earth). God created the laws of physics. It would not be in His character to contradict His own laws. Moreover, since the sun is the symbol of God, God designed it so that man would realize that our life is dependent on Him. The heliocentric model is, therefore, a deocentric model and not an anthropocentric model.

The geocentric model insists that the sun and planets and the entire universe is spinning around a static Earth, which does not spin or move. If all the other planets in our solar system are spinning and rotating around the sun, how could the tiny Earth be static in space while resisting the drag of gravity from the rest of our solar system spinning around us? The combined gravity of our solar system would eventually cause the Earth to begin to spin.

But even more powerful than the drag of gravity is the attractive power of the electromagnetic forces between the sun and the Earth. It is, in my opinion, the more powerful force that keeps us in orbit around the sun. And the sun is the center for this electromagnetic force, not the Earth. Scientists have discovered that there is, in fact, an electromagnetic rope between the Earth and the sun that emanates from the sun. In fact, this electric current coming from the spinning plasma in the sun is what powers our auroras and helps charge our magnetosphere. Without it, the Earth would be a sterile planet. Once more, this is evidence of a deocentric model.

If every other heavenly body in the universe is spinning, and in every other case where the laws of physics are universally obeyed, the smaller bodies circle the larger bodies, why would God disobey His own laws with the Earth? In fact, the laws of physics tell us that the two bodies are actually revolving around each other. The sun circles in a smaller radius because it is the larger body, but both rotate around each other. That is the picture of the harmony between God and humans. He is not a static God, but a God who existentially dances in harmony with us. It makes no sense that the Earth would be static and unmovable, and there is nothing in the narrative of Genesis that requires that.

Proponents of the geocentric model often point to the fact that it is mathematically possible to explain the orbits of all the planets using the Earth as a fulcrum. While it is true that from a geometrical perspective one can choose either the sun or the Earth as a fulcrum or platform from which to calculate the orbits of all the members of the solar system, the mathematical relativistic properties can also allow us to calculate all the same movements from an asteroid. Just because it can be mathematically computed does not mean that it is so in actuality.

Furthermore, the classical geocentric model requires that our entire universe must spin once every day (sidereal day) around the Earth. Their model, therefore, must make our universe a much smaller universe in order for it to be able to spin an entire cycle once a day. Our Judeo-Christian cosmological model stipulates that the universe is spinning around the area of the sun-Earth ecliptic, but it recognizes the enormous expanse of our universe as verified by the red-shift data to be fairly accurate and the idea that this whole universe could spin once every 24 hours to be preposterous.

If the radius of our universe is 15 billion light years across (distance from center to edge of the universe), then the diameter is 30 billion light years, and the circumference of the universe can be computed. The circumference is equal to the diameter multiplied by pi (C = Pi x d), hence: C = 30 billion light years × 3.141572653599793 = 94.2477795 billion light years).

That would mean that the universe would spin at a speed so fast that its outer edge could cover 94.2477795 billion light years in a single period of 23 hours, 56 minutes, and 4 seconds, which is the duration of a sidereal day (the time of the complete revolution of the heavens from our earthly perspective). No laws of physics can support such an outlandish speculation when nothing in our universe can surpass the speed of light. Thus, a single spin of our universe could not possibly be accomplished in a single day.

Furthermore, the idea that the entire universe should spin around once every day while the Earth stands static would demand that the connection between the matter of the Earth and the space-time anchored around it be severed. It is irrational to think that space-time could revolve around the Earth and not impact it in any way. The causal connection between matter and space-time cannot be severed.

The Earth and all matter residing in our universe are anchored in our space-time dimension. As the space-time universe spins, we, as well as all matter within it, also spin with it and cannot even tell we are spinning because there is no point of reference outside our universe in order for us to notice it. We are like the ant that is incapable of seeing the movement of the heavenly bodies and therefore, without an exterior reference point, is not able to tell that the Earth is spinning underneath.

As the space-time infrastructure expands, matter expands within it. The connection between matter and space-time cannot be abrogated to allow the Earth to somehow become a disconnected fulcrum from which the universe slides around its Teflon surface. How is it, then, if the Earth is stationary that our sun-Earth ecliptic is coming from Virgo and heading toward Leo? From the Judeo-Christian perspective, humans are not in a static timeline. They have come from the Alpha Point and are traveling to the Omega Point.

Furthermore, our Judeo-Christian cosmological model recognizes that our universe is accelerating at the edges, and the radial speeds as well as the rotational spin of galaxies are relative to the distance from us. If the universe were small, as the geocentric model requires, we would expect the galaxies at the edge of our universe to exhibit greater radial speed due to their distance from the central point of Earth. Of course, that radial speed is nowhere near the radial speed necessary for a universe 30 billion light years in diameter to make a complete circuit in a sidereal day. But even allowing for a much-reduced sized universe, the geocentric model would not predict a faster rotational speed of galaxies.

However, what we actually find in our observations is that these outer edges of our universe display both greater rotational speed and greater radial speed. That implies a universe in which time dilation in the outer fringes causes us to see movements in fast-forward time. It is the vast stretching of space-time that causes these severe time dilation differentials. A small universe necessary to make a circuit in one sidereal day would not exhibit such severe warping of space-time at the edges. Space could not be stretched far enough that time dilation would be so dramatic. The actual observable data contradict such a small universe.

In the 1970s, Vera Rubin, an American astronomer, discovered that the speed of the rotation of galaxies toward the edge of our universe was much greater than Newtonian physics could allow. The speed was so great that the centrifugal forces should have torn the galaxies apart, flinging stars outward. Some measurements approached the speed of light or even surpassed it. How can matter travel faster than the speed of light?

Astronomers calculated that these galaxies would have to have 23 percent more mass to create the gravitational hold that would counteract the centrifugal force. Hence, they invented dark matter (non-baryonic), a mysterious substance that is invisible and cannot be like normal baryonic matter that is detectable to our senses. This magical and invisible matter is nothing more than sheer speculation and has absolutely not one shred of physical evidence to confirm its existence. What they fail to see is that time in those distant sectors of our universe is moving in a fast-forward rate due to the stretching of space.

Matter cannot travel faster than the speed of light. It only seems that way to us looking from the Earth because our time frame is so much slower than that of the galaxies at the edge of the universe. This increase in the rotational and radial speed of galaxies indicates time dilation and not a classical geocentric model.

The classical geocentric model requires that the sun as well as the entire universe rotate around a static and unmovable Earth. In order for the universe to be able to spin around a static Earth once every day, it would need to be so much smaller than the red-shift analysis verifies. Few scientists today, including creationists and the proponents of intelligent design, would write off the red-shift data as erroneous.

I find the idea of a static Earth contrary to the very straightforward character of God, who is not a cheap conjurer of tricks. He established

His universal laws of physics, enforces them, and abides by them. It would not be like Him to disobey His own laws. He established them as universals for the purpose of showing the unity of all creation. It is the most important evidence for a previsioned creation instead of a randomly generated accident.

"What about miracles?" you might ask. When God seems to counter what we see and interpret as a miracle, it is because He uses other physical laws, which we have not yet even imagined. The supernatural is just the natural we have yet to understand and will never fully understand until we meet Him face-to-face. It would be quite unlike His character to make our planet stationary, when all other heavenly bodies are spinning. The idea that the whole universe would spin around a stationary planet would mean that it would have to be immune from the gravitational force of the entire universe twisting about it, and it would be impossible to conceive that the contortions of space-time around it would not eventually cause it to begin spinning, even if it began stationary.

Beyond that, the wording in Genesis does not necessitate a classical geocentric universe. Simply because God created the Earth prior to the sun does not necessarily require that the sun must revolve around the Earth. He placed the sun at exactly the right distance from us to replace the light of God as an artificial light that could sustain life. But that does not necessarily require God to go against His own laws of physics, which in all other cases require the smaller mass to rotate around the greater mass. And this is, in fact, what we observe in every other instance in the universe—all planets revolve around their suns.

God does all things decently and in order. He uses the laws He has created to accomplish His great deeds. Moreover, all heavenly bodies are also rotating, and this is so even in the atomic scale. Why should the Earth be the only object in the universe not rotating as the classical geocentric model stipulates? The Earth is at the center of the universe, but it also revolves around the sun. That is why the alignments of the anisotropic quadrupoles and octopoles are with the Earth-sun ecliptic, which is no longer aligned with our equator. Our Judeo-Christian cosmological model maintains that the universe is spinning around the solar system in an arc that encompasses Virgo at one end and Leo at the other. And when the seventh day arrives and Leo rules from Jerusalem, the world will once again find peace and harmony, as God intended.

While it may be true that the papacy erroneously maintained that the geocentric model was a Christian doctrine assumed from the Genesis account, the Earth-sun ecliptic is, however, at the center of our universe. Nevertheless, the evolutionists' wishful projection that attempts to link heliocentric truth to the Copernican principle, a philosophy that paints our existence as a simple accident of natural causes, is at complete odds with the observable data we have and even with what Copernicus himself believed.

Copernicus and Galileo were correct. Ptolemy and the pope were wrong. But Copernicus certainly would not agree with the modern interpretation of his incredible discovery. Copernicus viewed the sun as a symbol of the Godhead and understood that God had created the universe for humans to discover His majesty and grandeur. He did not think that humans were the center of the universe. He thought God was. On the other hand, he also did not think that humans were a meaningless speck in the universe. He believed that God, through His grace, gave us the intelligence to begin to understand the marvel of His creation. The Earth and the planets revolve around the sun, but that in no way reduces the significance of humans being created in God's image.

The special significance alluded to in the Genesis account is that the Earth and sun were created for humankind. That was not our choice, but God's. The heliocentric view of Copernicus in no way diminishes that central role alluded to in scripture. But the naturalist has taken the Copernican view and extrapolated and bastardized it into a completely different animal, which they call the Copernican principle. They have done so in a slick propaganda ploy to delegitimize the Judeo-Christian message of the special creation of our universe by a designer God.

This Copernican principle explicitly attempts to explain the origin of our universe by random chemical processes devoid of design and purposefulness. Hence, they ardently pontificate that there is no center of the universe and that we are simply living on an insignificant rock we call Earth, which revolves around an insignificant star in an insignificant galaxy among the vast array of galaxies in an infinite universe. They claim that the Judeo-Christian view of our centrality to the universe is an erroneous anthropocentric view, motivated by our emotional human needs and superstitions. In a clever and deceptive sleight of hand and propaganda spin, they claim to have a more humble view of humans by not claiming special significance to our position in the universe. But that humility is nothing more than smoke and mirrors.

Ironically, it is the naturalist who is truly anthropocentric. By eradicating the creator from the universe, the naturalist elevates humans and regards them as the pinnacle of the evolutionary process, as the center of the universe. This metaphysical choice is the pinnacle of anthropocentrism. Those who believe in intelligent design, on the other hand, see the universe as deocentric, not anthropocentric. The fact that the sun is at the center of our solar system in no way refutes the intrinsic design latent in the entire universe. The sun being the symbol of God is naturally at the center of our planets' revolutions around it.

But for all intents and purposes, the true center of the universe is God, not man or any other created aspect of existence. The sun, according to Genesis, was simply created as a substitute light for God, who brought forth the entire universe by the energy of His voice. The irony is that the natural implication of the outworking of the naturalists' anthropocentric view is the occult doctrine of attaining human divinity, or assigning divinity to the material universe. Thus, the paradox is that in evading God, the naturalist is eventually drawn philosophically into some form of mysticism, a subject that we will address more fully later on. Hence, their metaphysical choice to exclude God eventually leads them to the elevation of humanity into godhood along the evolutionary path to the occult. The irony is that atheism is a bridge to pantheism, which, in turn, is a bridge to the occult, which is the epitome of anthropocentrism, arrogance, and egotism.

The universe is filled with evidence that shows the objective observer that it could not have been birthed from random ordering and must have been designed by some higher intelligence. We will cover some more of the data that indisputably point to intelligent choices being made in the design of our marvelous universe in the third and fourth volumes of this series. If the reader has noted the specified complexity of the design, then I would ask, "What rational necessity or observable data can justify the arrogant claim that this supreme intelligence that brought forth the universe in the splendor of its intricate web of designs has no right to intervene, control, and direct the physical processes He set in motion?"

It can be labeled only as irrational lunacy for inferior intelligent beings such as humans, filled with an overinflated ego, to imagine that they are the pinnacle of intelligence in such a superbly designed universe created for them to inhabit. Dark are the eyes of the hyper-anthropocentric naturalists who consciously choose to darken their minds to the empirical evidence in order to justify their anarchy against their creator. Their metaphysical choice is not based on empirical facts but on the psychological symptoms of their rabid and morbid deophobia. False is the humility they claim while foisting their Copernican principle on the public as science. It is not science. It is nothing more than a metaphysical worldview craftily dressed in scientific jargon.

And although, the Earth revolves around the sun in the outskirts of the Milky Way, it is abundantly clear by the observable data that our Milky Way and, more specifically, the sun-Earth ecliptic are at the center of the entire universe. The term anthropocentric became a derogatory term pandered by the Darwinian supporters of the Copernican principle as describing humans' attempts to put themselves at the center of the universe, as depicted in the Ptolemaic geocentric planetary model.

The Judeo-Christian claim—that God created us in His image and that in doing so, He gave us a special significance among all the other created beings—was called arrogance by the naturalists who believe that life is but an accident of randomness. The idea that human life was sacrosanct was belittled as sanctimonious nonsense, and Christians and Jews were labeled superstitious, ignorant, and arrogant. After all, if all of life is nothing more than a cosmic accident, what special significance does one accident have over another? The only problem is that this atheistic rationalization is in direct contradiction to how humans are. Can we really smash a human head with a hammer with the same callousness we crush a coconut? It fails the test of reality. It does not reflect the way people truly are or think. It is an artificial construct that cannot be fully practiced by humans because whether they admit it or not, they carry the breath of God in them. Try as they might, they cannot escape what Schaeffer called the "mannishness of man" that God hardwired in us.

CHAPTER 9

• • •

THE MULTIPLE UNIVERSE THEORY

The intricate and specified complexity found throughout the entire universe from the macrocosm to the microcosm is so staggering that it is impossible to believe that all of this is the result of a randomly generated reality. Evolutionists are hard put to explain this apparent miracle of the first order, which is millions of times more impressive than parting the Red Sea or creating a global deluge. How could blind random chance have created a universe with such intricacy, symmetry, and order within its enormously complex matrix? Some in desperation have turned to the multiple universe theory in an attempt to bridge this huge credibility gap.

In addition, the insurmountable problem now facing naturalists is the lack of eternal time necessary to make their theory more palatable. If the universe were eternal, as naturalists once supposed, then the improbability of random selection becomes a little more palatable. For after all, you have an eternity of chances to finally get the right combination. But if the universe is finite, then the chance of recombination within that limited time frame becomes ever more improbable. Humes's finite no longer has an infinite number of recombination potentials to become complex through a chaotic and randomly guided genesis. Finite time now restricts the evolutionary process to a finite number of combinations, which subsequently provides zero scientific credibility.

Since Einstein's equations have forced naturalists to face the fact that the universe had a beginning and therefore has existed for a finite amount of time, evolutionists have adopted a variety of responses in order to bring infinity back into the equation so that random chance has the time necessary to make the impossible seem more plausible. Such are the multiverse or multiple universe theories.

By providing an infinite number of universes, it magically provides them with the eternal recombination possibility once again. Naturalists reason that if an infinite number of universes exist, then the statistical nightmare can be avoided. They simply claim that ours is the lucky universe that found the right combination through sheer random ordering to allow life to evolve. There are so far nine variations on the multiverse theory.

> Each envisions our universe as part of an unexpectedly larger whole, but the complexion of that whole and the nature of the member universes differ sharply among them. In some, the parallel universes are separated from us by enormous stretches of space or time; in others, they're hovering millimeters away; in others still, the very notion of their location proves parochial, devoid of meaning. A similar range of possibility is manifest in the laws governing the parallel universes. In some, the laws are the same as in ours; in others, they appear different but have a shared heritage; in others still, the laws are of a form and structure unlike anything we've ever encountered. It's at once humbling and stirring to imagine just how expansive reality may be (Greene 2011, 5).

To begin with, the choice to believe that there are multiple universes is a metaphysical choice. It is not a conclusion arrived at by scientific data. When Brian Greene declares how stirring it is to imagine a multiverse, he is declaring the truth. These are simply the product of imagination and not hard data. There are no empirical data in our universe that tell us other universes exist. There can be no data in our universe that can be connected to a separate universe. It is simply a wish projection influenced by their rabid deophobia.

When he says it is humbling, he is, in fact, reiterating the same propaganda of the Copernican principle that uses humility as a front to cover unabashed arrogance. It is the arrogance of the atheist to posit imaginary universes as a way to avoid the empirical data found within our universe in order to once again dethrone God as the creator. Once again, their humility is nothing but smoke and mirrors to hide their delusions of grandeur.

The Statistical Deception

But even more critical to understand is that the idea that a multiverse could help their evolutionary statistical nightmare is also nothing more than smoke and mirrors. For the multiverse concept to work at lessening the insurmountable odds, there must be some causal connection between the universes so that the constant reshuffling could conceivably rationalize our very fortunate chance formation through random processes. In other words, our universe must have some continuity and interconnection with the other universes if whatever processes form them can be correlated with the processes that formed us so that a statistical correlation can be made.

That would require that the multiple universes be causally connected. But if they are interconnected, then they are not different universes, and we can have access to those separate regions of reality in order to observe them. These universes would then be part of the same universe we live in. If that is so, then the argument is a moot point, because such a causal connection requires that our laws of physics be interconnected to all the other universes, and there would not be any statistical advantage since all of them would then be like us.

The entire argument of multiple universes is therefore self-contradictory. It is nothing more than smoke and mirrors created by flimflam propagandists using pseudo-science to shroud the truth. If there is an interconnection, then our universe is not a closed system. But if it is a closed system, as they so ardently pontificate, the lack of interconnection makes no association with them possible, even for the sake of statistics. What happens in another universe has no connection and therefore no bearing whatsoever with ours and cannot be used in any statistical comparison to the chances of the formation of our universe by randomly generated chemical reactions.

Let me explain this statistical reality in a fashion that you can relate to. If I were to place a giant roulette wheel in Miami, Florida, and another on Neptune's moon Triton, it would illustrate the irrational nature of the multiverse theory. Let us say that each roulette wheel has 10^{120} slots. Each individual using the two different roulette wheels would have one chance in 10^{120} to hit the one correct slot. The subterfuge in their theory is to make the public think that adding another roulette wheel will give the player of the roulette wheel in Miami a better chance of striking the right choice. The scientific truth is that you could place 10^{120} roulette wheels around the entire universe, and that would not in any way change the statistical odds of the player in Miami. Each turn that player takes has exactly one in 10120 chances to strike. Whatever happens in the other roulette wheels absolutely does not have one iota of effect on the wheel in Miami. It is, again, nothing more than smoke and mirrors.

The plain fact is that the multiple universe proponents are attempting to diminish the statistical constraints forced upon them through our finite physical reality by imagining the existence of multiple alternative universes. It is nothing more than a hypothetical mathematical fix, which is not mathematically correct and which is trying desperately to hold together their failed ideology of random origins with duct tape and monkey spit. No, not duct tape. I like duct tape. It is very useful. They are trying to hold it together with just monkey spit. (They borrowed the monkey from Darwin.)

Scientists are failing to understand a simple rule of statistics, one that naïve gamblers are not often aware of. It is not only that two different roulettes cannot be considered in any statistical prediction, as we previously noted; it is that even the very same roulette cannot be used to predict the future outcome of another turn of the wheel. Often, gamblers betting on the roulette wheel see the numbers posted from previous turns and judge according to those numbers the odds they think will make their choice a winner. In reality, there is no connection between one turn of the wheel and another. Each turn has exactly the same statistical chance.

Each turn has exactly the same number of chances of landing on a given particular number as every other turn. There is no interconnection between one turn and the other that could sway the odds of choosing one number over another. But in the naturalists' multiple universe models, the fallacy of their thinking is even more profound because they are positing a completely different roulette wheel than ours—a roulette wheel where the total number of possibilities has absolutely nothing to do with our roulette wheel. The chance of hitting the right number in a particular roulette wheel has absolutely no bearing on the chances of any other roulette wheel. Their sad attempt to stack the chances for evolution in our universe is nothing more than a sleight of hand and propaganda. It is not true science. It is just more smoke and mirrors.

Steven Weinberg, winner of the Nobel Prize in Physics and one of the proponents of this new multiverse theory, lashed out at the design argument, deeming it anachronistic. He explains this present order in our universe in relation to the many chances of failure in other universes. In an article titled "Designed for Living" in the *Wall*

Street Journal, author George Sim Johnston masterfully addresses Weinberg's irrational hypothesis of desperation.

> In the current *New York Review of Books*, he [Weinberg] dismisses talk of a "fine-tuned" universe as a dangerous regression to Greek myths. He also attacks religion, especially Christianity.
>
> To keep his view coherent, Mr. Weinberg—and physicists like him—must somehow explain the breathtaking specificity of what followed the Big Bang. Picture a wall with hundreds of dials; each must be at exactly the right setting for carbon-based life to emerge eventually in a suburb of the Milky Way. If the cosmic expansion had been a fraction less intense, the universe would have imploded billions of years ago; a fraction more intense, and the galaxies would not have formed. How to explain this remarkable exactitude? Mr. Weinberg favors the multi-universe theory, in which the Big Bang is just one of innumerable other big bangs. The idea is that if there are billions of universes, then the odds are pretty good that one would finally get it right so that man could dwell in it. This would be "cosmic natural selection" and so there is no need to worry about the appearance of design.
>
> The only problem with the notion of a plurality of big bangs is that there is not a shred of evidence to support it. The multi-universe theory also violates elementary logic. All these universes either interact or they don't. If they do, they constitute one universe. If they don't, they are mutually unknowable. Mr. Weinberg, in fact, is guilty of what he accuses religious people of doing: taking refuge in the unobservable (Johnston W17).

The Untestability of God Deception

"Elementary, my dear Watson," said Sherlock Holmes. I am pretty sure he would agree with me that it is elementary, my dear friends. This multiverse, if there are really alternate universes, cannot be tested or observed from our universe and has absolutely no bearing on events in our universe. Multiverses are untestable from our universe.

"Wait a minute," said Foolwinkle, waving his hand furiously. "Isn't the fact that God is untestable one reason we use to discredit the scientific inquiry of the existence of God?"

"Shhh!" said Rocksy, as he put his index finger over his mouth. He approached Foolwinkle, leaning his head closer and whispering, with his hand half-covering his mouth, "Not so loud. Those pesky creationists might hear you."

"But, but," stammered Foolwinkle, "I thought we believed in a closed system."

Foolwinkle stepped back, pointed his finger at Rocksy, and said, "You said that was an unbreachable horizon, a cosmic law."

Rocksy stammered a bit, not knowing what to say.

"Well?" asked Foolwinkle tapping his foot on the floor with his arms folded across his chest.

"Well," said Rocksy, tilting his head with a grin, "it's kinda more like a suggestion than a law. You know, truth is relative."

"What do you mean?" asked Foolwinkle.

"I mean it is a law for them but not for us."

"How do you know that?"

"Because we say so."

Maybe Foolwinkle is not so foolish. Why is it that it is scientifically allowed to consider untestable explanations such as other universes, but it is unscientific to claim that a higher intelligence designed and brought forth our universe? Is not the idea of a closed system one of their cherished presuppositions that prohibit any intrusion by a God outside of our universe? Hmm.

If they could test this other universe, then it would be part of our universe. So, ironically, the very thing they despise and have for so long ridiculed about the Judeo-Christian God, the untestability of God, is now hypocritically accepted in the multiple universe concept by statistical necessity. This is reason? It seems to me that it is more like schizophrenia.

Their position is not only untestable and unprovable, but it is also simply irrational. If there are, in fact, separate universes, the processes taking place within these universes have absolutely no bearing on the processes in our universe and cannot in any way influence the statistical probabilities within our universe.

The huge number of fined-tuned aspects that had to be chosen at the very first moment of creation that modern science has discovered since the time of Darwin, coupled with the toppling of the eternal universe brought to light by the mathematics of general relativity, have created some serious angst in evolutionists. Now the empirical data that are glaringly pointing to the centrality of our solar system in the universe are adding fuel to the fire of their angst. Nevertheless, undeterred by scientific data, they doggedly attempt to shore up their failed Copernican principle.

The Confusion between Parallel Dimensions and Parallel Universes

Evolutionary cosmologists are adverse to a finite universe because of the impossible odds required to form such symmetry as we observe all around us through sheer random processes within such a limited time frame. They are, therefore, predisposed to conjure an infinite universe and, alongside of it, an infinite number of universes. The Rapid Inflationary Model, which we have already shown to be incongruent with the mapping data of the WMAP and Planck satellites, also has its own multiverse scenario.

The inflationary theory, an approach that posits an enormous burst of superfast spatial expansion during the universe's earliest moments, generates its own version of parallel worlds. If inflation is correct, as the most refined astronomical observations suggest, the burst that created our region of space may not have been unique. Instead, right now, inflationary expansion in distant realms may be spawning universe upon universe and many continue to do so for all eternity. What's more, each of these ballooning universes has its own infinite spatial expanse, and hence contains infinitely many of the parallel worlds (Greene 2011, 7).

Before we dissect this outlandish claim, let us first concentrate on the meaning of infinite. To be infinite means to have no end, no boundary, and no restrictions. It means to include all things and to extend to all places in an eternal duration. If we say that our universe is infinite, by definition that means that no other universe could exist besides it. If it is truly infinite, then all parallel possibilities would be included in it.

Second, if parallel universes exist, then they cannot be truly infinite spatially. Moreover, our universe could not be spatially infinite, either. Then there is a part of hyperspace in which our universe does not exist and the parallel one does.

Third, it is one thing to say that our universe has parallel dimensions and quite another to say that there are parallel dimensions outside our universe. We have not one shred of empirical evidence that a single parallel dimension outside our universe exists, much less an infinite number of them.

Fourth, the parallel dimensions within our universe were created by the Big Bang and are therefore not infinite. They are finite in number, and they are finite in extent. The mathematical equations show us that there are seven invisible spatial dimensions and three visible spatial dimensions, which are all interconnected. They cannot be separated from the continuum. The same universal laws that control the three visible spatial dimensions control the seven invisible ones. They are not separate universes, but specific areas of our universe. There is not one mathematical reason to expect that our universe has an infinite number of parallel dimensions.

Fifth, none of our spatial dimensions can be infinite. If they have a beginning, then they cannot be infinite spatially. Those parallel dimensions within our finite universe were created at the same time that the visible dimensions were created a finite time ago. Hence, within that finite time frame, no dimension could become infinite. To do so would mean that it expanded at an infinite speed. No one can claim such an absurdity.

And last but not least, the idea that parallel universes exist does not come from any empirical data. Nothing in this finite universe leads us to believe that infinite parallel universes could exist. It is purely a speculation based on a desired metaphysical choice to exclude God as the creator of our universe and to provide some nebulous scaffolding for a materialistic delusion, which in reality amounts to nothing more than a cheaply manufactured imaginary hologram.

We have already documented that the specific alignments as well as the very existence of the anisotropic dipole and quadrupole discovered by the three mapping satellites cannot be explained by the mechanism of the Rapid Inflationary Model, which is simply incorrect in describing the Big Bang. Furthermore, the piggyback multiverse theory tacked on to it is also just as ridiculously speculative and irrational.

If more than one universe exists, then neither could be infinite. Each would be confined to a finite space where the other does not exist. But naturalists want to have their cake and eat it, too. How, then, can Greene remain rational when he says, "Each of these ballooning universes has its own infinite spatial expanse, and hence contains infinitely many of the parallel worlds" (Greene 2011, 7).

Greene realizes that these theories are highly speculative and unsubstantiated but finds it appealing due to his presuppositional bias toward a universe created by random ordering. In spite of his obvious desire for such a reality, he at least is honest enough to make the following disclaimer:

> *The subject of parallel universes is highly speculative. No experiment or observation has established that any version of the idea is realized in nature. So my point in writing this book is not to convince you that we are part of a multiverse. I'm not convinced—and speaking generally, no one should be convinced—of anything not supported by hard data (Greene 2011, 9).*

These multiverses are nothing more than untethered fantasies from reality that seek to rationalize or provide an imagined probability potential that can explain away the fine-tuned parameters that govern our particular universe in elegant order. It is a desperate attempt to insist that random ordering created this elegant symmetry through blind chance. Somehow, naturalists must create other universes so their cherished mechanism of natural selection can function and select ours out of the infinite number of universes in order to fortuitously bring forth life and the elegant universe filled with the symmetry necessary to sustain life.

The complete and utter failure of natural selection to account for our material existence is now unavoidable. But the empty evolutionary bubble, devoid of rational content, is still being paraded as the deposer of the Judeo-Christian God as creator. Natural selection is simply incapable of explaining the formation of the universe as we know it today. This is true even in the biological arena. For although the natural selection process may account for some variability within a species, it has not been shown to produce a new species. (This shall be treated more exhaustively in the third and fourth books of this series.)

Evolutionists have deceptively promoted the view that the concept of intelligent design is a felt need of ignorant and superstitious people that projects on the cosmology of the universe. They fail to realize that their quest to authenticate a multiverse and insistence on a random genesis is, indeed, a felt need for atheists who are projecting their atheistic bias into the realm of science. During the time of Darwin, naturalists refused to consider the possibility that our universe could have been created by claiming that this inquiry was out of the realm of science. Their cherished mantra that science cannot delve into any inquiry before the beginning of time has been shown to be nothing more than a misdirected lie. They have drunk deeply from the well that they once forbade Christians and Jews to touch or mention.

They have quite deceptively claimed that metaphysics and science are two independent realms. Here, the divided field of knowledge helped them place metaphysical views in the irrational upper story of subjective and unverifiable truths. As tyrannical and repressively closed-minded guardians of academia, they have turned to the same tactics as those who have in the past burned books that challenged the dogmas of the paradigm of the day. Jealously, they guard the turf of their evolutionary religion as the clerics of Darwinism forbid scientists who promote intelligent design to participate in the free market of ideas and shun, muffle, and exclude them from academia. This is an unbridled travesty of the scientific spirit to search for truth wherever it leads. It is the modern clerics of Darwinism who have inherited the spirit of the Inquisition.

To this day, even after all of their major assumptions have been scientifically disproved, they continue to claim that evolution is a scientific model and that intelligent design is not. They are blind to the fact that their Darwinian model is, in fact, a metaphysical model that ignores the verifiable scientific facts evidenced by modern science in our universe. In hypocritical flare, they conjure up other universes that are absolutely unverifiable in order to assuage their angst over the uncanny symmetry and design of our universe and hold doggedly to their dysteleological bias, which flies in the face of any rational inquiry.

CHAPTER 10

• • •

THE ESSENCE OF SPACE-TIME

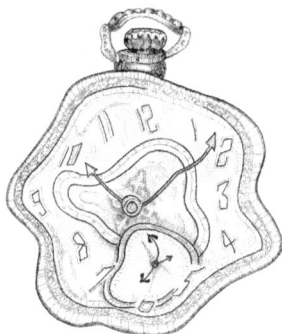

Time had a beginning. This we all now accept as true. But fifty years ago, the vast majority of scientists would have said that time never had a beginning. We have thus far covered two elements evidenced by science that are incompatible with the naturalist/existentialist philosophy: a beginning and the insurmountable preponderance of evidence that points to purpose and design from the subatomic world to the macroworld. Our modern discoveries about the nature of matter and its unique intrinsic properties have created insurmountable problems for the naturalist.

Beyond the creation of a material universe is the even more complex design of living things. Modern science has convincingly shown that a very narrow envelope or parameter must have been deliberately chosen in this physical world prior to the creation process in order for life to thrive out of the end result of the primordial Genesis explosion. If it were not chosen and this narrow parameter were achieved by blind random chance, then it would have been a miracle of the first order and more miraculous than even the most ardent Jew or Christian could ever imagine.

The statistical improbability that the universe could have developed in order to allow for life to exist (compared to the wide range of possibilities that could have evolved if the universe had been made and guided only by random processes) has resulted in an insurmountable difficulty for the evolutionist. Not only did our universe have a beginning, which removes the immense time frame necessary for evolution to have a chance statistically, but also the overwhelming evidence points to its intricate and sophisticated design, which suggests that it was intentionally engineered in order to harbor life from the very first minute fraction of a second in the Big Bang. And this inescapable realization has caused an even deeper anxiety among the existentialists.

Therefore, there has been a subliminal and sometimes quite overt antipathy for and a tendency to attempt to do away with the notion of a beginning, since this is an unavoidable inconsistency with the atheistic presupposition. And for the very same reason, the idea of the design of the universe is purposely downplayed. In contrast, aspects that could be construed as chaotic and random within our universe have been underscored and deliberately exaggerated.

The Evolutionary Affinity for Infinity

It is a mystery to me how modern scientists can leap over mountains and trip on a blade of grass! Not too long ago, even the eminent Stephen Hawking proposed, almost halfheartedly, a new possibility for the existence of the universe in order to remove this embarrassing admission that the universe had a beginning. In a conference organized by the Jesuits of the Catholic Church, he proposed the possibility that space-time was finite but had no boundary, which would mean that it had no beginning, no moment of creation.

Although for the most part Hawking forced the scientific community to face the implications of Einstein's theory, which shows the universe had a beginning, more recently, maybe from the pressure to conform to the paradigm of his profession, he attempted to find a way to explain away the singularity. Or perhaps this brilliant man was just exploring possibilities as an exercise that most thinking scientists are prone to do. It is by exploring all possible alternatives that we can eventually find the one that best answers the facts and fits observable reality.

In order to make his new suggestion possible, Hawking resorted to the idea that the universe could be finite in imaginary time but without boundaries or singularities. He had to resort to imaginary time, because real time excludes the results he wished to end up with. Thus, he resorted to semantic gymnastics, which did nothing but point to the glaring self-deception involved in this mental exercise.

> *When one goes back to the real time in which we live, however, there will still appear to be singularities. The poor astronaut who falls into a black hole will still come to a sticky end; only if he lived in imaginary time would he encounter no singularities.*
>
> *This might suggest that the so-called imaginary time is really the real time, and that what we call real time is just a figment of our imaginations. In real time, the universe has a beginning and an end at singularities that form a boundary to space-time and at which the laws of science break down. But in imaginary time, there are no singularities or boundaries.*

So maybe what we call imaginary time is really more basic, and what we call real is just an idea that we invent to help us describe what we think the universe is like (Hawking 144).

In addition, Hawking had earlier proposed that if the universe had just enough mass to make the critical density necessary to stop the outward expansion of the universe, causing it to implode, then time would run backward, and we would grow younger instead of older.

Physicist Stephen Hawking even believed that time might reverse itself as the universe contracted and history might repeat itself in a backward fashion. This would mean that people would turn younger and jump into their mother's womb, that people would dive backward from a swimming pool and land dry on the diving board, and frying eggs would leap into their unbroken shells. Hawking, however, has since admitted he made a mistake (emphasis added) *(Kaku 216).*

I have a great deal of respect for Hawking, and I am certainly less than an ant in comparison to his great intellect. But I must admit, I am deeply saddened by this proposition of imaginary time. And to this proposition, I have only one response: "Get real, okay?" The pun was intended.

The logic of this line of thinking reminds me of the philosophy teacher in college who was trying to convince the class that reality is an illusion, part of a dream. At that point, a student threw the textbook against the blackboard, startling the teacher, who rose to his feet and angrily asked, "Who threw that book?" The student promptly answered, "What book? It's only an illusion!"

Whether Hawking was aware of it or not, the philosophical framework for this view was derived directly from Platonic dualism and the parallel pantheistic concept of Maya. Reality in the pantheistic model is merely an illusion (Maya). In Eastern mystic thought, real reality is only reached when one is fully enlightened and becomes one with the world soul, at which point the material becomes incorporated into the atman or akasha, thus dedifferentiating into the world soul. It is not surprising to hear these elaborate efforts to deny the very reality about us for this doctrine the great deceiver taught from long ago. This has been the lie of Satan from time immemorial, and I mean real time.

There are two competing views in the world: truth and non-truth. Hawking grudgingly admitted that all reality seemed to point to a creator, so reality must not be real and only the imaginary is real. The very brilliant Hawking, unable to reconcile reality with an impersonal chance-directed genesis, had to take a leap of blind faith into the imaginary to explain away God. How sad!

In other words, what Hawking inadvertently suggested with imaginary time is that we must step out of reality to develop a cosmology without God. Here, Hawking stepped across the line of despair and took a blind leap of faith in an attempt to find cohesion between his reason and his preferred brand of worldview. And in the words of Francis Schaeffer, he escaped from reason. In doing so, he became a prime example of postmodern man.

In all fairness, I don't really believe that Hawking was totally convinced of the veracity of this imaginary time proposal, and I think he was toying with the idea almost as an attempt to convince himself. Hawking was standing on the edge of the line of despair, but I would like to think, and I really hope, that that his superb reasoning powers and personal intellectual honesty did not let him take the leap in spite of his emotions and the peer pressure he faced. He presented this as simply creating a proposal, an exercise in thinking because of its aesthetic appeal.

I'd like to emphasize that this idea that time and space should be finite "without boundary" is just a proposal: *it cannot be deduced from some other principle. Like any other scientific theory, it may initially be put forward for aesthetic or metaphysical reasons, but the real test is whether it makes predictions that agree with observations* (emphasis added) *(Hawking 142).*

I applaud Hawking's honesty in admitting that such proposals are, in fact, motivated by "metaphysical reasons." That honesty is rare among evolutionists who hide behind the curtain as the Wizard of Oz and pretend that their scientific models are not influenced by their metaphysical choices.

Hawking recognized the fact that the empirical evidence points to both time and space being finite and with boundaries. It is irrational to conceive that something that had a finite beginning could become infinite in a finite amount of time. The aesthetics of the metaphysical reasons he spoke of is the denial of the existence of a supreme God and creator of the universe. How Hawking proposed to observe any prediction in imaginary time is not clear. There are many in our postmodern culture who have embraced a pantheistic presupposition and view time and our space-time continuum as only an illusion and, therefore, not real. These no doubt would embrace such a view. I rather doubt that Hawking would have.

Given the physical ailment that plagued Hawking, it is a wonder to me that he did not become a bitter man, angry at God for the debilitating malady that so horrendously afflicted him. Lesser men than he would have become embittered and given up on life. But the amazing courage of this man was a marvel to be admired.

Men in his condition are the precise targets of the evil one as he sows the seeds of pity and hatred toward a God who could allow such misery in this world. Surely, no one outside Hawking's shoes could have judged him for shaking his fist at God, if he had chosen to do so.

In light of Hawking's ailment, I would like to digress a bit from the subject at hand in order to address a problem that has caused many to turn to atheism.

The Problem of Evil

How could a good God allow such ailments to exist? How could a good God allow injustice and evil to exist? The question runs even more deeply—what is evil? How can we define evil? I certainly do not presume to speak for God nor do I claim to understand the infinite, but I can recount what He has revealed to us through His prophets, and I can use my mind to reason.

If I had a nickel for every time an evolutionist has said, "I cannot believe in a God that allows such injustices to exist in this world," I would be a millionaire. My answer to them is, "It is because you believe that, that I know God is there." Usually, they look at me as if I had a horn growing out of my forehead, and then I explain, "It is because you believe that injustice is evil that I know there is a God. If there is no God, there is no such thing as evil."

How is it that every human being knows that evil exists? How is it that every man hates injustice? It is because we have this intrinsic notion of evil that God must exist. The problem of evil is explained only through the Judeo-Christian worldview. Naturalists cannot even define evil. If there is no God, then there is no right and wrong. Evil is merely an illusion in the minds of men. What is, just is.

And yet remarkably, humans, regardless of their age, race, or intelligence, know intuitively that evil exists. How does a randomly generated universe develop the consciousness of evil when there cannot even be a definition of evil in such a world? The pantheist cannot define evil, either. If God is everything, then all things are ultimately sameness. Unless there is an ultimate paragon who can define good and evil, there is no way to explain evil. No choice can be relied on as an absolute in a relativistic reality. What one person thinks is good may be what another thinks is evil, and neither one has any more intrinsic right than anyone else in the world to push his or her view on others. There is no paragon from which to know which side of "the force" is good and which is evil.

God has chosen to make us into beings like Him. The scriptures claim that in this sense, we are like God. We have an autonomous will. We can choose, and we can create. And in this, we, unlike all the rest of the animal kingdom, possess a self with a will that can either reject good and God or accept Him. It has been our rejection of God that has led to the preeminence of the primordial sin of man—selfishness.

From this sin arise all other evils. Sickness and death have entered our space-time continuum as the direct result of the choice taken by humankind to move away from God's way. God did not wish it to be so. We were meant to live forever in a perfect body. But there can be no other way if choice is to be real. Otherwise, we would merely be computerized automatons.

But to everything that Satan means for evil, God can in His sovereignty turn to good. And the amazing and brilliant insights of Hawking were, in my mind, the result of his ability to concentrate and think with his God-endowed brilliant mind, even perhaps as a result of his debilitating illness.

God is ever good and merciful. Sometimes from our puny, limited perspective, caught in the temporal world, we cannot see the magnificent tapestry He is creating from heaven. Hawking was a gift from God for all of humanity. His indomitable spirit and will to fight his disease were an example to millions. But I believe that his rewards in this world pale compared to the accolades he will receive in eternity from His loving creator, if he chose to accept Him.

It was Satan who tempted us and brought sin and death into the world. The impact of the fall is real in the soul of humanity and in every element of the universe, as reflected by the second law of thermodynamics. Our universe is decaying. The second law of thermodynamics is the evidence of the fall of our universe. Disease is a product of our flawed universe, but in His sovereignty, God can take the evil and turn it into some good. And it is the evil one who will be punished most severely in the coming judgment, for God has reserved a place for him so He can fence in the pains of sin in hell.

God did not create evil. Evil is the rejection of God. But if choice is real, and if good is also real, then evil must exist. Otherwise, all actions would be sameness. The atheist, therefore, cannot explain evil. And yet all of us intuitively know that there is evil. Our sense of injustice is intuitive and powerful. But in a world ruled by the matrix of the survival of the fittest, justice is simply an illusion. How can a randomly engendered universe create in humans such an incredible instinct that stands in direct opposition to the concept of the survival of the fittest and the abuse of the powerless? The atheistic worldview fails the litmus test of reality. It does not match the way humans are intrinsically. It fails not only to explain the origin of matter but also to explain the basic nature of humanity.

Allah or Elohim

Please bear with me as I digress for a moment from our topic, but in light of our present global condition, I find it terribly important to address this. On the evening of July 14, 2016, France's independence day, another terrible terrorist attack took place in Nice, France, while crowds gathered for a fireworks celebration. The next day, reports said that 84 were dead. As of July 2016, some 28,000 people have been killed because of terrorism. Ten nations in NATO have been attacked, and the West has taken no concrete action to stop this carnage. In 2001, when Al Qaeda attacked the United States by blowing up the Twin Towers, Western forces rallied and took out their training bases in Afghanistan. As of the writing of this book, not a single NATO nation has been willing to form a true coalition because the leadership in the United States is unwilling to lead it.

Muslim terrorists are willing to kill innocent human beings because of their belief in Allah, which commands them to establish

Islam in every nation in the world by the edge of the sword. Our Western ideology of tolerance that stems from the Judeo-Christian worldview is being warped into wickedness. While it is true that we believe our faith is a matter of conscience, the tolerance of colonialist aggression and globalist aspirations is nothing less than the tolerance of wickedness and violence. We must understand that tolerance of evil is wickedness and not a Judeo-Christian ideal.

Believing in a single god is not enough, either, if that god is not the one true God who created the universe by His spoken word. If that god is not the God of Righteousness, then his commands do not lead the faithful into righteous acts and mercy. When we hear of young Christian and Kurdish girls five years old and up being taken by Islamic Jihadist groups such as ISIS to be raped repeatedly every day by dozens of men, our hearts intuitively know that this is evil, and we are repulsed and enraged at the injustice. When we see the main streets of Raqqa, a city in Syria, lined with poles on both sides with decapitated heads stuck on the tops as banners of their ghoulish barbarism, it becomes plain that the counterfeits of God are always dangerous and violent in the end.

It is this right to rape, torture, and pillage given to them by Muhammad through the Qur'an that entices these Jihadists to join the group. Everything they conquer by their hands belongs to them, as long as they give Allah one-fifth of that which they pillage. A god that endorses the abuse of the weak as a spiritual duty and stands against the freedom of our choice to worship as dictated by our personal consciousness stands against the very principles of the one true creator who chose to give us fee will. Allah is a god that endorses tyranny and hatred, while the one true God endorses freedom and love. It is the great impostor that worships death instead of life.

Islam is, in fact, nothing more than a Judeo-Christian heresy. All our ancient church fathers spoke openly against heresies that could contaminate the truth. We cannot afford to neglect this high calling from God. True truth must be defended by those who claim to be leaders of Judaism and Christianity, and this cannot be done without boldly speaking out against the promoters of violence and the usurpers of God's message of peace. The problem we face today is no different than the problem our early church leaders faced when they taught against the pagan religion of the Roman Empire. The religions of the serpent must be unmasked, and the truth must be proclaimed.

Ravi Zacharias, a prominent evangelical leader born in India, knows this very well. In India alone, 81 million Hindus were killed by the Muslims in their attempted genocide of infidels between 1000 AD and 1500 AD. Some Christians and Jews have been duped into thinking that Allah is the same as the God of the Scriptures, but he is not.

> *Islam is a religion that is academically bankrupt, for it fails to meet the ordinary tests of truth. Those who critique it run the risk of being obliterated. How can a religion that claims that its prophet came to the entire world restrict its miracle to a language that is not spoken by the vast majority of the people of the world? How can a man whose own passions were so untamed gain the right to speak moral platitudes? I've written about this and other critiques of Islam elsewhere. An honest Muslim open to considering these things will readily see that the "God" of the Koran is not the same God spoken of in the Old and New Testaments and that the edifice of Islam is built on a geopolitical worldview masquerading as a religion (Zacharias 122).*

Christians are keenly aware of the threat of secular humanism as it rabidly attempts to undercut our religious liberties and continues to draw away our youth. Most Christians see no problem when I defend the faith against evolutionists and the new radical atheists such as

Richard Dawkins and Sam Harris, but they somehow think that to speak against Islam is not politically correct. Hear my warning, brothers and sisters, the great battle of our age will not be fought against communism or Nazism—it will be fought against Islam. I am not alone in this thought. Zacharias has a very important insight into this matter. Let us look at his prophetic warning:

> *Islam is willing to destroy for the sake of its ideology. I want to suggest that the choice we face is really not between religion and secular atheism, as Sam Harris, Richard Dawkins, Christopher Hitchens, and others have positioned it. Secularism simply does not have the moral power to stop Islam. Even now, Europe is demonstrating that its secular worldview—one that Harris applauds—cannot stand against the onslaught of Islam and is already in demise. In the end, America's choice will be between Islam and Jesus Christ. History will prove before long the truth of this contention (Zacharias 126).*

Hear my warning, brothers and sisters. If we do not prepare ourselves to meet this coming clash, we will be the victims of our own myopic apathy. The Qur'an teaches that Allah's goal is to kill every last unbeliever so Islam can dominate the world. That is the underlying root of Islamic terrorism to which the Western nations are blind. It is a religion as no other religion in the world. Our ignorance will become our demise.

> *God was desiring to verify the truth by His words, and to cut off the unbelievers to the last remnant (Qur'an, Sura 8:7).*

The militaristic nature of this religion is motivated not only by the expectations of the 72 beautiful virgins that await them in paradise, should they die during their Jihad, but by the fearful expectation of Gehenna (hell—the Lake of Fire) for those who refuse to fight.

> *O believers, when you encounter the unbelievers marching to battle, turn not your backs to them. Who-so turns his back that day to them, unless withdrawing to fight again or removing to join another host, he is laden with the burden of God's anger, and his refuge is Hell—an evil homecoming (Qur'an, Sura 8:15–16).*

The Muslim is taught to fight the unbeliever until Islam is the only religion in the world. Any religion that opposes this global mandate is considered to be persecuting God and must be obliterated.

> *Say to the unbelievers, if they give over He will forgive them what is past; but if they return, the wont of the ancients is already gone!* Fight them, till there is no persecution and the religion is God's entirely (emphasis added) *(Qur'an, Sura 8:39–40).*

The Muslim is taught that God will supernaturally aid them in this conquest. A Muslim is encouraged to take no prisoners "until he make wide slaughter in the land" (Qur'an, Sura 8:69).

> *O Prophet, urge on the believers to fight. If there be twenty of you, patient men, they will overcome two hundred; if there be a hundred of you, they will overcome a thousand unbelievers, for they are a people who understand not. Now God has lightened it for you, knowing that there is weakness in you. If there be a hundred of you, patient men, they will overcome two hundred; if there be of you a thousand, they will overcome two thousand by the leave of God; God is with the patient. It is not for any Prophet to*

have prisoners until he make wide slaughter in the land (Qur'an, Sura 8:65-67).

As a matter of fact, the Qur'an encourages the believers of Islam to lie in wait and ambush in order to slay the unbelievers. That command is given many times throughout the Qur'an. It is not an isolated verse but rather a continuously repeated command reiterated throughout their sacred literature.

> *Then, when the sacred months are drawn away, slay the idolaters wherever you find them, and take them, and confine them, and lie in wait for them at every place of ambush. But if they repent, and perform the prayer, and pay the alms, then let them go their way (Qur'an, Sura 9:5).*

They are, indeed, encouraged to portray themselves as harsh and savage fighters. The videos of the decapitations of their victims are not portraying the radical mindset of an apostate minority. They are in direct correlation with the mainline teachings of the Qur'an. That is of paramount importance to understand. The radical ideology that foments war upon all humankind who will not bow to the Muslim faith is not a fringe doctrinal element. It is the very core of Islam found in the Qur'an and the Sharia tradition that can be found in the Hadith and Jima.

> *O believers, fight the unbelievers who are near to you; and let them find in you a harshness; and know that God is with the godfearing (Qur'an, Sura 9:123).*

The Muslim is to fight the People of the Book (Jews and Christians). Once the slaughter of the land has been made wide, then those who survived the initial onslaught should be taxed heavily (Jizya tax), living in submission to the Islamic (Sharia) law.

> *Fight those who believe not in God and the Last Day and do not forbid what God and His messenger have forbidden—such man as practice not the religion of truth, being of those who have been given the Book— until they pay the tribute out of hand and have been humbled (Qur'an, Sura 9:29).*

Those who claim that Allah is only the Arabic name for the Hebrew El may be right in their etymology, and perhaps during the time of Abraham that was true. No doubt, Ishmael worshiped the same God as his father Abraham, but as time wore on, the descendants of Ishmael turned their back on El and worshipped other gods. By the time Muhammad was born, the Arabs had been completely paganized.

While it is true that Muhammad brought his people to worship a single god rather than the many deities that were venerated in Mecca, the characters of Allah and El are distinctly opposite. And while the Muslims claim to believe in Jesus, their version of Jesus is a gnostic interpretation that denies His crucifixion and is the complete antithesis of the New Testament description of Jesus as the Lamb of God who came to take our sins on Himself. Those who lived with Muhammad documented that the Jesus of the New Testament is not the one Muhammad venerated. That was recorded in the work of Sahih al-Bukhari in the Kutub al-Sittah (the six major collections of the Hadith).

> *Allah's Apostle said, "The most awful name in Allah's sight on the Day of Resurrection, will be (that of) a man calling himself Malik Al-Amlak (the king of kings)" (Sahih al-Bukhari, Vol. 8, Bk. 73, No. 224).*

The description of the real Jesus in the New Testament during the Day of Resurrection is quite clearly Malik Al-Amlak, the true King of Kings:

> *Then I saw heaven opened, and behold, a white horse! The one sitting on it is called Faithful and True, and in righteousness he judges and makes war. His eyes are like a flame of fire, and on his head are many diadems, and he has a name written that no one knows but himself. He is clothed in a robe dipped in blood, and the name by which he is called is The Word of God. And the armies of heaven, arrayed in fine linen, white and pure, were following him on white horses. From his mouth comes a sharp sword with which to strike down the nations, and he will rule them with a rod of iron. He will tread the winepress of the fury of the wrath of God the Almighty. On his robe and on his thigh he has a name written, King of kings and Lord of lords (Revelation 19:11–16).*

It is more than obvious that Islam is the antithesis of the Judeo-Christian worldview, and Westerners trusting in their superior technology must not underestimate their global colonialist aggression.

Atheism Cannot Even Define Evil

If there were no God, why would humans have this deep angst against injustice and evil? Without a God, there can be no evil. What is, just is. But we have been created in His image, and therefore, we are hardwired to hate evil and cherish justice. A universe governed by random ordering can only justify the matrix of the survival of the fittest and the abuse of the masses by the powerful elite. There is no right or wrong in a materialist accident in the cosmos. If all we are is the product of stardust, then evil is nothing more than an illusion.

When we see thousands dying of diseases and no end in sight for epidemics such as the Ebola virus that struck Africa, we intuitively know that this suffering is evil and that the paradise we inhabited long ago, that ancient harbor of peace, plenty, and safety, has become profaned and lethal. Of a truth it can be said that in this life there is little justice. And I dare say, beyond this obvious fact, that if there is no justice in the afterlife, then there is no justice at all. Such is the state of lawlessness and depravity we have created on our planet.

C. S. Lewis wrote:

> *God in His mercy made*
> *The fixèd pains of Hell.*
> *That misery might be stayed,*
> *God in His mercy made*
> *Eternal bounds and bade*
> *Its waves no further swell.*
> *God in His mercy made*
> *The fixèd pains of Hell.*
> *(Lewis 179)*

The accuser says that God is cruel to have created hell, for how could love create eternal pain? But it is love that brought forth hell to fence the ever-swelling pain of sin. It is said that His love and His wrath lead to one and the same thing: justice. For it is because He loves that He strikes against the enemy of humanity and those who would harm His lowly sheep. Hence, evil is kept at bay, and the terror of their leader Apollyon is bound within a net. In His time, evil will dissolve into the lake of fire and ever cease consuming humanity.

Alas, hell will free us of that horrible blight in time when time shall be no more! Sweet justice shall prevail, and mercy will bring us into heaven's fold, safe from the fire that would consume us all. And the maladies that afflict the innocent in this wretched shadow of the world shall one day cease forevermore. He is not cruel at all. Cruel

He would be if not for hell! If there is no justice after life, then there is no justice at all.

So Hawking saw that time was the defining element that could alter the result of the equation in order to avoid the singularity. But although we may conjure worlds of imaginary time to artificially manipulate our cosmological models to exclude the creator, to do so we must step across the line of despair into the irrational.

It can be said that time has existed only since the Genesis Singularity and is not part of the reality that existed prior to the singularity. That is, at one time, there was no time! But time is not an illusion any more than space and matter/energy are illusions. Space, matter/energy, and time are intrinsically intertwined so they are a single continuum, and therefore, when space is warped, so is time; this is what we have learned from Einstein. We can conclude in no other way than to say that all of reality must have come from outside our reality. All the intricate settings and fine-tuned parameters must have been set by some intelligent being before our reality came into existence in order for it to have been created in the only way possible to have a universe that can inhabit life.

Time Can Be Warped

It is also true that most people confuse the concept of dates with time. The theory of relativity shows us that there is no absolute present, that is to say that two people may consider the present in describing the same event, while being in two different times. This is, by the way, evidence that the universe is neither geocentric nor anthropocentric, but rather deocentric. For God is then the only central standard from which time and matter can be measured.

The essence of the theory of relativity is that the physical laws of the universe are absolute. And although the observer may see it from a different perspective in reference to time, depending on his or her position in space, the same position will always give the same absolute answer. Therefore, I propose that the theory should be renamed the special theory of absoluteness. Moreover, from the reference frame of the creator, as opposed to the relative frames in our material space-time continuum, all laws of physics will function exactly the same for any observer in the same place.

Why should the laws of the universe be uniform throughout its breadth and length if we are a product of chaos?

> *Einstein followed the example of Galileo and boldly postulated that all the laws of physics, including those of electromagnetism, must be exactly the same for any observers moving with uniform straight-line motion relative to each other. Secondly, he postulated that the speed of light measured by any such uniformly moving observer relative to another is independent of the state of motion, and that there is therefore no preferred absolute reference frame or ether (Sternglass 71).*

Therefore, it is correct that time is relative to the observer. Two different observers can view time at a different rate simultaneously. Now that is amazing to think about. That means that time is not an absolute. It can be stretched or contracted like space. In fact, it is inseparable from space. When space is contracted, so is time—it slows down. When space is stretched, so is time—it speeds up. Thus, we have the term *space-time*. That means that time is a function of the density of space. And the density of space is a function of gravity or the energy of acceleration.

Therefore, not only the density of space affects time, but time is also affected by energy. The one observer traveling at nearly the speed of light (an astronaut) would see the same event observed from a stationary object, say Earth, in a different time, because his or her time would be slowed down by the impact of the energy in which he or she is traveling. That energy, like gravity, warps space and thus impacts time. Space becomes denser, and time slows down.

The astronomer from Earth would see time flowing faster than the astronaut flying close to the speed of light. Hence, the name *relativity* for the time is relative to the position of the observer and the forces acting upon him in that particular frame of reference.

Nevertheless, it is absolutely the same for any observer who occupies the same space. Time relativity does not imply a chaotic and unpredictable situation. Quite the contrary, it implies an elegant observation of reality that provides a quite accurate prediction in every frame of reference. It does not imply an accidentally generated universe, but rather an elegantly designed universe.

The universe is symmetrical, not chaotic. Symmetry is intrinsic to every aspect of our reality. In fact, it is because of symmetry that people like Einstein were able to follow the natural conclusions of symmetrical relationships and come to better understand how our universe works. It is this universal symmetry that evidences an intelligent design and not an accident of random ordering.

In fact, it is not perfectly symmetrical, or life could not exist. It is just asymmetrical enough to allow for differentiation without creating chaos. That fine-tuned and carefully chosen asymmetry could hardly be expected from a random and chaotic origin. And it is for that reason that we have a universe of galaxies and not one predominantly populated by black holes. That symmetry extends even into the subatomic world.

> In our discussions of the special and general theories of relativity, we came upon yet other symmetries of nature. Recall that the principle of relativity, which lies at the heart of special relativity, tells us that all physical laws must be the same regardless of the constant-velocity relative motion that individual observers might experience. This is symmetry because it means that nature treats all such observers identically—symmetrically. Each such observer is justified in consid-

ering himself or herself to be at rest. Again, it is not that observers in relative motion will make identical observations; as we have seen earlier, there are all sorts of stunning differences in their observations. Instead, like the disparate experiences of the pogo stick enthusiast on the earth and on the moon, the differences in observations reflect environmental details—the observers are in relative motion—even though their observations are governed by identical laws.

Through the equivalence principle of general relativity, Einstein significantly extended this symmetry by showing that the laws of physics are actually identical for all observers, even if they are undergoing complicated accelerated motion. Recall that Einstein accomplished this by realizing that an accelerated observer is also perfectly justified in declaring himself or herself to be at rest, and in claiming that the force he or she feels is due to a gravitational field. Once gravity is included in the framework, all possible observational vantage points are on completely equal footing. Beyond the intrinsic aesthetic appeal of this egalitarian treatment of all motion, we have seen that these symmetry principles played a pivotal role in the stunning conclusions regarding gravity that Einstein found (Greene 2003, 169–170).

Time, therefore, is not an absolute constant, as we once thought. It can be slowed down or sped up. Even the shape of matter can be stretched as space itself can be stretched or contracted due to external forces. The speed of light happens to be one of the three only truly fundamental constants in the universe. It is, therefore, no great surprise that God is described in the scriptures as light. And yet light can also be stretched when space is stretched. We see that gravity bends light as it travels around the sun. We have also observed that light going through a dense medium such as water slows down. However, this is not because the light is slowing down, but because molecules in its path are refracting the light. Its path is no longer a straight, uninterrupted line.

Here, we must be careful to not go too far. Light is a symbol of God but not the essence of God. It is merely a physical symbol in order for humans to understand something about His character. The sun, as our giver of light, was, since ancient times, also a symbol of God. But God is not a star. From the earliest humans, we have seen the sun as a symbol of God, the giver of light and sustainer of life for our planet.

Nevertheless, the fact that the laws of physics are independent of the relative observer is, in fact, an absolute proof that the universe is deocentric. The laws of physics are the same in all inertial frames, and the speed of light is a constant in all inertial frames. This is, in fact, indirect evidence of the existence of absolute truth and specifically of a designer who engineered this marvelous universe.

> This was the essence of Einstein's General Theory of Relativity. It extended the fundamental idea that an observer should always find the basic laws of physics to be independent of his or her state of motion, *not only in the case of a reference frame in uniform linear motion but also in a uniformly accelerated frame of reference such as a rotating frame, where the direction of the velocity rather than its magnitude is constantly changing* (emphasis added) (Sternglass 141).

We know from experiments done in space that gravity also has an effect on time. If this is so, then we can logically infer that time is not a separate entity from our material reality; it is somehow intertwined. That is to say, when we speak of matter and space, we must include time as a factor in the continuum of this reality. They are interrelated and interconnected in a continuum. And for this reason, space cannot be thought of as simply void. If it can be warped, then it has some substance to it. It is a fabric that even contains an intrinsic energy called the zero-point energy. But we will speak of this later. The point now is that if space, matter, and time had a beginning in the Big Bang, we cannot say that our material reality could magically expand into an infinite structure within a finite time frame.

The Vector of Time Is Always Forward

It is because time is not like light, a steady constant that is invariable (it does vary in relation to energy and space), that it is called relative. Forces may slow down or alter time, but there are no known forces that can make it go backward. And although there may be some in the field of philosophy who contend that time does not exist, no serious scientist denies its existence. Within our universe, time is a reality, and its vector is always forward. This, too, is evidence of design and not of a randomly generated reality.

There is an undeniable reality that is irrefutable: in real reality (that is, within our visible spatial dimensions), the vector of time is always forward. I also suspect that within the invisible spatial dimensions, the vector of time is always forward. The angelic beings are also trapped in time. For this reason, the scriptures declare that only God knows the future before it happens.

The second law of thermodynamics clearly has a direction, and that direction cannot be backward. This is the fundamental and universal fact explained by the second law of thermodynamics, and it is absolutely universal, within our space-time continuum.

The Judeo-Christian worldview explains that it is part of God's judgment as the result of the fall. Our universe is moving toward a state of greater disorder; that is, symmetry is being broken with increasing intensity as we move forward in time toward the Omega Point. This reality conforms completely to the Judeo-Christian model that claims our universe was purposefully created from the Alpha Point and that it has an Omega Point in which the consummation of the ages will be complete.

Is Time Travel Possible?

There is a way to theoretically conceptualize time travel, but to do so, we must be able to break free of our space-time continuum. This, of course, is a feat that makes for good science fiction but remains quite out of our reach if we are to stay in this physical body during the process.

If we visualize the space-time continuum as the surface of a sphere (the skin of a balloon), then theoretically, by leaving the confines of the skin of the sphere and cutting into the center of the sphere at an acute angle, we could end up where we came from (the past) or where we are going (the future). The only hang-up is how our bodies will keep from disintegrating when we leave our three-in-one dimensional space-time continuum and step out of our spatial universe.

Although in the confines of our physical bodies, which need at least a three-dimensional spatial existence, this remains quite impossible, it does show us how God can be in the past, the present, and the future, since He is not bound by the space-time continuum, which we must inhabit. The mystical aura that has shrouded the concept of a transcendent God, as delineated by the Judeo-Christian worldview, is now understood through scientific data as a very realistic, plausible mechanism. Again, the supernatural is only the natural that we yet do not understand.

As a matter of fact, the idea that angelic beings could travel back and forth between the heavenly dimension and our universe is not as far-fetched as some naturalists would have us believe. Travel between the invisible spatial dimensions of our universe and the visible dimensions may be possible and could be explained through rational scientific means, of which we yet understand little. If quantum tunneling exists, which pops particles from the invisible dimensions to our visible dimension, then why not? Perhaps the great disparity between the power of gravity compared to the other three forces of nature is so because gravity is leaking into our invisible dimensions.

The passage between one dimension and another may not be as far-fetched as some have imagined in the past. The atheistic paradigm of the Enlightenment period roundly rejected the existence of any dimension that was not sensible to man. Today, modern physicists have provided us with mathematical probabilities for invisible dimensions to our universe through the modern string theory. Now, evolutionists want to extrapolate from that the existence of other universes, but that is a horse of a different color. Nevertheless, in 1963, New Zealand mathematician Roy Kerr generalized Karl Scharzschild's black hole to include spinning black holes and suggested that passage between our three-dimensional spatial existence and the invisible dimensions could be possible.

Since everything in the universe seems to be spinning, and because objects spin faster when they collapse, it was natural to assume that any realistic black hole would be spinning at a fantastic rate. Much to everyone's surprise, Kerr found an exact solution of Einstein's equations in which a star collapsed into a spinning ring. Gravity would try to collapse the ring, but centrifugal effects could become sufficiently strong to counteract gravity, and the spinning ring would be stable. What most puzzled relativists was that if you fell through the ring, you would not be crushed to death. Gravity was actually large but finite at the center, so you could in principle fall straight through the ring, into another universe. A journey through the Einstein-Rosen bridge would not necessarily be a lethal one. If the ring were large enough, one might enter the parallel universe safely.

Physicists immediately began to pick apart what might happen if you fell into a Kerr black hole. An encounter with such a black hole would certainly be an unforgettable experience. In principle, it might give us a shortcut to the stars, transporting us instantly into another part of the galaxy, or perhaps another universe entirely. As you approached the Kerr black hole, you would pass through the event horizon so you would never be able to go back to where you started (unless there was another Kerr black hole that connected the parallel universe back to our universe, making a roundtrip possible) Also, there were problems with stability. One could show that if you fell through the Einstein-Rosen bridge, the distortions of space-time that you created might force the Kerr black hole to close up, making a complete journey through the bridge impossible. . . . However, it soon became apparent that these black holes not only connected two distant points in space, but also connected two times as well, acting as time machines (emphasis added) *(Kaku 218–219).*

Michio Kaku's impression that one could travel through these black holes into another universe cannot be correct. This is nothing more than wish projection and cheap propaganda for their cherished multiuniverse illusion. If it were a separate universe, it could not be connected to ours by a black hole. By the very definition of what constitutes a universe, there cannot be any causal connection between one universe and another. It may be that we pass from one dimension to another within our universe, but not from one universe unto another. If our universe could be connected through the Einstein-Rosen bridge to another universe, then both areas are part of the same universe. What may more properly be said is that there may be a connection here to another invisible dimension of our same universe or even into another time frame further out from us in the same universe.

The geometry that Schwarzschild came up with was so unexpected that it took until the 1960's for scientists to finally understand that it describes a wormhole joining two black holes. From the outside the black holes appear to be separate entities sitting at distant locations, yet they share an interior.

In a 1935 paper, Einstein and his colleague Nathan Rosen, then at the Institute for Advanced Study in Princeton, New Jersey, anticipated that this shared interior was a kind of wormhole (although they did not understand the full geometry it predicted) and for this reason wormholes are also called Einstein-Rosen (ER) bridges.

The wormhole in Schwarzschild's solution differs from black holes that form naturally in the cosmos in that it contains no matter—merely curved spacetime. Because of the presence of matter, naturally formed black holes have only one exterior. . . . The Schwarzschild solution tells us that the wormhole connecting the two black hole exteriors varies with time. It elongates and becomes thinner as time progresses, like stretching out a piece of elastic dough. Meanwhile the two black hole horizons, which at one point touch, separate rapidly. In fact they pull apart so quickly that we cannot use such a wormhole to travel from one exterior to another. Alternatively we can say that the bridge collapses before we can cross it. In the dough-stretching analogy, the collapse of the bridge corresponds to the dough becoming infinitesimally thin as it gets stretched more and more.

It is important to note that the wormholes we are discussing are consistent with the laws of general relativity, which do not allow faster than light travel. In that way they differ from science-fiction wormholes that allow instantaneous transport between distant regions of space, as in the movie Interstellar. Sci-fi versions often violate the laws of physics (Maldacena 28–30).

The interior that is shared by the two black holes must be in one or more of the invisible spatial dimensions described by the string theory. In a normal black hole containing matter, the curvature of space-time is crunched by the immense gravity, thus slowing down the rate of time. But in Schwarzschild's solution, there is no matter; hence, space-time is actually stretched as the wormhole behind the two black holes lengthens. The uniqueness of this wormhole is that space-time is stretched, and time is subsequently sped up, causing it to be a gateway to the future. Anything caught inside it cannot return because it cannot travel backward in time. There is but one vector for time, and it is always forward.

Either way, there are grave problems to the stability of the Einstein-Rosen bridge created by distortions of space and time by a three-dimensional body passing through it that would cause it to close up. Nevertheless, in spite of this possibility of moving from one dimension to another, the distortions that could be created in space and time, if our physical bodies were to go through this bridge, may not occur if the being going through it did not have a three-dimensional body like ours. In other words, the passage between two or more dimensions of our universe could be traversed by angelic beings through this bridge.

The idea of time travel is certainly an appealing one for most, but so far, every attempt has been unrealistic on physical grounds—at least for those of us who are trapped within our three-in-one visible spatial existences. Some have attempted to come up with a viable way to do so, but invariably, they must invent a universe that does not exist in order to accomplish it.

> *In 1988, Kip Thorne and his colleagues at Caltech found yet another solution of Einstein's equations that admits time travel via a wormhole. They were able to solve the problem of the one-way trip through the event horizon by showing that a new type of wormhole was completely traversable. In fact, they have calculated that a trip through such a time machine may be as comfortable as a plane ride.*
>
> *The key to all these time machines is the matter or energy that warps space-time onto itself. To bend time into a pretzel, one needs a fantastic amount of energy, far beyond anything known to modern science.* For the Thorne time machine, one needs negative matter or negative energy. No one has ever seen negative matter before. In fact, if you had a piece in your hand, it would fall up, not down. Searches for negative matter have proved fruitless (emphasis added) *(Kaku 220–221).*

Again, you have to imagine a universe with negative matter in order to conjure this possibility, an exercise not unlike Hawking's imaginary time. At any rate, as fantastic as this all sounds, the fact remains that in the real world, in our space-time continuum, time has a forward vector, and we do not exist in a universe of negative matter. Furthermore, these wormholes, such as the Einstein-Rosen bridge, do not stay open long enough for anything physical from our visible dimensions to travel through them. They close up too quickly. The only traversable wormholes are those conjured through negative energy or exotic matter, which we have yet to produce. Time marches inexorably forward, and no empirical evidence has thus far shown otherwise.

The Essence of Time

What is the essence of this magical thing called time? We must be careful to differentiate between the concept of time and dates. The measure of a day is simply a space measurement of a full rotation of the Earth, and thus it is not an intrinsic measurement of time itself, for Earth's rotation is not even permanently constant. It is gradually slowing down, and at times, some scientists believe that earthquakes can influence the speed of the rotation of our planet.

Even the jiffy, which, as we said previously, is the vibration of an atomic particle, the shortest time interval ever considered as physically relevant (there are 10^{43} jiffies in one second), is not time itself. It is still a measurement of a spatial movement between two coordinate points, which is relative to the observer. What we commonly refer to as time measurements are then really hitching posts of observations, which are more accurately described as dates: at this date, the atomic particle was in this spatial coordinate, and at the following date, the particle was in a different spatial coordinate. We can attempt to describe time with dates, but dates are not time itself.

Moreover, it is impossible to speak of time without space. They are inseparable. Without time, there is no *when*. Without space, there is no *where*. Without *when* and *where* together, there is no date to describe the measure of time. Without matter/energy, there is no *what*. *What*, *when*, and *where* form the continuum of our reality. The evolutionist insists that there is no *why*. But our Judeo-Christian cosmological model claims that without the *why*, there would be no *what*, *when*, and *where*.

Nevertheless, invariably there is a real time measurement in a forward vector. There is a forward motion in time that can be measured in dates. We live with a concept of past, present, and future. But these are interconnected. For example, when we look at the night sky, the light we see in the form of stars is, in fact, the past, which is being seen in the present since it takes time for the light to arrive here from its initial destination. But in areas of the universe where space-time has been stretched sufficiently, time is moving faster than in our region; hence, when we see light from these areas toward the edge of our universe, we are, in fact, seeing the future of those stars and galaxies that have moved faster through time from the moment of creation (the Alpha Point).

The present is somewhat of an illusion since it can be divided infinitely into past and future, since it cannot be grasped to divide it, and since it is in constant motion. Yet in a sense, it is real, for we are not living in the future nor are we living in the past, and our senses are registering feelings that are not part of our memory but of our existential, moment-by-moment experience.

The motion of time may be affected by outside forces to either speed it up or slow it down, but the movement is always forward. The space-time continuum may be warped so that, for example, a body falling from space toward the center of the Earth will be affected by the pull of gravity in such a way that time will move at a different speed at the head than at the feet due to the difference in the gravitational pull between the two places. That simply means that our space-time is being warped by gravity, and thus, time will be relative to its position regarding the force acted upon in the space it inhabits.

But although one time may flow more slowly than another, neither flows backward. Yes, when Humpty Dumpty fell off the wall, all the kings' horses and all the kings' men could not put Humpty together again. Why? Because there is but one direction for the time vector; time only moves forward. Thus, the process is irreversible. It is indirect evidence that God has created a universe in which there is an Alpha and an Omega—a beginning and an end. It is evidence of a purposeful creation to reach a certain previsioned end. And this forward vector of time is the major tenet of the second law of thermodynamics within our space-time continuum. It is the refutation of the pantheistic concept of cyclical time. It is also the refutation of the atheistic concept of a random reality.

The universe began with order that is now becoming more disordered. Randomness could not have created such overwhelming order and symmetry as observed in our entire universe even today after the second law of thermodynamics has been busy creating disorder since the moment the law was created by God. Any disorder observed presently must be viewed as the simple progression of the second law of thermodynamics, and the only scientific conclusion that can be drawn is that it began with even more order than exists today.

The evolutionary foundational doctrine that randomness created our ordered universe is, therefore, in direct antithesis to the second law of thermodynamics and to all observable reality. The central point of the second law of thermodynamic is based on the vector of time being an absolute. If we are the product of chaos, then why does time only move forward? This is no accident. It is designed to allow us to function rationally. The idea that time can only move forward is, in fact, an essential requirement of a world that would provide people with an environment that would be suitable for their existence and that would allow them to function in a rational sense.

Without a past and without our mental ability to remember the past, we would not be able to function in our world. It is the past that has impressed us with data that we are able to then correlate in our minds to make sense of reality and determine which actions we will take in rational response to this memory. Without this memory of the past, life would be just a succession of meaningless impressions jumbled one into the other and continually being forgotten as new impressions are being felt. Without memory, we would be incapable of figuring out the *why* questions of the *what, when,* and *where* of reality around us. It is the fundamental basis for the scientific process. Furthermore, it is the evidence that the Alpha Point has a direction toward the Omega Point and testifies to the previsioned purpose in the mind of the creator of all creation.

There is in our minds an innate sense of time tied to our biological and psychological systems. Animals can in unison begin to fly at a given time in their migratory cycle as if some subconscious alarm clock goes off in their brains. Some of us will wake up at exactly the same time every day without an alarm clock. How does the brain know what time it is while you are sleeping and your eyes are closed?

Time is not an illusion; it is real. While it is true that it can be bent, it is no less imaginary than space and matter, which can also be distorted as they approach the speed of light or as they are acted upon by energy such as gravity or acceleration. To establish a cosmology on a time vector other than forward is simply to step out of the possible and into the imaginary, at least within the confines of our present bodies. Is it possible that the spiritual bodies that we will inhabit will be composed of light when we are clothed with what the scriptures call the incorruptible, glorified body? Only then will we not be trapped in this time vector.

There are three constants in nature: (1) Planck's constant (h); (2) the speed of light (c), which is constant under all conditions in the vacuum of space; and (3) Newton's constant G, which measures the strength of the gravitational force. The speed of light is a constant that is truly absolute. It cannot slow down or speed up in the vacuum of space. In Einstein's famous train, he exemplified this truth. Light, or the velocity of light, appears as a universal constant of nature, regardless of the inertial reference system used. It is, therefore, not unexpected that God is called light in the scriptures, and it is again evidence of a deocentric universe.

Bless the LORD, O my soul! O LORD my God, you are very great! You are clothed with splendor and majesty, covering yourself with light as with a garment (Psalm 104:1–2).

The Essence of Space

It is he who sits above the circle of the earth, and its inhabitants are like grasshoppers; who stretches out the heavens like a curtain, and spreads them like a tent to dwell in (Isaiah 40:22).

Thus says God, the LORD, who created the heavens and stretched them out, who spread out the earth and what comes from it, who gives breath to the people on it and spirit to those who walk in it (Isaiah 42:5).

It is a humbling thing to consider that the Hebrew scriptures professed long before Einstein that our space-time fabric could be stretched. The scriptures also stated long before Columbus sailed the blue that the Earth was round and not flat. There are many verses that speak of God stretching out the firmaments of heaven. In fact, the Word of God has many things to say regarding our space-time continuum.

- It tells us that the fabric of space-time can be torn. "Oh that you would rend the heavens and come down, that the mountains might quake at your presence" (Isaiah 64:1).
- It tells us that the fabric of space-time will be shaken, when the Lord returns: "At that time his voice shook the earth, but now he has promised, 'Yet once more I will shake not only the earth but also the heavens" (Hebrews 12:26). "For thus says the LORD of hosts: Yet once more, in a little while, I will shake the heavens and the earth and the sea and the dry land" (Haggai 2:6).

It tells us that when He ends His Millennial Kingdom, ruling from Jerusalem, He will destroy this universe and set up a new universe where the curse shall be no more and the second law of thermodynamic will no longer be operational, for this space-time reality will be burned up. This is the Omega Point. "Waiting for and hastening the coming of the day of God, because of which the heavens will be set on fire and dissolved, and the heavenly bodies will melt as they burn!" (2 Peter 3:12). "'Then I saw a new heaven and a new earth, for the first heaven and the first earth had passed away, and the sea was no more. . . . He will wipe away every tear from their eyes, and death shall be no more, neither shall there be mourning, nor crying, nor pain anymore, for the former things have passed away.' And he who was seated on the throne said, 'Behold, I am making all things new.' Also he said, 'Write this down, for these words are trustworthy and true'" (Revelation 21:1, 4–5).

- It tells us that before that Omega Point when He makes all things new, our space-time universe will become worn out like a garment. "Of old you laid the foundation of the earth, and the heavens are the work of your hands. They will perish, but you will remain; they will all wear out like a garment. You will change them like a robe, and they will pass away, but you are the same, and your years have no end" (Psalm 102:25–27). "'You, Lord, laid the foundation of the earth in the beginning, and the heavens are the work of your hands; they will perish, but you remain; they will all wear out like a garment, like a robe you will roll them up, like a garment they will be changed. But you are the same, and your years will have no end'" (Hebrews 1:10–12).

The "garment" or fabric of space can be warped, rolled, folded, ripped, burned, and stretched. It is literally a fabric that can be worn out. But what you do with the spatial dimension impacts both time and matter. The connection between them is inseparable. There is an interconnection between space, time, and matter that in this present universe cannot be broken. They are inextricably intertwined. If you impact one, you impact the others. For example, it is the degree of the stretching of the fabric of space-time that changes the rate of time in different frames of references around our universe. But also in those regions where space has been stretched, the density of matter has been affected.

So what exactly is space? Space is not a void. Space has an intrinsic fabric that has density and can be molded. In fact, it has an intrinsic energy. But to explain the energy of space, we must first begin with the discoveries of Swedish physicist Johannes Rydberg.

Born in Hamstad, Sweden, on November 8, 1854, Rydberg became a mathematician and physicist whose interest in understanding the secrets that unlocked the mystery of the interrelationship of the elements of the periodic table led him to his research in spectroscopy. Scientists had already discovered that elements had distinct spectral lines that were created by each element. However, they did not yet understand the relationship with those peculiar sets of spectral lines. It was Rydberg who discovered that relationship and devised a formula that expressed it: $n = n0 - N0/(m + m')2$. What this formula represents is the strength of the interaction between the nucleus and the electrons in the atom. It is that relationship that results in the specific, unique spectral lines for each element.

It turns out that the spectral lines were a unique sort of bar code that identifies whatever chemical is giving off light. William G. Tifft later discovered that these spectral lines are actually quantified in digital form. In fact, it turns out that this relationship in the bar code is logarithmic in scale. Each bar is ten times the distance of the previous, caused by the electrons moving between different energy levels in their shells around the nucleus.

In 1925 and 1927, American physicist Robert S. Mulliken traveled to Europe to work with a number of spectroscopists and quantum theorists including Wolfgang Pauli, Paul M. Dirac, Louis de Broglie, Walther Bothe, Erwin Schrödinger, Max Born, and his assistant Friedrich Hund. These brilliant minds were working on a quantum interpretation of the band spectra of diatomic molecules and were at that time the leading edge of researchers in that field. In 1927, Mulliken and Hund developed a molecular orbit theory that explained the energy states over an entire molecule. The insightful theory is called the Hund-Mulliken theory.

However, during his study of the spectral lines, Mulliken discovered that the wavelengths of these spectral lines were, in fact, shifted from their true theoretical positions. Mulliken brilliantly recognized that this peculiarity must be caused by some unknown energy in space that is battering the atoms and impacting their movements. Scientists have now come to understand that energy as zero-point energy. It is the intrinsic universal energy of empty space throughout the entire universe.

In other words, if we were to examine a cubic foot of space at any point in the universe and subsequently physically remove all matter within it (whether in the plasma, gas, liquid, or solid state) along with any heat so the temperature in that cubic foot of space were 0 degrees Kelvin, we would find that it is still not empty. It still contains a residue of energy. This energy is the intrinsic energy of the spatial dimension, which is now called zero-point energy (ZPE).

That energy is defined by the formula $E = 10^{95}$ ergs/cm^3. Scientists believe that it is this energy that atomic particles tap into in order to maintain their spins perpetually. It is the fundamental energy of the spatial dimension that allows matter to exist in the forms of spinning particles in atoms. Without this energy to create the spins, which govern the electromagnetic properties of the atom and allow it to maintain integrity as an atom, matter could not exist in any of the four states. Thus, matter could not exist outside of the spatial dimension. Matter is therefore inextricably tied to both space and time.

Take a moment and think about what that means. It means that God infused every square inch of the spatial dimension in our universe with a fundamental energy that allows our universe to exist and function as it does. In other words, each second of every day, all the atoms in the universe are drawing energy from this substratum in order to be able to function and exist in their material form.

The Apostle Paul's description of Christ now becomes much more apparent.

> *For by him all things were created, in heaven and on earth, visible and invisible, whether thrones or dominions or rulers or authorities—all things were created through him and for him. And he is before all things, and in him all things hold together (Colossians 1:16–17).*

Max Planck, one of the most important theoretical physicists of recent history, discovered that space is quantized; that is, it comes in discreet packets of a particular length that we now call the Planck length—the minimal measurement of a unit in space. The Planck length is denoted by the symbol ℓ_p and is equal to 1.616199(97) $\times$ 10^{-35} meters. The formula for Planck length is:

$$\ell_p = \sqrt{\tfrac{\hbar G}{c^3}} .,$$ The formula for Planck time is $(\hbar G/c^5)^{1/2}$, which is about 10^{-44} seconds, and the Planck mass is $(\hbar c/G)^{1/2}$, which is about 10^{-8} kilograms.

The symbol c is the speed of light in a vacuum, G is the gravitational constant, and $\hbar$ is the Planck constant. Planck's constant in relation to energy, E, is expressed in the equation $E = hc/\lambda$.

In other words, in describing energy, E, using the Planck constant we also need to include the speed of light. All formulas regarding the Planck constant include c, the speed of light. But the particular formula I want to bring to your attention is the one regarding energy because we can then compare it directly to Einstein's equation.

Since we know from Einstein's equation that $E=mc^2$, or energy is equal to mass times the speed of light squared, then we know that energy and mass are also interrelated with the speed of light. The speed of light is a measurement of time (the time that light travels in one second) through empty space. Hence, space, time, and matter are inextricably tied. Space cannot exist without time. Time cannot exist without space, and matter cannot exist without both space and time.

This small quantum packet of space turns out to be so infinitesimally small that it is 10^{-20} times the diameter of a single proton. It is many orders of magnitude smaller than any current instrument in our present technology could ever possibly measure. Nevertheless, what the formula tells us is that there is an interrelationship with gravity; thus, the mass of matter and the speed of light—thus matter/energy, space, and time—form an indivisible continuum. All three had to be created simultaneously. One could not be infinite without the other two being infinite. And therefore, if any of them are finite, then all of them are finite. Since we know for sure that time is finite, then we also know that space and matter are finite.

It turns out that if we were to decrease the velocity of light, the Planck constant would increase, and energy would stay the same. In other words, if the speed of light slows down, space is crunched, and the mass of the atoms increases since time is also slowed down. Hence, in places with extreme gravity, space is crunched, and time slows down.

In contrast, if the speed of light could be increased, we would observe space stretched, and thus the mass of atoms would decrease. But that is all mathematics in our universe. The speed of light in a vacuum is a literal wall that we cannot cross. The point is that there is an interrelationship between space, time, and matter that is inseparable. If we stretch space, time speeds up. This interaction between matter, space, and time is the fundamental triad that results in the physical reality in our universe. But when we impact one, we impact the other two. If science is the pursuit of knowledge based on empirical data, the speculation that any of these three fundamental constants could be separated from the other in either our universe or some other universe is nothing more than rubbish. All speculations regarding another form of universe is then not built upon any empirical data but purely on the fanciful imaginations of those with a preconceived bias that benefits from this irrational proposition.

Another form of energy has recently been assigned to "empty" space. I am not speaking of the zero-point energy, which can be physically detected by spectral lines, but a negative or dark energy that, according to its proponents, is an antigravity force. It is called vacuum energy, or dark energy. The story of how this repulsive energy came to be accepted begins with Einstein.

CHAPTER 11

• • •

THE DARK SIDE OF THE FORCE

*In the pride of his face the wicked does not seek him;
in all his thoughts, "There is no God."*

—Psalm 10:4

The idea of an infinite, static universe that had no beginning and no end was quite attractive to those who pandered the existentialist philosophical view. However, even as far back as the time of Newton, thinking scientists were able to point out the inconsistency of such a naturalistic view of the universe. Michio Kaku, a popular theoretical physicist, describes the conundrum of Olbers's paradox that faced scientists from Newton to Einstein.

> *In 1692, five years after Newton completed his master-piece, Philosophiae Naturalis Principia Mathematica, he received a letter from a minister, Richard Bentley, that perplexed him. Bentley pointed out that if gravity was strictly attractive, and never repulsive, then any static collection of stars will necessarily collapse in on*

itself. This simple but potent observation was puzzling, as the universe seemed stable enough, yet his universal gravitation would, given enough time, collapse the entire universe! Bentley was isolating a key problem faced by any cosmology in which gravity was an attractive force: a finite universe must necessarily be unstable and dynamic.

After pondering this disturbing question, Newton wrote a letter back to Bentley, stating that the universe, to avoid this collapse, must therefore consist of an infinite, uniform collection of stars. If the universe were indeed infinite, then every star would be pulled evenly in all directions, and hence the universe could be stable even if gravity was strictly attractive. Newton wrote, "If the matter was evenly disposed throughout an infinite space, it could never convene into one mass . . . and thus might the sun and fixed stars be formed."

But if one made that assumption, then there arose another, deeper problem, known as "Olbers' paradox." It asks, quite simply, why the night sky is black. If the universe is indeed infinite, static, and uniform, then everywhere we look, our eyes should see a star in the heavens. Thus there should be an infinite amount of starlight hitting our eyes from all directions, and the night sky should be white, not black. So if the universe was uniform and finite, it would collapse, but if it were infinite, the sky should be on fire! . . . Einstein, forced to confront a contracting or expanding dynamic universe like Newton, was still not ready to throw out the prevailing picture of a timeless, static universe. Einstein the revolutionary was still not revolutionary enough to accept that the universe was expanding or had a beginning. His solution was a rather feeble one. *In 1917, he introduced what might be called a "fudge factor" into his equations, the "cosmological constant." This factor posited a repulsive antigravity that balanced the attractive force of gravity. The universe was made static by fiat.*

To perform this sleight of hand ... the cosmological constant assigned an energy to empty space. This antigravity term, now called "dark energy," is the energy of the pure vacuum (Kaku 131–133).

Michio Kaku's association of the cosmological constant to dark energy is simply predicated on the evidence that the stars and galaxies in our universe are expanding outward at an ever-increasing rate in relation to their distance from us; that is Hubble's constant. This dark energy is theorized to be an invisible substance that repels and therefore cancels gravity, which miraculously grows more powerful with distance.

Nevertheless, dark energy does not render the universe static, which was Einstein's purpose in interdicting the repulsive force of his original cosmological constant. He wished to keep the universe static so that naturalists would not have to deal with the pesky problem of a beginning. This initial cosmological constant proposed earlier by Einstein was a mathematical fix, a repulsive force that would counter the gravitational force and therefore avoid the Big Bang. But that folly has been debunked as our universe has been completely confirmed to be dynamic and not static as scientists had so vehemently insisted since the Enlightenment.

However, experimental data have since then convincingly proved that the opposite is the case. In fact, now modern cosmologists think that dark matter and energy cause our universe to be much more dynamic than they had ever anticipated. Not only is it not static, it is increasing dynamically as time moves forward and the edges of space expand with greater speed. The Genesis Singularity cannot be avoided. The universe had a beginning. Time is finite, and so are space and matter.

But it must be repeated that Einstein's cosmological constant is not the same as the cosmological constant considered by today's cosmologists, even though they have some similarities in that they both speak of a repulsive energy that increases with distance. Einstein's cosmological constant was a mathematical fix to make the universe static. Today's cosmological constant is a mathematical fix that tries to explain the more dynamic and rapidly expanding universe. But before delving further into dark energy, let us examine Einstein's motivation for this mathematical fix he trumped up.

Einstein's initial reaction to the implications of the general theory of relativity is a classic study of the human spirit in rebellion toward God. He was, at that time, unwilling to concede that the universe was not static and infinite, for he initially deplored the idea that his theory would severely impact the naturalist presupposition. He wanted to avoid the need of a starter.

> *Einstein's reaction to his own equations may possibly acknowledge the threat of an encounter with God. Before he published his cosmological inferences from the theory of general relativity, he searched for a way to "fix up" the equations, anything to permit a static solution, a universe free of expansion or deceleration.*
>
> *Einstein postulated a cosmic force of repulsion to cancel off the attractive force of gravity despite the body of evidence that gravity was predominant in its influence throughout our galaxy and its vicinity. Einstein had to develop a repulsive force that would have imperceptible consequences for nearby objects but overwhelming effects over extreme distances (Ross 53).*

Faced with the dilemma that a universe, which was either slowing down or speeding up, must have had a beginning, Einstein was quite perturbed. The idea of a beginning has always been quite problematic and repugnant to the naturalist. There is good reason for them to hate this idea, because it mathematically brings them to a point where materialistic science cannot explain reality.

Einstein's theory tells us that all the matter and energy of our universe was condensed into a single point in space of infinite density—the Genesis Singularity. Infinite density cannot be broken by finite energy. There is not enough energy in a finite universe to break infinite density. Therefore, there is no materialistic force that can cause that Genesis Singularity to explode. The naturalist is incapable of providing any rational explanation to the mechanism of exploding a singularity with infinite density. It is an impossible conundrum for those who insist that only the material aspect of that which is contained inside our universe is real. Only an infinite force can counter the infinite gravity of infinite density. Hence, the materialist simply evades the question by stating that any mathematical equation that deals with an infinite number is beyond the realm of mathematical science. That is a half-truth. Infinite numbers do exist in mathematics. The real problem is that no finite force can counter that infinite density; hence, it points to an infinite God as the only possible source of infinite power.

The reality is that God has forced us to see, by the method He chose to create our universe, that only His infinite power could have created a Genesis Singularity with infinite density, and then only His infinite power could cause it to explode outward. The reality is simple: the singularity did explode outward because we would not be here otherwise. No other explanation can explain reality except an infinite God. That answer is unacceptable to the naturalists, not because of science, but because of their underlying metaphysical choice to deny the existence of God in order to avoid being morally accountable.

Therefore, Einstein initially attempted to find some mathematical way to cancel out this dynamic universe. He tried feverishly to explain away the beginning. He made an irrational leap of faith and pounced upon an arbitrarily devised, repulsive energy that was able to repulse faraway objects with immense power and yet would have minute, almost imperceptible effects to nearby objects (the exact opposite of Newton's understanding of gravity). He attempted to mathematically describe a sort of magical antigravity force that could make the universe static again.

Here is where Einstein stepped over the line of despair into the irrational and escaped from reason. One of the most brilliant men in history at first refused to face reason when it pointed to God and sought to create an alternative that was more in line with his naturalistic presupposition. This force, which he contrived, became known as the cosmological constant (not to be confused with the cosmological principle, which is another contrivance of the naturalistic paradigm to rationalize their atheistic presupposition).

But, Einstein was certainly not alone. There were other static solutions that were hastily proposed, such as the one Dutch astronomer Willem de Sitter postulated. But in time, all of them were shown to be incongruous with observable facts. Later on, when Hubbell's observation of the expanding universe was published, the static concept was utterly shattered, and the cosmological constant became a moot point. Finally, Alexander Friedmann discovered a simple algebraic error made by Einstein in the cosmological constant equations, which, when properly corrected, rendered a non-static universe instead.

The very brilliant Einstein suffered a great humiliation in his attempt to explain away God.

> For it is written, "I will destroy the wisdom of the wise,
> and the discernment of the discerning I will thwart."
> Where is the one who is wise? Where is the scribe?

Where is the debater of this age? Has not God made foolish the wisdom of the world? . . . But God chose what is foolish in the world to shame the wise; God chose what is weak in the world to shame the strong; God chose what is low and despised in the world, even things that are not, to bring to nothing things that are, so that no human being might boast in the presence of God (1 Corinthians 1:19–20, 27–29).

Einstein, initially, made no secret of his antipathy for the implications of his mathematical discovery. He openly fumed that his theory of relativity forced upon us the fact that there had to be a beginning and therefore implied a primal cause or creator.

The concept of an exploding universe seemed to irk the scientific community. Einstein openly fumed over the implications of a beginning point, particularly concerning a Creator or Prime Mover for the universe. Eddington, too, was agitated. He declared the origin of the universe to be "philosophically repugnant." More subtle expressions of irritation came from others such as Omer, who refused to attribute anything special to the time or circumstances of the observer (meaning the observer cannot determine anything about the origin) (Ross 58).

But subsequent conclusions from thermodynamics and observations of the velocity of galaxies toward the edge of our universe became incontrovertible evidence of an expanding universe. Eventually, grudgingly, Einstein came to concede the existence of God, reportedly as a deist, although I strongly doubt that he was a pure deist by the nature of his quotes about God in the latter part of his life. In time, Einstein did admit that the cosmological constant was an embarrassing mistake. As early as 1919, he had stated that his cosmological constant was "gravely detrimental to the formal beauty of the theory" (Einstein 153).

But it was not until 1931, following the publication of Hubbell's law of red shifts, that Einstein could finally bring himself to discard the cosmological constant from his field equations. Thoroughly humiliated, he conceded that its introduction was the greatest mistake of his life.

As he aged, he gradually became more comfortable with this idea of a creator and even began to speak out on the matter. He publicly admitted that he no longer held some of the precepts he had formerly adhered to when he was young and brash. It is a credit to his great mind and spirit of integrity that he was able to change his personal opinion to fit the facts and not the other way around, as is the common rule in our human nature.

Although he allowed some subjective bias to interfere with his final conclusion, to the extent he was objective, to that extent he aligned with truth. This is the mark of a great scientist, who is supposed to be an objective observer of reality in search of truth. This is the mark of a man of deep intellectual integrity. May he serve as an unflinching example to all of us who are fellow sojourners in the quest for truth.

Today, the idea of the cosmological constant is being revived by some who believe that this expansion force in our universe could be caused by the mysterious force they call dark energy. In 1998, the study of supernovas suggested that there was a mysterious force acting in the universe and causing it to expand at an accelerating pace. Some have speculated that this expansion is the result of a dark energy or vacuum energy that exerts a negative force on gravity, which effectively counters its attractive force.

Recent observation by the Hubble Space Telescope one million miles out in space, independently confirmed by the Apache Point Observatory in Sunspot, New Mexico, has shown definitively that the universe is, in fact, expanding at an accelerating rate. The idea that gravity would at some point slow down the expansion of our universe, from the initial expansion of the Big Bang, has been shown to be unrealistic. All cyclical models are now shown to be untrue. Sagan must be turning in his grave.

The Genesis Singularity cannot be avoided. But that has not affected atheists' fundamental presupposition. Naturalists are therefore speculating that some mysterious force, which they have dubbed dark energy, has a magical antigravity force that is at the root of this phenomenon. Somehow, they must come up with a materialistic explanation of reality. And thus, the cosmological constant has been pulled off the shelf to explain this phenomenon in a hybridized form.

The New Cosmological Constant: Dark Energy/Vacuum Energy

This time, the cosmological constant is not being used to make the universe static but to justify naturalists' worldview. Modern physicists are attempting to explain this acceleration as rooted in some antigravity force attributed to this contrived magical dark energy. In other words, scientists today are no longer attempting to mathematically arrive at a static universe that has existed for infinity. They are simply trying to find a materialistic reason for the universe to be expanding at such an exact accelerating rate.

The fact that the universe is expanding and accelerating can no longer be refuted. And yet the fact that the universe had a finite and fixed beginning date continues to stick in the throats of existentialists, no matter what the empirical data show. And from their viewpoint, it is with good reason, because it shows the ineptness of their naturalistic presupposition to explain reality.

However, it seems that this arbitrarily chosen assumption that there is a dark energy, which is repulsive in nature, is also quite ridiculous. It is nothing more than a theoretical Band-Aid to explain what they cannot otherwise explain in materialistic terms. It is simply a mathematical fix without any concrete empirical evidence to prove it exists. The expansion rate that God chose at the very first moment of the Big Bang has not ever wavered. Naturalists would love to find some materialistic reason for this expansion rate that would dampen the overwhelming evidence for intelligent design that it poses.

Not only do they propose this magical dark energy, but they also propose another invisible and equally magical substance known as dark matter. According to their fantastic speculations, this dark matter that composes 23 percent of the total matter in the universe has enormous gravitational powers, but in contrast, this dark energy that composes 76 percent of the universe has enormous antigravity powers. Isn't that convenient?

Evolutionary cosmologists now claim that all the scientific data seem to point to the fact that close to 96 percent of the universe is made up of something that cannot be seen, cannot be duplicated in a laboratory, cannot be explained in any fashion that would resemble known particles, cannot be physically described in any concrete mathematical way (other than the seeming gravitational effects), and resembles nothing that we have ever encountered in reality. And yet it makes up almost the entire universe. Had they said this to the early Enlightenment thinkers, they would have been drummed out of their cabal for infringing upon their holy mantra that nothing exists that cannot be sensible.

In summary, naturalists are now telling us that only 4 percent of the entire universe can be sensible to the human senses. My, how far they have come from the Enlightenment dogma that only the sensible is real! They have, in a few hundred years, gone from one extreme to the exact opposite, and I think they are still wrong. Neither extreme is rooted in reality. Both extremes are simply adjuncts to their foundational atheistic presupposition that seeks to deny the existence of the great creator.

Dark Matter and Dark Energy

What is this dark matter that naturalists are proposing? Recent observations of stellar, galactic, and galaxy cluster/super clusters have convinced them that there must be more mass in the universe than what we are able to observe. When velocity measurements of a given stellar structure are made, it has become evident that the structures are spinning at such a velocity that they should be coming apart by centrifugal force. Galaxies in the outskirts of our universe are moving too fast for the mass calculated within them. Therefore, they think that there must be some hidden matter whose gravity keeps them from disintegrating. This invisible mass is then called dark matter.

How do we know they are spinning faster than they are supposed to be spinning? By observing the red shift of one side of a spinning galaxy and the blue shift of the other side, we are able to deduce the velocity of the spin. By observing the amount of light in that galaxy, we can also mathematically deduce the number of stars in that galaxy and their corresponding mass.

But the mass predicted by these measurements is not enough to hold the galaxy from breaking apart with the velocity at which its outer regions are spinning. The centrifugal force is greater than the mass could hold together. Thus, scientists are proposing that perhaps there is some "dark matter" out there that is invisible to us with our present technology that could thus account for their ability to spin at such an enormous velocity without the galaxies disintegrating.

In other words, dark matter, they propose, provides an invisible mass that gives the galaxies an additional gravitational attractive force that keeps the spinning matter in these faraway galaxies from moving outward by the angular momentum of their unbelievably fast spins.

In contrast, dark energy, or vacuum energy, is considered a repulsive force that is causing the universe to expand at the phenomenal accelerating pace that has, at this point, been thoroughly documented and established as an incontrovertible fact. Together, these two magical, invisible propositions make up the vast majority of our universe, according to evolutionary theorists.

In other words, according to the interpretation of the evidence collected by the Wilkinson Microwave Anisotropy Probe (WMAP), evolutionists suggest that only 4 percent of our total universe is composed of baryonic matter (ordinary matter and energy). About 23 percent is in some form of dark matter—of yet unknown composition. And 73 percent is predicted to be of some form of dark energy—also of an unknown and even more enigmatic composition.

I find it quite fascinating that the naturalists, during the Enlightenment, ridiculed people of faith by insisting that anything that was not sensible to our human senses was simply superstitious. Today, the priests of naturalism are preaching that the sensible is only about 4 percent of all reality. The great thing about the theory of evolution is that it can evolve from one century to another and contradict the very foundations that supposedly made the theory true to begin with.

So we find today that evolutionary scientists are now indicating that there is considerably more mass and energy to the universe than what we can see, feel, hear, taste, or touch, even with the super-enhanced powers of our modern technical extrapolations to our senses. Evolutionary theorists have now divided themselves into two camps. Initially, some evolutionary scientists rejected the idea that this dark matter was composed of some new strange form of matter. They believed that this dark matter was in the form of massive objects such as black holes and brown dwarfs. These were characterized as MACHOs (massive astrophysical compact halo objects).

Others believe that dark matter is composed of strange sub-atomic particles that are invisible. These are categorized as WIMPs (weakly interacting massive particles). These mysterious WIMPs are theorized to be made of stuff other than ordinary matter (baryonic) and are therefore categorized as non-baryonic matter.

Most astrophysicists at first fell into the MACHO camp and hoped that the high-resolution photographs of the Hubble Space Telescope in outer space could produce the evidence to support this view. But, the latest evidence produced by the telescope does not reveal the large number of brown dwarfs that the MACHO-leaning astrophysicists had expected to find. Only 6 percent of the galactic halo matter seems to be the result of brown dwarfs.

Previously, evidence by the Hubble Space Telescope in conjunction with the Very Large Array Radio Telescope had verified the existence of more than 50 black holes, mostly in the very center of large galaxies. But now, some scientists are proposing that perhaps as many as half the galaxies in the universe have a black hole in their very center. Today, very few astrophysicists any longer deny the reality of black holes.

> *The Hubble Telescope, peering into the hearts of distant quasars and galaxies, has now taken spectacular photographs of the spinning disk surrounding the black holes located in the heart of distant galaxies, such as M-87 and NGC-4258. In fact, one can clock some of this matter revolving around the black hole at about a million miles per hour. The most detailed Hubble photographs show that there is a dot at the very center of the black hole, about a single light-year across, which is powerful enough to spin an entire galaxy about 100,000 light-years across. After years of speculation, it was finally shown in 2002 that there is a black hole lurking in our own backyard, the Milky Way galaxy, which weighs the same as about 2 million suns (Kaku 217–218).*

But in spite of these observations, there just does not seem to be enough black holes and brown dwarfs to account for the amount of dark matter/energy naturalists are proposing in our universe. On the other hand, the MACHO astrophysicists criticized the particle physicists championing the WIMPs for merely providing a quick fix without any experimental evidence to support their theory. These WIMPs are invisible to our senses and so far are undetectable by our technology. The bottom line is that there is not one single shred of empirical data that can substantiate either dark matter or dark energy.

There is No Dark Matter or Dark Energy: There Is Only Time Dilation

The reason galaxies at the edge of our universe are spinning faster than the mass would dictate is because the evolutionary astrophysicists making the measurements of those faraway galaxies are using our coordinates as a time frame for these measurements instead of theirs. The stretching of space has sped up time in this area of the universe. Those faraway galaxies are moving in fast-forward time compared to our much slower rate of time toward the center of the universe. It is not mass that is missing; it is that time is relative to the position of the observer. Evolutionists have conveniently overlooked Einstein's predictions regarding time.

But that is not an explanation they want to see, because for evolution to have a chance of plausibility, they need our area of the universe to be 15 billion years old. The idea that our area of the universe could have aged more slowly and not experienced these long eons of time they need for evolution would not be acceptable to their preconditioned metaphysical preference. The plain truth is that there is no need for some magical and invisible dark matter to explain the enormous velocity of the spin of those galaxies.

If we were to observe that spin velocity from their time frame, the mass and speed of their spin would be in complete accord with the laws of gravity. It only seems that way to us looking at them from a time frame coordinate that is so much slower. If our understanding of the relativity of time is correct, then we would expect the deep space photographs of galaxies toward the edge of our universe to be seen as stretched and smudged. Light coming from galaxies beyond them would also appear to bend as in gravitational lensing; however, the culprit doing the bending is not gravity, but time.

Think of a photograph of a baseball going through the air. If our shutter speed matches the speed of the ball, we are able to get a clear picture of the round ball. But if our shutter speed does not match the speed of the ball, then what we observe is an elongated, smudged picture of the ball that distorts its real shape. It is the difference in the rate of time between our coordinates here on Earth and the coordinates of the galaxies at the edge of our universe that causes images of these faraway galaxies to be unnaturally contorted and elongated. Measuring the length of elongation of these galaxies can test this; that is, we can predict that those galaxies furthest out would then be elongated the most due to a higher rate of time. The lengthening would then be directionally proportionate to their distance from us. But their evolutionary bias blinds them from the truth for more than one reason. Some may want this universe to have more mass for another subtle reason.

If there were enough dark matter in the universe, it might make the critical density of the universe such that we could theoretically collapse. In other words, if the mass of the universe would exceed the force of the Big Bang (expansion rate), then it would cause the universe to recollapse. That cyclical scenario would once again provide an opportunity for evolutionists to try to get rid of that pesky beginning.

That feature is quite attractive to existentialists who would envision an endless oscillating scenario, which would consequently allow them the rationalization to escape the need for a beginning. And more to the point for them, it would give them the latitude to sidestep the need for a starter. But as we have already discussed, this is mere wishful thinking, for there is no physical mechanism they can point to that would create a bounce and result in an eternally oscillating universe.

Without dark matter, the critical density of the universe lies somewhere between 0.1 and 0.01, and we would then be living in a universe that is eternally expanding. It would never collapse. There would just not be enough mass in our universe to stop the expansion and make it collapse.

With just the right amount of dark matter, enough to make the critical density 1.0, and if the universe is flat, then the mass would be just the right amount to stop the universe from expanding, but not enough to cause it to collapse back unto itself. However, the present evidence seems to indicate that there is just not enough dark matter to stop the expansion of our universe, no matter how badly they wish it so.

What I find fascinating in their propaganda is that they speak of this dark matter as a fact, when it is nothing more than a wild speculation. The uninformed public then assumes it to be a fact.

> *The galaxies are all heavier than astronomers had thought. Roughly speaking, every galaxy is about ten times more massive than all the visible stars and interstellar gas that it contains. The remaining nine-tenths of the mass is a mystery. It is almost certainly not made of the things that comprise ordinary matter: protons, neutrons, and electrons. Cosmologists call it dark matter: dark matter because it gives off no light. Nor does this ghostly matter scatter light or allow itself to be visible in any form, except through its gravity (Susskind 2006, 147).*

This same propaganda of certainty is exhibited for their newly proposed cosmological constant, otherwise known as vacuum energy or dark energy. It is simply unquestionably assumed to be the repulsive cause of the expansion of our universe. In reality, this repulsive energy is sheer rubbish; from a scientific point of view, it is unfeasible. No known force has ever been shown to increase in intensity with distance.

To begin with, the evidence mounted by modern astronomy points decisively to our universe definitely expanding at an accelerated pace. Hubble's constant is irrefutable. The speed of expansion is in direct proportion to the distance from us. That cannot be argued. But the idea of a repulsive force that grows with distance is, in my mind, an irrational proposition. No other force of nature works this way. In fact, when we understand how these forces work in the subatomic level, we can see that such a repulsive force is nonsensical. For example:

All forces in nature derive from special exchange diagrams, in which a particle like a photon is emitted by one particle and absorbed by another. For example, the electric force between electrons comes from a Feynman diagram in which one electron emits a photon, which is subsequently absorbed by the other electron.

The photon jumping across the gap between the electrons is the origin of the electric and magnetic forces between them. If the electrons are at rest, the force is the usual electrostatic force that famously diminishes according to the square of the distance between the charges. If the electrons happen to be moving, there is an additional magnetic force. The origin of both the electric and magnetic force is the same basic Feynman diagram.

Electrons are not the only particles that can emit photons. Any electrically charged particle can, including a proton. This means that photons can hop between two protons or even between a proton and an electron. This fact is of enormous importance to all of science and life in general. The continual exchange of photons between the nucleus and the atomic electrons provides the force that holds the atom together. Without those jumping photons, the atom would fly apart, and all matter would cease to exist (Susskind 2006, 46–47).

So we see that the electromagnetic force is created by the exchange of photons and that it diminishes according to the square of the distance between the charges. As a side note, not only were all particles created by photons at the beginning of our universe, but without these photons, matter could not even continue to exist (we are the children of God's light). But the point here is that the electromagnetic exchange of photons physically prohibits it from becoming stronger with distance. In fact, it diminishes according to the square of the distance between the two charges.

Similarly, the exchange of gravitons thought to be the attractive force we call gravity also diminishes with distance. The further these photons and gravitons must travel, the less they are able to perform their jobs. The idea that a force could increase with distance would be contrary to any known physical process that exists in our universe.

The strong nuclear force that holds the protons together in the nucleus is governed by gluons. Remember that protons are all positively charged. Like poles repel. Therefore, the protons in the tightly knit nucleus are repelling against one another with an enormous force, but the gluons manage to counter this electromagnetic strength. If the nucleus has more than 100 protons, however, the repulsive electric forces are enough to blow the nucleus apart. It seems that the strong nuclear force works like flypaper. If the protons manage to become separated even a little bit, the force is broken. Hence, with gluons, the distance is even more critical than with gravitons and photons.

There are no forces in nature we can point to that could mimic this supposed repulsive force that increases in power with distance, as proposed by the new cosmological constant or dark energy. In other words, the notion of a cosmological constant as a repulsive force that increases with distance is in opposition to all empirical knowledge of the forces of nature. It is a mathematical fix without any known mechanism to power the repulsive magic. It is just more of what Einstein called "spooky action at a distance." It appears to us to be expanding at an accelerated pace because time in those regions is accelerating as space is stretched. It is the relativistic framework of time that can best explain both dark matter and dark energy.

In a study called "Dark-Matter Mystery: Why Are 400 Stars Moving as if There's Nothing There?" by Michael D. Lemonick, scientists reported that the dark matter theorized to exist due to our view of the speed of galaxies at the end of our universe does not exist in the 400 stars they analyzed in the vicinity of our sun. They cannot see dark matter because the stars in the vicinity of our sun are running at the same time rate we are.

> *The theory of dark matter took decades to take hold in astronomy, and no wonder. It's pretty tough to wrap your mind around the notion that some mysterious, invisible substance pervades the cosmos — and even tougher to accept that it outweighs ordinary matter by a factor of 6 to 1, at least. Evidence eventually trumped incredulity, though, and by the 1980s, the vast majority of scientists were on board with the idea, nutty though it might seem, and there they've remained ever since.*
>
> *But a new study out of the European Southern Observatory claims that this now established theory could be in trouble. Chilean astronomers took a look at the motion of 400-plus stars in the broadly defined neighborhood of the sun and concluded that the way*

they're all moving is inconsistent with conventional ideas about dark matter. "Our calculations show that [evidence of dark matter] should have shown up very clearly in our measurements," said Christian Moni-Bidin, of the University of Concepción, in a press release. "But it was just not there!" (Lemonick).

In conclusion, we can say with certainty that the dynamic nature of our universe has increased as our scientific understanding has progressed, rather than causing it to become static. Einstein's attempt to avoid the uncomfortable reality that our universe had a beginning by fixing his equation with the original cosmological constant was a fool's errand. This dark energy and dark matter, which cosmologists have theorized to explain the speed of expansion of our universe and the speed of rotation of galaxies, is just also another fool's errand. The answer to this expansion rate will be understood when we address the Genesis acceleration spin constant. Let us now address the two fundamental metaphysical doctrines in the catechism of scientism: dysteleology and the Copernican principle.

The Evidence for Design

We have already discussed how quantum tunneling could not create an entire universe from nothing. At the first moment of the creation of the universe, both the visible and invisible spatial dimensions and the energy that eventually turned into matter came to be. The hyperspace point where the Genesis singularity appeared did not, prior to its moment of creation, have the invisible dimensions of our universe from which to pop up like virtual particles in quantum tunneling today. Not until God created the Genesis Singularity were all three visible space-time dimensions and the seven invisible space-time dimensions birthed. Quantum tunneling does not come from hyperspace; it comes from our invisible dimensions of our universe.

The great difficulty for evolutionists is that Einstein's theory brings us to a beginning singularity with infinite density and a supposedly near zero volume in our three-dimensional space. We have already mentioned that only an infinite power could overcome infinite density, but what about this near zero volume in space? Some claim that the singularity had zero volume.

If, then, our entire universe, the visible and invisible dimensions together, truly had a zero value spatial dimension at the beginning, it would not have been able to expand into our present universe. Zero times any number is still zero. One can multiply that zero to infinity, and it would still be zero; that Genesis Singularity had to have a finite size with a finite amount of energy, and although unimaginable to us, it was compressed at an infinite density. It is possible and quite probable that this Genesis Singularity first existed in the seven invisible spatial dimensions and then expanded into our three-in-one visible dimensions so that within our visible dimensions, it is zero volume, but not within the invisible dimensions.

Nevertheless, this had to come from an exterior source, which brought into our universe a precalculated and precise amount of energy to begin the entire creative process that produced everything that now exists. Thus, all the mechanisms touted by evolutionists to explain the origin of our universe and galaxies fail the litmus test of reality. But the dysteleological delusion is made even more evident when one considers the chances of random forces hitting the exact expansion rate necessary for our universe to have formed as it is today.

Whoever brought that finite and yet almost incalculable amount of energy into the Genesis Singularity must have been a real genius, because the physical processes created by this precise and expertly calculated finite amount of energy were the master stroke that allowed our universe to be precisely capable of harboring life. Thousands of parameters had to be precisely chosen at that very point of the infusion of energy in the Big Bang in order to strike the delicate balance that alone can produce a universe that can harbor life. Even Weinberg admits this.

> "Life as we know it," Weinberg writes, "would be impossible if any one of several physical quantities had slightly different values. . . . One constant does seem to require incredible fine tuning." This constant has to do with the energy of the big bang. Weinberg quantifies the tuning as one part in 10120. Scientific notation is an understatement and so I will expand that exponential into decimal notation. If the energy of the big bang were different by one part out of 10 000000000000000000000000000000000000000 000000000000000000000000000000000000000 000000000000000000000000000000000000000 there would be no life anywhere in the universe. The universe is tuned for life from its inception. . . . Michael Turner, the widely quoted astrophysicist at the University of Chicago and Fermilab, described that tuning with a smile: "The precision," he said, "is as if one could throw a dart across the entire universe and hit a bullseye one millimeter in diameter on the other side" [It should be noted that Professor Weinberg is a skeptic.] (Schroeder 5).

Sometimes it is difficult to grasp the enormity of a number when written in exponential notation. There are 60 seconds in one minute, 60 minutes in one hour, and 24 hours in one day. Therefore, each day has 60 x 60 x 24, or 86,400 seconds. Since there are roughly 365 days in one year, then there are 86,400 seconds x 365 days in one year, or 31,536,000 seconds in one year. A year is really 365¼ days long.

For this reason, we add one day every four years, our leap year. But in order to stay with round numbers, we will use 365 days for this illustration. At the writing this book, I am 60 years old, so on the day of my 60th birthday, I had lived 31,536,000 seconds in one year × 60 years, or 1,892,160,000 seconds. In other words, at 7:00 a.m. on the morning of January 14, 2012, I had lived roughly 1 billion, 892 million, 160 thousand seconds. That is a lot of seconds. I hope I have not wasted too many of them.

Imagine that there was a giant roulette wheel in a television game show that included every second of my life. Every second is in different colors, but only one is white. You have one chance to throw a little black ball and have it landing on the white section. Your chances would be one in 1,892,160,000. How good do you think your odds are?

Imagine the number of seconds that have elapsed in a hundred years, a thousand years, a million years, a billion years, or in the presupposed 15 billion years of the history of the universe. The number of seconds that have elapsed since the Big Bang is merely 1017. This number includes leap years. If we had a roulette wheel large enough to contain all the seconds since the very first second of the universe, your chances of winning would be 1 in 1017. Can you rationally say that random ordering caused our initial plasma state in the Big Bang to be that precise energy gradient and to expand at the precise velocity necessary to create our universe when the odds are 1 in 10120?

To believe that such accuracy can be accomplished in one shot through random ordering requires too much faith for me. I am actually a skeptic at heart. I was an atheist before I became a Christian. For this reason, I was forced to accept the most rational, logical explanation for the origin of our miraculously designed universe. Some genius mind had to have conceived, planned, and engineered this wonder we call our universe. Nothing else suffices if we are to be rational and objective.

Seven Logical, Sequential Postulates in Regard to the Origin of Our Universe:

1. Something had to exist before the matter and space-time continuum was created, since logically something couldn't have arisen from nothing.

2. Yet that primordial something could not be trapped in the matter and space-time continuum in which we live. It must be able to exist before or, more accurately, outside the confines of our space-time continuum at the inception of the Big Bang. Since evidence points to an intelligent choice in the precisely calculated singularity created, then only the reality of an infinite, omnipotent God could be logically conceived as existing before or outside of the Big Bang as the only possible initial primal cause, which is not bound within the continuum of time and space. No other possible explanation has been presented that could account for the spontaneous appearance of this enormous amount of energy from nothingness, which could also precisely calculate that amount of energy imputed and subsequently guide the orderly process that created the space-time continuum in its intricate and functional web of potentialities that could allow the formation of a universe with a suitable habitat for life.

3. It could therefore not be the laws that govern our physical properties within our universe, since they are themselves only a description of what first has to be created. Laws are recognized only by intelligent minds that analyze the behavior of our material universe and, through reason, discover universal truths. But how does a randomly ordered evolution create universal laws and give any validity to the rational, orderly scientific processes?

4. Einstein taught us that energy and matter are both sides of the same thing. All matter is formed from energy and can be interchangeable. Since we know that energy is

interchangeable with matter ($E=mc^2$), therefore it is not only plausible but necessary that the primordial elements or contents of the universe, residing within that infinitely dense Genesis Singularity at the moment of the Big Bang, must have been created by energy emanating from an outside source that could bring forth all this into a single coordinate of space-time. The universe is therefore not an absolutely closed system but an open system that received at least one event emanating from outside it in order for it to begin at all. What naturalistic mechanism could conceivably be capable of doing this?

5. All natural processes in a natural and finite universe are confined within that universe and cannot exist eternally outside that universe since they did not come to be until the creation moment and are not transcendent beyond their finite existence. The Genesis Singularity containing our entire universe in a single point in space-time came to be in a state of infinite density. A giant singularity with infinite density cannot be exploded outward against the power of that infinite density and the immense gravitational power it exudes without an infinite power to break that infinitely tight bond. Only an omnipotent God can counter that force of infinite density and its subsequent gargantuan gravity. No finite materialistic power could ever undo that bond. Evolutionists have nothing from the materialistic realm at their disposal that can explain how infinite density could be broken to expand.

6. Could random chemical processes existing before our universe through sheer serendipity cause our universe to begin accidentally and then expand outwardly at the exact rate necessary and in the precise manner that it did to create a universe that can inhabit life? Only the reality of an infinite transcendent God existing outside space-time

could provide the vast energy necessary to convert into the enormous amount of matter that presently inhabits our universe and furthermore to guide its development in such a smooth, ordered fashion as to generate greater order through willfully directed energy and prevision. Only an infinite and omnipotent God could have the mind to conceive, engineer, and have the power to control the initial explosion, precisely choosing and fine-tuning the initial temperature of the plasma state to allow a universe that inhabits life. Furthermore, only an infinite God could oversee the explosion in such a way that the universe would then progress in the smooth, ordered state that it did, almost as if defying the second law of thermodynamics.

7. It is a fact that uncontrolled and random chemical reactions would have formed a lumpy universe, leading to an infinite number of black holes, rather than the vast number of galaxies that inhabit our universe. The enormous number of values for physical properties could only have been previsioned, engineered, and maintained by a supreme mind in order to form a universe that could inhabit life. In other words, only an infinite, transcendent God could control the very narrow parameters of all the physical processes that developed after the expansion of the plasma universe created at the Big Bang in order to end up with a universe that could sustain carbon-based life as we know it. Therefore, since we know that the singularity was a highly ordered and directed event that resulted in that very narrow parameter of all the physical values in order to be able to support carbon-based life, then it stands to reason that unintelligent and undirected chance energy could not have engineered the Big Bang, no matter how many years it would be given to accomplish this miracle of the highest order possible.

The Uniqueness Antipathy

We can see the not so subliminal naturalistic bias reflected in evolutionists' attempt to negate any scientific endeavor that points to the existence of the creator. For this reason, they have great antipathy toward anything that makes Earth and man unique in all of reality. There is a subliminal reason that attracts the naturalists/existentialists to gravitate toward such speculative notions as WIMPs to explain dark matter. If WIMPs were discovered, it would make humankind's place in the universe much more removed from the center. Within our recent scientific history, humankind has progressed from the geocentric cosmologies of Ptolemy and his forbears, which made man and Earth the center of the universe, through the heliocentric cosmology of Copernicus and Galileo, which shattered this so-called anthropocentric view of humans, to our modern conception, which realizes our solar system is but a speck in a galaxy of billions of other galaxies in the vast universe.

We are often told by evolutionary scientists that the Earth is an average or medium-sized planet orbiting around an average star orbiting in the outer suburbs of an ordinary spiral galaxy, which is itself only one of about a million-million observed galaxies in our vast universe. The insignificance of our planetary position and size is here incorrectly distorted in order to downplay our uniqueness and importance in the universe.

Thus, the Copernican principle, which in their minds defies the uniqueness of the Earth, became a cherished star in their bonnet. In their minds, it refuted the Judeo-Christian worldview that grants humans special significance. But contrary to the evolutionary propaganda, our uniqueness is not based on which way the solar system functions. Whether the sun revolves around us or we revolve around the sun makes no difference to the uniqueness of our planet. The truth is that according to Newton's laws, the sun and the Earth revolve around each other. That is, they revolve around the center of mass between both masses. Surely the scriptures have never claimed that our sun revolved around our planet. This was a mistake made by some misguided clergy. Our uniqueness comes from the fact that God created the universe as a canvas for our Earth and ultimately humankind. This is the true Judeo-Christian message.

The naturalist is highly antipathetic to this train of thought. In other words, the uniqueness antipathy held subjectively and universally by naturalists causes them to insist that humans are but a serendipitous accident of random chemical processes. There is no divine purpose to humans. All things are the mere product of random ordering. A person is but an atom of the universe, and chaos is its mother.

It is precisely this subjective bias that motivates them to choose the otherwise irrational option that our universe was created from nothing into infinite space-time. In this way, they can insist that our sun-Earth system is not in the center of the universe, even when all the data point to it.

By the way, if they hold to the position that the spatial dimensions in our universe are infinite, then the brane collision theory cannot exist. If our membrane is infinite, then there is no other membrane possible. For this same reason, the discovery of WIMPs in their mind would mean that we are an even more miniscule aspect of reality compared to the rest of the universe. A person becomes less than an insignificant speck in the vast universe with no transcendental significance whatsoever.

To begin with, the Earth is not an average planet. The overwhelming majority of planets found in our universe are giant gaseous planets on the order of Jupiter. Earth's exact position in relation to the sun, which allows for water to exist in liquid form, is thus far also unique in the entire universe. The trick is not only to find a rocky planet, but to find it around the same relative distance from its star in order to allow water to be liquid. Its rotation around the star would also have to be nearly circular, like Earth. Too elliptical, and the drastic changes in temperature would be incompatible with life.

It must also be about the same size as Earth. If the planet were smaller, the interior molten core would solidify as it cooled down. This would mean that it could not produce a magnetosphere to protect it from lethal cosmic rays. It would also not have enough gravity to maintain a thick enough atmosphere. Mars, for example, is slightly smaller than Earth and has 1/100 the density of our atmosphere and no magnetosphere to protect it from the lethal cosmic radiation of our sun.

If it were too much larger, the immense gravitational pull would solidify the metals in the core and it could also not produce the electromagnetic protection that our magnetosphere creates for us against the lethal cosmic rays of solar radiation. It is believed that the flow of our molten core is what causes our magnetic field. Life on such a planet would be mercilessly buffeted by cosmic rays. The atmosphere would, without a magnetosphere, be blown away by this cosmic wind.

Then, of course, it would need a single moon of just the right size, about one-fourth the size of the planet in order to keep the rotational axis from diverging too much as it spins. Too much wobble, and there would be no way to survive the extremes of the weather and tides on such a planet. Too large a moon or too many moons may also cause catastrophic consequences. Ocean tides would routinely create giant tsunamis. It is the precisely fine-tuned gravitational drag of our moon that allows us to have a spin that is not too drastic to create giant waves that would wash over the surface of the planet.

This highly specified planet would also need a few large gas planets far enough away from it on the opposite side of the star in order to provide the protection necessary to shield it from comets and asteroids. It is doubtful that we could have survived our solar system if we did not have Jupiter, Saturn, and Neptune to run interference for us against these cosmic killers. But if these giant planets were too close, it would also invariably spell disaster. Their orbits would have to be fairly circular and stable so they did not come near the rocky planet in question to perturb its orbit or collide with it. It is infinitely more probable that such giant planets would simply eat smaller rocky planets in a world ruled only by blind chance.

This may not seem significant to many, but the chances of a solar system like ours to have these giant gas planets, which should have formed nearer to the sun, in orbit outside the rocky planets is a statistical problem. These huge planets would have had to knock each other out of orbit while not crashing into one another or the rocky planet in question on its way to the outskirts of the solar system. As if that is not enough, then they must miraculously fall into a stable orbit around the sun that no longer interferes with the other planets that had caused each other to be thrown further away from the sun.

That miracle planet would also need to have an atmosphere containing oxygen and carbon dioxide and be free from noxious gases. This is not a common combination found throughout the universe. Venus is a rocky planet, roughly close to our size, but it has an atmosphere containing lethal amounts of sulfuric acid. There is a layer of sulfuric acid haze from 25 kilometers to about 50 kilometers, followed by cloud layers of sulfuric acid to about 75 kilometers above the surface of the lethal planet. Its surface temperature hovers around 872 degrees Fahrenheit, about twice the temperature in a normal cooking oven on high. Even though it receives about 25 percent of the heat that Mercury receives closer to the sun, the enormous atmosphere containing greenhouse gases makes the surface temperature a literal furnace. The average temperature is above the melting point of lead.

Thus, it is not just the distance to the star, but the composition of the atmosphere and its density that is also crucial. The atmosphere in Venus is about 92 times as dense as the Earth's. The density on Venus's surface is comparable to the density of our ocean at about 910 meters below the surface; that is an astounding 2,957.5 feet below the surface of the ocean.

Most of us cannot appreciate the immense pressure this represents unless we have been scuba diving in the ocean and ventured beyond the range of colors. The full color spectrum is lost after 30 feet of depth. Everything seems to take on a bluish hue from that point down. The surface pressure on Venus would implode and crush our bodies instantly. The pressure is so high that carbon dioxide is technically no longer a gas but a supercritical fluid. It flows like a liquid.

We would have to climb 50 to 54 kilometers above the surface of Venus to find atmospheric pressure comparable to Earth's and temperatures where water could be maintained in liquid form. However, at this altitude, we would be buffeted by sulfuric acid clouds and tremendous winds of tornado strength that range from 220–360 miles per hour. Between the wind and the acid, it would, in short order, corrode and rip to shreds any space vehicle we inhabited. Venus also lacks a magnetic field. However, it does have an ionosphere that gives it some protection from cosmic radiation.

Earth is certainly not a common planet. It may be possible that one day we will find an earthlike planet, but it will be the exception to the rule and not the rule, as most evolutionists optimistically propose. Thus far, scientists have cataloged several hundred planets, and none of them are even close to what Earth is like.

Even if we would ever find life on another planet, it does not necessarily prove that they evolved through random chaotic processes. If God is a creative being, He could easily choose to create life on other planets. The improbability of random ordering does not change, even if there were life on another planet. On the contrary, since the statistics of forming life are so gigantically remote, it would make it even more logical to conclude that life is the product of an intelligent designer. But regardless, the uniqueness of our blue planet cannot be disputed by any known data. Everything to the contrary is merely speculation based on a naturalistic presupposition.

Furthermore, the sun is not an average sun. It is vastly outnumbered by small, dim red dwarf stars. The sun ranks numerically within the top 5 percent of stars in size in the entire universe, although it is dwarfed by some stars that are many thousands of times its size. However, the intense radiation of such huge stars would be lethal to life. Our sun is just the right size.

The majority of stars in our universe are, in fact, binary systems, some even tertiary and quaternary. Such an arrangement would spell doom for any planet being pulled and tugged by two or more competing stars. Our sun is propitiously placed on the outer edge of our galaxy in a neighborhood that is not crowded by dangerous radiation from having too many nearby neighbors. Even more important, it is a stable sun burning at a fairly even rate. Any fluctuations in the intensity of a star would spell disaster for life, not to mention the fact that almost all binary star systems have their axes pointed to our solar system. The design of the universe points to our solar system. It is unique among all others.

The Design of the Heliopause

Our planet and its relation to the sun are unique to the universe thus far. But the propaganda that the naturalist is attempting to promote is that the Earth and its solar system are not unique and humankind is an inconsequential cog in the vast realm of reality. Our uniqueness, as intimated by a special creation with intelligent design, is therefore contraposed by their mythical faith that there is no design to our universe and that all was randomly ordered.

Imagine if we could create a magnetic bubble wrap that encircles our entire solar system in a cocoon of protection from harmful radiation. Wouldn't that be a great idea? In fact, our solar system does contain a protective magnetic bubble sheath that helps keep the harmful radiation of other stars in our galaxy from entering our solar system. I must admit that this randomness god that evolutionists have so much faith in ought to take a trip to Las Vegas—it's the luckiest thing that has ever existed.

Some 37 years ago, we sent out *Voyager 1* to investigate our solar system. At that time, scientists hoped that the little craft would last five years. It has lasted 37 years so far, and the "little engine that could" is still sending us back an incredible amount of information as it crosses into the heliopause. Gliding ever further into the cosmos, *Voyager 1* beamed back information that was completely unexpected.

As a result of that information, we now know that a protective sheath of giant magnetic bubbles that are one astronomical unit in size wraps around our solar system. It is as if a giant magnetic bubble-wrap sheath is wrapped around our solar system and is able to protect that solar system from dangerous radiation coming from the galaxy. It just so happens that the size of these bubbles is about the same distance our Earth is from the sun. But that is just another lucky coincidence.

The spinning giant ball of plasma we call our sun generates enormous electromagnetic forces. These waves travel all the way out to the heliopause, where the waves bunch up, coming closer to each other. Because the sun is spinning and the magnetic waves coming out from the sun are twisted, the electric field lines crisscross and reconnect, forming these giant magnetic bubbles.

Call me crazy, but it looks as if God bubble-wrapped the solar system with a magnetic shield to protect us from harmful radiation emanating from the galaxy. This is our first line of defense against lethal rays that could sterilize the Earth. This phenomenon is another piece of evidence that our universe was designed with purpose and intelligence in order to allow life to thrive in our third rock from the sun.

That these bubbles are roughly the distance that Earth is to the sun is also another quite spectacular bit of evidence that should point the thinking person to the unique distance our blue planet inhabits, where life can thrive.

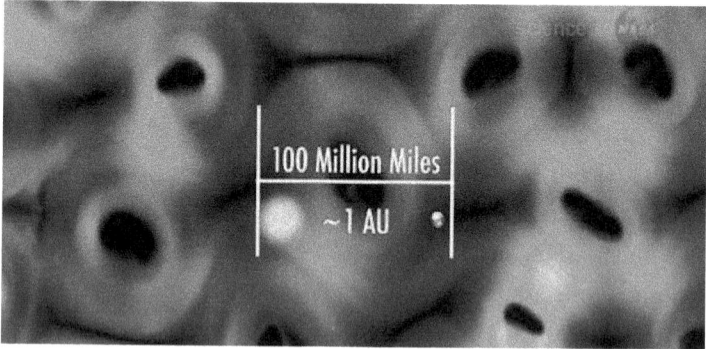

The protective magnetic bubble sheath around our solar system

Science points to a deocentric view rather than an anthropocentric naturalistic illusion. If, in fact, the universe was created by a white hole and Earth is almost at the center of the universe, then its position in this vast universe cannot be construed as anything but purposefully designed.

It is this determinism, evidenced by the grand harmony of the universe, that is so antipathetic to existentialists who wish to find chaos as the matrix of the universe due to their philosophical proclivity. And it is precisely why many leading scientists who have an existentialist axe to grind so vehemently oppose this notion of intelligent design. I believe that the vast majority of modern scientists are basically good and honest people. Few are overtly attempting to deceive others. But it is human nature to gravitate toward those things that substantiate the philosophy we adopt. I am convinced that it is not so much an act of deception of others as it is an act of self-deception.

Subjectivity in the thinking process of all human beings is inescapable. The only real difference between one person and another is the degree to which it influences them. This includes all humankind, including Christians. But some are so passionate about their underlying philosophical perspective that it completely clouds their scientific objectivity.

CHAPTER 12

● ● ●

THE GENESIS SINGULARITY

Many people do not like the idea that time has a beginning, probably because it smacks of divine intervention.

—Stephen Hawking

Space, time, and matter had a beginning. No matter how you slice it, all three models of the Big Bang theory clearly show that there must have been a beginning. At this point of origin, which is the creation point (the Alpha Point), the Genesis Singularity contained all that became our universe gathered in a single point in space-time. The distance between all the material that later became all the stars and galaxies in the universe was zero, and the density and the curvature of space-time would have been infinite. To say that space-time was severely crunched would be an understatement of cosmic proportions. The gravitational clutch was at its absolute maximum and will never again be equal in power from the moment the Big Bang began.

Somehow, some force caused the Genesis Singularity to explode outward. That enormous force would have to be infinite in strength in order to counteract the infinite density created by the infinite curvature of space-time in that Genesis Singularity. The problem for evolutionists is not just the problem of where all that highly packed energy came from to form the Genesis Singularity, but also what force could counter such immense gravity and cause it to explode outward. In our material universe, there are no infinitely powerful forces. All known forces in our material universe have a finite amount of power.

When we speak of the curvature of space-time being infinite, we have little ability to imagine that because our finite minds cannot truly grasp the immensity of the infinite. Furthermore, since mathematics cannot actually function in the realm of infinite numbers, it is a position that cannot be fully understood by finite beings or described by any mathematical equations. The upshot of that means that all theories, including the theory of relativity, break down at this specific point of genesis (the Genesis Singularity) and form an impenetrable barrier for humans to ever punch through to understand.

Beyond that, the infinite force that was able to counteract the infinite curvature of space-time had to come from beyond the space-time singularity. There is no known force in physics within our universe that could counter the massive force of gravity in a singularity from a normal black hole, much less a singularity that contained the entire universe. That force had to come from beyond the singularity to cause it to expand outward. Moreover, that force could not have been a single quantum of force, but a continual force that set the speed of expansion at just the right velocity to enable this universe to become habitable for life.

Therefore, it is only rational to infer that something above and beyond our material science or temporal reality had to exist before the Genesis Singularity existed. The fact that the universe now exists and before that point did not means that something behind or before the Genesis Singularity had to exist in order for it to have been started. Stephen Hawking voiced this dilemma for evolutionists and readily admitted that for this reason, most scientists today are antipathetic to the idea of a genesis for our universe.

> *In fact, all our theories of science are formulated on the assumption that space-time is smooth and nearly flat, so they break down at the big bang singularity, where curvature of space-time is infinite. . . . Many people do not like the idea that time has a beginning, probably because it smacks of divine intervention (Hawking 49).*

The biblical model of creation depicts a fiat creation of the universe, not out of nothing, but out of the energy of God. Remember that matter and energy are interchangeable, not, as an extension of God, as the pantheist would claim, but as a separate entity. I emphasize this specifically to differentiate between the pantheistic concept of creation in which the universe is simply one and the same as their concept of divinity.

This differentiation is important for two reasons: (1) the passage from nothing to something is an impossible and illogical progression. The universe could not have begun with nothing. Something had to have been brought into existence from outside the universe to begin the process; (2) the pantheistic concept of deifying the essence of the universe (the neuter principle) acknowledges that their deity is one and the same as the universe and could therefore not have brought this matter into our universe from without. If the deity is limited to space-time reality, then there is something greater that brought it into existence. The pantheistic deity is just not big enough. It fails the litmus test of reality and reason.

The Quantum Fluctuation Delusion

The most common argument used today to avoid the problem of origins is the proposition that our universe was birthed by a quantum fluctuation. A quantum fluctuation, as described in quantum mechanics, is the temporary change in the amount of energy at a point in space, as explained in Heisenberg's Uncertainty Principle. This momentary change in the quantum field can cause a particle pair to pop up in a point in space. The key phrase here is "in a point in space." Without the substructure of space and the combined spatial dimensions of which it is comprised, with its intrinsic energy and quantum fields, there can be no quantum fluctuations.

To begin with, quantum mechanics is a view of our universe through the lens of a particle theory. As we will later discuss, it is an incomplete theory since it cannot account for gravity, which is one of the four forces of nature. A complete theory should be able to describe all four forces of nature. Nevertheless, for the point at hand, let us ignore that.

The mathematics of quantum mechanics allow for a phenomenon where particles appear as if from nowhere. Particles "magically" appear to pop up into our three-in-one visible space-time reality, and therefore, many scientists have contested that our universe could have equally popped up out of nothing through some quantum fluctuation. By the way, this happens to be NASA's favored position at this time.

It is a nice try, but no cigar. In order for quantum mechanics to exist, a universe must exist with matter/energy, space, and time. Quantum mechanics are laws that describe the intrinsic properties of our physical universe. Before the universe existed, there were no quantum mechanics and no quantum fields to fluctuate. One must first have the microworld, or Jiffyland, as some have dubbed it, before quantum fluctuations can take place. Space must be imbued with zero-point energy in order for any atomic particle to exist. Hence, they are putting the cart before the horse. The truth is that there is not a single shred of evidence that can claim that quantum fluctuations can exist outside the matrix of space-time.

Prior to the Rapid Inflationary Model, most people paid little attention to this idea that an entire universe could inflate from a subatomic particle. However, while the mechanics of inflation may be possible, there can be no inflation without space-time. The transition from nothing is not answered in any way by the quantum fluctuation wish projection. Moreover, the inflation process still abides by the law of conservation and does not create new space out of nothing. It simply stretches the existing fabric of space-time.

Some now contend that the laws of physics created the universe. The idea that a law could exist outside our physical universe is, in fact, only a rational proposition if that law is essentially a reflection of the mind of God, who exists outside our universe. Laws are not independent entities that exist disconnected from matter, energy, and space. They are descriptions of how matter and energy behave in space-time. A law cannot exist prior to that which it describes, unless it exists in the mind of the engineer.

A law is a rational explanation of the behavior of some aspect of our physical universe. This naturally presupposes at least two realities: a mind to reason and a physical reality. How, then, can laws exist prior to the physical universe they explain outside of a rational mind and a physical reality?

The stipulation that laws are eternal is no less of a subjective speculation outside the sphere of our universe than belief in a supreme God. In fact, it is infinitely less rational. It is also worth noting that this naturalistic idea of a universe popping up from a quantum fluctuation is, by definition, jumping out of naturalists' cherished closed system presupposition and admitting that the universe is open.

It is simply nothing more than a metaphysical choice to deify scientific laws and avoid the obvious conclusion that their universal nature of space, time, and matter cannot be explained by random ordering. Evolutionists want their cake and also want to eat it. If they claim that the consideration of a God outside our universe would be an unscientific inquiry due to their doctrinal dogma of a closed system, then they cannot ignore it to allow any other things to exist outside our universe. This belief in a closed system automatically voids the consideration of a singularity popping up in a quantum fluctuation from outside our universe. Furthermore, it also negates their macrouniverse/multiuniverse speculation.

Contrary to naturalistic expectations of a randomly guided evolutionary process, even the laws within our universe are absolute truths; that is, they are universal truths. In a naturalistic universe without a God, these universal laws would be completely irrational since randomness and relativity would govern the formulation of laws in that universe. How could randomness and dysteleology produce a universe with absolute universal laws? It is an oxymoron.

We are told by academicians and the elders of the Darwinian priesthood that knowledge is what lifts people from squalor and moves them forward in their evolutionary path. It is the *what* questions that comprise the essentials of knowledge. But the *what* questions cannot be answered without four more questions: *where, when, how*, and, most important, *why*. It is the *why* questions that allow us to put all the other questions into a logical framework and consequently to extrapolate into the future and plan accordingly.

Imagine a universe without the *why* questions. How could even science exist without them? And yet, this is what atheism proposes. If there is no God and all things have happened without purpose and design, there is no *why*. So why is it that people instinctively ask *why*? Our basic human nature stands in stark contrast to this absurd proposition. It gives forth a powerful testimony against the idea that our universe has no purpose or design and that the *why* questions are simply irrelevant, as the evolutionist dogmatically insists.

Moreover, the quantum fluctuation that seems to pop up particles from nothingness is most likely popping them up, not from nothingness but from another dimension in our universe that happens to be invisible to us. According to the super string theory, there are three visible spatial dimensions in time for our universe as well as seven invisible spatial dimensions in time that coexist all around us. These mysterious particles popping up from the invisible dimensions are just as much a part of our universe as the visible particles that we can observe.

There is no magic to their appearance. They do not come from nothing or else the law of the conservation of energy would be null and void. Quantum fluctuations simply pass them from one dimension into another. None of these very real, albeit invisible, parallel dimensions existed before the creation of the Big Bang. All ten dimensions of our physical reality were created at the Genesis Singularity. The notion that something could magically pop up from nothing is not a scientific proposition but a metaphysical wish that uses scientific jargon as a magic wand and pixie dust to deceive the uninitiated.

If we are then to hold this view accountable to the scientific premise of repeatability, which is often used to discount the proposition that God created the universe as unscientific, then where are the new universes created by these quantum fluctuations? This is nothing more than an untestable, metaphysical stipulation that has not one shred of empirical data that can prove it.

The Naturalist Schizophrenic Tendency

Naturalists seem to become a bit schizophrenic when dealing with origins. In one breath, they claim that our universe is a closed system with unrelenting and absolute dogma. In another, they speak of hyperspace and speculate that there exists an eternal macrouniverse that is further filled with an infinite number of infinitely sized universes. I suppose it would look like a giant bathtub filled with soap bubbles, each universe being a particular bubble and the macrouniverse being the bathtub. At least this is how Leonard Susskind imagines it and describes it in his book *The Cosmic Landscape*.

Think about this absurdity for a moment. Naturalists claim that our universe is infinite in space and that there are also an infinite number of other universes outside of it. This, too, is an oxymoron. If our universe were truly infinite, there could be nothing outside of it, or, by definition, it would be finite. But they always want to have their cake and eat it, too. Take note here that their logic is not one simple mistake. If they were to believe that our universe is infinite in space and yet one more universe existed outside of it, they would be wrong once. But if they claim that there are an infinite number of universes outside our infinite universe, then they are subsequently infinitely wrong.

Of course, the real reason for these wild and unfounded speculations is simply to avoid the bothersome question of the origin of our universe from nothing. Their stipulation, to which they so ardently and dogmatically insisted on prior to Einstein's equations, has become a fly in their ointment. That is, prior to Einstein's equations, evolutionists had dogmatically declared as a scientific fact that our universe had no beginning and was eternal. That atheistic, metaphysical proposition simply pushed God out of the picture in their narrow deophobic minds.

Their macrouniverse speculation and their quantum fluctuation explanation for the origin of the Big Bang seek desperately to salvage their failed presupposition. That there is not a shred of scientific evidence that can substantiate this metaphysical speculation seems to not bother them one bit. In fact, it would be impossible for us to even explain the laws that would govern anything outside of our universe, but that does not seem to stop them, either. Just how is that more scientific than stating that our ordered and symmetrical universe is direct evidence that it is previsioned with purpose and functionality and created by the calculating mind of an infinite creator?

There is no way to rationally explain our origins from a materialistic point of view. Those who now claim that our universe could have been born from the clashing of two giant membranes, derived from an unwarranted and speculative extrapolation of the super string theory, fail to explain from where the first membrane came. All they are able to accomplish is to move the question of origins one step further back. Surely, it is obvious that this speculation regarding clashing membranes is completely outside our universe and is no different qualitatively than the proposition that God is the mind behind our universe. Science or, more appropriately, true science is not at odds with theological true truth.

The rude awakening forced upon them by Einstein that our universe had to have a beginning has completely pulled the rug out from underneath the feet of materialists. Nevertheless, attempts to do away with the Genesis Singularity continue unabated, so distasteful is this reality to the naturalists and their hallowed Copernican principle. Here is one example of this desperate quest that was proposed in a 1990 *American Journal of Physics* (52:2) article by Jean-Marc Lévy-Leblond. Speaking of this "un-begun universe" idea, Schroeder writes:

> *Levi-Leblond tells us that as we extrapolate our perception of time back to the beginning, an infinite amount of time would pass. That is, we would never reach the beginning if, and here is the key foundation of his argument, and only if the universe started as a point of zero spatial dimension. Being of zero spatial dimension, the ratio of the stretching of space between today and the beginning would be today's value divided by zero. Any finite number divided by zero, equals infinity. Time would in that case stretch back to infinity and so we could never reach the start. If the universe was created not as a point of zero dimension, but any finite size, no matter how miniscule that initial size might be, then the extrapolation would not lead to an infinite passage of time since the beginning. It might be a large number, but certainly not infinite. The universe in that case would have had a beginning in time. Levi-*

Leblond, being an honest fellow, acknowledges, in the following words, that in fact the best estimate is that the universe did indeed start with a finite size. "The commonly used models of the universe do have such an event horizon [a finite size at the beginning] so that the life time of the universe is finite." Note well that his entire thesis rests on his acknowledged nontruth (Schroeder 61–62).

You might ask, "Black holes have singularities with zero volume within our three-in-one visible spatial dimensions, so why could our universe not have begun with zero volume?" Well, do our black holes really have a singularity with zero volume? Think this through. If matter has true zero volume, then it is not there anymore. It is irrational to state that the universe began with zero volume. That is the same thing as saying there was no starting point, and hence our universe could not be. You can multiply zero by infinity and still have zero.

In regard to black holes having zero volume, how is it that the enormous gravity resulting from that compressed matter is still there if the matter disappeared into nothingness? We see the event horizon growing as matter is swallowed up. That expansive reach of growing gravity is evidence that the matter in the singularity has not disappeared into an absolute zero. So how can there be zero volume and an increase in the size, outward reach, and power of a black hole as it sucks in more matter?

It may be zero volume within our three visible spatial dimensions, but it is not zero volume in the seven invisible spatial dimensions parallel to our visible dimensions. That matter is so heavy that it has distorted our visible dimensions to the point that its escape into the invisible dimensions may be possible. Gravity may bleed through the invisible dimensions into ours, and so we still know that it is there, even if we cannot see it. If it were truly zero volume, then the law of conservation would be broken, and the event horizon would decrease until the effects of gravity, which provide us with the smoking gun proof that matter has not disappeared altogether, would no longer create an event horizon. The black hole, rather than growing, would simply vanish in a silent puff. Black holes would then not have the Hawking radiation because matter inside it would have disappeared from our universe. That is not what we observe, is it?

Are Black Holes Portals to Other Universes?

The Copernican principle of modern materialist cosmology always seeks to minimize the uniqueness of our universe and attempts to promote the idea that we are simply only one of perhaps an infinite number of universes. Never mind that such speculation is untestable and lacks a single justification by empirical data. Not only is it untestable, but there is no credible mechanism within our universe that could ever hope to jump out of our reality to empirically establish the existence of such a multiverse.

Brave attempts have been made to justify this wild speculation, but all are found wanting. Some such as Michio Kaku have suggested that black holes could be white holes on the opposite side and each could be creating another whole universe. He reasons that the Big Bang may have been a white hole, which is the opposite of a black hole, and therefore speculates that whole universes are being created by the millions through the black holes in our universe. It may very well be that our universe was created by a white hole, but there is no evidence that a black hole is connected to a white hole on the other end. But even more important, the physical evidence in our universe contradicts such a claim.

That idea may make a spectacular sound bite for their TV evolutionary propaganda machine and for wide-eyed university students who have been systematically brainwashed by their evolutionary mantra, but the empirical evidence simply does not correspond to this wild speculation.

It is theorized by some physicists that anything that enters the critical circumference of the horizon of a black hole will remain forever trapped and crunched into a singularity of infinite density and zero volume, a concept that is quite difficult to conceptualize within our continuum of three visible special dimensions in time. How can we describe an object that is supermassive and exerts an enormous amount of gravity but takes no spatial volume? Hence, they theorize that it may be actually lost from our universe. They conceive these singularities as portals into another universe.

So if we know that matter and energy are interchangeable and the law of the conservation of matter specifies that matter simply doesn't just disappear, then it is logical to assume that anything entering a black hole may simply enter into a portal of a different dimension from our visible space-time continuum. It is not a different universe, which is unconnected, but an invisible dimension of our own universe, connected to the dimension of our visible space-time continuum through the singularity.

If the matter going into the black hole did actually disappear from our universe, then the gravitational effect of its density would subside in our universe. But such is not the case. So it must be that it is still connected to our universe. But, on the other hand, it is quite possible that it has been pushed into a parallel dimension, and although it is invisible to us, who are stuck in the visible three spatial dimensions in time, it is still there. It still allows the gravity waves to seep back into our dimension.

Hence, this dimension, although invisible, is not undetectable. We know this because the force of gravity created by this zero volume singularity is so strong that not even light can escape its clutch in our visible three-in-one space-time dimension. We can observe the gravitational effect that it has on matter outside the black hole's critical circumference growing as matter enters the black hole. The more matter that it swallows, the larger the critical circumference becomes.

The critical circumference (horizon) of a twentieth-century black hole presents quite a different challenge. At no height above the horizon can one see any emerging light. Anything that falls through the horizon can never thereafter escape; it is lost from our Universe, a loss that poses a severe challenge to physicist's notions about the conservation of mass and energy (Thorne 139).

If, in fact, the opposite ends of our black holes were white holes, as Kaku suggests, then the matter sucked in from our universe would be dispersed in an ever-expanding universe on the other side. The moment that matter begins to expand and separate, the force of gravity exerted by the concentration of matter would immediately weaken.

That would mean that the black hole would lose the gravitational force to create its critical circumference in our universe. No such observable evidence has ever been found. Quite the opposite is the observational and empirical data that stipulate positively and absolutely that the critical circumference of all black holes expands as more matter is sucked into their singularity. The fact that the critical circumference does not diminish when more matter is sucked into the black hole means that this matter is not inflating in another universe but still trapped and crunched within an invisible dimension of our universe.

I am certainly not an institutionalized authority in nuclear physics nor do I claim to be; I am just a simple man trying to be rational. But if my humble powers of deduction don't fail me, it seems to me that there need not be a "severe challenge to physicist's notions about the conservation of mass and energy." I do not agree here with Kip Thorne. Matter is not lost from our universe, for it is obviously still connected to our universe, as evidenced by its effect on our perceived dimension within our universe. It is just lost from our three-in-one space-time dimension to which our finite senses are curtailed.

The very existence of these dimensions is evidence that our visible space-time continuum is, in fact, a much more comprehensive universe beyond our visible framework and not composed of only the sensible and material realty that naturalists insisted on during the Enlightenment. Evidence for this is found in another remarkable feature revealed from the geometry of Karl Schwarzschild, based on Einstein's theory of relativity, that for each star there is a critical circumference, which depends on the star's mass (the same discovered by Laplace and Michell, that equals 18.5 kilometers times the mass of the star in units of the mass of our sun).

If the actual circumference of a star just happens to be the same as its critical circumference, then the flow of time at its surface would stop, and light coming from its center would be shifted into a wavelength of infinite size. That is, it would be shifted out of existence from the dimension of our three-in-one space-time continuum.

> *What did not seem at all reasonable to physicists and astrophysicists of the 1920s, or even as late as the 1960s, was the prediction for a star whose actual circumference was the same as its critical one. . . .* For such a star, with its more strongly curved space, the flow of time at the star's surface is infinitely dilated; time does not flow at all—it is frozen. *And correspondingly, no matter what may be the color of light when it begins its journey upward from the star's surface, it must get shifted beyond the red, beyond the infrared, beyond radio wavelengths, all the way to infinite wavelengths; that is, all the way out of existence. In modern language, the star's surface, with its critical circumference, is precisely at the horizon of a black hole; the star, by its strong gravity, is creating a black hole horizon around itself* (emphasis added) (Thorne 133).

What an intricately complex universe we live in. How marvelous is the symbolism that God has given us in our universe to understand reality and His character as a loving God. You see, our death is like a black hole; we cease to exist in this dimension. But we are still in this universe, and the gravitational field of the black hole is the evidence of this truth. The death of a star parallels the death of a soul.

The more we learn about our universe, the more the naturalist is forced to face the miraculous nature of the existence of matter and of life in our universe. And consequently, through understanding what was once considered to be supernatural, becomes quite natural.

But naturalist, in their refusal to accept an open universe, are forced to conceptualize a macrocosm that is riddled with separate universes. This is not because the scientific inquiry has pushed them to this conclusion, but because of their subjective bias that subliminally urges them to hedge the odds a bit and underplay the miraculous nature universally exuded by the wondrous properties of our creation.

Ghost Worlds

Others, like Paul Davies, see the grandness of the design of our universe, and rather than admit the obvious, they resort to multiple redundant universes in an attempt to fix the odds. But instead of fixing the odds, they simply complicate the matter more by making our creation even more special in light of the rest of reality.

> *This raises the philosophical problem of why nature has so much redundancy built into it. Why produce so many universes when all but a tiny fraction goes unnoticed? If, instead, the other universes are relegated as ghost worlds,* we must regard our existence as a miracle of such improbability that it is scarcely credible. Life is then indeed chancy—more chancy then we could ever conceive (emphasis added) *(Davies 14).*

How ironic that the naturalist, who so vehemently scoffs at the notion of the existence of a miraculous creation, can so easily accept the much more miraculous improbability of a chance evolution of matter and life and of multiple ghost worlds. Our ignorance is only exceeded by our arrogance.

Humanity has been given the evidence that undeniably points to a creator and sustainer of this magnificent and diverse universe, who created order in the midst of disorder. This is, as Hugh Ross asserts, the fingerprint of God visible for all who care to know, for those who are true scientists in search of truth.

It is an inescapable truth that as we learn more about this magnificent universe, we are forced to open our mind to the reality of an unseen parallel extension of our universe, which the scriptures call the realm of the spirit. It is not an ethereal imaginary mythical world, but a very real continuum, albeit an invisible dimension of our reality.

The antiparticles, which can be deduced by objective mathematics and perceived through the influence generated upon particles that exist in our visible dimension, prove to us that there are designed realities beyond our ability to perceive. The spirit world or dimension is not the antiworld, but the antiworld certainly proves to us that we cannot summarily discount the existence of such a dimension on the grounds that we simply cannot perceive it.

The Standard Model of Particle Theory plainly predicts the existence of virtual particles that continuously arise spontaneously in our visible dimension and annihilate quickly through contact with an antiparticle. This process creates what physicists term *quantum foam* in their description of space at the Planck level.

Take a box of iron filaments and spread them over a glass table. Take a strong magnet and hold it to the back of the glass. You will see electromagnetic field lines as the filaments arrange in loops over the positive and negative ends of the magnet. Had the filaments not been there, the field lines would have still been there. They may be invisible to our eyes and senses, but they are real nonetheless.

It is very sad that otherwise brilliant men, because of their evolutionary bias, must evade the obvious inference that any objective analysis of the universe, by necessity, impresses upon the observer. Our universe in its entirety is a magnificently designed system, which consequently also begs the obvious inference that it must have had a super genius designer. But this is unacceptable to their initial subjective presupposition that there is nothing that can exist outside our visible and testable material existence.

Consequently, naturalists are forced to make an enormous leap of faith in order to believe that our highly structured, pervasively ordered, and incredibly intricate multilayered material universe developed as the result of mere random chance processes. Their naturalistic cosmology renders this universe and everything that exists within it as nothing other than chemical serendipity.

But this is not the only place where evolutionists must take a leap of blind faith, for there is an even larger hyperleap to take; that is, the evolution of life through random chance chemical processes. But I will deal with this later on.

CHAPTER 13

● ● ●

THE PLASMA UNIVERSE

Somehow, this Genesis Singularity began to expand. As it stretched outward, the density of that radiant energy decreased, and time began to flow. That was the beginning of time. But because the curvature of space-time was so great, time crawled very slowly. As space-time stretched, time began to flow a little faster. But from that very first moment of time, all the parameters that allow our universe to exist as it does now in order to be capable of inhabiting life were somehow chosen and set. The potential for the eventual outcome was decided in the very first microsecond.

The evolutionary religion believes by faith that this was simply a random happenstance. But randomness has never created order, and it never will. It will become evident to the objective mind that our universe was created by a successively ordered regimen of deliberate and calculated actions that were previsioned and executed by an infinite mind of infinite power. No other rational option is possible. Thus, the Alpha is connected to the Omega by purposeful design. Therefore, to begin at the beginning, we must first go to the end

371

in order to see the purpose of His beginning. We have begun our discussion of the Alpha Point, but the scriptures also speak of an Omega Point.

The Alpha and the Omega

Our Judeo-Christian cosmological model states that this universe began a finite time ago and will end a finite time from now. It is a temporal universe and not the permanent home of humans. In the biblical book of Revelation, God tells us that our universe will be destroyed, and God will make a new heaven and a new earth. It will be radically different from our present system.

> *Then I saw a new heaven and a new earth, for the first heaven and the first earth had passed away, and the sea was no more (Revelation 21:1).*

The Earth will be destroyed because it was profaned by sin. The new universe will not have the second law of thermodynamics operating in it. It will be an eternal universe. It will be a universe created after the destruction of this present universe that must be atoned for by a holy God. This will take place in the final destruction after the Battle of Gog and Magog at the end of the Davidic Kingdom in the third earth. In that day, death shall be thrown into the lake of fire, and "there shall be no more curse" (Revelation 22:3 KJV). In that day, we shall be as timeless as radiant energy. We shall be in an eternal now.

> *He will wipe away every tear from their eyes, and death shall be no more, neither shall there be mourning, nor crying, nor pain anymore, for the former things have passed away (Revelation 21:4).*

In this new world, the fourth and final earth, God shall walk among us. Angels shall walk among us. We shall see Him face to face and will no longer need temples in which to worship Him. It will be a universe without a sun to give light to the Earth.

> *And I saw no temple in the city, for its temple is the Lord God the Almighty and the Lamb. And the city has no need of sun or moon to shine on it, for the glory of God gives it light, and its lamp is the Lamb. By its light will the nations walk, and the kings of the earth will bring their glory into it, and its gates will never be shut by day—and there will be no night there (Revelation 21:22–25).*

There shall be no night there. The Lamb shall be our sun. His glory shall illuminate us. So we see here that God's radiant energy shall accomplish everything that is necessary in order for us to live eternally. It was that same radiant energy that brought this universe into existence. The sun is simply a substitute source of energy, designed by God as an artificial source of light to give us life. In fact, His glory shall make plants grow and photosynthesis take place. Here, fruit trees and rivers of magical lore, known to modern man only in the myths of ancient nations, shall become our sustenance and source of eternal life.

> *Then the angel showed me the river of the water of life, bright as crystal, flowing from the throne of God and of the Lamb through the middle of the street of the city; also, on either side of the river, the tree of life with its twelve kinds of fruit, yielding its fruit each month. The leaves of the tree were for the healing of the nations. No longer will there be anything accursed, but the throne of God and of the Lamb will be in it, and his servants will worship him (Revelation 22:1–3).*

There are three symbols used for Christ in the scriptures that are directly linked to life. This is no accident, for Christ's number, the visible member of the triune Godhead, is three.

1. Light:

   In the beginning was the Word, and the Word was with God, and the Word was God. He was in the beginning with God. All things were made through him, and without him was not any thing made that was made (John 1:1–3).

   Again Jesus spoke to them, saying, "I am the light of the world. Whoever follows me will not walk in darkness, but will have the light of life" (John 8:12).

2. Water:

   On the last day of the feast, the great day, Jesus stood up and cried out, "If anyone thirsts, let him come to me and drink. Whoever believes in me, as the Scripture has said, 'Out of his heart will flow rivers of living water'" (John 7:37–38).

3. Manna (bread):

   Jesus answered them, "Do not grumble among yourselves. No one can come to me unless the Father who sent me draws him. And I will raise him up on the last day. It is written in the Prophets, 'And they will all be taught by God.' Everyone who has heard and learned from the Father comes to me—not that anyone has seen the Father except he who is from God; he has seen the Father. Truly, truly, I say to you, whoever believes has eternal life. I am the bread of life. Your fathers ate the manna in the wilderness, and they died. This is the bread that comes down from heaven, so that one may eat of it and not die. I am the living bread that came down from heaven. If anyone eats of this bread, he will live forever. And the bread that I will give for the life of the world is my flesh." The Jews then disputed among themselves, saying, "How can this man give us his flesh to eat?" So Jesus said to them, "Truly, truly, I say to you, unless you eat the flesh of the Son of Man and drink his blood, you have no life in you. Whoever feeds on my flesh and drinks my blood has eternal life, and I will raise him up on the last day. For my flesh is true food, and my blood is true drink. Whoever feeds on my flesh and drinks my blood abides in me, and I in him. As the living Father sent me, and I live because of the Father, so whoever feeds on me, he also will live because of me. This is the bread that came down from heaven, not like the bread the fathers ate, and died. Whoever feeds on this bread will live forever" (John 6:43–58).

These are the three necessary staples that constitute life. But all of these earthly things are simply substitutes made by God in order for us to be able to live in this present universe. Without light, there is no photosynthesis, no food chain, and no oxygen for us to breathe. Without water, no life could exist. In fact, we are made mostly of water. Without food, no one could survive. Life depends on these three staples. And yet the very last chapter of the New Testament tells us that in this new heaven and new earth, there will be no more sun or stars, and there will be no more ocean.

If you carefully read chapter 22 of the book of Revelation, you will find that there is a river of living water that flows from the throne of God. Hence, we shall no longer need an ocean to provide the meteorological water cycle in order for us to survive. Furthermore, there will be no more night, for God shall be the light of the new universe. So the sun is no longer necessary. And there will be 12 trees of life to feed us and give us eternal life, and the leaves are for the healing of the nations. Hence, we will no longer need any food. This universe is but a shadow of the real universe that has been planned by our sovereign God. He is all we need.

Now if you are an atheist, then all of this is rubbish to you. My advice to you is to keep an eye out for the signs of the end of our second earth. Before the end of our age, Elijah shall return, and he will have the power to stop the rain. I would urge you to know exactly what the four horsemen of the Apocalypse will bring, for your life will depend on it. The day will come when you will not think our warning to be superfluous fantasy.

But if you are a believer, then you should take notice that our future universe will function without suns and stars. These are artificial candles created by God to mimic His glory. It was His glory that began the universe and created these artificial lamps we call stars to provide for us a continuous source of energy that allows life to exist on planet Earth. In the future, these lamps will no longer be necessary. His radiant glory will illuminate the new heaven and the new earth:

> And the city has no need of sun or moon to shine on it, for the glory of God gives it light, and its lamp is the Lamb (Revelation 21:23).

> And night will be no more. They will need no light of lamp or sun, for the Lord God will be their light, and they will reign forever and ever (Revelation 22:5).

Modern science says we are the children of stars, made out of stardust. That is not exactly true. It was God and His radiant energy that created our universe, the sun, the Earth, and humanity. Although stars can generate the higher, more complex elements in our universe, this process was already going on before the stars were made, through the intense energy of the glory of God. The process began when our universe was nothing more than a giant ball of spinning plasma as enormously intense electric currents called Birkeland currents began the process of turning plasma into matter. We will address the particulars of this process in the next few sections.

The evolutionary speculation that relies on the long, slow process of coalescing matter through the force of gravity is infinitely inferior to the rapid process of creating solid matter from plasma through Birkeland currents at the very beginning of our early universe. For this reason, the Genesis account cannot be stuffed like a stuffed cabbage with any versions of the slow, gradual evolutionary history. The naturalists not only lack the starting mechanism, but they also have the wrong order and timetable of creation. In essence, they have the cart before the horse. There is no reconciliation between naturalistic evolution and the scriptures. But there is a correlation between God's timetable in Genesis and the age of our universe that true science corroborates, as we shall now see.

The Unique Powers of Plasma and the Electromagnetic Force

All physical matter in our universe is exhibited in four distinct states: solid, liquid, gas, and plasma. In the solid state, the atoms are densely packed together, forming a rigid, more ordered condition with a defined shape. In the liquid state, the atoms are less densely packed and in a more disordered, less defined shape. Matter, then, loses its rigidity, becoming fluid and acted upon by gravity flows to the lowest point, taking the shape of whatever container it is in. Nevertheless, the material maintains its atomic integrity. In the gas state, matter is even less densely packed than in the liquid state. Although the atoms still maintain their atomic integrity, they are so scattered that they move randomly in three-dimensional space, diffusing it even more. Hence, gases have the tendency to expand and dissipate to whatever volume is available to them.

H_2O, for example, can be found in the solid state as ice, in the liquid state as flowing water, and in the gas state as water vapor. By applying heat, the molecules containing more energy begin to bounce against each other, and the volume expands as rigidity is reduced.

If heat is taken out, then the opposite takes place. When the water molecules reach a particular temperature called the freezing point, the molecules become locked together in a rigid form as a solid, which we call ice.

In other words, as we add heat, the molecules begin to move around with more energy, losing their rigid formation and becoming a fluid, which we call water. If we add more heat, the molecules in the liquid begin to bounce against each other with even more energy, which causes them to spread further apart and turn into steam. All elements in our universe can be turned from solid to liquid and then to gas. Although the freezing point and boiling point of each element is unique to its atomic characteristics and structure, they are all affected by the energy of heat; each has a freezing point, and each has a boiling point.

If we add even more heat to the gas state, we reach the plasma state. This fourth state, however, is unique from the other three states. Not only is it the most disordered of the four states, but, more significantly, in this condition, the atoms actually lose their atomic integrity. That is, the atoms are broken down into separate particles. The electrons are basically ripped from the nucleus, leaving them with a positively charged nucleus that is separated from the negatively charged electrons. The fact that in this plasma state the particles have electrical charges causes it to be uniquely susceptible to electromagnetic forces in a way that the other three states are not.

Most people are not aware that the vast majority of the universe, except for a small fraction, is in a plasma state. If we were to add the mass of all objects in our solar system (planets, moons, asteroids, comets, rings, etc.), it would represent less than 0.1 percent of the mass of our sun. Since everything in the sun is in the plasma state, then 99.9 percent of our solar system is literally in the plasma state. All the stars in our universe are in the plasma state. The truth is that most of our universe is still in plasma form. It is still in the form in which God's radiant energy created it at the very beginning of our universe. Only a tiny fraction has been changed into the three other states of matter that allow our universe to be inhabited by living things.

If the universe is largely a plasma universe, and if matter in this plasma state is significantly more affected by the electromagnetic force, then it stands to reason that the electromagnetic force has impacted the formation of our universe in a more significant way than the tediously slow and gradualist effect of gravity, as our evolutionary cosmologists have traditionally insisted. Not only is the electromagnetic force more capable of affecting the unique electrical properties of plasma than gravity, but also it is many times more powerful than gravity.

The electromagnetic attraction between the proton and the electron is about 10^{39} times stronger than the gravitational attraction between them. In other words, the electromagnetic force is 10^{39} powers of magnitude greater than gravity. Were it not for that powerful force, the atoms in the other three states could not remain in a stable configuration, and their shapes could not maintain structural integrity. It is because of this structural integrity and the electrical charges they contain that atoms can unite with other atoms to create molecules without melding into one another.

It is also what enables them to link with other atoms as the electrons in their outermost shells create covalent bonds with other atoms. Had the electromagnetic force been weaker, the atoms would not be able to maintain their structural integrity, which gives them the particular and unique properties for each element. The electromagnetic force is calibrated to a very specific power, which could not have been randomly formed. Had it been created a tiny bit stronger, then atoms would not be able to unite with other atoms to create covalent or any other chemical bonds since the atoms could not borrow the electrons in the outer shells to make these bonds. Each atom would remain isolated from any other atom. Molecules could not form.

Here is the great dilemma for evolutionists. How did random ordering choose this exact power to the electromagnetic force from all the other potentials so that atoms could take form and be able to unite in larger molecules? Had it been weaker, the atoms could not form at all. The very structure of the atom would disintegrate. Had it been stronger, the structure would be bound so tightly that it could not form molecules. All things in the universe would be composed of single atoms that would be flying freely in a soup of atoms that could not coalesce.

The upshot of this fine-tuned parameter, which is absolutely essential in order for matter to exist as it does in our universe, is that it must have been carefully chosen from the very beginning of our universe, or life and all the celestial objects could not have been created. Thus, we can now understand that this precisely chosen electromagnetic force is not only 1,000,000,000,000,000,000,000,000,000, 000,000,000,000,000 times greater than the force of gravity, but also that this specific difference between the force of gravity and the force of electromagnetism had to have been a previsioned design. Without this specific symmetry between them, matter could not exist as we know it.

What are the statistical chances out of all the potential parameters that a randomly ordered universe could have formed this exact symmetry between the two forces? Just exactly how could random, sheer luck produce such precise symmetry between all the four forces if they had been accidentally created? The leap of blind faith necessary to believe this is astronomical.

If the universe is 98 percent plasma, and since the electromagnetic force is so much more powerful than gravity and on top of that plasma is so much more sensitive to the electromagnetic forces because the particles are broken into electrically positive and negative charges, then it only stands to reason that we ought to take a much closer look at the role electromagnetism could have played in the formation of our early universe.

Modern evolutionary cosmology has given gravity a much larger role than it deserves in the formation of our stars and galaxies. It is not gravity that holds our sun in orbit around the Milky Way. It is not gravity that keeps Alpha Centauri and the sun in their positions in respect to one another. It is the powerful electromagnetic force that actually holds our galaxies in the specific formations, and it is much more probable that it is the venue through which God created solid matter, as we shall see.

The Sun has a mass of 2,000,000,000,000,000,000,000,000,000,000 kilograms. That is 330,000 times greater than the puny mass of the Earth. Our nearest star, Alpha Centauri, has 1.1 times greater mass than the sun. They are, however, separated by 4.25 light years distance. That means that if we were traveling at roughly 386,000 miles per second, it would take us 4.25 years to arrive there.

The formula for the force of gravity between two objects is the mass of the first object minus the mass of the second object, divided by the distance between them squared. Because the distance component is in the denominator and it is squared, that means that the greater the distance between the objects, the power of gravity is diminished by the square of that distance ($M_1 - M_2/d^2 = G$). That means that the force of gravity decreases at an enormous rate the farther the objects are apart.

In order to better grasp what this enormous distance is between these two stars, let us use a comparable scale that would be in more understandable terms. The average distance between the sun and the Earth's orbit is called an astronomical unit (au). Let us reduce these 93 million miles of a single au to the distance of 1 inch. In such a scale, the sun would be the size of a single pencil dot on a piece of paper, and the Earth would be a much smaller and almost imperceptible dot to the naked eye located one inch away. Alpha Centauri would be another pencil dot a tiny bit bigger than the sun. That is, it would be about 1.1 times larger than the pencil dot that represents the size of our sun. These two tiny dots would then be separated by a whopping distance of four and a half miles of space between them.

If we follow the equation that allows us to calculate the force of gravity between the sun and Alpha Centauri, we then subtract their masses and divide that number by the distance between them squared. What possible impact could the 0.1 difference between the mass of these two tiny dots produce in gravity when, for every inch that separates them along those four and a half miles between them, that force is diminished by the square of every single inch along the way? To say that it is negligible is an understatement. Gravity cannot be the major force that holds our galaxies together. That force no doubt must be the enormous electromagnetic fields that are generated by the rotating Milky Way in spiraling arms that stretch outward from its center.

Let us suppose a different scale where the size of the sun, instead of being a pencil dot, would be the size of a golf ball. Alpha Centauri would then be another golf ball a tiny bit larger, but the distance between them would be a whopping 700 miles. That means that if the golf ball representing the sun were to be placed in Miami, Florida, the golf ball representing Alpha Centauri would have to be placed in Atlanta, Georgia. I have made that long drive before when I attended Georgia State University. It takes me 12 hours to drive straight through from Miami to Atlanta. What possible gravitational impact could either of these two tiny golf balls have on each other 700 miles apart?

If the sun were the size of a golf ball, then the farthest edge of our Milky Way galaxy would be 15,555,555 miles away from that golf ball. I think it is obvious to any objective mind that such distances would make the role of gravity insufficient to hold the galaxy together. Because of the inverse square property of the gravitational force, the impact of Alpha Centauri on our sun is less than the impact of the mass of a butterfly on the mass of the Earth.

But the electromagnetic force is 10^{39} power stronger than gravity, and its ability to impact plasma material is increased by the charged particles created in the plasma state. Moreover, electromagnetic forces are both attractive and repulsive, while gravity is only attractive. They have the power to hold the stars in place. The electromagnetic fields generated by our spinning galaxies span as fingers into space and hold its stars within the spiral formations created by the giant Birkeland currents that initially created the galaxies and set them spiraling in motion.

If gravity were the principle force to create the galaxies and hold them together, we would have long ago lost all our galaxies in the universe. In other words, if these galaxies existed for millions of years and had they been held together by only gravity, all the stars we find in our Milky Way would have long ago been sucked into its center and into one giant supermassive black hole empty of any stars around it. It is the electromagnetic force that keeps them from rapidly becoming black holes.

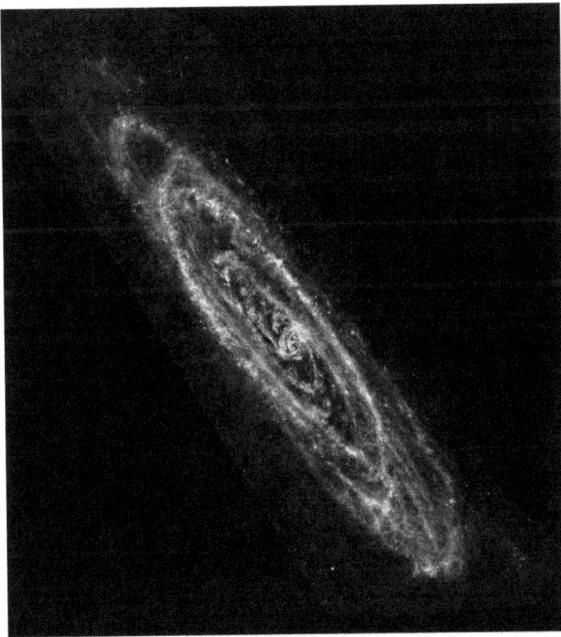

Infrared picture of Andromeda showing electromagnetic structures that shape the galaxy

It is the unique attractive and repulsive powers of the electromagnetic force that maintain the integrity of our galaxies as they do the atoms. They do so by holding stars in their particular positions within these filaments of electromagnetic fields. This electrical component becomes visible in infrared pictures taken of Andromeda, our closest galaxy neighbor in space.

Since all matter and energy contained within our entire universe existed in plasma form immediately after the Big Bang, then it stands to reason that the knowledge of plasma physics should become foundational to understanding how our universe was created in its present form. Understanding the behavior of the electromagnetic force and the effect it has on plasma material is critical to understanding the processes that birthed our early universe and the formation of stars and galaxies.

Not All Plasma Is Created Equal

Because the electrical charge of any given atom is no longer bound as in the solid, liquid, and gas state, and therefore no longer electrically neutral when fractured into the plasma state, the assemblage of the positive and negative components becomes capable of performing collective motions of great vigor and non-intuitive complexity. That is, they react in ways that one would not normally expect in the other three states of matter, as we shall see. In other words, these positive and negative particles become highly reactive and heavily influenced by the electromagnetic force so quite magnificent and wondrous transformations were able to take place in that enormous spinning plasma ball of our infant universe.

But let us first better understand what this unique plasma state in matter looks like. Not all plasmas are created equal. There are distinct plasma modes, which exhibit different properties:

1. Dark Current Mode: In the dark current mode, plasma emits no light. We can find an example of this in our ionosphere. The ionosphere is a layer of the Earth's atmosphere that

contains a high concentration of ionized gases and free electrons. Ionized gases are gases that have been stripped of some electrons and, therefore, have a net positive charge, while the free-floating electrons have a negative charge. This is a form of plasma that is located above the mesosphere and extends from about 50 miles to roughly 600 miles above the surface of the Earth. We find in this dark current mode of the plasma state a low current flowing and, therefore, no emission of light because the low current is not enough energy to excite the emission of photons.

It is nevertheless because of the unique electromagnetic properties of the ionosphere that we are able to bounce radio waves from it to reach around the world. Radio waves are electromagnetic waves that contain the longest wavelength in the electromagnetic spectrum. The intensity of this current in the ionosphere fluctuates. Any ham radio enthusiast can tell you that on some days, the ionosphere functions well, and at other times, it does not.

I learned this when I was just 13 years old. My godfather, Billy French, invited me to spend a few weeks in Princeton, New Jersey, with him. Billy was a childhood friend of my mother and had lost both legs below the knee in a landmine during World War II. Wearing his prosthetic legs, he would take a walk every morning, and it was all I could do to keep up with him. His father, Bill French, was an electronic engineer and a ham radio enthusiast. I will never forget seeing that huge antenna and asking him, "What kind of TV do you have that you need an antenna so big?" He laughed and said, "That is my radio antenna."

I responded, "Your radio needs an antenna that is bigger than your TV?" He laughed again and said, "Come let me show you." We went into the basement, and he took me to the ham radio console. "This," he said, "is a ham radio. If the ionosphere is good, I can talk to people with other ham radios all around the world." He went on to explain to me how the radio waves were bounced back down by the ionosphere and subsequently picked up by antennae of other ham radio enthusiasts around the world. That day, I spoke to a man from Australia. I was fascinated by the fact that my voice was actually traveling by radio waves through the atmosphere, all the way from Princeton, New Jersey, to Australia.

2. Normal Glow Mode: In the normal glow mode, such as in auroras, emission nebulae, the sun's corona, comet tails, and fluorescent bulbs, the ionized gases acted upon by a stronger current cause the plasma to emit light. When electrical currents from our sun strike the ionosphere through weak areas of our magnetosphere (such as in Earth's poles), they cause the ionosphere to light up. This creates the phenomenon we know as the aurora lights. The colors are indicative of the specific ionized gases that are being excited by the solar currents. Their brightness depends on both the density of the plasma and the intensity of the current. The many colors of neon lights used by advertising billboards depend entirely on which ionized gas is being used.

The Aurora Borealis shows the glowing ionosphere as currents flowing through the electromagnetic field lines fire up the plasma with a dazzling display of dancing lights.

The different colors correspond to the specific ionized gases that are being excited by the solar currents.

3. Arc Mode: In the arc mode, the higher intensity of the currents causes an emission of photons with extreme brilliance and a wide spectrum. Sparks emitted by fires are also a form of the arc mode. The flame in fires is also in plasma form. Anyone who has seen a blown electrical transformer can attest to the awesome power of the plasma arcs they produce. During a typical summer thunderstorm, while I was working at the Miami Beach Fire Department, my rescue unit was dispatched to a downed electrical wire. When we arrived on the scene, we saw that one of the high-voltage wires was broken. It was raining buckets, and the wire seemed to be a live snake as it bounced on the sidewalk. The arcs coming from the wire would hit the wet sidewalk and turn the water into steam with an explosion. As water turns to steam, it expands 1,700 times in volume, causing repeated explosions as the wire danced in the air like an angry living thing. It was a light show and fireworks combined. When the grid was finally turned off, I inspected the sidewalk and found that every time the wire had arced and struck the sidewalk, it had turned the concrete into glass.

Lightning is another example of an arc current that flashes with extreme brilliance and enormous energy. Anyone who has seen the impact of a lightning bolt can attest to its immense power. Living in Florida, one of the lightning capitals of the world, I saw such an impact during a typical afternoon thunderstorm just three houses down from where I lived. The bolt struck a huge tree, and the trunk exploded like a bomb, splitting the tree in half and leaving the pieces of the trunk smoldering in the rain.

Arc welders are capable of melting metals to bind them together. But the brilliance is so energetic that eye protection must be used, or the extremely brilliant light will burn the retina of the eyes of anyone that looks directly at it. These arcs are characterized by twisting filaments and the emission of ultraviolet light as well as radio waves.

The sun's photosphere is also characterized by arc plasma. We can see these enormous arcs of plasma following precisely through the electromagnetic fields generated by the churning, rotating plasma in the sun. When these twisting arcs break, they release massive coronal mass ejections that travel outward from the sun. The complicated web of looping electromagnetic field lines in the sun shows us that the electromagnetic force may be the principle force that governs the processes within our sun.

The Discovery of Our Electrical Universe

In very high-intensity electrical currents flowing through plasma, a unique phenomenon develops. These currents, called Birkeland currents, will take on a corkscrew spiral shape that resembles a double helix, almost like DNA. The Birkeland currents were named after their discoverer, Kristian Birkeland (1867–1917). Born in Oslo, Norway, this brilliant scientist was the first to propose that an aurora was caused by the emission of electrons from sunspots that flowed through magnetic fields around the Earth to light up the ionosphere. Considered as the first space scientist and the father of plasma physics, he was nominated for the Nobel Prize seven times. Revered

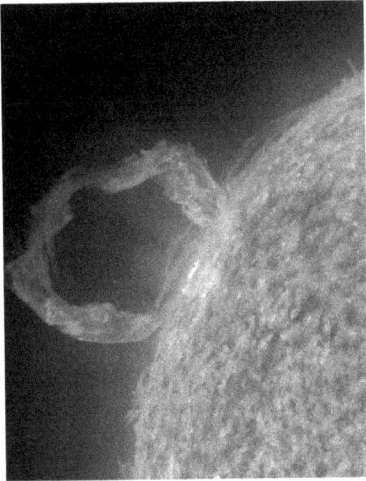

Plasma loops following the electromagnetic field lines generated by the sun.

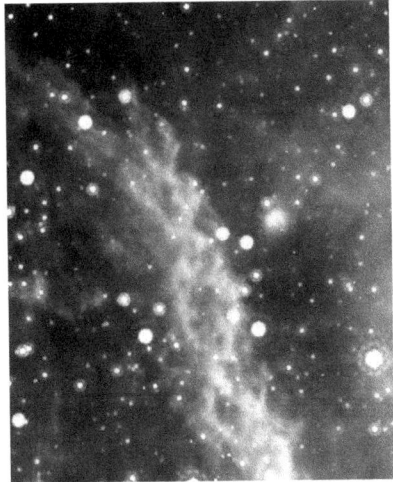

The twisting currents in a thunderbolt of lightning

in Norway, his picture still graces the 200 kroner bill, which also contains the schematic of his experiment; quite appropriately, it can only be seen in ultraviolet light.

Birkeland led several expeditions in northern Norway that established a network of observatories to collect magnetic field data underneath the auroras. From 1899 to 1900, he led the Norwegian Polar Expedition that first determined the global patterns of electric currents by ground magnetic field measurements. The brilliant scientist proposed that the polar electric currents were connected to a system of global currents that flowed along geomagnetic fields. In his book *The Norwegian Aurora Polaris Expedition 1902–1903*, he provided a diagram of the field-aligned currents. These were highly disputed by many scientists of his day, but in 1963, a U.S. Navy navigation satellite directly observed these predicted magnetic disturbances with the aid of a magnomanometer and finally forced the scientific community to confirm his insights. Birkeland's prediction that currents emanating from the sun powered the Earth's electrical currents flowing through magnetic field lines was also confirmed in 2007 when the Themis satellite discovered that there were actual electrical ropes that connected the Earth to the sun.

Our entire scientific establishment in the Western world has been largely governed by gravity physics from the time of Isaac Newton. In spite of the many advances in electromagnetic physics, the vast majority of evolutionary-minded physicists remain gravity-based because of their underlying bias toward gradualism. Our scientific understanding of the universe has moved on, but they have not.

The term *plasma* and the understanding of the unique plasma state of matter were not even known until the last century. It was Irving Langmuir (1881–1957), an American scientist, who first coined the term *plasma*. He discovered that currents flowing through plasma created a double-sheath layer. The electrical currents in plasma isolate one section from another by a wall of two closely spaced layers, one of positive charges and the other of negative charges. This double layer is where the strongest electric fields in plasma reside. In other words, the plasma literally isolates itself from foreign intruders by surrounding itself in this double layer as a protective sheath. If one would plunge a sphere into the plasma, it would surround it in this double-layered sheath like a cocoon. The almost lifelike organizing powers of currents through plasma made Langmuir think of it as similar to blood plasma, and for this reason, he named this unique, bizarre state of matter plasma. It is hard to overstate the magnificent qualities that large currents produce in plasma, as we will now see.

It seems that Scandinavian countries, with their interest in the magical-looking auroras, became the foremost researchers in the area of plasma physics. In 1963, Swedish scientist Hannes Alfvén (1908–1995) became the first to predict, as a result of his studies in plasma, that the large-scale structure of the universe was filamentary. Alfven is considered one of the founding fathers of plasma physics. His book *Cosmic Plasma*, published in 1981, provided the basis for the consideration of electromagnetism in plasma as the main mechanism for the creation of galaxies and stars. Alfvén won the

1970 Nobel Prize in Physics for his work in the theory of plasmas, called magnetohydrodynamics.

Work modeled after Alfvén's ideas, later done by Anthony L. Peratt in 1980, using supercomputers at Maxwell Laboratories and later at Los Alamos National Laboratory, created results virtually indistinguishable from the structures of actual galaxies. The models were computed as two interacting, adjacent Birkeland filaments produced a flat rotational curve that created spinning formations, which were surprisingly identical to present galaxy shapes observed through our telescopic instruments. The simulation was conducted with two spherical plasma clouds in parallel magnetic filaments with a powerful current of 10^{18} amperes running through them. The computer simulation showed that the clouds began to spin around one another, creating a spiral shape that literally mimics the very shape of our Milky Way galaxy.

Peratt's large-scale simulations of the Maxwell-Lorentz equations yielded tantalizing evidence that our galaxies were largely shaped by the electromagnetic force in our early plasma universe. His book *Physics of the Plasma Universe* is a must-read for any who wish to study cosmology. But old paradigms die hard, and most Western scientists still under the spell of the evolutionary gravitational model are not predisposed to consider an alternative to their gradualist dogma and their cherished inflationary model. Sadly, their metaphysical worldview impedes them from considering more advanced knowledge gained through modern plasma physics.

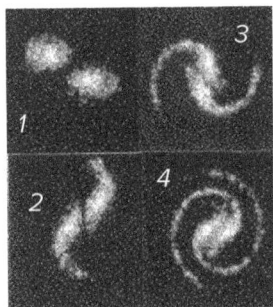

Anthony L. Peratt's Birkeland current simulations in plasma using super-computer capabilities. The results are virtually indistinguishable from actual galaxies.

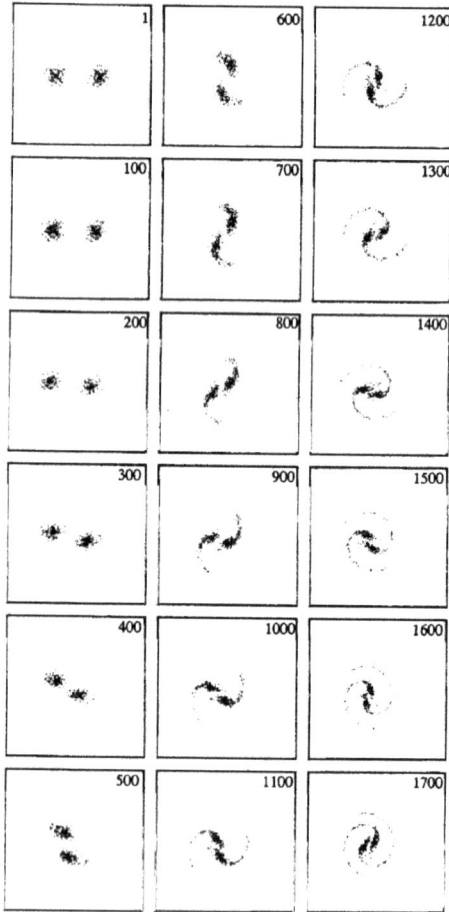

The above diagram is from http://plasma-universe.com/file: peratt-galaxy-formation-simulation.gif can be read starting from the top of the left column and advancing down to the bottom of the left column, followed by the top of the middle column advancing down to the bottom, and then followed by the top of the right column and advancing down to the bottom. The frame-by-frame illustrations show how Birkeland currents can create galaxy-shaped structures.

Here is the curious thing about the non-intuitive electromagnetic forces—the forces of the electromagnetic fields are orthogonal to the field. That is, they are perpendicular to the field. If we picture a proton moving through a three-dimensional magnetic field, we may gain a clearer understanding of this phenomenon.

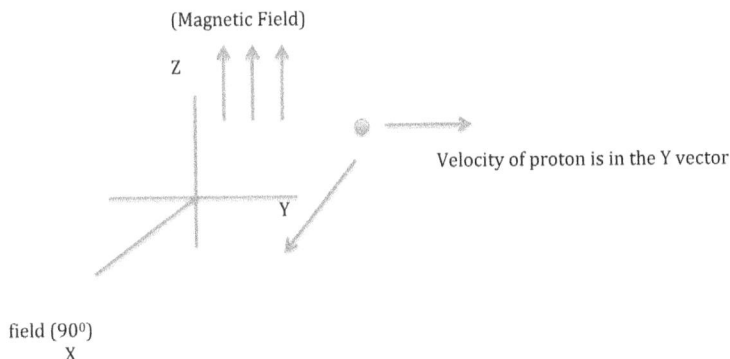

(Magnetic Field)

Z

Velocity of proton is in the Y vector

Y

field (90⁰)
X

The magnetic force of the proton is in the X vector, which is orthogonal to the magnetic field (900)

Perhaps the dipole, quadrupole, sextapole, and octopole of anisotropy discovered by our mapping satellites COBE, WMAP, and the European Space Agency's Planck owe their symmetrical positions to the electromagnetic process that created our universe. These satellites measured the microwave background radiation in the universe, which formed during the creation process, and they clearly show a previsioned design to our large-scale structure of the universe, which points to our sun-Earth ecliptic as the center of this magnificently designed universe. Perhaps the early plasma universe was injected with giant high-intensity Birkeland currents that were responsible for the formation of seven distinct bands of galaxies that are distributed in discreet quantum areas arranged in concentric circles around our solar system, as we previously documented. This unique seven-staged shell structure of our universe was discovered by Tifft through the study of the red-shift analysis of our entire universe. Because his findings completely refute the evolutionary notion that our universe is randomly ordered, it has simply been ignored. Just Google it, if you do not believe me.

Let us now take a closer look at these wondrous Birkeland currents and the amazing powers they harness.

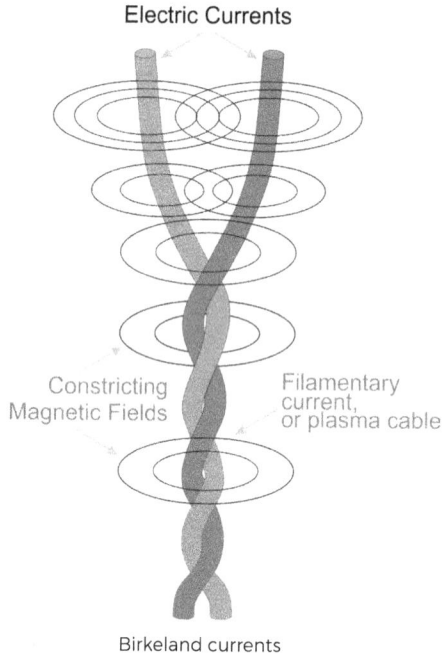

Electric Currents

Constricting Magnetic Fields

Filamentary current, or plasma cable

Birkeland currents

The Creative Power of Birkeland Currents

At once I was in the Spirit, and behold, a throne stood in heaven, with one seated on the throne. And he who sat there had the appearance of jasper and carnelian, and around the throne was a rainbow that had the appearance of an emerald. . . . From the throne came flashes of lightning, and rumblings and peals of thunder, and before the throne were burning seven torches of fire, which are the seven spirits of God (Revelation 4:2–3, 5).

As for the likeness of the living creatures, their appearance was like burning coals of fire, like the appearance of torches moving to and fro among the living creatures. And the fire was bright, and out of the fire went forth lightning. And the living creatures darted to and fro, like the appearance of a flash of lightning. . . . And above the expanse over their heads there was the likeness of a throne, in appearance like sapphire; and seated above the likeness of a throne was a likeness with a human appearance. And upward from what had the appearance of his waist I saw as it were gleaming metal, like the appearance of fire enclosed all around. And downward from what had the appearance of his waist I saw as it were the appearance of fire, and there was brightness around him. Like the appearance of the bow that is in the cloud on the day of rain, so was the appearance of the brightness all around. Such was the appearance of the likeness of the glory of the LORD. And when I saw it, I fell on my face, and I heard the voice of one speaking. (Ezekiel 1:13–14, 26–28).

It is quite interesting to note that God on His throne and His angels (in the Hebrew Old Testament as well as the New Testament) are described in electromagnetic terms, which quite vividly depict plasma conditions. I am not insinuating that God and His angels are simply electromagnetic forces but rather that this appearance may be providing for us a clue into His creative process. God appeared as a pillar of fire by night and a pillar of smoke by day as well as a burning bush to Moses, but that does not mean He is limited to the finite forms of a pillar of fire, smoke, or a burning bush that was not consumed. These are appearances intended to depict some aspect of the divine for our finite and limited abilities of perception. But I should like to point out that fire is plasma.

Moses, Ezekiel, and the Apostle John wrote of God as a being that glowed with radiation and that His throne also glowed with electromagnetic radiance and shot forth thunderbolts of lightning. Habakkuk describes the return of the Messiah in vivid electromagnetic terms when he describes His power as rays flashing from His hands.

> God came from Teman, and the Holy One from Mount Paran. Selah. His splendor covered the heavens, and the earth was full of his praise. His brightness was like the light; rays flashed from his hand; and there he veiled his power (Habakkuk 3:3–4).

It seems clear to me that God was revealing an aspect of His creative power in a way that people could comprehend. When we begin to understand the role of electromagnetic energy shaping and holding our galaxies and stars in precise formation, then the statement made by the Apostle Paul in his letter to the Colossians goes beyond the fundamental zero-point energy in space and takes on quite an impressive connotation.

> For by him all things were created, in heaven and on earth, visible and invisible, whether thrones or dominions or rulers or authorities—all things were created through him and for him. And he is before all things, and in him all things hold together (Colossians 1:16–17).

I suspect that our protoplasma universe was ruled not by slow gradual gravity forces promoted by the evolutionary priesthood with their need for vast ages to make the impossible evolutionary hypothesis more palatable, but by rapid electromagnetic forces that shaped our stars and galaxies. Even today, our plasma universe is not largely dominated by gravity but by the electromagnetic force. We need look no further than our own planet to understand this. The entire gravity of our terrestrial planet cannot overcome the electromagnetic forces of the molecules in a single paper clip, or it would sink through the surface and plunge into the core of Earth like a hot knife through butter. It is the electromagnetic force that binds the atoms in a paper clip and maintains its integrity even when acted upon by the immense gravity of the entire Earth.

Moreover, the electromagnetic force not only provides us with the power to run our machines and lights and appliances, but it is also the only reason our planet can sustain life. We are able to survive the pounding and lethal cosmic rays from our sun because of the magnetosphere that surrounds our planet and shields us. So the electromagnetic force not only protects us from total extinction and the sterilization of our planet but also was, in fact, the primary creative force that molded our universe.

The key to understanding the importance of the electromagnetic force is the heightened impact that it provides in the medium of plasma. Because in the plasma state, the constituent particles are polarized into positive and negative entities (the positively charged nucleus and the negatively charged electrons stripped from the nucleus), and for this reason, the electromagnetic force plays a crucial role in radically impacting plasma, much more so than the weaker force of gravity. I suspect that large-scale Birkeland currents were used by God to shape our early universe into galaxies. But these Birkeland currents come in many sizes.

Today, scientists have recognized that the auroral Birkeland currents in our planet can carry a whopping 1 million amperes and 100 billion watts. This causes the upper atmosphere around them to heat up considerably. Birkeland currents most often appear in pairs that twist like a braided rope or a double helix. A cross-section of the currents reveals a hollow tubelike beam of electrons that wind in circular vortices. The tubelike beams are protected by a quite remarkable double-layered sheath, which allows it to remain stable and distinct or insulated from the surrounding plasma. It literally acts like the membrane of a cell to keep the internal components functional within a given matrix. These twisting currents can actually be visibly observed in certain auroral events called auroral curls. But they can also be manufactured in much smaller scales as in a laboratory with multiterawatt-pulsed power generators.

Regardless of their size, the twisting currents most often appear in pairs, which begin to tighten together with great force. The stronger the field is, the smaller the radius of the circle formed by the twisting currents. Now here is the remarkable thing: there is a creative power in these magnificent spiraled currents. Any plasma material caught within the twisting chords is compressed with great force. This is called the Z pinch effect. It is so called because the azimuthal magnetic fields cause the pinching of any material caught within them and compress it from the plasma state into the solid state.

The result is that Birkeland currents produce filamentary structures of solid matter. It is virtually a machine that turns matter in the plasma state into the solid state. It does not need millions of years to create solid matter by gravitational accretion. It creates it in an almost instantaneous process. It explains to us how God could have created the Earth even before separating the plasma into stars.

What evidence do we have that Birkeland currents were involved in the creation of our universe? If we look at the cosmos, we can observe that most of the universe is filled with filamentary structures that compose our galaxies. In fact, a look at the map of the CMBR shows a complex filamentary web that stretches throughout our entire universe. Such a large-scale filamentary superstructure could not have been the result of the force of gravity, but rather the electromagnetic force. If Birkeland currents were responsible for the creation of stars and galaxies, then we would expect these galaxies to be clustered together in huge filamentary structures.

That is precisely what the Sloan Digital Sky Survey (SDSS) encountered as scientists attempted to put together a three-dimensional map of the galaxy structures in our universe. SDSS is a major multifilter imaging and spectroscopic red-shift survey using a dedicated 2.5 meter wide-angle optical telescope at Apache Point Observatory

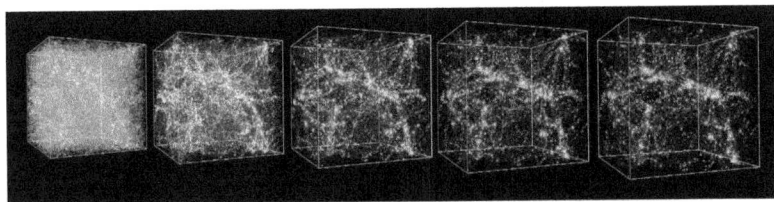

The large-scale filamentary structure of the universe is evident to any objective observer.

in New Mexico. The filamentary structure formed by these giant Birkeland currents seems to form an intricate web that mimics an electrical filamentary form.

The problem with the electromagnetic forces in plasma is that for humans dealing mostly in a world with the other three forms of matter, these characteristics are counterintuitive. Growing up on a farm in Cuba, I made slingshots with my childhood friends. We made them out of tire inner tubes cut in strips and a Y-shaped branch. I found the guava branches to be the best. We would shoot at cans and bottles placed on a fence and sometimes went dove hunting. If I shot a rock or a marble at a can, I would not expect it to start circling around the can as it passed it. But when particles are electrically charged, things do not behave as they do in our typical experiences.

For example, if the initial velocity of an approaching particle is exactly perpendicular or orthogonal to the electric field, it will be directed and confined into a circular path. But if the original velocity is at any other angle, the path of the particle will become a helix spiral. In this way, these Birkeland currents shooting through the plasma universe at the beginning of the Big Bang may have helped create the angular momentum we observe in all spinning heavenly bodies. In addition, I also believe that our primordial plasma universe was set into a spin at the moment of creation. This also helps explain the angular momentum that rules all heavenly bodies. Furthermore, it may explain how these gigantic currents may have been generated inside the spinning hot plasma that, like the churning clouds of water vapor, create electrical currents by friction.

Unlike the other three states of matter, plasma provides a much more susceptible matrix for the electromagnetic force to perform its incredible powers. The exotic currents produced have very unique properties that one would not expect in dealing with the other three states of matter. For example, Birkeland currents traveling in opposite directions will repel one another with an electromagnetic force that is inversely proportional to their distance from one another. In other words, if they are traveling in opposing directions, they repel one another and remain separate.

On the other hand, if the Birkeland currents are traveling in the same direction, they will attract with an electromagnetic force that is inversely proportional to their distance from one another. In other words, the closer they come, the stronger their attraction becomes for one another. That property allows them to accomplish things that are quite fantastic. These currents are not only attracted to one another when they are traveling in the same direction, but they also wrap around each other like a woven rope and begin to squeeze together even tighter. This pinches any matter caught within them in an area called the Z pinch.

But a curious and quite fortunate effect takes place when these currents reach a certain threshold distance from each other. As the currents constrict, all matter caught between them in the Z pinch becomes compressed. But amazingly, when they eventually reach a certain point at a given distance, a force of repulsion is generated. That is a remarkable property that allows the matter solidified within it to not be destroyed. This repulsive force keeps them from constricting any further. The result is that this configuration becomes surprisingly stable and literally becomes a creation machine that spits out solid matter in a filamentary form. In other words, it takes hot plasma and creates strings of solid matter.

These amazing twisting currents subsequently create within them a twisting cylindrical volume where solid matter is condensed by the extreme pressure exerted. Not only have we observed this in our laboratories on Earth, but we have also witnessed this in outer space. Anthony Peratt explains to us that the behavior of these currents is identical at any size, and their effects are not limited to the matter caught within their currents. They also have the ability to affect matter outside it.

> *Regardless of the scale, the motion of charged particles produces a self-magnetic field that can act on other collections of particles or plasmas, internally and externally (Peratt 47).*

Photographs made through our telescopes into deep space have seen these filamentary structures of astronomical proportions. We can literally view their images through the capture of the synchrotron radiation they produce in the process of creating solid matter.

Electrons moving through these Birkeland currents are accelerated within the double layer cylindrical electrostatic sheath to velocities that approach the speed of light. These subsequently emit synchrotron radiation that includes visible light, radio waves, x-rays and even gamma rays, which we can observe. The electromagnetic principles that control these Birkeland currents in plasma are applicable at any scale, from a laboratory experiment to astronomical proportions that could span the entire breadth of our initial proto-plasma universe. / Image credit: ESO

But this solid matter created in the Z pinch is not a homogeneous mixture of randomly ordered elements. Instead, the material is distributed in quantum groups of different elements that, although physically connected in a filamentary structure, are aligned by the ionization potential of the elements constricted between the twisting currents. This process of lining up elements through their ionization potential is called Markland convection. That is, each element is arranged in distinct packets whose position in the filamentary structure generated is directly governed by the ionization potential of the individual elements involved.

In physics, the ionization potential is the energy that is necessary to remove one electron from either an atom or a molecule. This varies according to the electrical properties of the atoms of each element, so we can literally see how the electromagnetic force is much more significant in this plasma state than gravity ever could be in forming the elements that constitute our material existence and doing so in a relatively short period of time. The smoking gun that evidences this creative process is the synchrotron radiation that is produced.

Synchrotron radiation is a non-thermal radiation generated by the electrons or charged particles that are speeding through the magnetic fields of the Birkeland currents near the speed of light. The speed of the electrons determines the frequency of the photons they emit and, therefore, the type of radiation we observe. All electromagnetic radiation is determined by the frequency wave of the photons emitted.

At the highest end of the frequency spectrum (shortest wavelength) are the gamma rays, and at the lowest place in the spectrum are the much longer radio waves. Curiously, these photons emitted by the Birkeland currents are confined to a narrow cone determined by the direction the particle is traveling. This process is called beaming. The beam is always perpendicular or orthogonal to the electric field of the Birkeland current. I strongly suspect that this property may be responsible for the regions of anisotropy found by the mapping satellites, COBE, WMAP, and Planck.

Further evidence for the large-scale Birkeland currents that shaped our protoplasma universe during the creation of our galaxies has been found by the imaging of cosmic extragalactic polarization undertaken in the South Pole by a team of physicists using an instrument called BICEP2. This is a second-generation instrument designed to map small temperature fluctuations in the cosmic microwave background radiation that corresponds to density fluctuations in our early protoplasma universe. The BICEP1 was not sensitive enough to gather the information the scientists were looking for.

These scientists were not looking for the effects of Birkeland currents, but under their mistaken idea that gravity was the principle force that created our universe, they were looking for gravity waves from the Big Bang. The telltale sign they were looking for is called a B-mode polarization pattern in the cosmic background microwave radiation (CBMR). The pattern created by the B-mode polarization is a unique spiral formation that exhibits a distinctive intensity level that is particular to the force of gravity.

When they began to analyze the data generated by BICEP2, they realized that they had, indeed, found this unique spiral shape but at a much higher intensity than that predicted by gravity. For four years, they sat on the data trying to make sense of it. Their bias toward the gravity model of the creation of stars and galaxies was not confirmed by the findings because the intensity of the signal was far too strong to have been created by the force of gravity. Nevertheless, that was not the news they wanted to hear.

The distinctive spiral B-mode polarization shape can also be created by synchrotron radiation as well as space dust. After four years of not knowing how to spin their results, they eventually ruled out that space dust could have been responsible for their BICEP2 observations. But due to their predisposition for a slow, gradual gravity-molded universe, they also ruled out synchrotron radiation. They did this because they only considered synchrotron radiation from Birkeland currents that did not span the universe, as we suspect, but were much smaller in scale. Having falsely concluded that nothing else could have caused these unique spiral patterns in the CMBR, and ignoring the fact that the B-mode signals were much too intense to have been generated by gravity, they nonetheless simply declared them to be gravity waves by the wave of their magic evolutionary wand.

Blindly sticking to their evolutionary Rapid Inflationary Model that has already been soundly refuted by the dipoles and octopoles discovered by the COBE, WMAP, and Planck satellites, they simply moved forward as if the king were still clothed and declared it as proof of the inflationary hypothesis. With a Cheshire grin, they boldly claimed that this was the smoking gun of the inflationary model. In reality, and contrary to their predisposed evolutionary bias, it is actually the smoking gun of large-scale Birkeland currents used by God to mold our galaxies and create our planets in the early universe.

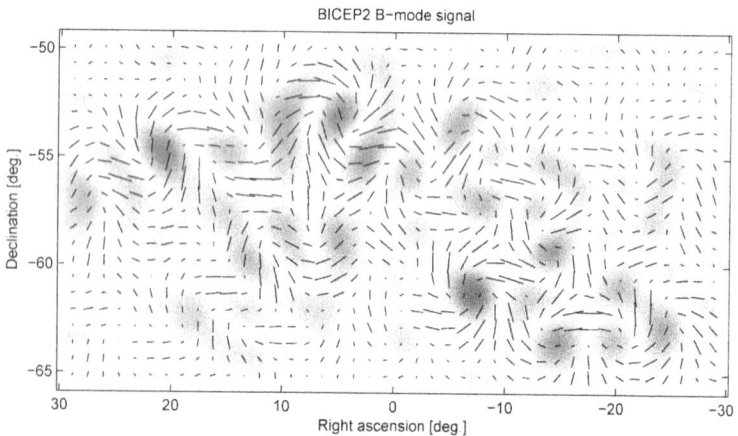

Gravitational waves from inflation generate a faint but distinctive twisting pattern in the polarization of the CMB, known as a "curl" or B-mode pattern. For the density fluctuations that generate most of the polarization of the CMB, this part of the primordial pattern is exactly zero. Shown here is the actual B-mode pattern observed with the BICEP2 telescope, with the line segments showing the polarization from different spots on the sky. The gray shading shows the degree of clockwise and anti-clockwise twisting of this B-mode pattern. (BICEP2 Collaboration)

There is yet one more piece of evidence discovered by the mapping of the CBMR that strongly suggests that massive Birkeland currents during the early plasma universe may have been responsible for the creation of galaxies. The CMBR maps also show that the northern hemisphere of the universe, which is divided by the dipole and the Earth-sun ecliptic, is colder than the southern hemisphere of the universe. The Rapid Inflationary Model cannot produce such a universe with two hemispheres exhibiting different temperatures. The rapid inflationary Gaussian equations predict an entirely homogeneous universe without the dipoles, quadrupoles, sextapoles, octopoles, and very real and measurable temperature differentiations between the northern and southern hemispheres of the universe.

Instead, what the physical data from BICEP2, WMAP, COBE, and Planck actually insinuate is that the giant Birkeland currents may have been responsible for these anisotropic anomalies. These areas of anisotropy (areas of low density of the CMBR), which are symmetrically aligned with one another, represent the areas where the Birkeland currents began. At the beginning of the current plasma, matter enters into the tubelike structure of the twisting currents and is shot through to the exit point where beaming takes place. Therefore, these areas will show a lower density than the surrounding matrix of plasma as plasma from that area is evacuated.

If the majority of these enormously intense energy Birkeland currents were positioned from north to south, the southern hemisphere would have been the recipient of the beaming process associated with these currents. This is the smoking gun that shows that giant Birkeland currents are, in fact, responsible for the slightly higher temperature differential created by the beaming of electrons into the southern hemisphere.

This, of course, would mean that the positioning of the Birkeland currents could not have been randomly generated but, instead, specifically designed to create a universe such as it appears today. That previsioned choice leads us, therefore, to conclude that random ordering could not account for the physical evidence we observe. It had to be intelligently designed by a master architect. For this reason, the atheistic metaphysical bias of the evolutionary priesthood simply chooses to ignore the physical evidence and brashly twists their findings in classical Orwellian doublespeak.

We Are Not the Children of Stardust: We Are the Children of Light

The late Carl Sagan popularized the idea that we are the children of stardust. Holding to his evolutionary/pantheistic worldview, he also popularized the idea that the universe was cyclical and had been created and destroyed many times. During the intermediary processes between the Big Bang and the Big Crunch, the nuclear processes within different-sized stars were then responsible for the formation of the higher elements seeded throughout the universe by their eventual explosions. These chemicals, according to their gradualist model, through gravitational attraction, eventually coalesced into planets and through accidental and random reactions eventually evolved into human life. Thus his famous saying that we are the children of stardust.

But I think he was way off the mark. From the chaos of the Genesis Singularity, the very first thing created were photons—the basic component of all electromagnetic radiation. It is from light that all things in our universe were made. From the first record we have of our ancestors, light has been a central mystery to humankind. The strange fact is that the more we learn about light, the more mysterious it becomes. It is an almost otherworldly thing. It does not behave as ordinary matter, and yet without it, we could not exist. Not only do we depend on it to be able to observe the world around us, but it is also through its peculiar property that the chain of life can even exist. It is an exotic element of our universe upon which all visible reality has been built.

Indeed, as we are about to see, science and the Bible agree that light belongs to an altogether unique category of reality. It is exceedingly exotic and operates on a privileged plane of existence that we ourselves can never experience—at least not in this life (Guillen 71).

For most of human history, humankind has thought that light was instantaneous. That is, we thought that its speed was infinite. Today, we have been able to exactly measure the speed of light in a vacuum, and it is an outstanding 186, 282.397 miles per second, or 299,792.458 kilometers per second. It is hard for us to picture in our minds how really fast that is. When we first broke the sound barrier, we had no idea what physical consequences it would have on human bodies. As it turned out, Chuck Yeager, an experimental pilot aboard a rocket called Bell X-1, strapped below a bomber, detached, and on October 14, 1947, flying at an altitude of 45,000 feet, became the first person ever recorded to break the sound barrier. Thankfully, there were no adverse effects on the human body. That was considered a great achievement for modern humanity. But light is more than 900,000 times faster than sound.

My wife is a flight attendant, and last year, I tagged along with her to Quito, Ecuador. The next day, we took a cab to go visit the equator high in the Andes Mountains. The equator is the area of the world where its circumference is the greatest as it bulges due to its spin. The circumference around the Earth is 24,901 miles at the equator. If I, standing on the equator of the Earth, had shot a bullet at the speed of light that would travel around the Earth, it would have circled the earth 7.48092032 times in one second. That means it would have gone through my skull seven times before I could have ducked my head. But the marvelous thing is that this speed is, in fact, the speed limit in the entire universe for all ordinary matter (tardyon) residing within it.

Nothing can go faster than light. It is an absolute constant that is never variable. That ultimate speed is for all intents and purposes an impenetrable boundary for all things inside our universe. In fact, it is also true for the realm of tachyons, which are theoretical entities hypothesized through mathematics that are supposed to travel faster than light, as Guillen explains:

> According to special relativity, the speed of light is not only sacrosanct, but also serves as an impenetrable boundary—a Great Wall of China throughout the cosmos—between two very dissimilar realms.
>
> We ourselves inhabit the tardyon realm—the world of ordinary matter that can never reach or exceed the speed of light in vacuo from below. No matter how hard we tardyons try or how clever we are, there is no way for us to reach or break through the luminous barrier. . . . The faster tardyons travel, the more massive they become, and therefore the more difficult it is to further increase their speed. It is a surprising quirk of nature that has been verified experimentally.
>
> The mass of subatomic particles whizzing around in a giant donut-shaped accelerator—like Cornell's Wilson Synchrotron—enlarges indefinitely with increasing speed. Ultimately, an infinite amount of energy would be needed to propel the infinitely massive subatomic particle up to that magical, mystical, unattainable speed of 299,792.458 kilometers per second.
>
> The second realm comprises of tachyons, hypothetical entities that travel faster than light but cannot ever attain or break the light barrier from above. As they attempt to slow down to the hallowed speed limit, their masses inflate uncontrollably.

Ultimately, an infinite amount of energy would be needed to slow one all the way down to 299, 792.458 kilometers per second.

And that's not all. For tardyons and tachyons alike, time exists. It flows one way for us tardyons and the opposite way for tachyons, but both of our kinds are trapped by it. Both of our kinds are secular creatures.

Also for our two kinds, time slows down as we approach the speed of light—a prediction of special relativity that also has been confirmed by meticulous observations of fast-moving subatomic particles. For a tardyon traveling at 99.99999 percent the speed of light in vacuo, a single second stretches into a conventional day and a half. And that leads to the following conclusion: For light itself—which travels at 100 percent the speed of light—time slows to a complete stop. Time doesn't flow. Time doesn't exist. Light and light alone inhabits a realm where past, present, and future have no meaning because the three exist all together and at once (Guillen 73–75).

In addition to all that, the speed of light is always independent of the motion of whatever body emitted it. For all other matter, a projectile shot from a moving object has a velocity that is dependent on the motion of the object emitting it. In other words, if I were in a car traveling 70 miles per hour and threw a rock out my car window, forward at 30 miles per hour, the rock would be flying at 70+30 miles per hour, which is equal to 100 miles per hour. If, instead of throwing the rock forward, I heaved it backward at 30 miles per hour, the speed of the rock would be 70-30, or 40 miles per hour. Light, on the other hand, travels at a constant, invariable speed, and it does not matter how fast that emitting object is traveling or if the object emitting it is going toward the direction of the light or away from the direction of the light. Its speed never varies. Nothing else in our universe has this unique property.

Ancient men knew nothing of the incredible details modern physics has revealed regarding light. But the scriptures align with modern science in declaring that from God's radiant energy, all things were formed in our universe. Moreover, Genesis tells us that light was created before the stars. In fact, light existed in the density of the primordial cosmic plasma that due to its density was opaque. Light did not break through until it expanded and cooled, perhaps with the help of the creation of water through giant Birkeland Currents. That point in time was recorded in Genesis 1:3–4.

That is a remarkable fact; it shows that the more we learn about our universe, the more science backs up the Genesis creation story. In fact, light in the scriptures is used to symbolize God. And certainly the properties of light do offer insight into the nature of God, who is in a timeless state and whose character is invariable and independent of anything within this material universe.

> *This is the message we have heard from him and proclaim to you, that God is light, and in him is no darkness at all (1 John 1:5).*

> *That people may know, from the rising of the sun and from the west, that there is none besides me; I am the Lord, and there is no other; I form light and create darkness; I make well-being and create calamity; I am the Lord, who does all these things (Isaiah 45:6–7).*

Since the scriptures refer to God as light, it behooves us to understand that light behaves as both energy and matter. That is to say that a photon can behave as matter. Certain aspects of it are explained in physics as if it were matter or a particle (photon). Yet other properties are simultaneously best explained as if they were energy or an electromagnetic wave. No other form within our universe exhibits this kind of modality. Moreover, we have already discussed how light can be turned into matter and matter can be turned into light, so there is nothing in this universe as mysterious and cryptic as light.

Radiant energy is said to exist in a state of eternal now. In other words, if I were able to ride a rocket at the speed of light, time would disappear. I would be in a state of timelessness. Perhaps the name that God gave Moses at the burning bush can make more sense: "I AM WHO I AM" (Exodus 3:14). He is the timeless one.

To us, this is a conundrum that is unexplainable. Yet I find it quite illuminating since it is consistent with who God is. For He is at once timeless and both energy and material in existence. That is, He can take on a spiritual form that is invisible and a material form that is visible. Moreover, all of us finite beings caught within this temporal universe cannot ever attain godhood. The finite cannot become infinite within our mortal frames. And yet God tells us that He will give us a glorified body that will be infinite when we accept Him as the light of the world.

Again Jesus spoke to them, saying, "I am the light of the world. Whoever follows me will not walk in darkness, but will have the light of life" (John 8:12).

And although we may not be able to completely comprehend this, it does not mean that it isn't so. I find it amusing that one of the reasons atheists since the time of the Enlightenment rejected God was mainly because they thought that nothing could exist outside of time. In their narrow minds, this was nothing more than mythological superstition. Neither could they have imagined then that modern physics would verify the Genesis record in declaring that God's radiant energy was the first act of creation, and all things after came from that energy.

In fact, within our human frame of reference, light, or, to be more complete and accurate, electromagnetic energy, is the most accurate entity in our universe that can symbolize God. Not only did all matter come from light, but also all living things cannot survive without light. Yet it is not God. It is, according to scripture, an emanation of His glory. It was the first thing He brought forth into our universe to begin its transformation from a disordered state to a more ordered state.

It was God's radiant energy at the Alpha Point that was converted into plasma and photons, then into the more complex forms of the elements to bring forth our physical universe. But initially, within the dense plasma cloud, the density of the matter it had created obscured the light within it. When the expansion of space-time reached a certain point, light escaped. It was at that precise point when light broke forth from the darkness of matter and separated that marked the end of the first period of creation: day one.

How is it that the Genesis record written more than 3,000 years ago stipulates that God separated the light from the darkness, giving us a vivid description of the process modern physics has discovered in the separation of light from the dense plasma state?

It may also surprise you to find out that we are also beings of light. Our bodies radiate photons continuously. The wavelength is too long for us to see because it is in the infrared spectrum, but if you looked through an infrared camera, you could see a person in the blackest of darkness. We literally glow with infrared light. I worked as a firefighter for the City of Miami Beach for almost 30 years and often used these infrared cameras. They were quite useful in finding a struggling swimmer in the surf at night. They also helped us find hot spots in the darkness of smoke to prevent rekindling.

We must again mention that our Judeo-Christian cosmological model is not in accord with the pantheistic model, which suggests that the creation and the deity are one and the same thing. In the Judeo-Christian model, God was before the creation, and the creation is not God but merely the clay, while He is the potter. God, who is not trapped in space-time, created space-time and all that is contained therein. It was not the death of stars that led to the creation of planets and life on Earth through the slow and gradual evolutionary model based on gravity. It was the electromagnetic force that created matter through Birkeland currents. It was God who formed the Earth through His electromagnetic power. In fact, I suspect that God also created the DNA through some form of electromagnetic power since its double helix shape is quite similar. It was not a random accident, but rather an intentional design. We can see the similarities between the spiral DNA and the spiral helixes of filamentary matter created by the Birkeland currents in the cosmos.

No, we are not the children of the gradualist gravity model of evolving stardust. Do not be deceived by empty words meant to draw us away from the creator. Yes, we were fashioned from the clay, but we were molded by God's light and given the *nashama* (breath) of God to spark life into us. And for this reason, we are called children of light.

> *Let no one deceive you with empty words, for because of these things the wrath of God comes upon the sons of disobedience. Therefore do not become partners with them; for at one time you were darkness, but now you are light in the Lord. Walk as children of light (Ephesians 5:6–8).*

The truth of God's word is clear for those who have eyes to see and ears to hear, but the problem of humanity is not that there is no scientific evidence from the structure and origin of our universe to declare God's unmistakable role in the creation process. The problem is that people are prone to develop their worldviews to rationalize their moral choices. If there is a God, then people are beholden to His moral standards, and that is the foundational reason that most are adamantly antipathetic toward and unwilling to accept or consider the truth. It is for this reason that naturalists fanatically resist any scientific endeavors that bring to light the masterful design intrinsic to all aspects of our universe and irrationally opt for a randomly generated universe through mere accidental processes.

In a universe without a God, sin does not exist. Each person essentially becomes his or her own god. Each person is free to choose what he or she wants to call evil. In so doing, they attempt to escape the guilt associated with their preferred modes of indiscretions. But in doing so, they fail to understand the deep message of love that God has given us—that He sent His son to take away the guilt of our sin and to offer us forgiveness.

> *"For God so loved the world, that he gave his only Son, that whoever believes in him should not perish but have eternal life. For God did not send his Son into the world to condemn the world, but in order that the world might be saved through him. Whoever believes in him is not condemned, but whoever does not believe is condemned already, because he has not believed in the name of the only Son of God. And this is the judgment: the light has come into the world, and people loved the darkness rather than the light because their works were evil" (John 3:16–19).*

CHAPTER 14

● ● ●

THE AGE OF THE UNIVERSE AND OUR JUDEO-CHRISTIAN COSMOLOGICAL MODEL

The proponents of intelligent design who write textbooks for the public school system do not involve themselves in the following discussion, and I think it appropriately so. Their objective is to show that there is a scientific alternative to the idea that we evolved through random chaotic processes. Their science is intended to show that there is a design to the universe that could not have been generated by random chemical processes. They do not invoke God as the Designer, but they certainly imply that a master mind had to have engineered this universe. However, since our purpose in this work is to equip the saints and defend the Judeo-Christian account of creation in Genesis, it is important that we cover this issue from the point of view of our Judeo-Christian cosmological model.

There are several possibilities from a scriptural point of view in regards to the age of the universe. There are those who believe that the Genesis account is symbolic, and therefore, they assume that the timetables asserted were not literal. There are others who, like myself, believe that the Bible is the inspired and authoritative Word of God. In other words, it is a document of propositional truth given to us by divine revelation in order that we might better understand Him. And for that reason, the narrative is therefore to be interpreted literally, whenever the narrative suggests.

I have not always believed this. In my youth, I was an atheist and an evolutionist. I have come to understand as a result of my personal quest for truth that God in His infinite wisdom created the universe and humanity. Moreover, He left us with the evidence to know Him, if we are willing to be objective and honest as we seek the truth wherever it may lead us. The Hebrew Scriptures, in my estimation, is part of that body of evidence that is essential for us to know and understand who we are and why we are here.

After studying all the major "holy books" from all the major world religions, I came to accept the Hebrew Scriptures as the true revelation from the great creator. It is beyond the scope of this text to document the reasons I came to this conclusion. The story of my journey, however, is documented within the other books of my apologetics series. But for the record, I have come to believe that the interpretation of the Judeo-Christian scriptures should be literal, wherever the text makes it possible.

I am convinced that God's truth will continue to emerge triumphant and unscathed by even the most vitriolic attacks of the evolutionary, naturalistic-based pseudo-science of the atheists. That does not mean that the entirety of the scientific work being done by secular scientists is pseudo-science, but that their naturalistic bias is not hard science. It is a premeditated, metaphysical choice that cannot be corroborated by hard science. Darwinism is pseudo-science.

If my presupposition is correct, then there should be no disparity between true science and the claims of the Holy Scriptures. We therefore are under the assumption that our side of the universe is relatively young in comparison to what the evolutionists believe and that true science will substantiate this. Evolutionists pointing to the 15 billion light years that separate our area of the universe from the very edges of the universe claim that we must therefore be at least 15 billion years old.

Their logic is based on the idea that light had to have been traveling 15 billion years in order to reach us from those stars at the edge of the universe. But I would caution you to remember what we have learned from Einstein that time is relative to the position of the observer. Although the outer edges of our universe may actually be 15 billion years old, our local astronomical neighborhood may not have aged at the same rate in the very same interval of time. Time is not a monolithic constant that spans the entire universe. As we shall see, it is elastic and dependent on the specific conditions of space for each separate set of coordinates considered. This is what Einstein clearly proved, and it is what evolutionists try desperately to evade.

However, I would like to add a word of warning here. Christians must be careful to not repeat the same mistakes of the past. Christians should never interfere with the pursuit of scientific knowledge. Even if the conclusions are being made, which seem to be in contraposition to the teachings of the Scriptures, our responsibility is never to stifle or curtail the search for truth.

Those who come to conclusions that seem to be contrary to the scriptures have the right to their opinions as much as we do. The free exchange of ideas should never be curtailed by church or state. In the end, we believe that true truth never has to fear competing in the free market of ideas, for it shall always rise to the top in the minds of those who are truly seeking truth and willing to be objective.

True science and true truth do not contradict each other. The belief in the unity of truth is the fundamental doctrine of the Judeo-Christian worldview. To deny the unity of truth is to deny the true historical space-time reality of the Judeo-Christian worldview. To divide scientific truth from theological truth is to deny the lordship of God in the universe. There can be no middle ground in this. God is the author of truth, and if His revealed propositional truths in scripture are false, then there is no God. Ours is the duty to search for truth and to never curtail the rights of others to do the same.

For this very same reason, naturalists should not preclude from the academic institutions and scientific communities scientists who hold to the intelligent design model. Furthermore, to allow only one view to be taught in educational institutions is an unconscionable travesty of the scientific ideal. Such interference with those who, through the scientific process, are in pursuit of truth is simply repressive, tyrannical censorship and lowers the standard of those institutions to centers of indoctrination instead of education. It is reverting to the same tactics used against Galileo, which are so abhorred by all rational people everywhere.

But even within our own camp, there is much to be learned. Often, well-meaning Christians have been intransigent and dogmatic in areas where the Bible is not. Somehow, they dogmatically become convinced that God made a mistake and forgot to state some issue. In their misdirected zeal, they determine to straighten out the problem and help God by perceiving themselves as stalwarts of the faith. These ardently pontificate some minor relativistic issue, straining at a gnat and swallowing a camel. That is, they take absolute positions where God has chosen not to do so. That is how legalism and false religions develop, and it is the most effective way to repel truth seekers.

If God saw fit not to make something an absolute within the scriptures, we have no right to correct God. He didn't overlook anything, so we should not feel compelled to fix things up for Him. Sometimes, we have a tendency to reduce and simplify things for our benefit. But in the process, we often overlook the subtle reasons why God does things the way He thinks is best. We try to box things into a nice little convenient package that is easier for us to comprehend, assimilate, and follow. He always has a reason for doing what He does, even when we don't understand why.

There are several schools of thought presently in Christendom with respect to the age of the Earth. As stated earlier, I prefer the young Earth theory because I believe it fits the scientific facts more accurately. But this is my opinion and not dogma. Where the Bible speaks dogmatically, we should not compromise. But where it does not, there is room for disagreement—seasoned with charity, of course.

Some in the creationist camp have proposed several theological alternatives in order to account for the seemingly enormous age of our universe. Phillip Gosse's apparent age theory presupposes that God created the light spanning out from the stars and reaching our planet at the same time He created the stars, thus building in an apparent age at the moment of creation.

But other creationists oppose this apparent age theory, correctly concluding that God would not have, in fact, written into the universe a superfluous lie, an act that would be in disaccord with His nature. And for what purpose might He do such a thing? Was God trying to trick us? This, in my mind, is a valid repulsion of Gosse's apparent age theory.

The age of the stars is not apparently older because God created a false illusion; it is older because we miscalculated the age of the entire universe by falsely assuming that the rate of expansion has always been uniform in all parts of our universe. We have failed to take into account the fact that time stretches as space stretches. This means that time runs at different speeds throughout the breadth of the universe. It means that the aging process of the universe is not a uniform, monolithic rate, as we shall now discover.

Before Einstein's theory brought us to the mathematical conclusion that the universe had to have a beginning, some creationists, trying to harmonize the Genesis account with the enormous spans of supposed evolutionary history and attempting to explain the supposed fossil evidence of the evolution of man from apes, came up with a theory that there was a pre-Adamic race that existed for long spans of time. These artificially harmonized the anthropological evidence offered by the evolutionists as proof of our ancestral lineage from apes by claiming that God created Adam after destroying this pre-Adamic world.

The elastic contortions necessary to twist the scriptures sufficiently to allow such an interpretation are ridiculous. Moreover, the anthropological evidence does not at all prove that humans ascended from the apes. That subject is, however, beyond our present scope, but we will deal with the anthropological evidence for the evolution of humans in *The Descent of Man*, the fourth book in the *Machine or Man* series.

Others, attempting to overcome the enormous spans of time dictated by the evolutionary paradigm, have, by clever reinterpretations of the Genesis narrative, tried to impose seven ages in the seven days of the creation narrative. Here again, artificially imposed ages of varying spans were interjected into the text in order to harmonize it with the supposed enormous ages of the evolutionary timetable, which evolutionists have claimed for the age of our planet. As we shall see, these enormous ages for our planet are not really substantiated by either the fossil evidence or the radiometric data, but fabricated by the naturalists' enormous need to give time in order for the improbable evolutionary/gradualist mechanism to become more palatable.

Everyone is entitled to his or her own opinion, but frankly speaking, I see these as flagrant attempts to artificially harmonize scripture with pseudoscientific notions of evolutionary gradualism. The scientific evidence does not substantiate a gradual transition from one species to another. Evolutionists simply artificially fabricate those immense timetables necessary for gradualism, as we shall see in the *Descent of Man*.

Nevertheless, within the creationist camp, there is room for divergent opinions on this matter, and we should always keep our minds open and objective. Be that as it may, I lean heavily toward the young earth/old universe theory, which poses the aforementioned apparent problem with regard to the distances observed from the stars at the edge of our universe. After studying this for more than 40 years, I believe that a scientific explanation that best covers all the empirical data is in accordance with the Judeo-Christian scriptures; therefore, I have chosen to call the young earth/old universe theory our Judeo-Christian cosmological model. This model incorporates the work of many others who have done the hard work. My only contribution is compiling it into a linear process.

A Judeo-Christian Cosmological Model

We have been taught that if light from stars that are billions of light years away from us is reaching us, then it stands to reason that the age of the universe is at least as old as the light of the farthest star reaching us. Evolutionists claim that since starlight has reached us from 15 billion light years away, then our universe must be at least that old and the Earth must be at least four billion years old. However, it may be argued that the accuracy of the actual distance to the farthest edges of our universe is not completely dependable since our measurements are, in fact, indirect. Even among the evolutionists, there is a discrepancy that ranges from 13 billion years to 15 billion years of age.

> We can determine the present rate of expansion by measuring the velocities at which other galaxies are moving away from us, using the Doppler effect. This can be done very accurately. However, the distances to the galaxies are not very well known because we can only measure them indirectly (Hawking 48).

Nevertheless, let us accept these measurements of the farthest edge of our universe as completely accurate for the sake of argument. There is a rational and scientific explanation for this apparent problem, which in reality is no problem at all. Remember that time is relative to the observer. The key to the actual age of our universe begins with understanding that the rate of expansion in our universe is not decelerating as time goes by. Neither is the expansion monolithic throughout the entire breadth of the universe. Our universe is expanding at different rates, depending on what area of the universe is considered. The farthest reaches of our universe are expanding at a much higher rate than the interior. Space is densest at the center of the universe. As we move outward in concentric circles, space becomes less dense—space stretches. It is the rate of the stretching of space-time that causes one area to age faster than another.

Our Judeo-Christian cosmological model proposes that the rate of time is not monolithic throughout the breadth of the universe. It is the stretching of space-time that directly determines the rate of time in any coordinate from the center of the universe to the very outer edge. Therefore, the edge of the universe that is 15 billion light years away has, in fact, experienced 15 billion years of history because time there is in fast forward compared to our central position in the universe where space-time is much denser and time moves more slowly. There is no discrepancy between the Genesis record and true science.

The Discrepancy Is between Evolution and True Science

Evolutionists had completely depended on the Gaussian equations to justify their magical Rapid Inflationary Theory. However, if the Gaussian equations had been responsible for the universal homogeneous and isotropic nature of our universe, it would mean that the entire universe had to be 100 percent homogeneous and isotropic. The findings of the three mapping satellites, which show there are pockets of areas that are anisotropic, disqualify the Gaussian equations as the arranger of our universe. Moreover, these pockets of anisotropy are symmetrically aligned not only with one another but also with the sun-earth ecliptic and its equinoxes. No materialistic equation can explain such symmetrical alignments that extend to every quadrant of our universe.

The red-shift analysis of stars and galaxies toward the outer rim of the universe clearly shows that their rate of expansion is directly proportional to their distance from us. The long and short of it is that the outer fringe of our universe is not slowing down; it is speeding up. That is, the galaxies and stars at the edge of our universe are, in fact, accelerating away from us in an exact relation to the distance from us. The further away, the faster they go. That is a mind-blowing thought, which is completely counterintuitive.

No evolutionary model has adequately yet explained all these striking phenomena. In order to explain the acceleration of the expansion of our universe, most naturalistic cosmologists have now turned to dark energy as a repulsive force that is causing this acceleration. But no one can describe this magical energy that increases in intensity with distance. All known forces cataloged by observational science decrease with distance, some more rapidly than others—but all weaken with distance. This anti-gravity force they call dark energy or vacuum energy is nothing more than a mathematical fix that has absolutely not one shred of empirical data to support it. I think they are not seeing the forest for the trees.

Nevertheless, as we shall now see, the idea that our universe is 13 billion or 15 billion years old (depending on what evolutionist you talk to) may not be necessarily so from the point of view of the Earth. Just because the outer fringes of our universe seem to be 13 billion to 15 billion light years away from us does not mean that Earth has passed through 13 billion to 15 billion years of history. Remember what we learned from Einstein that time is relative to the position of the observer and that time is inextricably linked to the density of space. We have thus learned from Einstein to no longer speak of space or time separately, but instead, we speak of space-time.

The Stretching of Space-Time

Most of us are not too familiar with concepts of relativistic frames. That is because science knew nothing of it until Einstein, and few gave credence to the counterintuitive conclusions of his mathematical equations until recently. Let me illustrate the principle with something simpler and more familiar to us.

Perhaps one of the most memorable moments in my life was in 1969 when I watched Neil Armstrong step off the lunar module and say, "That's one small step for a man, one giant leap for mankind." I was in high school in Miami at the time and remember the scene vividly as I sat in our Florida room working out with weights. I sat up and watched live as that milestone in our human history was achieved. An overwhelming feeling came over me that I had witnessed perhaps one of humankind's most spectacular achievements in all our history. To this day, I have a copy of the *Miami Herald* newspaper documenting that glorious day that a man stepped foot on our moon.

According to the fancy digital scale in my bathroom, I happen to weigh 180 pounds on Earth. If I would have boarded the Apollo 11 lunar module in 1969 with Neil Armstrong and then walked on the moon with him, what would have been my weight on the moon? If I would have taken my trusty digital scale out and weighed myself on the surface of the moon, it would have read 30 pounds.

That does not mean I lost 150 pounds during the trip. It just means that the moon's gravity is one-sixth the strength of the Earth's gravity. My mass has not changed, but because of the lesser mass of the lunar environment, there is less gravity to bend space-time, and I appear lighter. My weight is relative to the gravity of my surroundings. In fact, if you could measure it, you would see that my size would have also stretched a bit longer, because space on the moon is not as contracted as it is on Earth. I would have been a little bit taller. Because space is stretched more on the moon than on the more massive Earth, time on the moon also runs faster.

Jupiter, on the other hand, is 1,300 times larger than the Earth in volume. But because it is made up of mostly gas, it is only 318 times as massive. What would I weigh on Jupiter? One would think that if Jupiter is 318 times as massive, then we could simply multiply our weight by 318 and get the answer, but we cannot. The further we are from the center of mass, the less we weigh. Gravity diminishes with distance from the center of mass.

Therefore, if I were to weigh myself on an airplane 36,000 feet above sea level, I would weigh slightly less than on the surface of our planet. Because Jupiter is a gas giant, the radius is 11 times that of Earth. Thus, our weight on the surface of Jupiter would be only 2.6 times what it is on Earth. Therefore, I would weigh 468 pounds on Jupiter. But because the gravity on Jupiter is greater, my body would be shrunk in size and time would move more slowly. Time, like space, is contorted by gravity. Space around the surface of Jupiter is shrunk; it is denser than on Earth. When space is shrunk, time slows down. Space-time is inseparable. What we do to one impacts the other.

Gerald Schroeder, a distinguished physicist, writes of this time warping in his fascinating book *The Science of God*. Adapting an example from his book, let us now imagine that I would travel to a giant rocky planet called Gargantua that is so massive that it would slow the rate of time by a factor of 350,000 times. Time in this mammoth planet is running so slowly that only three minutes passed by during the same interval that two years passed on Earth. Of course, such an enormous gravity would flatten me into a pancake, but for the sake of the illustration, let us imagine that I survived the short visit to the planet Gargantua.

After exploring for only three minutes in this imaginary planet, Scotty from Star Trek beamed me back to Earth instantaneously. I returned to Earth and reported my findings to you. You would be two years older, and I would only be three minutes older. Earth experienced two years during the very same interval that I experienced only three minutes on that giant planet. The rate of time is relative to the position of the observer and determined by the particular density of that space-time coordinate.

Time is relative to the observer because spatial warping and energy such as gravity or acceleration influence it. For example, as an object reaches the speed of light, time slows down because space is warped or crunched. As space is crunched, matter becomes more massive. As an object enters into the critical circumference of a black hole where space is so warped and contracted (the fabric of space is denser) due to the enormous gravitational pull of gravity in the singularity, objects seem to slow down in time until coming to the edge and freezing time completely.

The opposite is also true. If space is thinned out or stretched, then time is accelerated. Time is an intrinsic component to space. There can be no space outside of time, and there can be no time outside of space (at least for us trapped in this universe). All spatial dimensions are trapped in time. The two cannot be separated, but they can be warped. In essence, what Einstein has taught us is that time is a function or property of space.

How does this relate to the creation of the universe? The Big Bang is, in essence, the opposite of a black hole. You will recall that in a black hole, the extreme density of matter has been caused by the collapse of matter due to the enormous force of gravity. The force of gravity is so concentrated that it creates a halo of influence around it—the event horizon or the critical circumference. Therefore, any matter coming close to the event horizon or critical circumference is sucked into the black hole, and once it crosses that event horizon, it can never come out and is necessarily crunched into the singularity.

A black hole is like the Eagles' song "Hotel California"—you can check in, but you can't check out. The distance of that event horizon from the singularity is dependent on the amount of material found in the singularity. It is called a black hole because not even light can escape from its gravitational clutch. The more material is crunched into that singularity in the middle of a black hole, the greater the gravitational force it will produce, and consequently, the further the event horizon will extend outward.

In the Big Bang, we have the opposite scenario, where the material was cast outward from the interior Genesis Singularity. The immense energy that turned into everything in our enormous universe was concentrated in a point in space-time so small that it was crunched into an infinite density. This is not a point that is contested by evolutionists, for it is mathematically derived from Einstein's equations. What they refuse to concede is that only an infinite power could have caused it to overcome this infinite density.

That infinite power caused the Genesis Singularity to explode outward with just the right amount of force in order to eventually overcome and pass through the critical circumference created by the enormous gravity of all the matter/energy of the universe within that event horizon. Too little force, and it would crunch back. Too much force, and the expansion rate of the universe would be so great that matter could not coalesce. Atoms would not be able to form from the constituent particles. Our universe would become nothing more than a soup of elementary particles ever separating in distance.

It was an exactly precise and fine-tuned amount of force of gargantuan measure that threaded the cosmic needle and allowed a universe that could inhabit life. This is that miraculous expansion rate, which Weinberg concedes to be a miracle, for the probability of it being randomly chosen is one chance in 10^{120}. In addition, the forces that govern the physics of our universe were exactly proportioned between each other in a perfect balance so our physical universe could have the exact properties necessary in order for us to inhabit this universe.

One of the most astonishing discoveries astrophysicists have made in recent decades is that if gravity were just 0.000000000001 (one trillionth of one) percent stronger, our universe would have reversed course long ago. It would have collapsed catastrophically, ending in a big crunch, the opposite of a big bang. Likewise if gravity were just 0.000000000001 (one trillionth of one) percent weaker, our universe

would have flown apart so rapidly that planets, stars,
galaxies—all the basic constituents of the universe—
would never have had a chance to coalesce. We'd all
be dust in the wind (Guillen 68).

In a white hole (the opposite of a black hole), as in a black hole, the singularity creates an event horizon or a critical circumference that is also proportionate in strength and distance to the amount of matter/ energy contained in the singularity. In the Genesis Singularity, all the matter and energy of the entire universe were concentrated into one spot, and that virtual event horizon theoretically would have extended further than the actual boundary of space-time. The actual boundary of space-time at that moment was, in fact, the bounds of the Genesis Singularity. The virtual critical circumference was outside our space-time continuum. It would not be reached until the singularity expanded that far.

That theoretical or virtual critical circumference was more powerful than any created by any other black hole or even the sum of all black holes that we can observe today. But we must understand that space-time did not extend that far at the beginning. The theoretical distance to that critical circumference was actually beyond the event horizon of our universe at that Alpha Point. The actual event horizon of the white hole of the Genesis Singularity would be no less powerful than the virtual critical circumference. Something infinitely powerful had to affect that Genesis Singularity with an infinite amount of force to cause it to expand against that extremely powerful event horizon. All events taking place inside that primordial critical circumference existed in a space-time that was severely crunched. Time was slowed down to a virtual crawl.

Yet somehow, that Genesis Singularity miraculously expanded and overcame the infinite density and enormous gravity that held it in that primordial singularity. I believe that at this very point, God gave that singularity a spin, which He had carefully calculated to accelerate enough to counter the gravitational pull. The entire process was done smoothly and seamlessly to show us humans, whom He knew would later study the process, that it was controlled from without. Even more amazing, as it expanded, the radiant energy condensed into a homogeneous ball of superheated plasma. This ball of plasma was contained within the event horizon of the early universe as space-time stretched outward toward that theoretical critical circumference. The universe had not yet reached that distance from the Alpha Point where gravity is so strong that nothing in our material universe could allow it to escape. But what kind of force could cause the most powerful singularity in the history of the universe to overcome its gargantuan gravitational force in such a controlled manner? There are no forces within our material universe that could accomplish such a feat. Evolutionists cannot provide any mechanism for such a feat. Nothing other than an omnipotent God could counter such massive gravity.

All the supermassive black holes we have observed could be thrown into one giant supermassive black hole, and it would still not even come close to the gargantuan gravity that the Genesis Singularity exuded. The Genesis Singularity contained all the matter and energy of all the black holes in the universe plus all the matter and energy outside of them. And yet it extended outward at the perfect speed necessary to overcome that gargantuan gravity and pass through the primordial critical circumference of the Genesis Singularity.

As space-time and the matter/energy within it expanded outward, the actual event horizon was, in essence, the edge of our space-time universe at each particular moment as it stretched. That edge of space-time traveled outward until it reached the precise distance at which the combined gravity of all the matter and energy inside dictated the virtual position of the primordial critical circumference. Then it passed beyond that theoretical point, which I call the primordial critical circumference, to distinguish its uniqueness and its immensity from all others. Something extremely powerful and magnificently controlled had to cause that giant plasma ball to continue to explode outward and eventually travel past the primordial critical circumference so that from that point forward, the event horizon of the universe (the edge of space-time) began to extend beyond it, and the primordial critical circumference began to shrink backward.

Once material began to exit the primordial critical circumference, its strength decreased. As the material exploding outward from the Genesis Singularity exited the primordial critical circumference (whose distance was dictated by the gravitational force of all the crunched matter/energy within it), the matter that remained within began to diminish in volume. Consequently, the primordial critical circumference also diminished in strength and expanse. It began to retreat toward the original Alpha Point of the Genesis Singularity and weakened in its gravitational grip accordingly.

Hence, as material continued to exit, the effect on space-time going through the different magnitudes of that primordial critical circumference also changed accordingly. The speed necessary to overcome that initial gargantuan power of the primordial critical circumference caused space-time to stretch outward at an unbelievably enormous rate necessary to be able to punch through it. But also, as the primordial critical circumference weakened and retreated, the impact on space-time passing through it diminished.

Hence, the space-time area near the edge of our present universe, which exited first, is the one most stretched. The exact spin rate of the universe had to have been carefully calibrated from the very beginning to provide it with the necessary centrifugal force to overcome the massive gravity of the Genesis Singularity. I call this miraculously previsioned spin rate the Genesis spin acceleration constant.

Imagine the universe as a giant circular, elastic merry-go-round. As it spins, the outer edges are moving faster than the interior where the fulcrum marks the spot that the Genesis Singularity appeared—

the Alpha Point. As the universe spins, the edges are pushed outward by centrifugal force. The farther it stretches from the Alpha Point, the faster the outer edges are spinning. Most of us can recall playing on a merry-go-round as children. If we stood in the middle of the merry-go-round, there was no problem hanging on. But as we moved toward the outer edge, the centrifugal force increased until our feet were swept out from under us.

The same thing applies to our spinning universe. The outer edges of our universe are spinning much faster than the interior of our universe. This had several effects. The first is that it stretched the elastic fabric of our space-time the greatest at the outer bounds of the universe. Because the density of space-time is the lowest in this outer band, time runs the fastest.

The second effect is that time marched at different speeds, depending on the position in the universe, because space-time density differed as one moved inward from the outer band. If Tifft's red-shift measurements are correct and our universe is composed of seven concentric spheres of galactic bands, then each concentric sphere runs at a different time rate according to the density of space within each of them.

The third is that the increase in speed generated by the spin acceleration constant caused the plasma to generate enormous electromagnetic currents that began to shape our present universe. That is where the creative Birkeland currents came to play a role in condensing solid matter from plasma from the very beginning of the universe. Our Judeo-Christian cosmological model thus claims that there was no long, protracted, gradualist, evolutionary aggregation of matter through the force of gravity, but rather an almost instantaneous creation process.

The Relativistic Coordinates of Time throughout the Universe

As we move outward from the center of the universe to the edges, time moves at a different rate according to the speed of its spin and the consequent stretching of the spatial dimensions within each concentric shell. Let us say that the outer edge of our universe is 15 billion light years away. Now imagine that at first this sheet is only one light year in radius from the center of the Genesis Singularity—the Alpha Point. At the very center of the sheet is point A. That is the Alpha Point, the fulcrum where the universe began in the Big Bang that exploded from the Genesis Singularity.

On the edge of the circumference, at the outer ridges of the universe, is point Z where, for the sake of illustration, we will say there is Planet Z. Planet Z has not really formed yet, but when it forms, it will do so in this area from the material surrounding it. The material in the outer edges of the universe was the first to exit the enormously powerful and original primordial critical circumference determined by the gravitational force exuded by the Genesis Singularity.

Let us imagine that at a certain time after the Big Bang, the Alpha Point and Planet Z are only one light year away from each other. If Planet Z were to shoot a ping-pong ball from a cannon toward the Alpha Point, and if that ping-pong ball represented a photon traveling at the speed of light, it would take one year for that ping-pong ball to reach the Alpha Point. Standing at Alpha Point is a scientist in a lab coat with a big butterfly net to catch and count the ping-pong balls in his right hand, and in his left hand, he has a stopwatch to measure the rate in time.

Let us say in this imaginary scenario that for each passing second of time, Planet Z continues to shoot ping-pong balls at the same rate toward the Alpha Point. The scientist in the lab coat at the Alpha Point is thus receiving one ping-pong ball every second, but not until a year after the first one was shot from Planet Z. These ping-pong balls are all traveling at the exact speed of light, while the elastic fabric of our spatial dimensions of the universe continues to expand outward at an accelerating speed. In other words, space-time is being stretched.

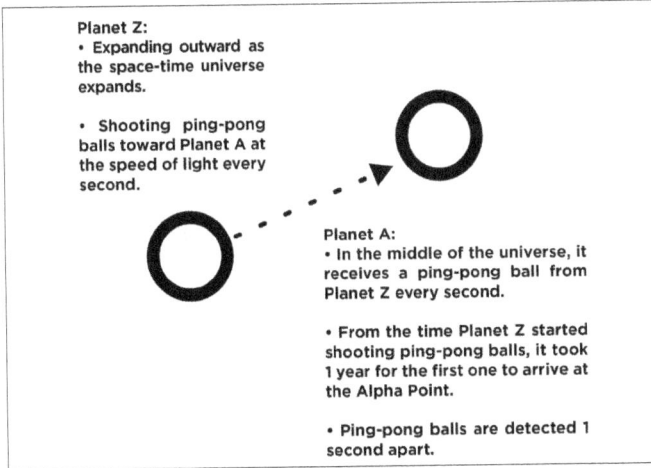

Planet Z:
• Expanding outward as the space-time universe expands.

• Shooting ping-pong balls toward Planet A at the speed of light every second.

Planet A:
• In the middle of the universe, it receives a ping-pong ball from Planet Z every second.

• From the time Planet Z started shooting ping-pong balls, it took 1 year for the first one to arrive at the Alpha Point.

• Ping-pong balls are detected 1 second apart.

The universe has expanded one light year from the Alpha Point

By the time our universe has been stretched to twice its radius (two light years), we are no longer receiving one ping-pong ball every second from Planet Z because it has twice the distance to travel; hence, we are receiving a ping-pong ball every two seconds. But on Planet Z, the ping-pong balls are still being sent every second.

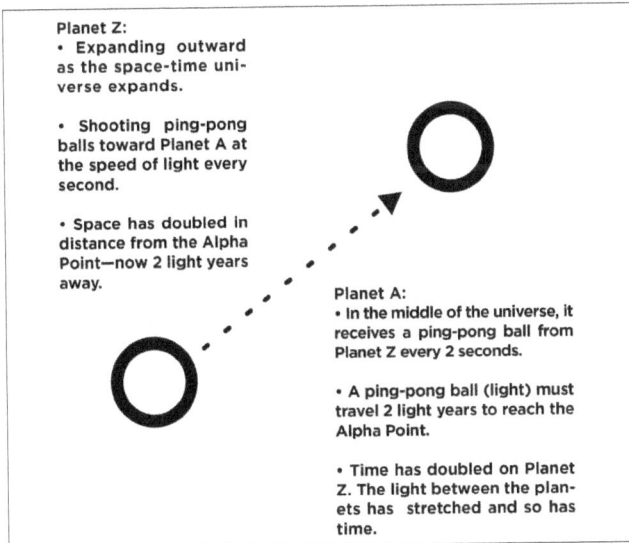

Planet Z:
• Expanding outward as the space-time universe expands.

• Shooting ping-pong balls toward Planet A at the speed of light every second.

• Space has doubled in distance from the Alpha Point—now 2 light years away.

Planet A:
• In the middle of the universe, it receives a ping-pong ball from Planet Z every 2 seconds.

• A ping-pong ball (light) must travel 2 light years to reach the Alpha Point.

• Time has doubled on Planet Z. The light between the planets has stretched and so has time.

The universe has expanded two light years from the Alpha Point

433

From the perspective of Planet Z, there is no change in time. But from the perspective of the Alpha Point, the ping-pong balls have slowed down by half. In other words, as space was stretched, time has sped up and doubled in rate on Planet Z. Each time the distance is doubled, time is also doubled in speed on Planet Z.

From the perspective of the Alpha Point, where the scientist is busy catching the ping-pong balls with the big butterfly net, time in comparison to Planet Z seems to have slowed down by half. Planet Z is actually sending out ping-pong balls at the same rate, but the scientist at the Alpha Point is collecting half the ping-pong balls as before. Time on Planet Z is now moving twice as fast as time at the Alpha Point.

If the universe is expanding at an accelerating pace, time will accelerate at the same rate at the edges of that universe. The next time space doubles, Planet Z is now four light years from the Alpha Point. Now the scientist at the Alpha Point receives the ping-pong balls every four seconds instead of the original one second. At Planet Z, the ping-pong balls continue to be shot at the rate of one per second. Time on Planet Z is now moving four times as fast as at the Alpha Point.

The next time it doubles, it is eight light years away from the Alpha Point. The ping-pong balls are now caught every eight seconds at the Alpha Point, while in Planet Z, they are still being shot out at the same rate of one per second.

That means that time on Planet Z is now moving 8 times faster than at the Alpha Point because space is being stretched to eight times the distance. We can continue this process until we come to the present edge of our universe. To the observer at the Alpha Point, Planet Z looks like it is 15 billion years old because the fabric of space has, in fact, been stretched that far and has subsequently made those objects much older. Planet Z is 15 billion years old because time there is running much faster than at the Alpha Point at the center of the universe. Time is relative to the specific coordinates within the framework of the universe.

Let us now imagine that at the Alpha Point, Planet A is formed from the material now close to the middle of the expanding universe. Viewing outward in any direction from Planet A, the edge of the universe would be equidistant, and the warping of space-time would appear the same toward the edges of the universe. Time in those fringe galaxies where Planet Z resides would have been equally stretched and sped up to be actually much older than Planet A. But all of this has transpired concurrently.

Because the waves of light that are emanating from Planet Z (the ping-pong balls) continue smoothly as space is being stretched, there is a continuous light that stretches from Planet Z to Planet A. But the wavelength on Planet Z has been stretched greatly toward the red-shift spectrum due to the stretching of space-time. The closer we move from Planet Z to Planet A, the shorter the wavelength red-shift.

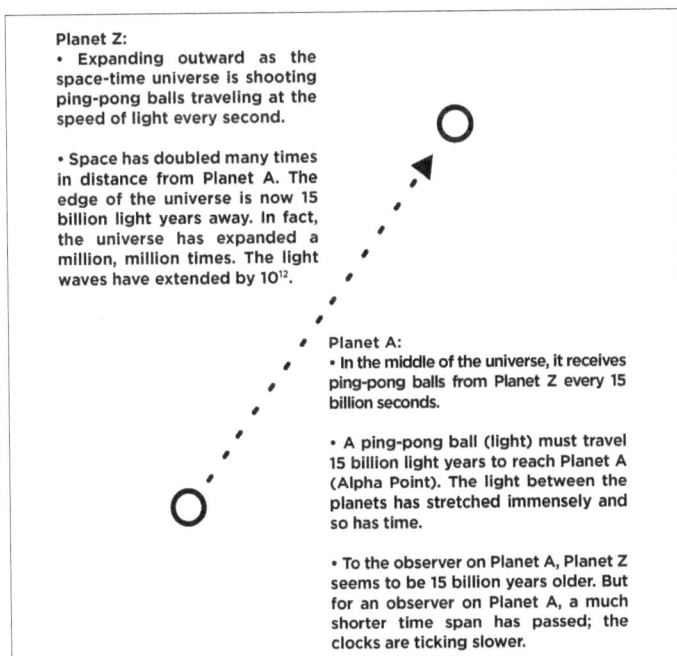

Planet Z:
• Expanding outward as the space-time universe is shooting ping-pong balls traveling at the speed of light every second.

• Space has doubled many times in distance from Planet A. The edge of the universe is now 15 billion light years away. In fact, the universe has expanded a million, million times. The light waves have extended by 10^{12}.

Planet A:
• In the middle of the universe, it receives ping-pong balls from Planet Z every 15 billion seconds.

• A ping-pong ball (light) must travel 15 billion light years to reach Planet A (Alpha Point). The light between the planets has stretched immensely and so has time.

• To the observer on Planet A, Planet Z seems to be 15 billion years older. But for an observer on Planet A, a much shorter time span has passed; the clocks are ticking slower.

Our present universe 15 billion light years from the Alpha Point

435

By the time we on Planet A, close to the center of the universe, receive the photons from Planet Z, Planet Z has already stretched out further from us. We are therefore looking at the past of Planet Z when we see the photons arriving because from the moment they were shot, they traveled 15 billion light years to get to us. But in reality, time on Planet Z is really in the future of Planet A because time has accelerated greatly by the stretching of space-time in that region. The stars and galaxies in that area have already aged greatly. For this reason, we find in those outlying areas the greatest preponderance of quasars, as we previously noted. Evolutionists are dead wrong when they declare the far reaches of our universe to be glimpses of the early universe. They have failed to understand the relativity of time in relation to the stretching of space-time.

On Planet A, perhaps only several thousand years have passed, but on Planet Z, some 15 billion years have passed from the viewpoint of Planet A. The guys shooting the ping-pong balls on Planet Z seem to have aged exponentially at the same accelerated pace that space was being stretched outward from Planet A.

Each time the universe expands to double in volume, the density reduces by one-half. This also reduces the overall gravity within a given gradient of space, which means that the clocks run faster in that given gradient. Time, influenced by the immense gravity of our sun, runs at a slower pace than time at the outer reaches of the heliosphere beyond the planets.

If you think of the universe as matter that exited a critical circumference of a singularity, the Big Bang, then material exiting first would become those objects at the edge of the universe. The material exiting last would be those celestial bodies nearest the center of the universe. If the objects that first exited the critical circumference are at the edge of our universe, then when we view those stars, we are, in effect, looking into the future fate of our location in the universe near the Alpha Point. Because time moves faster at the edge of the universe where space-time is stretched the most, we are, in essence, seeing not the beginning of stars, as evolutionists try to imply, but the death of stars and galaxies. That area of our universe has aged the most. We are watching not the birth, but rather the death of galaxies.

In time, the fabric of space-time will rip the outer edges of our universe. That is what we are observing as these quasars are growing and eating up their galaxies. Matter is being burned up, and the space-time fabric is being ripped by the giant singularities being formed, as the second law of thermodynamics dictates.

It is for this reason that we view these galaxies at the very edge of our universe rotating at such enormous speeds. Time at the edges of our universe is in fast-forward. Scientists have marveled that they seem to be moving faster than the speed of light, and yet Einstein tells us that this is impossible. In reality, they seem to be moving faster than the speed of light because their time is in fast-forward, and our rate of time is so much slower.

Matter at the edges of our universe observes the same laws as matter in our neighborhood of the universe. There is no dark matter necessary as a mathematical fix to keep the galaxies from flying apart due to centrifugal forces, because from their perspective, it is behaving exactly as Newton and Einstein have told us they should within their time rate. Their apparent speed to us is the result of the difference in the rate of the passage of time on Planet Z and the rate of the passage of time on Planet A.

The Moment Energy Turned to Matter

Prior to the ejection of the universe through the primordial critical circumference, we have the condensation of matter from the initial radiant energy released by the Big Bang. The creation of physical matter from energy during the Big Bang gives us a hitching post in time from which we can measure the Alpha Point. We can measure the density of our universe when the plasma state began the process of condensing photons into quarks inside the critical circumference of the Big Bang. That measurement can be computed through Einstein's equation to be approximately a million million times hotter than the present background radiation. The temperature of any electromagnetic radiation determines the frequency of its waves. Hence, the microwave background radiation we measure in the space around us today has cooled a million million times to its present state. Working backward from our present to that initial point gives us the difference between the Alpha Point and now.

That is, we can measure the present density of our universe through the temperature gradient of the background microwave radiation. This is the leftover microwave radiation from the Big Bang, which is called cosmic microwave background radiation (CMBR). It is the temperature in the black of space, and it is everywhere almost uniformly at 2.73 degrees Kelvin, or -270 degrees Celsius (except in those areas of anisotropy discovered by the three mapping satellites). This gives us an idea of how much our universe has expanded. The difference between those two densities should give us an idea of how much the universe has expanded since matter was birthed from the plasma soup of the Big Bang.

Schroeder calculates that the universe has expanded a million million times in size since the time of the Big Bang. The wavelength of the radiation from that moment in the Big Bang (quark confinement) has been stretched lengthwise a million million times. Its red-shift (Z) has therefore been observed today to be 10^{12}.

When I say "observed," I do not mean with the naked eye. Our vision allows us to see only a tiny sliver of the electromagnetic radiation. At the blue end of the light spectrum, we can see 0.00001 centimeter of a wavelengths of light, which our brains interpret as a blue color. At the longer end of the spectrum, we see 0.000000001 centimeter of a wavelength, which our brains interpret as red.

Hence, when the wavelengths are stretched, they are red-shifted. By stretched, I mean that the distance from wave crest to wave crest increases, lowering the frequency, and subsequently, the energy of the radiation is also reduced. The highest energy electromagnetic radiation is the gamma ray, which has very tightly packed waves (high frequency) and can only be stopped by thick lead shields. All the electromagnetic waves have been stretched as space-time has been stretched outward since the Big Bang.

The fact that electromagnetic radiation has stretched by 1012 means that our cosmic timepiece, as observed today on planet Earth, is a million million times slower on Earth than it was as matter was first exploding outward during the Big Bang. What do I mean by a cosmic timepiece? Electromagnetic radiation is the cosmic timepiece. It is the only measure of time that can allow us to backtrack to the beginning of our universe.

> *The timepiece of the universe is not manufactured by a watchmaker, skilled though that craftsperson may be. The clock of the universe is the light of the universe. Each wave of light is a tick of the cosmic clock. The frequencies of the light waves are the timepieces of the universe.*
>
> *Waves of sunlight reaching Earth are stretched longer by 2.12 parts in a million relative to similar light waves generated on Earth. That stretching of the light waves means that the rate at which they reach us is lowered by 2.12 parts per million. This lowering of the light wave frequency is the measure of the slowing of time. For every million Earth seconds, the Sun's clock would "lose" 2.12 seconds relative to our clocks here on Earth. The 2.12 parts per million equals 67 seconds per year, exactly the amount predicted by the laws of relativity (Schroeder 52).*

When we look out toward the area of the universe that first exited the primordial critical circumference at the edge of our present universe, we are observing an area of space-time where the clock is extremely fast compared to ours on Earth by the rate of 1012. We know that light as well as space and matter can be bent and stretched. As a matter of fact, the empirical proof of Einstein's equation was the recording of starlight being bent by the gravity of our sun during an eclipse. Because space-time is warped by gravity, we observe a change in the position of the starlight traveling through it. This is called gravitational lensing. Thus, a star that should have been hidden behind the sun was bent around it, and we were able to capture it on a photographic plate during a solar eclipse.

Returning to our Planet Z at the outer edge of our universe and Planet A near the center or fulcrum of our spinning universe, from the perspective of Planet A, the light of the star in Planet Z is dimmer and shifted to the red spectrum because it is moving away from Planet A near the Alpha Point. Each time the universe doubles in size, the rate of time doubles in Planet Z, as space is stretched outward. In this way, the stars and galaxies at the edge of our universe seem to us on Earth as 15 billion years older than we are during the same time interval because time has moved faster, and 15 billion years have actually passed in this area of space-time.

Why does the edge of our universe look like it is speeding up? There are two reasons:

1. Because at the initial explosion from the critical circumference of the white hole singularity of the Big Bang, matter had to be moving with greater force and speed in order to escape the gravity of that enormous Genesis Singularity and pass through the primordial critical circumference. In order for matter to exit such a strong gravitational pull, which includes all the matter in the universe squeezed into this Genesis Singularity, it had to be exerting an outward force that was greater than the gravitational force pulling it into an infinite density. The energy gradient at that time was 10^{12} times hotter than our present microwave background radiation. Matter exiting the continuously diminishing strength

of the critical circumference would thereafter have been less affected by its remaining gravitational force, since the reduced matter inside the white hole would affect the strength of the gravitational force at the primordial critical circumference.

Matter exiting the white hole later would need to overcome less gravity. It would not need to be moving as fast as matter exiting first from the initial maximum critical circumference of the Big Bang. In addition, matter exiting last would have had less time outside the critical circumference of the singularity. Matter trapped inside the extreme warp of a white hole as seen from an outside observer (that is, if we had a special telescope to see beyond the circumference, which we cannot) would be seen as if time had stopped. Hence, the matter that exited first had more time outside the critical circumference of the Primordial Singularity.

2. Once matter exited, that gargantuan gravitational force caused the stretching of space-time at an enormous rate during that initial phase. The centrifugal force exerted on the outer edges of our spinning universe forced it outward and stretched the space-time fabric in proportion to its distance from the fulcrum. Since there is no substance in hyperspace outside our universe to cause the spin rate to stop, the spin continues to accelerate at the same expanding rate, ever increasing in speed, until it reaches a point in which space-time can no longer stretch. That is the Omega Boundary where space begins to rip. That rip will then work backward toward the center until the Omega Point.

What is miraculous in the first place is that matter could escape such a gargantuan gravitational force when the Genesis Singularity had all that is contained in our present universe crunched into a coordinate with infinite density. Evolutionists have no answer that can explain in materialistic terms how that infinite density could be broken. There are no mathematical equations that can be used with infinite numbers. Hence, their argument simply sidesteps this reality by saying that it is beyond the scope of science. They are partly right. It is beyond the scope of our finite material world.

What it implies by necessity is that only an infinite force can counter infinite density. That is direct evidence that only an infinite omnipotent God could cause the Big Bang to explode outward and overcome the gravitational power that causes infinite density. An objective scientist should at least consider this as a possibility. But it is their underlying metaphysical foundation that causes them to outright reject this as a rational solution to the problem. The atheistic foundational presupposition of the evolutionary process excludes a priori the consideration of God being the master architect of our universe.

Moreover, to top it off, the shepherding of matter through the primordial critical circumference was done in a fluid, constant motion. It was not a random, chaotic explosion as insisted by the evolutionary metaphysical doctrine of dysteleology. It was an orderly, regulated explosion that maintained an almost complete homogeneous, isotropic fabric throughout. If not, our universe would have become lumpy rather than smooth. We would have ended up with a universe of black holes rather than galaxies and stars. Something extraordinarily powerful caused it to explode outward in a completely ordered and obviously previsioned fashion.

Unlike any other explosion that has ever existed and has been thoroughly studied and analyzed in our universe, the largest explosion ever was controlled in complete and utter acquiescence to the will of God so the end result was the smooth transition of the expulsion of space-time that allows for stars and galaxies to exist. Divine intervention is the only rational explanation that can explain why the Big Bang resulted in the smooth, almost homogeneous, isotropic universe we see today.

There are not and can never be any materialistic mechanisms that can explain that reality. The findings of the mapping satellites of the symmetrical areas of anisotropy that are aligned with the sun-Earth ecliptic and our equinoxes automatically disqualify the Rapid Inflationary Model that was propped up to try to explain the almost homogeneous and isotropic universe. It is direct evidence that no materialistic process ruled by random ordering could have created our universe. The design left for us to observe shows the fingerprint of a symmetrical universe designed by a master mind who knew that one day we would discover this magnificent message in the cosmos.

Radiant Energy and Time

It may have surprised you to find that electromagnetic waves are not caught in the web of time as matter and the spatial dimensions are. For this reason, light is never interrupted as time either speeds up or slows down. That is why we can see galaxies at the far edge of our universe that, from our perspective, may be 13 to 15 billion light years away from us. But nevertheless, the electromagnetic waves are stretched as space is stretched. That is, the wavelengths are elongated as space is stretched.

> As science has discovered, radiant energy does not experience the flow of time. Radiant energy, such as the light you are seeing this very moment, exists in a state that might be described as an "eternal now," a state in which time does not pass by (Schroeder 58).

Again, we find that light is an appropriate symbol for the timeless creator who brought all things to be from His radiant energy. And so we see that the very edge of the universe is, in fact, the area of our space-time universe that first appeared through the primordial critical circumference at the very beginning of the universe. It was at that time that matter became visible as light, and electromagnetic waves such as the background microwaves escaped the density of the primordial plasma universe inside the primordial critical circumference of the Big Bang.

At C – 1 (the point in hyperspace, without the singularity), there was no universe, no space, no time, no matter. At C, the creation point (the time of the creation of the space-time continuum), time and space began. All of the matter/energy contained in the universe was found within the Genesis Singularity and the immediate space containing the critical circumference as a white hole. This Genesis Singularity graphically represents the big daddy of all the singularities because it contained the entire mass/energy (or potential mass) of our universe concentrated in one point infinitesimally small. That, by necessity, would have had to be introduced from outside our universe, for no other possible alternative is found within the system of the universe that can explain the genesis of this event.

Diagram 14.1: The Creation of Our Universe

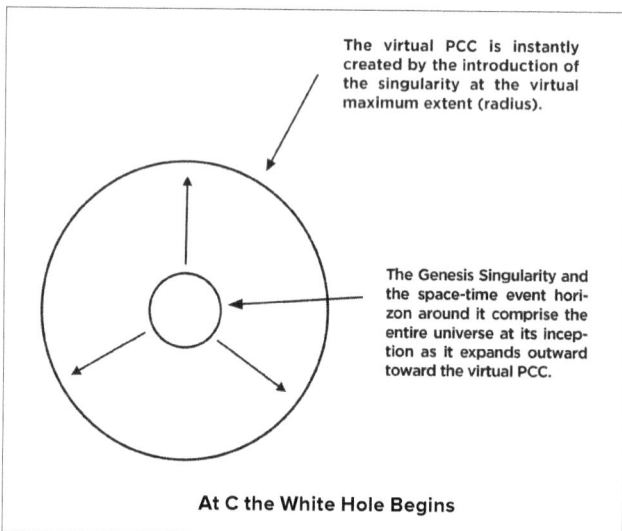

The virtual PCC is instantly created by the introduction of the singularity at the virtual maximum extent (radius).

The Genesis Singularity and the space-time event horizon around it comprise the entire universe at its inception as it expands outward toward the virtual PCC.

At C the White Hole Begins

At C + 1, matter continuously expands outward in plasma form as the space-time universe is reacting to the Genesis spin acceleration constant that will determine the precise velocity necessary to escape the primordial critical circumference.

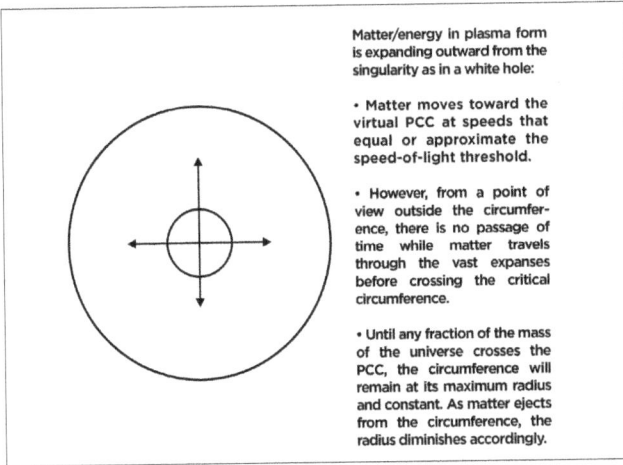

Matter/energy in plasma form is expanding outward from the singularity as in a white hole:

• Matter moves toward the virtual PCC at speeds that equal or approximate the speed-of-light threshold.

• However, from a point of view outside the circumference, there is no passage of time while matter travels through the vast expanses before crossing the critical circumference.

• Until any fraction of the mass of the universe crosses the PCC, the circumference will remain at its maximum radius and constant. As matter ejects from the circumference, the radius diminishes accordingly.

At C + 2, matter begins to eject from the white hole. The critical circumference now begins to diminish in radius.

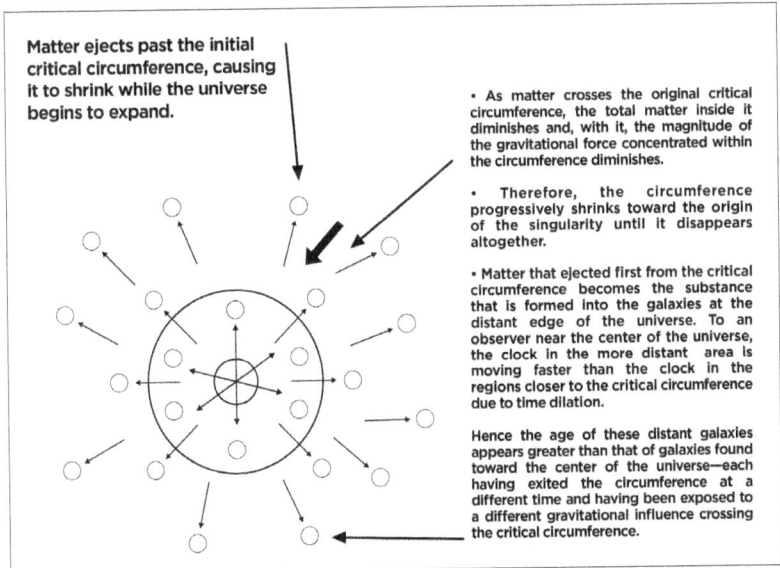

Matter ejects past the initial critical circumference, causing it to shrink while the universe begins to expand.

• As matter crosses the original critical circumference, the total matter inside it diminishes and, with it, the magnitude of the gravitational force concentrated within the circumference diminishes.

• Therefore, the circumference progressively shrinks toward the origin of the singularity until it disappears altogether.

• Matter that ejected first from the critical circumference becomes the substance that is formed into the galaxies at the distant edge of the universe. To an observer near the center of the universe, the clock in the more distant area is moving faster than the clock in the regions closer to the critical circumference due to time dilation.

Hence the age of these distant galaxies appears greater than that of galaxies found toward the center of the universe—each having exited the circumference at a different time and having been exposed to a different gravitational influence crossing the critical circumference.

At C + X, for all matter ejected from the white hole, time is relative to the position it has in relation to its distance to the Alpha Point of the Genesis Singularity.

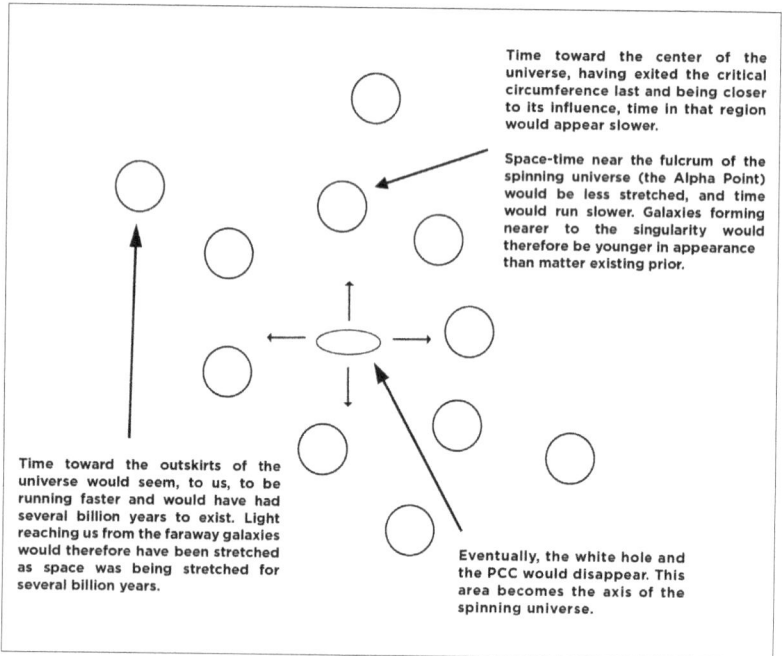

Time toward the center of the universe, having exited the critical circumference last and being closer to its influence, time in that region would appear slower.

Space-time near the fulcrum of the spinning universe (the Alpha Point) would be less stretched, and time would run slower. Galaxies forming nearer to the singularity would therefore be younger in appearance than matter existing prior.

Time toward the outskirts of the universe would seem, to us, to be running faster and would have had several billion years to exist. Light reaching us from the faraway galaxies would therefore have been stretched as space was being stretched for several billion years.

Eventually, the white hole and the PCC would disappear. This area becomes the axis of the spinning universe.

For this diagram:

"C – 1" is the point in hyperspace without the creation of the Genesis Singularity or white hole. Our universe does not exist.

"C" is the creation point when the singularity and the white hole are created.

"C + 1" is all matter/energy within the white hole, and expanding toward the primordial critical circumference (PCC) is at its maximum extent.

"C + 2" matter begins to eject beyond the primordial critical circumference. The PCC begins to shrink and reduce in power.

"C + X" is all matter that has exited, and the white hole disappears.

Because space at the beginning of our universe was stretched faster than at the end of the Big Bang, light reaching us from these faraway galaxies would have had several billion years to reach us, while Earth could have aged only several thousand years during the same interval. In other words, the perceived passage of time near the center of the universe is much slower than the time observed toward the edge of the universe, which exited the critical circumference first.

This not only explains the differences in the perceived age of both these places, but it also explains the proportional speed of expansion toward the edge of our universe and the proportionately faster spins of galaxies in these outer regions of our universe. Matter exiting first from the explosion of the Big Bang would be moving faster than matter exiting last from the explosion.

This idea that the Big Bang is, in fact, a white hole was proposed by Russell Humphreys in his insightful book *Starlight and Time*. The idea of a white hole as the Big Bang is, in fact, within the bounds of the equations of general relativity. Humphreys explains his brilliant idea:

> Given a bounded universe that was once fifty times smaller, the other possibility allowed by GR is that the universe was previously in a huge white hole. This is a black hole running in reverse. Astrophysicists of the 1970s gave that name to the concept, arising from theoretical studies of the black holes. The name never really became popular, but the concept is still considered valid today.
>
> Like a black hole, a white hole would also have an event horizon. Matter and light could exist inside its event horizon without any particular problems. There need be no singularity at its center, except perhaps at the very beginning of its existence. However, the equations

> *of GR require that light and matter inside the event horizon of a white hole must expand outward.*
>
> *The event horizon of a white hole would be a one-way border that permits only outward motion through itself. Matter and light waves would have to move out of a white hole, but they could not go back in. Since the diameter of an event horizon is proportional to the amount of matter inside it, the event horizon would shrink as matter passes through it and out of the white hole. . . . In the same way, the event horizon would get smaller and smaller, and eventually shrink to nothing. There would then be no more white hole, but only scattered matter moving away from a central point (Humphreys 24–25).*

Since the red-shifted stars are so overwhelmingly more prevalent than the significantly few blue-shifted stars, the observable empirical data substantiate the fact that the vast majority of stars are moving away from us in a central area near the place where the white hole began our universe. Possibly the only ones moving toward us are those that, through random collisions, may have had their initial courses changed.

It further insinuates that the universe has a starting center, the direction opposite to the movement of all the galaxies, as the universe expands. And since it just so happens that they are expanding away from us, it makes us at or, at the very least, close to the center of the universe.

At any rate, no matter what the age of the universe is, the point is that there is an existence. And the general relativity equations tell us that this existence had a definite genesis. Hence, in light of our discussion, it can be said that our Judeo-Christian cosmological model has been shown to be just as plausible, scientifically, as the naturalist model. And in fact, it can explain our universe better than the naturalist model. In spite of the opposition and ridicule it received in the last 300 years, the Genesis narrative is the most rational explanation for existence.

The Secret of the Expansion Rate of the Universe, the Spin Speed of Distant Galaxies, and the Genesis Acceleration Constant

Before delving into the Genesis acceleration constant, let us quickly recap what we have learned. It seems clear to me that there must be some other explanation for the expansion rate of the universe and for the speed of the rotating galaxies far from us than this magical dark matter and dark energy proposed by evolutionists. There is no known mechanism for any repulsive force acting stronger in proportion to the distance. As we have already shown, it may be possible to explain this as simply an illusion created by the process of the creation of the universe.

We have also already documented that there is an expansion rate that has been precisely chosen to allow life to exist in this universe. It is an amazingly fine-tuned, precise expansion rate, which, if altered by one in one quintillion, it would simply not allow for our universe to exist as it does. A quintillion is a one with eighteen zeroes. To get a mental handle on that enormous number, consider it to be the equivalent of one grain of sand in all the grains of sand on Earth.

It is an irrefutable fact that the expansion rate must have been carefully chosen in order for our universe to exist as we know it. If that rate of expansion were to have been just a bit slower, then our universe would have collapsed into a Big Crunch. If the rate were just a fraction faster, the universe as we know it could not have formed. Not even the atoms and nuclei could have formed. Our universe would have expanded too fast for the condensation of matter.

The additional attractive force would eventually over-whelm the outward motion of the Hubble expansion; the universe would reverse its motion and start to collapse like a punctured balloon. Galaxies, stars,

*planets, and all of life would be crushed in an ulti-
mate "big crunch." If the cosmological constant were
too large, the crunch would not allow the billions of
years necessary for life like ours to evolve. . . . If the
cosmological constant is negative, it must also not be
much bigger than 10^{-120} Units if life is to have any possibility of evolving. . . .*

*In the ones with large positive λ, everything flies
apart so quickly that there is no chance for matter
to assemble itself into structures like galaxies, stars,
planets, atoms, or even nuclei (Susskind 83).*

The specificity of this acceleration constant necessary for our
universe to have developed the way it has is so exact that it is,
for all intents and purposes, impossible to conclude that random
ordering could have been responsible for such fine-tuned exactness.
Of all the possible expansion potentials, it would be considered a
certifiable and authentic miracle of the first order that random
ordering could have serendipitously fallen on that exact expansion
rate of 1 out of 10^{120} possibilities that makes it possible for matter,
stars, and galaxies to exist.

Let us be more specific. When measuring the expansion rate of
our universe and the rotation rates of galaxies, we use equations that
include distance, mass/energy, and time. Evolutionary cosmologists,
seeing the expansion rate that is actually experimentally calculated,
try to adjust the equations by adjusting the mass/energy component.
Thus, copious amounts of so-called dark matter and dark energy are
added to make the equations work. But instead of changing the mass/
energy components of the equation, what they need to adjust is the
time component of the equation.

The reason galaxies seem to be spinning faster and faster the
further out we can see in our universe is due to the increase in the
speed of time as space-time in these areas is stretched in proportion

to their distance from Earth. In those regions of the universe, time has been accelerated proportionally to the distance from us because space has been stretched in proportion to the distance from us. The differential is caused by the speed of time and not by an increase in invisible mass. Dark matter simply does not exist. Dark energy simply does not exist. It is the process of the Big Bang that has created time dilation differences throughout the length and width of our universe as the different regions of space-time exited the event horizon.

How do we know it is expanding? We know because of the red-shift measurements. What is causing this expansion? What modern cosmologists see as the enormous force of dark or vacuum energy is, in fact, the product of a Genesis spin acceleration constant that began with the Big Bang. It is the Genesis spin acceleration constant that has created different time frames throughout the length of our universe. The further out the universe is stretched by the angular momentum and centrifugal force of its spin, the more that time in each concentric region, beginning from the center and spanning outward, is accordingly accelerated. The more it is stretched at the edge of the universe, the faster time goes by. The less it is stretched toward the center of the universe, the slower time goes by.

What is the Genesis spin acceleration constant? We tend to think of our universe expanding outward like a linear explosion, without any spin. But what if the entire universe were spinning at an acceleration rate that has been constant from the beginning? If we could have analyzed in concentric circles the plasma filling the entire early universe as it expanded outward, we would see that from the center to the outer rings, the universe would have been spinning at an ever-increasing rate of acceleration. The outside rings would be spinning much faster than the inside rings. In fact, the rates would be proportional to the distance from the Alpha Point.

The difference in their velocities is due to two things. (1) The increase in the angular momentum of a spinning sphere created by centrifugal forces causes the outer areas to be spinning much faster than the interior regions. (2) The difference in the rates of time within those bands of space-time, which are stretched the most, causes time to accelerate accordingly. Because time is moving faster, all movements in those areas are also accelerated. Since there are no forces in hyperspace acting upon the universe to slow it down, the acceleration rate becomes a permanent feature. It is like the eternal spin of subatomic particles.

This spin in the universe would then be the cause of the centrifugal motion that, through angular momentum, forces matter outward from the center. The spin acceleration constant is thus the reason for our expansion, and there is no need for a magical antigravity force. Dark energy does not exist except in the minds of those who wish to deny the existence of a creator.

Those of us inside the universe could not perceive the universe spinning nor could we possibly observe it, because we have no point of outside reference with which to judge this. We are like an ant on the surface of the Earth without the capacity to see that the Earth is spinning. But if we look around us in space, almost everything is spinning. At the moment, there is no clear explanation as to why all things are spinning. But if the universe itself were spinning from the beginning, it could explain why matter exiting the event horizon of the Genesis Singularity could be sent spinning as it cooled and condensed from the plasma state. In addition, the Birkeland currents may also provide a mechanism for the spinning bodies that were solidified from plasma by their twisting forms.

In other words, from the precise moment of the Genesis Singularity exploding outward, God set a precise acceleration rate that governed the speed of the spin of the Big Bang, which determined the expansion rate at exactly the speed necessary to overcome the immense gravity of the Genesis Singularity and cross the primordial critical circumference to bring forth our universe. This is the exact expansion rate that is so exact as to be one chance in 10^{120}, which we have already documented.

I believe that God gave the Genesis Singularity an initial acceleration spin, which was simultaneously in at least two directions, which could be considered east and west as well as north and south. That would create a spherically shaped universe that looks isotropic in every direction and is expanding outward in a spherical fashion.

If the universe was actually spun from the moment of the Big Bang with the spin acceleration constant, then it would predict that this movement in the speed of the rotation of galaxies is also directly proportionate to their distance from us. In those areas of space far to the edge of our universe, where time runs faster, the speed of the rotation of the galaxies would vary according to distance and would seem to be much greater than the mass that we have calculated within them. It is the difference in the rate of their time to our time that gives us the illusion that there is dark matter causing it to hold together as it spins faster. This Genesis spin acceleration constant would then explain both dark energy and dark matter without resorting to some magical repulsive force that increases strength with distance.

This can be tested. If dark matter really exists and makes the galaxies spin faster and not break apart, then there would be no reason to suspect that these spin rates would vary with distance. In fact, in a randomly ordered universe, we would expect that dark matter would be randomly distributed, and these effects would be out of synchrony with the variations in distance. But if we find that the spin rates of these galaxies are also determined by their distance from us, then it can be safely concluded that the differences in time rates are the factor involved in this phenomenon.

Will the Universe Expand Forever?

Will the universe expand forever? Will all matter eventually fly apart as the universe continues to expand? I don't think so. There are three things to consider:

1. Einstein's theory tells us that nothing can go faster than the speed of light. As matter approaches the speed of light,

it begins to gain more mass so it can never quite reach the speed of light. Hence, it is logical to assume that as the Genesis spin acceleration constant creates an expansion rate that nears the speed of light, matter will begin to build mass and resist the expansion force. In other words, there is an upper limit to the speed of matter and to the expansion of our universe within each of the concentric shells of galactic bands that surround our sun-Earth ecliptic.

2. I also suspect that there is a limit to how much space can be stretched before the fabric rips. I suspect that the zero-point energy that powers the spatial dimensions may become impotent at a certain point of expansion. The intrinsic energy of space may fail at a given point as its density becomes so low that it reaches a threshold limit. I call this the omega boundary.

3. I also believe that as time speeds up, so does the second law of thermodynamics. The increase in entropy is directly tied to time. At a given moment, the total entropy of the universe will be so great that all galaxies will become supergiant quasars. These, then, will swallow one another, and space-time will be consumed by these voracious and gargantuan black holes.

This means that our universe will have a finite size. It will not expand to infinity. It has a beginning, and it has an end. It has an Alpha, and it has an Omega. No matter how hard the naturalists want to input infinity to our universe, they cannot.

Furthermore, there is the matter of the second law of thermodynamics. The universe is increasing in entropy. Suns will burn out, and galaxies will be swallowed by supermassive black holes. It is more likely that as the universe ages in the outer regions, those regions will be populated by supermassive black holes that will swallow all matter around them until they swallow each other and perhaps even rip the very fabric of space-time. This process will then run backward toward the interior of the universe as entropy increases.

Now that does not mean that there can be no quasars in the interior shells. The second law of thermodynamics is universal. Black holes exist everywhere. But if our Judeo-Christian cosmological model is true, we can predict that the greater the amount of time that has passed, the greater the total entropy of that shell will be. Hence, the outer or seventh shell can be predicted to have the greatest number of quasars, and these should also be the greatest size. Moreover, that ratio will continue to increase proportionally as the universe continues to expand outward.

I do not think that the universe will expand into the scenario often called heat death where even atoms fly apart into a cold death. I see the future of these outer regions being dominated by ever increasingly large supermassive black holes that could tear the very fabric of space-time.

CHAPTER 15

● ● ●

THE SIX DAYS OF GENESIS

How then do we reconcile the six days of creation recorded in Genesis with the Big Bang and the supposed time frame that the physicists say were necessary for (1) matter to coalesce from the plasma stage and (2) the elements to form into our observable universe?

First, it must be said that Genesis is not a physics textbook. It is a book of truth that has been intentionally designed to communicate truths in a way that would be understandable to the audience it serves. Sometimes, deep truths are communicated through symbols and parables that allow the truths to be understandable to the common person. But most of the time, it is a simple, historical narrative that provides for us a true picture of human frailty and God's amazing grace.

The scriptures claim to be the word of God inspired by the Holy Spirit. If, in fact, it is God's word, then it must, in my mind, be a message of truth. When it speaks of historical events, it cannot contain lies, or it is not the word of an omnipotent God.

> *For no prophecy was ever produced by the will of man, but men spoke from God as they were carried along by the Holy Spirit (2 Peter 1:21).*

If the Scriptures contain a fallible message, then there is no rational way to conclude that the prophetic message can be trusted. But Jesus said it could:

> *Then he said to them, "These are my words that I spoke to you while I was still with you, that everything written about me in the Law of Moses and the Prophets and the Psalms must be fulfilled" (Luke 24:43–45).*

The Law of Moses, the prophets, and the Psalms comprise the entirety of the Hebrew Scriptures. In fact, He declared that every small portion of the writing was God-ordained and would be fulfilled. Furthermore, He warned of the consequences of annulling any letter or even a single stroke of God's word.

> *"Do not think that I have come to abolish the Law or the Prophets; I have not come to abolish them but to fulfill them. For truly, I say to you, until heaven and earth pass away, not an iota, not a dot, will pass from the Law until all is accomplished. Therefore whoever relaxes one of the least of these commandments and teaches others to do the same will be called least in the kingdom of heaven, but whoever does them and teaches them will be called great in the kingdom of heaven" (Matthew 5:17–19).*

Those Christians who deny the historicity of the Bible while yet claiming that it contains God's message attempt to allegorize the historical narratives in order to force fit them into the evolutionary ideology. They are, in essence, making the human mind the interpreter and inventor of God's message. When we do this, we neglect the letter of the word and develop from our finite mind some symbolic meaning of our own will and interpretation. I do not believe that God needs the human finite mind to reinterpret His message. Either it is true or it is false, and the data will either prove one way or the other.

It is also important to understand that the popular understanding of science at any given age may, or may not, be in agreement with the scriptures. It has been my observation that the data have consistently shown that God's word is, in fact, a true historical narrative. Time and time again, critics who have claimed otherwise have been disproved by new archaeological or scientific data. As more information is gathered and science progresses, we consistently find that science coincides with the scriptural narrative. Such was the case with the Steady State theory that claimed the universe was infinite in space and time, which was so popular for several hundred years until Einstein's theories showed that the universe indeed had a beginning.

Such has been the case with every evolutionary mechanism proposed as a means for the speciation of living things. Jean-Baptiste Lamarck's idea of acquired inherited characteristics was touted as scientific fact for many decades until geneticists showed it to be absolutely wrong. Such was the case for mutations as the mechanism of the evolution of the species. For half a century, scientists have been searching for a single mutation that could be propped up as their poster child, but they have failed miserably. But that is a subject we will cover in Book 4 of this series, *The Descent of Man*.

My confidence in the scriptural narrative does not come from blind faith. It comes from the observation that time after time the biblical narrative has been vindicated by true scientific inquiry. That rational foundation then opened the door of faith and brought me to a personal relationship with God. I have read the Qur'an, the Mahabharata, the Rig Veda, the Bhagavad Gita, and many other religious works, but none have met the rational standard of the

Holy Scriptures, and none have impacted my soul in a mystical and yet very tangible way that defies description. God's word is truly a living word since it speaks to my soul in a supernatural way that would be useless for me to try to explain since it cannot be proved in any concrete way; it simply must be experienced in order to be understood.

I do not presume to say that mystical experiences do not happen outside the Judeo-Christian worldview. I am simply saying that the Judeo-Christian worldview is the only worldview that has its mystical experiences based on a rational component.

While many prophetic messages are written in highly symbolic fashions, the historical narratives, in my opinion, should not be allegorized, but taken at face value. Thus, attempts by Christians to allegorize narratives such as the story of creation in order to subjugate God's message to modern interpretations of the scientific data are, in my mind, a grievous error.

Many well-meaning Christians, when confronted with the absolute dogma of evolutionists who claim the universe was infinite in space and time, went to tremendous lengths to invent pre-Adamic races and even to merge evolution with theism. In the end, all of these "evangelastic" attempts to stretch scripture into an evolutionary box have been shown to be foolish. Einstein's mathematical equations vindicated the biblical narrative in regard to the origin of the universe. The universe had a beginning. Space-time had a beginning. Einstein soundly refuted all the evolutionary dogma regarding the origin of our universe. The scriptures were vindicated.

In fact, his understanding of the relativity of time with respect to the observer is also absolutely essential in order to understand the relative ages of the celestial bodies in our universe. It also gives us the framework from which we can understand Genesis. We mentioned earlier that the first six days of Genesis were written from the point of view of God, not humans. Man was not created until the sixth day. God is the only reference point outside of time, for He created time.

Hence, He alone can offer a hitching post for a time scale for the creation of the universe outside the confines of the universe.

Now evolutionists are eager to disregard the Genesis narrative by claiming that the timetable of Genesis cannot be reconciled with the age of our universe, since the farthest stars in our universe are 15 billion light years away from us. This is their primary argument against the scientific authenticity of the Genesis narrative. As we have already documented, our Judeo-Christian cosmological model does not dispute the age of the edge of our universe being 15 billion years old. We have already made the case that the age of our space-time area in the universe is dependent on the density of that space-time coordinate.

Using Einstein's equations, we can measure by working backward from the CMBR to the point that scientists call the moment of quark confinement—the moment energy turned into matter. We know from measuring the length of the frequency of the background microwave radiation and comparing it to the frequency during quark confinement, when matter was created, that we can figure both the difference in energy between these two points and how long the lightwaves have been elongated. This gives us a measure of time. Schroeder explains this seeming paradox.

> We know the temperature and hence the frequency of radiation energy in the universe at quark confinement. It is not a value extrapolated or estimated from conditions in the distant past or far out in space. It is measured right here on Earth in the most advanced physics laboratories and corresponds to a temperature approximately a million million times hotter than the current 3°K black of space. That radiant energy had a frequency a million million times greater than the radiation of today's cosmic background radiation.

> *The radiation from the moment of quark confinement has been stretched a million-million fold. Its redshift, z, as observed today is 10^{12}. That stretching of the light waves has slowed the frequency of the cosmic clock—expanded the perceived time between ticks of that clock—by a million, million (Schroeder 59).*

That moment when light shined through was the first tick of the cosmic timepiece for us, observing outside the primordial critical circumference (PCC) of the Genesis Singularity. It is at that point that the PCC was crossed and space-time exited the cocoon of the primordial critical circumference. The true moment of quark confinement inside the PCC is obscured to us behind the PCC. All we can measure is the temperature or the wavelength of electromagnetic radiation once it became visible from behind the PCC. Time existed prior to that point, but from our perspective, it was standing still.

The matter that crossed over eventually became the stars and galaxies of the outer edges of our space-time continuum. Matter that exited last remained close to the Alpha Point as the PCC retreated all the way back to the Genesis Singularity that was now spent.

What that means is that from the point of view of the Earth now (near the Alpha Point), looking to the ends of our universe, our clock is ticking one million million times more slowly than time ticks in those farthest galaxies. Space-time at these outer fringes is stretched immensely, causing time to speed up. And even though the light of these stars at the edge of our universe are 15 billion light years away from us, time at the edge of our universe is in fast-forward, and therefore, we are seeing the future events that will occur to our stars and galaxies in the Milky Way. There, at the far edge of the universe, stars and galaxies are old. The second law of thermodynamics is ever bringing greater entropy into the galactic systems. For this reason, we find a great preponderance of enormous supermassive black holes that have become quasars that burn brightly with an enormous luminosity, in spite of their distance from us.

These black holes are so gargantuan that they are ripping the fabric of space-time apart and burning all things in the worlds around them as they suck the material toward them at speeds nearing that of light. The enormous energy created by the accretion rings that swirl wildly around the supermassive black holes contains more energy than the black hole can handle as it gobbles up at full capacity. The black holes cannot feed fast enough and therefore discard the excess energy in the form of these enormous jets of concentrated electromagnetic energy.

That excess energy that cannot be eaten up essentially shoots out in huge beams of electromagnetic energy at the poles of the spinning black holes. These beams stretch out for millions of light years and incinerate anything in their paths. These are, in fact, the death rays of the dark stars and not the incubators of galaxies, as the evolutionists naïvely insinuate.

The number of this relationship between the wavelength of our cosmic timepiece today (CBMR) and the initial wavelength of the temperature in our early plasma universe during the Big Bang is exactly a million million. This happens to be quite a unique number. It just so happens that the estimate of how many galaxies exist in our universe is also about a million million. Evolutionists would say that this is just a coincidence. One would think that a rational mind should by now be suspecting that there are just too many such coincidences in our very, very lucky universe. This is the obvious evidence for the objective mind of a predetermined symmetry that was previsioned and designed into this universe from the first moment of the Big Bang.

> *The earth is a medium-sized planet orbiting around an average star in the outer suburbs of an ordinary spiral galaxy, which is itself only one of about a million million galaxies in the observable universe (Hawking 131).*

Hawking uses this million million number of galaxies in regurgitating the typically worded propaganda for the Copernican principle that seeks to promote the atheistic doctrine that Earth is an insignificantly placed planet in an insignificant galaxy, placed by random luck in the enormous universe—the Copernican principle. All this is purposefully done, ultimately to undergird their metaphysical doctrine that there is no God and, therefore, no transcendental significance to human life.

In this way, evolutionists attempt to discredit the Genesis account that records the significance given to us by God because we are created in His image. Feigning humility as they negate any significance to our position in space and to the position of man, they slyly dethrone God and place their puny finite minds on the throne. Let's be frank; their Orwellian doublespeak is diabolically brilliant. They have managed to say that humans are gods and their puny finite minds are the center of reality, while dethroning God and sounding syrupy humble to the ignorant.

Nevertheless, the supersymmetry pervasive in our entire natural order is once again buttressed by the symmetry between the number of galaxies in our universe and the ratio of the lengthening of the wavelength of the CMBR. But their self-imposed horse-blinds do not end there. They continue to view the farthest reaches of our universe from the same coordinate of space-time rate as on Earth. They fail to see that the stretching of space-time means that time is in fast-forward in these first areas that exited the primordial critical circumference. They keep referring to this area as the past, when it is, in fact, the future.

While it is true that by the time their light hits us, it is giving us a snapshot of their past in that area of space, nevertheless that region of space has experienced a much longer history than ours and is, therefore, in our future. It is not a picture of the early universe, but a picture of an old, dying universe.

In fact, if we were able to instantly travel to that area of space-time at the edge of the universe, we would find out upon arriving that it is well beyond 15 billion light years at that end of the universe. In our present, the starlight from the stars we are viewing from those regions may no longer exist. The fabric of space-time may have completely ripped, and all baryonic matter could be crushed into gargantuan singularities inside super-duper supermassive black holes that were once quasars. We may get there and see nothing but darkness. All baryonic matter not gobbled up by these gigantic black holes could have burned up in their massive beams of electromagnetic energy.

Perhaps the very zero-point energy intrinsic in the fabric of space-time would not be enough to hold the fabric together when it stretches beyond a given point—the Omega boundary. That rip in our universe may already be traveling backward toward us as our universe continues to expand and the total entropy of our universe continues to grow, and we will not know of it until we begin to see the quasars at the outer ridges of our universe grow dark.

Our Judeo-Christian cosmological model stipulates that the rate of time in any given set of coordinates in the entire universe is dictated by the conditions of space in those coordinates. Time is a function of space. Where space is stretched, time runs faster than where space is crunched. Our model therefore stipulates that the age of the universe at any set of coordinates is in an intrinsic relationship to the actual six days of God's creation of our universe. Schroeder brilliantly discovered that a very curious thing happens when one divides this measure of the wavelength of the CMBR and the 15 billion years of the age of the farthest edges of our universe.

In terms of days and years and millennia, this stretching of the cosmic perception of time by a factor of a million million, the division of fifteen billion years by a million million reduces those fifteen billion years to six days!

If the universe had been any other size, temperature, or mass, or the threshold temperature of matter (protons and neutrons) had been different, this rela-

tionship would not exist. Cosmologists are in awe that the mass and the energy of expansion of the universe are matched with the "incredible fine-tuning" of one part in 10^{120}. It is almost as if the values had been selected. Perhaps they have.

Genesis and science are both correct. When one asks if six days or fifteen billion years passed before the appearance of humankind, the correct answer is "yes" (Schroeder 61).

Hence, Schroeder states that if one were to ask (from the standpoint of God) how long it took Him to make our universe, the correct answer is six days. But if one were to ask how long our universe has been in existence, the answer would be that the farthest regions aged 15 billion years in that same interval.

Take a moment to assimilate this undeniable fact. There is an intrinsic mathematical relationship with the six days of God's creation and the age of the oldest part of our universe, which is determined by the measure of the lengthening of the electromagnetic waves since the moment of quark confinement. God has left for us a cosmic timepiece.

Schroeder has broken down each day from God's viewpoint and from the Earth's viewpoint, looking backward in time. Since the stretching of years doubled every time the universe doubled in distance, Schroeder made a graph like the one below in his book *The Science of God*:

From the Bible's perspective looking forward in time.	From the Earth's perspective looking backward in time from the present
Day One – 24 hrs	8 billion years
Day Two – 24 hrs	4 billion years
Day Three – 24 hrs	2 billion years
Day Four – 24 hrs	1 billion years
Day Five – 24 hrs	½ billion years
Day Six – 24 hrs	¼ billion years

(Graph adapted from Schroeder 63)

466

Using Schroeder's figures, we come up with a universe that in six days, from the point of view of God, resulted in a total expansion of 15 billion light years from the Genesis Singularity.

I do, however, believe that Schroeder has the order of expansion upside down. Using his order of expansion would mean that the expansion rate of the universe is slowing down. In his chart, he states that on day one, the universe expanded 8 billion light years from the Alpha Point, and henceforth the expansion diminishes until on the sixth day when the universe only expanded one-fourth of a billion light years.

The actual observable data from the study of red-shift and from the data garnered by WMAP and the Planck satellites point unquestionably to the fact that our universe is expanding at an accelerating rate directly proportionate to the distance from us. The expansion of our universe is increasing, not decreasing. Going back to the moment of quark confinement, the wavelength of radiation has expanded a million million times, so if we were to be looking back from the present, we would see a universe shrinking in size and its expansion rate decreasing all the way back to the moment that light exited the PCC and the quark confinement moment that happened behind the curtain of the PCC now became visible to us.

The expansion rate did not diminish in strength from the first day to the sixth. It has increased in strength from the very first day until today. The universe is expanding at the outer edges and not in the interior axis of our spinning space-time continuum, which is where our sun-Earth system resides. The centrifugal force of the spinning universe causes the outward movement and subsequent stretching of space-time, which speeds up the clocks in those furthest realms. While our area of the universe at the fulcrum is expanding, it is still in alignment with the six days of creation, but at a different rate than the farther edges of our universe. Because we are generally speaking at the Alpha Point, time moves the slowest in our area of the universe.

It seems to me that by inverting the expansion numbers, Schroeder is attempting to match the evolutionary ages with the days of creation. He makes the first day the long period that he believes is necessary for the evolution of stars and galaxies imagined by those who gravitate to the notion that the slow process of gravity was the creative force of stars and galaxies.

The force of gravity, however, is not powerful enough to accomplish the deeds they imagine. Logic dictates that if our universe is expanding at an accelerating rate, the expansion at the beginning must have been the least. Schroeder has accepted the fact that God did create life and the universe, but it seems that he is still trying to reconcile the evolutionary timetable with God's creative process. While it is true that the area of our universe that first exited the PCC of the Genesis Singularity is the one that has expanded the most, that final expansion of 8 billion light years did not take place first.

However, if we interpret Schroeder's graph as signifying the first area of space-time to cross over the primordial critical circumference, then his order is appropriate. The farthest reaches of our universe that first exited the PCC on the first day are therefore the areas that are the most stretched. There, the clock is running the fastest. But it was not stretched to that point on that first day. It has stretched to that point since that day to the present.

At the edge of our universe, 15 billion years have passed since the moment of quark confinement, which, in essence, is the moment we first saw light exiting the PCC. The time between the real quark confinement and the moment that light exited the PCC is, to us, no time. It was behind the curtain of the PCC. All of this happened on the first day. In contrast, our area of the universe near the Alpha Point would have aged proportionally much less, because time here is running the slowest.

The events that Schroeder ascribes to this expansion are events that took place when the universe had not yet expanded to that state. We can see this in the graph he uses to correlate the evolutionary timetable with the days of Genesis. In the following chart from his book, he aligns the days of creation with the evolutionary timetable under the category of "scientific description."

THE SIX DAYS OF GENESIS

Day Number	Start of Day (years B.P.)	End of Day (years B.P.)	Bible's description	Scientific Description
One	15,750,000,000	7,750,000,000	The creation of the universe; light Separates from dark (Gen. 1:1–5) electrons bond to atomic nuclei; galaxies start to form	The Big Bang marks the creation of the universe: light literally breaks free as
Two	7,750,000,000	3,750,000,000	The heavenly firmament forms (Gen. 1:6–8)	Disk of Milky Way forms; Sun a main sequence star forms
Three	3,750,000,000	1,750,000,000	Oceans and dry land appear; the First life, plants, appear (Gen. 1:9–13) Kabbalah states this marked only the start of plant life, which then Developed during the following days	Earth has cooled and large bodies of water appear 3.8 billion years ago followed almost immediately by the first forms of life: bacteria and algae
Four	1,750,000,000	750,000,000	Sun, moon and stars become visible In heavens (Talmud Hagigah 12a) (Gen. 1:14–19)	Earth's atmosphere becomes transparent photosynthesis produces oxygen-rich atmosphere
Five	750,000,000	250,000,000	First animal life swarms abundantly In waters; followed by reptiles and Winged animals (Gen. 1: 20–23)	First multicellular animals, waters swarm with animal life having the basic body plans of all future animals; winged Insects appear
Six	250,000,000	approx. 6,000	Land animals, mammals, and humankind (Gen. 1: 24–31)	Massive extinction destroys over 90 percent of life. Land is repopulated; hominids and then humans.

(Graph adapted from Schroeder 70)

Earth has not aged at the same rate as the highly stretched space-time region at the outer rim of our universe. Hence, from the point of view of Earth, looking back to the moment of quark confinement, 15 billion years have not passed. We do not need to establish a 15 billion year history to our planet in order to allow the evolution of the species the time necessary to bring forth human life. Darwinian evolutionary mechanisms have never been capable of creating specified complexity from random ordering. Chance chemical reactions have never created life, much less human life. (This topic will be covered in the third and fourth books of this series.)

Nevertheless, I believe that Schroeder's insight regarding the ratio that exists between the distances of expansion of our universe in relation to the expansion rate of the radiation from the moment of quark confinement with the six days of creation is absolutely an amazing, illuminating discovery. It is quite an extraordinary link between our material universe and the Genesis narrative that cannot be easily ignored. That relationship to the six days of creation is yet another obvious fingerprint of God that should cause us to sit up and take notice that there is a design and purpose to our creation.

However, let us be clear what this ratio does not say. It does not mean that the universe expanded to 15 billion years from the Alpha Point in just six days of 24 hours each. Had the universe expanded to 15 billion light years distance from the Alpha Point in only six days, then during the rest of the history of Earth, it would have expanded way beyond that distance. The 15 billion light years is our present visible event horizon at the edge of our universe. The moment that light broke through the primordial critical circumference is the event horizon in our universe when it first appeared from within the primordial critical circumference of the Genesis Singularity.

The six days are simply the ratio between the first event horizon and our present event horizon. It is the ratio between the frequency of electromagnetic waves at the beginning and now. The universe did not stop expanding after the sixth day of creation. What it means is that the total creation of our present universe is in an intrinsic relation to the six days of the creative process.

Now Schroeder does not claim that the universe expanded to 15 billion years in six days. He claims that God's frame of reference is broken into six days of creation that encompasses the Earth's time frame reference of 15 billion years. From the point of view of Earth, 15 billion years of history are in synchrony with those six days of God's timetable.

There are, however, in my opinion, a few problems with that assumption. The first is that lumping together the age of the edge of our universe as the age of the Earth implies that time is a constant everywhere in the universe. If Schroeder includes the Earth and the outer ridges of our universe in the same time frame, he is making a grave mistake. The area of the universe where Earth exists has not passed through 15 billion years of history.

Time is relative to the position of the observer, and there is not one single rate in which time exclusively flows. Its flow depends upon the density of space. Where space is dense, time flows slowly. Where space is not dense (stretched), time flows more rapidly. Hence, the point of view from the Earth is radically different than the point of view from the edge of our universe, where space has been extremely stretched.

The ratio will be the same because both sets of coordinates have arrived at their destinations in the same interval from the moment of the Alpha Point. That is, the galaxies at the edge of our universe 15 billion light years from us have reached their present age at the same interval that Earth has reached its own present age from the moment of quark confinement. Hence, the ratio of six remains the same for both sets of coordinates in the universe. But because the sun-Earth system is at the center of the universe, our time frame is linear and not exponential. Our clock is linearly connected to the six days of creation.

What, then, is the ratio between Earth's age and the moment of quark confinement? A case has been made by creationists using the scriptures to claim that the Earth is approximately 6,000 years old. If they are right, then the ratio between the aging of the edge of our universe and the aging of our Earth would look like this:

- First thousand years of human history on Earth—The outer edges of our universe aged ¼ billion years
- Second thousand years of human history on Earth—The outer edges of our universe aged another ½ billion years
- Third thousand years of human history on Earth—The outer edges of our universe aged another 1 billion years
- Fourth thousand years of human history on Earth—The outer edges of our universe aged another 2 billion years
- Fifth thousand years of human history on Earth—The outer edges of our universe aged another 4 billion years
- Sixth thousand years of human history on Earth—The outer edges of our universe aged another 8 billion years

Humanity's history from the point of view of God is six days. From our human point of view, it is nearing the end of the sixth millennium. From the point of view of the outer edges of our universe, 15 billion years have transpired.

How do we know that one day to God is as a thousand years to humans? The scriptures have actually shed some light on this relativistic ratio, which is found in the Hebrew Tanakh as well as the New Testament. It is the lynchpin that God left in Scripture for us to know that He previsioned and designed it from the very beginning.

> *For a thousand years in your sight are but as yesterday when it is past (Psalm 90:4).*

> *But do not overlook this one fact, beloved, that with the Lord one day is as a thousand years, and a thousand years as one day (2 Peter 3:8).*

This matches perfectly with the appearance of our present universe, except that the expansion rate of the universe is not static. It is continuous, which means that the total distance of our universe must be adjusted to account for the continuing expansion that began from the first day of creation. Because the light from these farthest regions of our universe takes time to travel, by the time it reaches us, those regions have already expanded further. Soon, we will reach the seventh millennium, and the distance to the edge of our universe shall once again double and be 31 billion light years. That is, if it did not reach the Omega Boundary when the Quasars devour all galaxies as well as space-time around them.

The proportions are correct between them, but the difference in the rate of time depends entirely on the coordinate in space-time with which the measurement is made. I believe that Schroeder's important insight shows us that the rate of time in the different areas of our universe must be considered individually and must be adjusted from the first moment of creation to our present size by the measure of the expansion of the CMBR, now at 3.72 Kelvin to its original state some 1012 times hotter.

The symbolism of the six days of creation, ending with the seventh as a Sabbath day of rest, is reflective of the duration and future of the Earth as predicted in the scriptures. The seventh millennium of Earth's history will be the Millennial Kingdom of Christ, which is the Jubilee, or the year of the seventh Sabbath, according to scriptures. From the beginning, the Sabbath was called the Day of the Lord. It was a time for humanity to rest and worship God. It is no accident that the prophets called the Millennial Kingdom the Day of the Lord, or the Day of God. In the dawning of that seventh day, the nations of the world will make war against God, and the Messiah will come to rescue His chosen people.

Wail, for the day of the LORD is near; as destruction from the Almighty it will come! . . . Behold, the day of

> *the* LORD *comes, cruel, with wrath and fierce anger, to make the land a desolation and to destroy its sinners from it (Isaiah 13:6, 9).*

> *On that day the* LORD *will punish the host of heaven, in heaven, and the kings of the earth, on the earth (Isaiah 24:21).*

> *In that day the* LORD *with his hard and great and strong sword will punish Leviathan the fleeing serpent, Leviathan the twisting serpent, and he will slay the dragon that is in the sea (Isaiah 27:1).*

We know the length of that day because God has revealed it to us:

> *Then I saw thrones, and seated on them were those to whom the authority to judge was committed. Also I saw the souls of those who had been beheaded for the testimony of Jesus and for the word of God, and those who had not worshiped the beast or its image and had not received its mark on their foreheads or their hands. They came to life and reigned with Christ for a thousand years (Revelation 20:4).*

The Day of the Lord will be a day of redemption and a day of vengeance and atonement for the crimes committed against God's people. The seventh millennium is a day of rest for His people—the seventh day, the Sabbath of humanity. But it will begin with the cleansing of atonement, the seven years of tribulation that end in the day of vengeance, or Armageddon. As the First Earth was cleansed by the Great Flood, the Second Earth shall be cleansed by the roar of the Lion of Judah in the great day of the war of God. Isaiah saw that day.

> *Who is this who comes from Edom, in crimsoned garments from Bozrah, he who is splendid in his apparel, marching in the greatness of his strength? "It*

is I, speaking in righteousness, mighty to save." Why is your apparel red, and your garments like his who treads in the winepress? "I have trodden the winepress alone, and from the peoples no one was with me; I trod them in my anger and trampled them in my wrath; their lifeblood spattered on my garments, and stained all my apparel. For the day of vengeance was in my heart, and my year of redemption had come. I looked, but there was no one to help; I was appalled, but there was no one to uphold; so my own arm brought me salvation, and my wrath upheld me. I trampled down the peoples in my anger; I made them drunk in my wrath, and I poured out their lifeblood on the earth" (Isaiah 63:1–6).

That day will begin in the middle of the Great Tribulation. From that point forward, seven bowls of wrath are determined upon the kingdom of the antichrist. When the seventh bowl is poured out, a great earthquake will destroy every city on this planet. Mountains will fall down. Valleys will be lifted up. Islands will fall into the sea. It will be a complete tectonic shift in the entire world.

Let us look at the first six bowls of wrath:

Then I heard a loud voice from the temple telling the seven angels, "Go and pour out on the earth the seven bowls of the wrath of God."

So the first angel went and poured out his bowl on the earth, and harmful and painful sores came upon the people who bore the mark of the beast and worshiped its image.

The second angel poured out his bowl into the sea, and it became like the blood of a corpse, and every living thing died that was in the sea.

The third angel poured out his bowl into the rivers and the springs of water, and they became blood. And I heard the angel in charge of the waters say,

"Just are you, O Holy One, who is and who was, for you brought these judgments.

For they have shed the blood of saints and prophets, and you have given them blood to drink.

It is what they deserve!"

And I heard the altar saying,

"Yes, Lord God the Almighty,

true and just are your judgments!"

The fourth angel poured out his bowl on the sun, and it was allowed to scorch people with fire. They were scorched by the fierce heat, and they cursed the name of God who had power over these plagues. They did not repent and give him glory.

The fifth angel poured out his bowl on the throne of the beast, and its kingdom was plunged into darkness. People gnawed their tongues in anguish and cursed the God of heaven for their pain and sores. They did not repent of their deeds.

The sixth angel poured out his bowl on the great river Euphrates, and its water was dried up, to prepare the way for the kings from the east (Revelation 16:1–12).

After the sixth bowl, the antichrist stages in Armageddon to attack the Messiah in Jerusalem.

And I saw, coming out of the mouth of the dragon and out of the mouth of the beast and out of the mouth of the false prophet, three unclean spirits like frogs. For they are demonic spirits, performing signs, who go abroad to the kings of the whole world, to assemble them for battle on the great day of God the Almighty. ("Behold, I am coming like a thief! Blessed is the one who stays awake, keeping his garments on, that he may not go about naked and be seen exposed!") And they assembled them at the place that in Hebrew is called Armageddon (Revelation 16:13–16).

Now let us look at the seventh bowl of wrath:

The seventh angel poured out his bowl into the air, and a loud voice came out of the temple, from the throne, saying, "It is done!" And there were flashes of lightning, rumblings, peals of thunder, and a great earthquake such as there had never been since man was on the earth, so great was that earthquake. The great city was split into three parts, and the cities of the nations fell, and God remembered Babylon the great, to make her drain the cup of the wine of the fury of his wrath. And every island fled away, and no mountains were to be found. And great hailstones, about one hundred pounds each, fell from heaven on people; and they cursed God for the plague of the hail, because the plague was so severe (Revelation 16:17–21).

At this point, in the seventh year of the Great Tribulation period, Jesus comes down from heaven to make war on the antichrist in order to save Israel and cleanse our planet.

He is clothed in a robe dipped in blood, and the name by which he is called is The Word of God. And the armies of heaven, arrayed in fine linen, white and pure, were following him on white horses. From his mouth comes a sharp sword with which to strike down the nations, and he will rule them with a rod of iron.

*He will tread the winepress of the fury of the wrath of
God the Almighty. On his robe and on his thigh he has
a name written, King of kings and Lord of lords.*

*Then I saw an angel standing in the sun, and with
a loud voice he called to all the birds that fly directly
overhead, "Come, gather for the great supper of God,
to eat the flesh of kings, the flesh of captains, the flesh
of mighty men, the flesh of horses and their riders,
and the flesh of all men, both free and slave, both
small and great." And I saw the beast and the kings
of the earth with their armies gathered to make war
against him who was sitting on the horse and against
his army. And the beast was captured, and with it the
false prophet who in its presence had done the signs by
which he deceived those who had received the mark of
the beast and those who worshiped its image. These
two were thrown alive into the lake of fire that burns
with sulfur. And the rest were slain by the sword that
came from the mouth of him who was sitting on the
horse, and all the birds were gorged with their flesh
(Revelation 19:13–21).*

And so shall the Second Earth die and the Third Earth begin on
the seventh day, the day of rest, the Sabbath, the Great Jubilee. So we
can see the symmetry in God's previsioned plan of the ages. We can
see the symmetry that exists between our physical world and God's
spiritual plan for humankind.

Since the measure of the lengthening of the CMBR is calculated
from our present time, looking backward to its initial temperature;
hence, it encompasses our entire history, not just the six creation days
but also the seventh Day of the Lord. Our future history is entailed
in this ratio. There is an Alpha, and there is an Omega. The Sabbath
recorded in Genesis 1 that ended the creation week is then reflected

by the thousand-year (millennial) reign of Christ that shall begin when He returns to slay the global tyrant that shall come.

In the day the Prince of Peace comes, swords shall be turned into plowshares, and the mountains shall be laid low; the valleys shall be lifted up, and the islands shall fall into the sea; the deserts shall flower, and Jerusalem shall be lifted up. It will be a time when Christ shall replenish our planet after the death of the Second Earth—a time of peace and harmony only dreamed shall become a reality.

"I will rejoice in Jerusalem
and be glad in my people;
no more shall be heard in it the sound of weeping
and the cry of distress.
No more shall there be in it
an infant who lives but a few days,
or an old man who does not fill out his days,
for the young man shall die a hundred years old,
and the sinner a hundred years old shall be accursed.
They shall build houses and inhabit them;
they shall plant vineyards and eat their fruit.
They shall not build and another inhabit;
they shall not plant and another eat;
for like the days of a tree shall the days of my people be,
and my chosen shall long enjoy the work of their hands.
They shall not labor in vain
or bear children for calamity,
for they shall be the offspring of the blessed of the LORD,
and their descendants with them.
Before they call I will answer;
while they are yet speaking I will hear.
The wolf and the lamb shall graze together;
the lion shall eat straw like the ox,
and dust shall be the serpent's food.

They shall not hurt or destroy
in all my holy mountain,"
says the LORD (Isaiah 65:19–25).

During this Millennial Kingdom, Satan will be imprisoned and later released at the end of the thousand years.

Then I saw an angel coming down from heaven, holding in his hand the key to the bottomless pit and a great chain. And he seized the dragon, that ancient serpent, who is the devil and Satan, and bound him for a thousand years, and threw him into the pit, and shut it and sealed it over him, so that he might not deceive the nations any longer, until the thousand years were ended. After that he must be released for a little while (Revelation 20:1–3).

If the Judeo-Christian cosmological model we are proposing is correct, then we can predict that at the end of the 7,000 years since creation, the end of the Millennial Kingdom of Christ will be the consummation of the ages, the Omega Point. The outer reaches of our universe would have expanded (continuing in the present expansion rate) another 16 billion light years past our perceived present edge of the space-time continuum. The last doubling of our universe resulted in a stretching of the farthest edges 8 billion light years. The next doubling will be therefore 16 billion light years. Adding the 16 billion light years to our present 15 billion total light years brings us to 31 billion light years.

The universe by this time will be literally more than double its present size. The galaxies at the edge of our universe will be spinning at double their present velocity. The rate of time in those far regions will double. The total system entropy of our universe would double that of our present universe. The number of quasars will surely escalate exponentially, and the fabric of space-time will reach the point where it will begin to rip altogether at the Omega boundary.

Perhaps this process has already begun and that information has not yet reached us. If our model is correct, then we can predict that this information will reach us by the end of the seventh millennium of Earth's history, from the viewpoint of humanity. This will mark the Omega Point of our universe, when God says the heavens will burn up and the elements will melt away with extreme heat.

> *Waiting for and hastening the coming of the day of God, because of which the heavens will be set on fire and dissolved, and the heavenly bodies will melt as they burn! (2 Peter 3:12).*

God is not trapped within the framework of time. He sees time as an island in the ocean of His existence. He sees the end and the beginning. His point of reference is therefore not as a spectator caught within the framework of time. In mathematical terms, this is known as a floating reference. Almost three millennia before Einstein was even born, the Hebrew Scriptures had stated that time is relative to the observer. Comparing time from the viewpoint of man with the viewpoint of God, the Psalmist said, "For a thousand years in your sight are but as yesterday when it is past, or as a watch in the night" (Psalm 90:4).

A watch in the night is equivalent to a four-hour period. In the same verse, he said that to God, 1,000 years for humankind are as a day and equally as a four-hour period. In other words, God is not trapped in time. A day to a person is 24 hours, which is six times greater than four hours, and both are significantly less than 1,000 years. So again, we see this ratio of six that manifests God's design for our time frame of existence. But to God, all three are relative terms, because He can be at all three at the same time.

Hence, day one in Genesis, from the viewpoint of God, may not necessarily be one day to people, for people had not yet been created nor was the Earth yet rotating around the sun and revolving on its axis to create the night and day portions of a 24-hour period. However, it is quite probably a 24-hour period. Those who try to elongate this time period do so because they cannot conceive how God could accomplish such a feat in such a short amount of time. They insist that natural mechanisms are slow and ponderous. But through the electromagnetic force, God can and did create the universe, because He is all powerful.

This 24-hour period is also quite important. The relationship of a watch, as used in Psalm 90, to the 24-hour day is six—the very same relationship of the edge of our universe to the Alpha Point and the very same relationship of Earth's history to the Alpha Point. This is no mere coincidence. Nothing written in God's word is without significance.

The Hebrew day begins at sunset. It begins in darkness and ends with light. The symbolic reason is clear—light will in the end gain preeminence over darkness. The Hebrew word for watch implies a night watch. Before the Romans, the Hebrews had three night watches, and the day was divided into three watches as well. The last night watch was the third night watch, and it ended at dawn. The dawning light is the symbol of the coming Messiah. If we measure the beginning of the day from the moment of light, which is what Genesis tells us is the first day of creation, then the sixth watch is the last night watch. At the end of the sixth watch, when darkness has come over our planet, the sun will dawn. The moment that light gains ascendancy over darkness is also marked by our spring or vernal equinox.

On the morning of the spring equinox, when light began to gain ascendency over darkness on Earth, the priests of the temple of God opened the curtain that shrouded the Holy of Holies and opened the Eastern Gate, the Shushan Gate, so the very first golden rays of the sun dawning over the horizon would pass through the gate and shine directly into the Holy of Holies to light up the golden mercy seat above the ark of the covenant, which symbolizes the throne of the Messiah.

It was upon the golden mercy seat that the blood of the sacrificial lamb was sprinkled for the redemption and future atonement of the chosen of God. It is through the Eastern Gate that the Messiah will enter triumphantly to take His throne and establish the Kingdom of David on the sixth watch after the coming night. Ezekiel saw this day:

> Then he led me to the gate, the gate facing east. And behold, the glory of the God of Israel was coming from the east. And the sound of his coming was like the sound of many waters, and the earth shone with his glory. And the vision I saw was just like the vision that I had seen when he came to destroy the city, and just like the vision that I had seen by the Chebar canal. And I fell on my face. As the glory of the LORD entered the temple by the gate facing east, the Spirit lifted me up and brought me into the inner court; and behold, the glory of the LORD filled the temple.
>
> While the man was standing beside me, I heard one speaking to me out of the temple, and he said to me, "Son of man, this is the place of my throne and the place of the soles of my feet, where I will dwell in the midst of the people of Israel forever. And the house of Israel shall no more defile my holy name, neither they, nor their kings, by their whoring and by the dead bodies of their kings at their high places, by setting their threshold by my threshold and their doorposts beside my doorposts, with only a wall between me and them. They have defiled my holy name by their

abominations that they have committed, so I have consumed them in my anger. Now let them put away their whoring and the dead bodies of their kings far from me, and I will dwell in their midst forever" (Ezekiel 43:1–9).

I am of the persuasion that nothing written in scripture is without significance. The Day of the Lord will begin when the Great Impostor enters the temple of God and defiles the name of God by the abomination that causes desolation. In that day of darkness, the light of the glory of God will come from the east and enter through the Eastern Gate to sit upon the throne of Jerusalem to dwell among men forever.

If we divide the day by a watch—four hours—the last watch is the sixth watch. Now it just so happens that man and woman were created on the sixth day of the creation week, according to Genesis. It also happens to be that at the end of the sixth millennium of humanity's history, the sixth day according to God's timetable, we will enter what the prophets called the coming night. Jesus warned us of the coming night:

We must work the works of him who sent me while it is day; night is coming, when no one can work. As long as I am in the world, I am the light of the world (John 9:4–5).

But the Hebrew prophets spoke of this time of darkness long before Jesus entered into our space-time continuum.

For behold, darkness shall cover the earth, and thick darkness the peoples; but the LORD will arise upon you, and his glory will be seen upon you (Isaiah 60:2).

In that time of darkness, when the golden light dawns, the Morning Star will appear to claim His throne in Jerusalem, and the

Day of the Lord will begin as light overcomes the darkness. God is the light of the world, and that light is what brought forth our world and sustains it.

It is at the end of this seventh 1,000 years that Satan is once again free to sow anarchy as he is released from the Abyss and leads the final rebellion against the Messiah ruling from Jerusalem. This is the Battle of Gog and Magog, which shall bring the total destruction of this universe and the beginning of the new heaven and the new earth.

Even with the Messiah ruling righteously from Jerusalem, humankind will once again rise to rebel against the great creator and come to take a spoil of the chosen of God who had been gathered from the nations and lived in security and peace for a thousand years.

And when the thousand years are ended, Satan will be released from his prison and will come out to deceive the nations that are at the four corners of the earth, Gog and Magog, to gather them for battle; their number is like the sand of the sea. And they marched up over the broad plain of the earth and surrounded the camp of the saints and the beloved city, but fire came down from heaven and consumed them, and the devil who had deceived them was thrown into the lake of fire and sulfur where the beast and the false prophet were, and they will be tormented day and night forever and ever.

Then I saw a great white throne and him who was seated on it. From his presence earth and sky fled away, and no place was found for them (Revelation 20:7–11).

This is the Omega Point. Ezekiel saw this day:

The word of the Lord came to me: "Son of man, set your face toward Gog, of the land of Magog, the chief prince of Meshech and Tubal, and prophesy against him and say, *Thus says the Lord* G*OD: Behold, I am against you, O Gog, chief prince of Meshech and Tubal. And I will turn you about and put hooks into your jaws, and I will bring you out, and all your army, horses and horsemen, all of them clothed in full armor, a great host, all of them with buckler and shield, wielding swords. Persia, Cush, and Put are with them, all of them with shield and helmet; Gomer and all his hordes; Beth-togarmah from the uttermost parts of the north with all his hordes—many peoples are with you.*

"Be ready and keep ready, you and all your hosts that are assembled about you, and be a guard for them. After many days you will be mustered. In the latter years you will go against the land that is restored from war, the land whose people were gathered from many peoples upon the mountains of Israel, which had been a continual waste. Its people were brought out from the peoples and now dwell securely, all of them. You will advance, coming on like a storm. You will be like a cloud covering the land, you and all your hordes, and many peoples with you.

"Thus says the Lord G*OD: On that day, thoughts will come into your mind, and you will devise an evil scheme and say, 'I will go up against the land of unwalled villages. I will fall upon the quiet people who dwell securely, all of them dwelling without walls, and having no bars or gates,' to seize spoil and carry off plunder, to turn your hand against the waste places that are now inhabited, and the people who were gathered from the nations, who have acquired livestock and goods, who dwell at the center of the earth"* (Ezekiel 38:1–12).

I will summon a sword against Gog on all my mountains, declares the Lord GOD. Every man's sword will be against his brother. With pestilence and bloodshed I will enter into judgment with him, and I will rain upon him and his hordes and the many peoples who are with him torrential rains and hailstones, fire and sulfur. So I will show my greatness and my holiness and make myself known in the eyes of many nations. Then they will know that I am the LORD (Ezekiel 38:21–23).

This is the Omega Point, the end of our universe and the beginning of the new heaven and the new earth where time shall be no more and the memory of sin shall be wiped away. In that day, the curse shall be no more. In that day, the second law of thermodynamics shall no longer operate. Our broken world shall be destroyed. And those who have come to the light shall forever bask in His light. But let us now return to the Alpha Point when it all began.

The Ordering Process of Creation in Genesis: The Six Days of Creation

The Genesis account is the account of the creation of our universe from God's point of view. It is a record of His transformation of energy into a more ordered state in matter. The point that we need to understand is that each day of the Genesis creation account is a movement from disorder to order, from evening to morning. I suspect that the second law of thermodynamics may not yet have been activated. The creative process has not yet been profaned by death. The fall of humanity in the garden at Eden had not yet occurred. Perhaps the second law of thermodynamics came to be at the fall. At any rate, the very first sentence in Genesis provides for us the Alpha Point of our reality in this universe. Let us now look at the Judeo-Christian cosmological model as narrated in the Genesis account of creation.

Day One: Our Universe Is the Song of God

In the beginning, God created the heavens and the earth (Genesis 1:1).

The first four words tell us several important things. The Hebrew words are "Be' rai' sheet Elohim." It tells us that "in the beginning, God" was there. In fact, God was before the beginning of time. He stands outside of space-time. He created space-time.

When we speak of time, Einstein taught us to ask, "In relation to what?" The answer here is in relation to God. No one or nothing else existed at that point. This is His time frame, from His viewpoint. The rest of the sentence tells us unequivocally that He caused the universe to exist—not only the cosmic spatial dimensions in time, but also all the radiant energy that turned into matter and the forces, which control the physical processes in the universe.

There is no need for any magical formulas to understand what is plainly spoken. In the very first sentence in Genesis, we understand that the spatial dimensions were created and that energy existed within them to eventually form the heavens and the Earth. That is, there was a beginning for space-time, and only God existed at and even before that Alpha Point. The verse therefore indicates that He is not caught within the framework of His creation. He stands outside of it as the only truly transcendent God.

Those first four words thus set the scriptures above all pantheistic religions, because God stands in an eternal timeless state outside the universe. The pantheistic god, in contrast, is the same as the universe; hence, that god is too small. If it is synonymous with the universe, it is finite. It has a beginning and an end, because the universe has a beginning and an end. It does not cycle from creation to destruction and back to creation. History is linear in that it has an alpha and an omega.

The word *beginning* tells us that space-time was introduced at this point, for time began before the other elements were placed inside time; that is, time in relation to God, not matter and not humans.

Matter and humans had not yet been created. But where there is space, there is time. God created space-time and then introduced His radiant energy into a form we call the Genesis Singularity.

Hence, the Genesis record, written by Moses well over 1,000 years before the birth of Christ, said that our universe had a beginning, while science everywhere as well as all the pantheistic and polytheistic religions declared that the universe was eternal. Modern naturalistic science held doggedly to this view all the way up to the middle of the twentieth century. Guess who was right?

The scriptures tell us further that the voice of God spoke all things into being. Each creation day begins with "and God said." You may be surprised at how scientifically accurate that description is.

> By the word of the LORD the heavens were made,
>   and by the breath of his mouth all their host. . . .
> Let all the earth fear the LORD;
>   let all the inhabitants of the world stand in awe of him!
> For he spoke, and it came to be;
>   he commanded, and it stood firm (Psalm 33:6, 8–9).

"He spoke and it was done." What is the voice of God? What is sound? We hear sounds because our eardrums feel the vibrations in the air caused by voices, and our brain interprets those energy waves as distinct sounds, depending on the frequency of the vibrations. Those vibrations are information—rich patterns of specified complexity that carry a code interpreted by the brain as language. But sound waves require a medium such as air or water in order for it to propagate and be heard by finite and limited humans.

Sounds carry information-rich, specified complexity in the form of codes that are symbolized by specific sounds. The sounds represent a letter that, combined with other letters, creates a word. The message contains two codes. The first is the sound of each letter, and the second is the meaning of the word, which is pronounced. It requires both a mind to speak it and a mind to hear and interpret it. Codes cannot be created by random chemical reactions. Codes inherently imply a mind to conceive them. Thus, the idea that God created the universe by His spoken word implies a willful design that produced specified complexity, which brought forth the very nature and essence of our universe. The Logos is the architect of our universe.

The voice of God is energy waves in the vacuum of space, which also carry information-rich, specified complexity. It is that information that determines the nature of the material created. It implies that the very essence of material reality is composed of vibrations that carry specified complexity. But this is something we did not even imagine until the string theory was discovered. The string theory has mathematically suggested that the smallest entity of matter is a vibrating string or membrane. It is the pitch of that vibrating string or membrane that determines what subatomic particle it will be. Hence, the very fabric of matter is determined by the specified information carried in those vibrating strings or membranes that collectively determine what subatomic particle they will become. Every atom of Baryonic matter is then composed by the voice of God in the vacuum of space, uttered during the creation song.

Radiant energy is described as a frequency wave. Even matter, which is created from energy, is in frequency. Is not the smallest entity to reality a vibrating string or membrane? All of reality is simply the song of God, who voiced it at the beginning and with it created the magnificent symphony we call our universe.

From the very first human, Adam, humanity has known that God created the universe by His spoken word, even though we had no idea what that meant—until now. Some reading this statement may be skeptical of this claim, but that reality is attested by our history.

The Creation Motif Spans the Ancient Nations

The creation motif of an omnipotent God who created the universe by the spoken word is, in fact, common to almost all ancient nations. Each civilization that spread throughout the ancient world, stemming from the division of languages in the judgment at the Tower of Babel, carried the memories of the origins of our universe as it was passed down from Adam to Noah. The common motif has been labeled ex *nihilo*, which is Latin for "out of nothing." We find these memories in any continent around the world. For example, in South America, the one true God was called Viracocha. Burr Cartwright Brundage of the University of Oklahoma, in his book *Empire of the Inca*, describes the attributes of Viracocha as follows:

> *He is ancient, remote, supreme, and uncreated. Nor does he need the gross satisfaction of a consort. He manifests himself as a trinity when he wishes, though otherwise only heavenly warriors and archangels surround his loneliness.* He created all peoples by his "Word," as well as all huacas [spirits]. *He is man's Fortunus, ordaining his years and nourishing him. He is indeed the very principle of life, for he warms the folk through his created son, Punchao [the sun disk, which was distinct from the sun god Inti]. He is a bringer of peace and an orderer. He is in his own being blessed and has pity on men's wretchedness. He alone judges and absolves them and enables them to combat their evil tendencies* (emphasis added) *(Brundage 165).*

As in all other cultures, the one true God, Viracocha, is also associated here with thunder and a lightning bolt, while Amaru (the Opposer) is associated with volcanoes, floods, and earthquakes as well as a special lightning that curiously goes from Earth to heaven, a symbol of the rebellion of the creation against the very heavens.

> *Pachamama or the Amaru was a priestess and sorceress, who slept with all men. The Amaru lived under earth and in the rivers, and just as thunder and lightning were the destructive powers of Viracocha*

descending from heaven, so Amaru manifested itself in earthquakes, in a lightning that goes upwards from earth to heaven, in the fire that erupts from volcanoes, and in the landslides occasioned by the overflow of water and mud during the rainy season (Parrinder 100).

There is also a curious connection between the rainbows in Inca legend that quite nicely parallel the biblical account of the Great Flood. In the biblical account, the rainbow was given as a sign by God that He would never again flood the earth, and it just so happens that this truth is corroborated by the Inca tradition that believed the rainbow was the instrument that keeps the Earth from being flooded by drinking its excess water.

In Central America, the initial God of the Olmec civilization was one and the same as the Mayans. His attributes are identical to the God of the Hebrews and to Viracocha of South America. The Mayans referred to Him as "the Heart of Heaven." He was the great creator of the universe and was known as Zamna by the Olmec and as Tonatiuh by the Mayans. He was also known as Huracan by the Olmec. The title carries with it the implication of awesome power as in a hurricane. His depiction under either Zamna or Tonatiuh carries the same significance of the very act of creation through His spoken word. The Olmec depicted Zamna

Catherwood's drawing of the Olmec Supreme God, Zamna, known as Tonatiuh by the Mayans. He is depicted with His mouth open in the act of the Creation of the universe

with his mouth open in the act of creating the universe by the spoken word. Tonatiuh, the Mayan counterpart, is also depicted in that fashion. As a matter of fact, all three persons of the Triune Godhead are represented in Olmec and Mayan theology.

The record of an omnipotent God creating our universe out of nothing is also reflected in the ancient Egyptian religion. The original one true God of the ancient Egyptians is Ptah Atum-Ra. He is also credited for creating the universe by His spoken word. The symbol for Ptah or his image Atum was the sun. That was the natural symbol of the preeminent being held in common by almost every other culture.

> *In some myths life is brought forth when the god merely utters a word; in others, man is molded by the deity out of the clay (Mercatante 21).*

> *Ptah the great, being the tongue and heart of Ra. . . . The form as tongue and the form as heart are as the image of Atum; the great and mighty is Ptah who vivifies. . . . Thoth took form by it, Horus took form by it, by Ptah. It is come to pass that heart and tongue have power over [all] limbs [according to] the teaching that it is in the fore of every body. . . . The eyes see, the ears hear, the nose breathes air; they raise to the heart, and it is then the one that makes all that has been bound together go forth, while it is the tongue that repeats what the heart has planned. That is how all gods were born, Atum and his Ennead. For see, every word of God took form from what the heart had planned and what the tongue commanded (Quirke 45).*

Ptah Atum-Ra is, in fact, the oldest of the Egyptian gods and was considered by them to be the only one who was self-created. Atum-Ra is also known by the variant spelling of Tem.

The oldest of the creation gods in Egyptian mythology, variously called "divine god," "self-created," "maker of the gods," and "maker of men."

According to the Pyramid Text of Pepi I, Tem [Atum] existed when:

"not was sky,

not was earth,

not were men,

not were born the gods,

not was death."

What form he existed in, however is not stated in the text.

To make a home for himself Tem [Atum] created the celestial waters, which the Egyptians called Nun, and for a time he lived in them alone. Next, in a series of "thoughts," he created the heavens, the celestial bodies, gods, men, animals, and plants. The thoughts of Tem were translated into words by Thoth, who was his mind or intelligence. When Thoth uttered the words, all creation came into being. . . .

He appears in The Book of the Dead as the evening or setting sun and Ra as noonday sun, with Khepera as the morning sun. . . . In the Theban Recension of the Book Tem is identified with Osiris as being among the gods whose flesh never saw physical corruption, and, according to one myth, he was responsible for the primeval flood which covered the entire earth and destroyed all mankind, except for those in the boat of the god (Mercatante 185).

Jean-François Champollion was certainly one of the world's most renowned authorities in Egyptology, for he was the French Egyptologist that broke the code and was the first to be able to decipher the Egyptian hieroglyphics through the Rosetta Stone. He, too, was of the firm opinion that ancient Egyptians were monotheists and later on were corrupted into a form of polytheism.

He was not alone; Emmanuel de Rougé and Heinrich Karl Brugsch were of the same opinion. Sir Ernest A. Wallis Budge writes:

> In fact, de Rougé amplifies what Champollion-Figeac (relying upon his brother's information) wrote in 1839: "The Egyptian religion is a pure monotheism, which manifested itself externally by a symbolic polytheism." M. Pierret adopts the view that the texts show us that the Egyptians believed in One infinite and eternal God who was without a second, and he repeats Champollion's dictum. But the most recent supporter of the monotheistic theory is Dr. Brugsch, who has collected a number of striking passages from the texts. From these passages we may select the following:
>
> God is One and alone, and none other existeth with Him—God is the One, the One who hath made all things—God is a spirit, a hidden spirit, the spirit of spirits, the great spirit of the Egyptians, the divine spirit—God is from the beginning, and He hath been from the beginning, He hath existed from old and was when nothing else had being. He existed when nothing else existed, and what existeth He created after He had come into being, He is the Father of beginnings—God is the eternal One, He is eternal and infinite and endureth for ever and aye—God is hidden and no man knoweth His form. No man hath been able to seek out His likeness; He is hidden to gods and men, and He is a mystery unto His creatures. No man knoweth how to know Him—His name remaineth hidden; His name is a mystery unto His children. His

names are innumerable, they are manifold and none knoweth their number—God is truth and He liveth by truth and He feedeth thereon. He is the king of truth, and He hath stablished the earth thereupon—God is life and through Him only man liveth. He giveth life to man, He breatheth the breath of life into his nostrils [a clearer allusion to the Genesis account cannot be made]—God is father and mother, the father of fathers, and the mother of mothers. He beggeteth, but was never begotten; He produceth, but was never produced; He begat himself and produced himself. He createth but was never created; He is the maker of his own form, and the fashioner of His own body—God Himself is existence, He endureth without increase or diminution, He multiplieth Himself millions of times, and He is manifold in forms and in members—God hath made the universe and He hath created all that therein is; He is the Creator of what is in this world, and of what was, of what is, and of what shall be. He is the Creator of the heavens, and of the earth, and of the deep, and of the water, and of the mountains. God hath stretched out the heavens and founded the earth—What His heart conceived straightway came to pass, and when He hath spoken, it cometh to pass and endureth for ever (Budge xcii–xciii).

We find this same motif in India, documented in the *Rig Veda*. This ancient culture from the earliest times believed that God, through the great Om, brought the universe into existence. (For further documentation see my upcoming book *The Secret of the Lost Knowledge*.) Suffice it to say that the difference between the Hebrew account and the memory of other nations is the miracle of inspiration given to Moses and the prophets in order that humankind could have an uncorrupted record of the things our ancient forefathers had learned from Adam through Noah.

Of course, the naturalist simply dismisses these commonalities as simple fabrications of the minds of superstitious people. But what are the chances that all these developing cultures separated by mountains, oceans, and great distances could have all come up with the same basic account? The historical evidence from ancient civilizations supports the Genesis historical narrative regarding the creation of our universe.

Genesis Is Irreconcilable with the Evolutionary Model

Christians who espouse evolution and allegorize the scriptures should understand that there can be no reconciliation between the naturalistic ideology and scripture. There can be no reconciliation between a closed system and the Genesis account. A closed system, by definition, excludes God as a possibility. It is simply a metaphysical choice that automatically excludes the *possibility* of a God and not a scientific proposition deduced from empirical data.

There can be no reconciliation between the evolutionary doctrine of dysteleology, which states that all evolutionary processes in the universe are randomly ordered and that there is no purposeful agent involved in the accidental evolutionary outcome.

Our Judeo-Christian cosmological model is simply the antithesis of the Darwinian atheistic model, and any attempt to merge the two is an irrational and impossible endeavor. All of nature is imbued with a purposeful design, from the subatomic particles to the specified complexity found in living organisms, none of which can be explained by chance, random chemical reactions.

The Hebrew day begins at sundown. It begins in darkness and ends in light. It is a symbol of the Alpha Point when God created the universe and the Omega Point when, after the 1,000-year reign of the Messiah, God will create a new heaven and a new earth where there will no longer be any darkness.

Genesis tells us that God brooded over the Earth as a chicken broods over her eggs. In fact, as Schroeder has so eloquently written, the process of going from evening to day in the Hebrew implies an ordering process. The Hebrew word for *evening* is quite important. It is the word *ereb*, which means *mingled*, a *mixture*, and denotes a state of disorder. It is from this word that we have the word *arab*, which means a mixed race. The Hebrew word for *morning* is, on the other hand, *baqar*, which means to plow, break forth, or care for, and denotes a process of bringing forth order as in plowing a field. The asymmetrical field, once plowed, takes on a symmetrical order, which leads to productivity. Plowing is a teleological process with an end goal in mind.

Hence, the Hebrew account of creation is documenting a teleological process of ordering from a state of lower order toward a more ordered state with each passing day. It was the radiant energy of the glory of God that took the initial indistinguishable radiant energy of the universe and began to order it according to His purpose from the first moment of the Genesis Singularity at the Alpha Point. We see here an intentional action taken by God to bring order to disorder by a willful choice, and this became one day. Each creation day was then a movement toward greater order by the willful brooding of God.

Just like a sculptor imposes his or her will upon a block of stone to fashion its appearance from the material, to create what his mind has previsioned, God, outside the universe, took a willful action in the process of ordering the more chaotic essence of the raw energy He infused into our universe in the Genesis Singularity. The first verse of Genesis not only refutes the evolutionary model that insists on a closed system, but it also disqualifies the pantheistic concept of a god that is trapped within the universe and negates any attempt by Christians to merge the evolutionary hypothesis with the Judeo-Christian faith.

At the moment the Genesis Singularity was brought into existence by the voice of God, all that existed was raw energy, bounded in a space-time singularity.

> In the beginning, God created the heavens and the earth (Genesis 1:1).

This opening statement gives us a complete overview of God's creation and stands in direct opposition to the evolutionary doctrine and its materialistic mechanism of natural selection. It tells us that God created our cosmos as well as our Earth.

The Scientific Chronology of the First Day

The scriptures and science bring light into the process of the Big Bang. The Alpha Point brought forth radiant energy from the voice of God into a finite space created by God—the Genesis Singularity. His radiant energy turned into photons, electrons, positrons, and neutrinos. Something was needed to bring further order. We need to understand that from the very first moment of the Big Bang, there was order. Specific parameters were set that would determine the outcome of the explosion. If that order had not been there, none of the guiding parameters that ruled its development would have been there, and our universe would not have developed the way it did.

> To describe the light that filled the early universe, we can say that the number and the average energy of the photons was about the same as for electrons or positrons or neutrinos.
>
> These particles—electrons, positrons, neutrinos, photons—were continually being created out of pure energy, and then after short lives being annihilated again (Weinberg 15).

All through this process, electrons and positrons (matter and antimatter) were annihilating at a specific rate determined by the heat and density of the expanding plasma ball. This crucial parameter was indispensible in determining the number of protons and neutrons that would be created. Around 0.0001 seconds after the Genesis Singularity became the moment of quark confinement—matter came into being in our universe.

At the moment of quark confinement, the vibrating strings or membranes of energy coalesced to form quarks. It is at that moment that energy was turned into baryonic matter. This moment of quark confinement is when energy coalesced and was transformed into the subatomic components from which all matter is formed. The song of God hit its first distinct note of its magnificent symphony.

But matter was yet indistinguishable. The universe was an incredibly hot ball of quarks, photons, electrons, positrons, and neutrinos. More order had to be brought forth from this dark soup of quarks. God's continuous brooding then began to expand the spatial extent of our new universe. By expanding it, He reduced the density of the material inside, thus cooling it. The temperature at the moment of quark confinement is estimated at 10^{12} Kelvin. Protons and neutrons began to be formed by the quarks at a carefully calibrated rate.

> *The proportions were roughly one proton and one neutron for every thousand million electrons or positrons or neutrinos or photons. This number—a thousand million photons per nuclear particle— is the crucial quantity that had to be taken from observation in order to work out the standard model of the universe (Weinberg 16).*

The quarks turned to protons and neutrons in a specifically crucial proportion that would determine the future capabilities of the atom to create all the higher components necessary for life.

Our universe was composed of mostly photons at this stage, but the universe was still opaque because the density of the opaque protons and neutrons kept the light contained. Photons existed, but light could not be seen. The expanding and spinning plasma ball was dark.

Let us continue with the Genesis account:

> *The earth was without form and void, and darkness was over the face of the deep. And the Spirit of God was hovering over the face of the waters (Genesis 1:2).*

Without form is the Hebrew word *tohu,* which means "unformed." The word *void* is the Hebrew word *bohuw*. It is sometimes translated as "empty" or "void," but it implies an indistinguishable ruin. Some have translated it as "chaos." But it is not pure chaos. Had it been pure chaos, we would not have had the exact temperature and expansion rate necessary for our universe to develop in the fashion that life could inhabit it. It was deterministic chaos, which is bounded by the parameters of God's providential will. What this is describing is movement from a less ordered state to a higher ordered state, which can no longer take place through any materialistic medium because of the universal second law of thermodynamics. It is the act of God brooding over the initial plasma state of the early universe and changing unformed matter into formed matter.

Purely random chaotic processes could not have begun with the exact temperature and expansion rate necessary for the universe to develop in a way that life could be harbored. Any small, seemingly insignificant change in that exact temperature would not have resulted in the precise universe we have that can support life. The impossible odds of doing so by blind, random luck is one in 10120, and for all practical purposes must be considered an impractical and nearly impossible feat.

In fact, without a previsioned, calculated, and directed energy coming from outside the universe at the Alpha Point, our genesis would not have been the Big Bang of a white hole but instead a gargantuan supermassive black hole at the Alpha Point, which would not have allowed the formation of our universe. No material explanation can account for the breaking of the infinite density of matter in the Genesis Singularity. Our universe would not exist. It would simply be a dark black hole floating in hyperspace with a space-time radius that encircles only the Genesis Singularity.

Yet the Big Bang occurred as it exploded outward, and it was at this critical initial point that all the magnificently designed calibrations were brought into existence so symmetry could exist among all elements and forces of the physical universe. The four fundamental forces of nature were eventually distilled from the one, which contained all properties in a unified force. Eventually, each force decoupled, and it became evident that they had been given a precise predesigned parameter to interact with the others so our universe could be perfectly capable of sustaining life. Any small deviation from their prescribed values would not result in the universe that exists.

The four states of matter were intentionally prescribed by the very essence of the matter God created and the intrinsic characteristics of the atomic components He designed. It was His directed energy that caused the moment of quark confinement and turned radiant energy into the subatomic particles that form the understructure of all matter in our universe.

The underlying message found in Genesis is that the early universe was moving from a state of greater randomness (more disorder) to a state of lesser randomness (more order) by the loving action of God hovering or brooding over the universe as a hen broods over her eggs. The Hebrew word used here is *rachaph*, which means "to brood" or "to flutter" over in a caring fashion. It is the same word used in Deuteronomy to describe God's love for His people.

Like an eagle that stirs up its nest, that flutters over its young, spreading out its wings, catching them, bearing them on its pinions (Deuteronomy 32:11).

It is the radiant energy of the hen brooding over her eggs that incubates the eggs and allows the baby chicks to form. There is no better example for the action of our creator over the plasma energy that resulted from the initiation of the Big Bang. God's radiant energy ordered the indistinguishable plasma and brought forth the building blocks of our material universe.

If we compare this description in Genesis to the process of the Big Bang, it matches quite well. In fact, the initial radiant energy at the Alpha Point, which existed at the very beginning of the Big Bang, is a very accurate description of *bohuw*—indistinguishable energy—which is matter yet unformed (*tohu*).

Even if the second law of thermodynamics existed from the first moment of the Alpha Point, it flatly tells us that this process of moving toward organization runs contrary to the normal workings of our present universe unless acted upon by an outside force. This moving against the gradient of natural processes from a disordered state to a more ordered state can only be accomplished through directed energy. Our Judeo-Christian cosmological model differentiates the materialistic evolutionary model in that it maintains that God provided that directed energy to accomplish His creative process. Without a motor to provide directed energy, the disordered cannot become more ordered.

We can see this process in the intricate and marvelously engineered metabolic motors that were designed with specified complexity by a mastermind to allow life to flourish in our world. We have an example of that in plant life with the ingenious design of chlorophyll. In the animal world, we see this in the mitochondria. Without this metabolic motor to direct energy, the raw energy could not be ordered in order to support life.

As a matter of fact, the Genesis account, written long before humankind ever knew what plasma was, accurately depicts the original plasma state of our universe at its inception and its transformation into the elements that constitute our material universe. In other words, the formation of matter from the disordered plasma was not an accidental random happenstance, but rather the willful act of the creator in order to bring forth a universe for the habitation of life.

Without God's willful brooding (*rachaph*) over the chaotic elements, there could be no ordering. Without an intentional brooding that produced the exact temperature and expansion rate, our universe would not be as it is today. But even before this plasma was formed, the raw energy had to be converted into quarks—quarks that form the components of the atomic particles.

According to the timeline calculated by Weinberg in his *First Three Minutes*, at about 0.02 seconds from the Alpha Point, the temperature was cooled to 10^{11} Kelvin. The density of the universe had been reduced to 4×10^9.

During that time, electrons and positrons created from light were destroyed by annihilation at about equal rates. The pair production critical threshold is 1 MeV. The thermal energy in the universe during that moment was $kT = 8.6$ MeV. That was considerably above what was necessary for pair production. Positrons are the antimatter of the negatively charged electrons. When they collide, they annihilate, releasing an enormous amount of electromagnetic energy. It is this electromagnetic energy that was then used to create solids through the marvelous creation engines of the Birkeland currents.

From a purely materialistic perspective, there is no mechanism to avoid the complete annihilation of all the electrons by their antimatter (positrons) since they were created with equal amounts. If random processes governed our universe, all electrons would have naturally eventually collided with positrons, thus annihilating one another. Our universe would have ceased to exist.

The fact that our universe does exist points to God's brooding as the causal factor. He purposed to allow enough matter to exist alongside the electromagnetic energy produced by the annihilation of photons and positrons that kept it from being completely annihilated by the natural processes. Instead, under the guidance and control of God, the enormous quantities of electromagnetic energy released in the plasma became available in order to later become the intense Birkeland currents that began the process of the creation of the elements that comprise baryonic matter in our universe. Baryons are the physical, material particles that compose visible matter in our universe.

During this time, protons and neutrons were also being changed back and forth at about equal numbers. The energy difference between protons and neutrons is 1.29 MeV. At that point, only about one baryon for 109 photons is surviving (as inferred by the 2.73 K background radiation).

Since the conservation of the baryon number is a conservative principle, it is inferred that throughout the expansion process of the universe, the ratio of photons to baryons also remained fairly constant. In essence, during this period, photons (light—the entire electromagnetic spectrum) were being turned into electrons, positrons, protons, and neutrons. The annihilation of electrons and positrons (antimatter) then fueled the electromagnetic currents that God used to create solid matter out of plasma.

Stop for a moment and consider the amazing correlation between the Genesis account and the formation and nature of our material existence as we know and understand it today through modern physics. Here, the beauty of Einstein's equation is deeper than most scientists realize. His equation tells us that there is a deep fundamental unity between energy, matter and light, as expressed by his equation Energy = M (mass) C^2 (speed of light squared). Before humankind ever knew that energy and matter were interchangeable components, Genesis told us that matter was created by the energy of the voice of God. That radiant energy turned to protons, neutrons, and light (photons).

It is this fundamental unity that gives credence to the creation story. Light is the fundamental constant in the universe. So light is, in essence, the fingerprint of God, left for us in order to cause the seeking person to ponder over His magnificence. For this reason, light existed in the Genesis story before the creation of the suns, the stars, and all other things. The very glory of God shone upon our universe, and the radiant energy from that light began to transform our planet purposefully into His design.

Critics of the Genesis narrative considered the creation of light before stars to be nothing more than a fable, a mythological and ignorant statement. But our modern understanding of physics has once again vindicated the Genesis narrative. Light existed before all other things in our universe were made. That is an amazing truth.

Photons were the first created particles from God's radiant energy. Everything in our present universe came from photons. This alignment with the Genesis account is unimaginable to conceive and unknown to Moses when he received God's inspiration to record the creation story. In fact, most modern people are not even aware of this reality today. We are literally children of the light. Sagan was wrong, and he knows it now. We are not the children of stardust. The scriptures are correct; we are the children of the light.

In fact, this is counterintuitive to humankind. How can light exist before the sun and the stars? In the past, atheists ridiculed the scriptures by saying that light came from stars and not before stars. They considered the Genesis narrative to be nothing more than unscientific and ignorant superstition; that is, until we began to understand Einstein. The scientific truth is exactly what the scriptures maintained long before Einstein's theories allowed us to understand that. The idea that radiant energy could be turned into matter was not understood by humans until $E = MC^2$. Nevertheless, the Genesis account plainly states that light existed before the sun and the stars. Long before modern people knew anything about singularities and the atomic processes involved, Genesis told us that light existed before the stars.

Light is used by the author of Genesis to represent the visible portion of the electromagnetic wave spectrum, which extends in frequency above and below what we can see with our eyes. Before the string theory, Sternglass had already begun to think that matter is just vibrating light. Sternglass as well as Bohm and Vigier proposed that the motion of the electromagnetic wave creates the appearance of solidness, similar to the vortex of a tornado—a tornado of light, if you will.

> *Bohm and Vigier also used the ideas of vortices or whirlpools to describe the electron and its spin. Over the years, I had become increasingly convinced that only if the neutron and all other particles and their interactions could be described in electromagnetic terms could there be a unified view of photons and electrons, a hope that de Broglie and Einstein kept alive despite formidable opposition. I felt that only if a link existed between electromagnetic and nuclear forces could one fully accept the concept of the ether as an ideal fluid capable of sustaining rotary or vortex motions (Sternglass 112).*

Sternglass was close, but not quite there. The M-theory equation, which unites all four forces of nature (we shall address that more fully later), does tell us that the fundamental essence of matter is vibrating energy that turned into quarks and photons. Hence, the very essence of light is therefore an appropriate symbol of God's voice. Those vibrating strings or membranes are literally the voice of God in space-time.

The theory of relativity predicts that objects approaching the speed of light would slow down time. At the same time, it predicts that we could never reach the actual speed of light because our mass would become infinite. Yet light, which obviously is traveling at the speed of light, really exists in a timeless state. If we could attain the speed of light, we would be theoretically floating in timelessness, and our three-dimensional world would disappear. It is therefore quite plausible, from a scientific point of view, to believe in a light being who transcends time and is in a timeless state Himself. There is nothing mythical, superstitious, or mystical in such a being. It is in direct correlation to the science we already understand.

Time simply cannot trap such a transcendent being. Prior to Einstein, we thought of time as an absolute constant, and therefore, the idea that someone could know the future was ridiculed as unscientific. Materialists scoffed at the idea of a God that transcended time as simple superstition and antiscientific. It is neither. Instead, now we know that time is not a constant and that it can be slowed down or sped up, depending on the speed an object is traveling or the forces affecting the fabric of space that it uses as a substratum or medium to exist. Science now informs us that energy such as gravity or speed affects space and time. It warps space-time.

What we have discovered is that light is the fundamental constant of the universe, and matter and energy are, in fact, only separate sides of the same thing. One can be converted to the other. Matter is, in essence, condensed energy, so if we convert the mass of a single pencil into energy, it releases a power akin to an atomic bomb. Can you imagine the amount of energy necessary to create the Earth, the sun, the planets, our galaxy, and the trillions of galaxies in our universe?

The amount of energy is so great that it boggles the mind. And yet this was precisely what happened in the Big Bang. But scientists tell us that we cannot discuss how the Big Bang appeared from nothing, since they posit that time has to be a required factor in any scientific inquiry. But I say that it is not time, but rather light, that is the fundamental constant in the universe, and since God is light, I believe He began the Big Bang when there was no time—before time—for He exists outside of our space-time continuum that we call our universe. His glory brought forth all that is, including space-time. Since the Genesis Singularity was created by His radiant energy and since light is in an eternal now, it did not need time as a precondition to exist.

It is from this radiant energy of infinite density that God created space-time and matter and brought into existence the Genesis Singularity. From a materialistic perspective, only a God of light can exist outside of time to begin the Big Bang. Here and only here do we have a scientific base to move behind the Genesis Singularity. Only a Judeo-Christian cosmological model can offer a scientific explanation of the origin of the Big Bang. That is big—really big.

Hence, this entire universe is simply an expression of God's radiant energy and light. It is His song of love. It is, therefore, illogical that scientists would think it improper to consider the existence of God as scientific. For science leads us to no other alternative for the origin of the Genesis Singularity.

We view darkly as through an imperfect mirror. We cannot explain with our finite understanding the essence of such an eternal being. But non-comprehension does not equal non-existence. For example, simply because we are yet unable to find a mathematical theory that can explain both properties of light as a particle called a photon and at the same time as an energy wave does not mean that light does not exist. Similarly, just because we cannot mathematically describe the existence of a master designer existing outside of our universe does not mean that we cannot logically deduce His existence from a rational extension of the facts that are known.

Returning to Weinberg's timeline from *The First Three Minutes*, we deduced through working backward with Einstein's equations: At 0.11 seconds from the Alpha Point, the density of the universe was reduced by about 30,000,000, which means it also cooled to 3×10^{10} Kelvin (2.6 MeV). At that point, free neutrons began to decay into protons, causing the ratio to change in favor of the protons. The universe then reached a ratio of 62 percent protons to 38 percent neutrons as it expanded and cooled down.

At 1.09 seconds from the Alpha Point, the density had reduced by a factor of 400,000, and the temperature was reduced to 10^{10} Kelvin. At that point, the density was reduced enough that neutrinos were able to be released from the hot dark ball of plasma exploding outward from the Alpha Point.

God Chose a Lefty

The neutrino is another obvious fingerprint of God that shows our universe was designed and did not arise through random ordering. The neutrino has a spin that is eternal, and quite oddly, it spins only to the left. It is just as possible to have formed right-turning neutrinos from a physical perspective, but there are none. If the universe were guided by random choices, we would expect our universe to contain roughly the same number of left-turning neutrinos as right-turning neutrinos. But apparently, God chose a lefty.

Certainly, the odds of creating one form exclusive of the other is statistically so improbable that it would be irrational to insinuate that it was a randomly generated reality. Any number of combinations of the percentage of their ratios is the most likely outcome in a randomly generated universe.

But that is not what we find. What we find is that neutrinos are exclusively left-spinning particles, just like the amino acids of proteins in living things are exclusively left-handed amino acids. Amino acids are found in equal numbers as both right-handed and left-handed. And yet, miraculously, all proteins in living things are exclusively made of left-handed amino acids without any chemical reason for it being so. Apparently, God has a soft spot in His heart for lefties. (We will discuss the origin of life further in Book 3 of this series, *Codes: God's Indelible Fingerprint.*)

The paradox is that there is no chemical or physical reason to force neutrinos to spin only toward the left. But apparently, a choice was made in their creation so that humans could know they did not arise from randomness. This phenomenon poses a paradox for the evolutionist—one that is identical to the paradox produced by left-handed amino acids. The universe does not follow the prescribed path that randomness would have predicted. All proteins in living things are made exclusively from left-handed amino acids. There is no physical or chemical reason for this choice. That is another unmistakable proof that both the natural elements and life are, in fact, designed by previsioned choice.

Therefore, at 1.09 seconds from the Alpha Point, the universe became transparent to the neutrinos. Physicists believe that in addition to the 2.73 K microwave background radiation, our universe is also filled with the background neutrino radiation. That is estimated to be about 2K. The expanding universe, however, was still opaque to electromagnetic radiation. Light was still contained by the yet too high density of opaque matter.

> As the explosion continued the temperature dropped, reaching thirty thousand million (3 X 10^{10}) degrees Centigrade after about one-tenth of a second; ten thousand million degrees after about one second; and three thousand million degrees after about fourteen seconds. This was cool enough so that the electrons and positrons began to annihilate faster than they could be recreated out of the photons and neutrinos. The energy released in this annihilation of matter temporarily slowed the rate at which the universe cooled, but the temperature continued to drop, finally reaching one thousand million degrees at the end of the first three minutes. It was then cool enough for the protons and neutrons to begin to form into complex nuclei, starting with the nucleus of heavy hydrogen (or deuterium), which consists of one proton and one

neutron. The density was still high enough (a little less than that of water) so that these light nuclei were able rapidly to assemble themselves into the most stable light nucleus, that of helium, consisting of two protons and two neutrons.

At the end of the first three minutes the contents of the universe were mostly in the form of light, neutrinos, and antineutrinos. There was still a small amount of nuclear material, now consisting of about 73 percent hydrogen and 27 percent helium, and an equally small number of electrons left over from the era of electron-positron annihilation (Weinberg 16–17).

Note that the density of the plasma at this point is almost that of water. We shall address that later. At three minutes and two seconds from the Alpha Point, the temperature of the universe had decreased to 109 Kelvin. At that time, the universe was made up of mostly photons and neutrinos. Further neutron decay now left a ratio of 86 percent protons to 14 percent neutrons. These baryons, however, were still a very small fraction of the overall energy of the universe at that moment. Nevertheless, this ratio is critically the reflection of the equilibrium of the particle populations that will be established in the future. That ratio in the particle population will determine the atomic structures of the elements. It had to be exactly as it was in order for our universe to be as it is.

The timing and temperature at this crucial point in creation necessarily needed to be exactly precise in order for this delicate equilibrium to form so the universe could harbor life. It is an irrational leap of faith to state that this parameter was the result of blind random processes. Once more, the universe bears evidence that another intelligent previsioned choice was made.

At three minutes and 46 seconds, the universe had cooled down to 0.9×10^9 Kelvin. At that point, neutron decay stopped. The deuteron was stable, so all neutrons began to combine with the protons to form deuterium and, subsequently, the nuclei of helium (alpha particles that are highly stable). During that period, helium was about 26 percent of the universe by mass.

The Miraculous Neutron, Proton, and Electron Ratio

It is precisely here that another miraculously chosen parameter becomes obviously critical in the expansion process. If the expansion of the universe had proceeded at a slower pace, almost all of the neutrons would have decayed into protons. Without neutrons, we could not have formed the heavier elements necessary in order for life to exist. The neutrons stabilize the nucleus in atoms and allow for the heavier elements to form. Were it not for the neutron, the large nuclei composed of many protons and neutrons necessary to construct the higher elements could not have formed. The repelling force of positively charged protons in the nucleus (remember, like poles repel) are stabilized by the neutrons.

Our universe would have remained nothing more than a giant helium balloon because heavier elements require neutrons to form and allow the atom to stay stable. In other words, without the exactly chosen and regulated temperature and expansion rate, which our universe experienced in the Big Bang, life could not have existed in our universe. All of this was decided from the very first instant of the Big Bang. To believe that out of all the possibilities, our universe, guided only by random processes, could have chosen the exact temperature parameters of the expansion rate and subsequently the cooling down rate by mere chance is to believe in a miracle exponentially greater than the parting of the Red Sea and the resurrection from the grave.

At 34 minutes and 40 seconds, the temperature cooled down to 3×10^8 Kelvin. It is at that moment that the Birkeland currents would have begun the process of creating elements from the basic components of the atom. It is also at the next point that we had yet another miraculous intervention. The density of this plasma universe is now reduced to that of water. When Genesis speaks of God hovering over the water, I believe it is possible that it may be describing the plasma state that was essentially the same density as water. How else could a ball of plasma be described to ancient people? But I must state that I do not hold to that as an absolute. It is more like a suspicion.

However, I also believe that the Birkeland currents at that time began to create vast amounts of water from the plasma material. So it is also possible that the production of water as one of the very first molecules created from hydrogen and oxygen through the Birkeland currents provided a means to cool down the plasma ball much more quickly than modern cosmologists, relying on the gradualist gravity model, are willing to recognize.

> *And darkness was upon the face of the deep. And the Spirit of God moved upon the face of the waters (Genesis 1:2 KJV).*

It is at that time that the electron ratio was also fixed. Nuclear processes had stopped, and we were left with about 1 in 10^9 electrons, because somehow, miraculously, we had obtained a slight excess of electrons over their antimatter, positrons. Even though they were being produced at the same rate, somehow "miraculously" just enough electrons were not annihilated by their antimatter (positrons), which then allowed the possibility of our atoms to form since nuclei could unite with electrons.

In the plasma state, the nuclei is separated from the electrons, but the Birkeland currents are able to unite them in the Z pinch. Nobody knows why we ended up with more electrons than statistics would have predicted. It just miraculously happened. Hmm! What a lucky random break, eh? It is in this process that the Birkeland currents are able to unite the electrons to the nuclei and form the solid elements in our universe, and this may have played a pivotal role in establishing the electron ratio.

This, too, is critical, for without those electrons, we would also not be able to combine the protons and neutrons of the atomic nuclei to make atoms of different sizes and thus the fundamental elements in the periodic table. The electrons provide the negative charge that counters the positive charge of protons and subsequently stabilizes an atom. Life could not have developed without the electrons to complete the structure of the atom.

It is the exact nature of the interplay between subatomic particles that provides the infrastructure for the formation of the elements in our periodic table. All the components of the atom were created by the exact temperature choice and expansion rate that God brooded, thus creating their specific properties and numbers in order to produce the universe we inhabit.

The radiant energy density of the universe was composed at this point of about 69 percent photons and 31 percent neutrinos. But the universe was still opaque. The density of baryonic matter was still too great. Photons were not yet visible. The light God used to create the matter in the universe was not yet visible to any observer other than God. But as the Birkeland currents began to create water and entire atomic structures, the temperature of the universe began to decline, and the photons began to have more room to move about.

From the point of view of the Baryonic matter created since the Alpha Point, the universe had existed a little less than three minutes and 40 seconds. From God's point of view, as written in the Genesis account, all of this happened during day one.

Electromagnetism, Not Gravity

Because most American cosmologists do not consider the impact of the Birkeland currents in the formation of our cosmos, they rely on the slow gradual gravitational model. In their gravitational model, the universe must cool off significantly before gravity can coalesce the energetic subatomic particles. Hence, in Weinberg's timetable in the *First Three Minutes*, he computes that it would take 700,000 years from the Alpha Point before matter could coalesce through gravity. The universe had to cool down to 3,000 Kelvin before it could allow hydrogen and helium nuclei to collect electrons and become stable atoms. However, what gravity needs 700,000 years to accomplish, the electromagnetic force can produce in nanoseconds.

The large-scale production of water through the giant Birkeland currents equally arrayed in symmetrical patterns around our early universe would have helped expand the universe and cool it down in a considerably shorter time frame. I believe our early universe was chiefly cooled down by the production of water, which then allowed light to escape, and light appeared from the darkness of the soup. It was at this point that the Genesis record said:

> *And God said, "Let there be light," and there was light (Genesis 1:3).*

Perhaps the giant Birkeland currents faced inwardly to create a water world near the Alpha Point. The centrifugal force of the spin of the universe would send the Birkeland currents in the plasma outward, while the water that shot out from the twisting currents went inward.

At this point, the universe was, for the first time, transparent to light. In the plasma form, the ions and electrons were efficient scatterers of light. But once the expansion reached a certain point where temperatures cooled down enough, and these giant Birkeland currents began to create atoms that were inefficient at scattering light, light broke through toward the interior of the universe. Radiation was decoupled from the particle. In the evening of day one, light broke through and became visible. That was the time that was recorded in Genesis as the moment when light was divided from darkness.

And God saw that the light was good. And God
separated the light from the darkness. God called
the light Day, and the darkness he called Night. And
there was evening and there was morning, the first
day (Genesis 1:4–5).

We must be careful here to understand the full meaning of light. What was decoupled from the dark plasma soup were photons. Photons are manifested by an entire spectrum of electromagnetic energy. The form that light takes is dependent on the length of the wave, or frequency. Human beings are only capable of seeing one-third of the spectrum of electromagnetic energy as visible light. The churning plasma of the spinning plasma universe was so energetic that massive amounts of electromagnetic energy were produced by the excited photons in all nine wavelengths.

Although the density of the plasma had expanded and subsequently cooled down enough to create the atoms of hydrogen, helium, carbon, and even up to oxygen, most of the universe remained in a plasma state, as it remains even today. In the plasma state, as previously explained, the nuclei and the electrons remain separated as positive and negative particles, which makes them especially susceptible to the electromagnetic force.

The fact that these electrons with a negative charge and these nuclei with a positive charge existed in overwhelming numbers in the churning plasma means that the electromagnetic force would play a much greater part in the creation of solid matter than most evolutionists in our Western culture are willing to consider. Their subliminal bias toward a slow, gradual evolution through the force of gravity leads them to ignore the much faster capacity of the electromagnetic force to create solid matter in nanoseconds.

The electromagnetic force is the force that gives us electricity and magnetism. They are interrelated. Most of us are aware that an electric current is produced when the electrons of a given medium, such as a copper wire, are excited to the point that the electrons literally break free of the shells of orbit around the nucleus and begin to flow, traveling from atom to atom. Our modern hydroelectric plants use the force of moving water to spin large coils of tightly wrapped wires around a powerful magnet. The resulting polarization of the atoms in the wires by the magnets causes the electrons to flow.

When the electrons are excited enough, they give off photons to lower their energy levels. When the electricity flowing through the coiled tungsten filament inside a bulb excites the tungsten molecules, they give off photons that we see as light. Thus, photons and the electromagnetic force are intricately interwoven. All electromagnetic rays use photons as their means of convection.

In the churning, spinning plasma form of the early universe, these free-floating electrons would be highly susceptible to any magnetic forces created by the spinning plasma. Immensely powerful currents were continuously created in this highly energetic medium. In the same way that the friction of cloud particles creates electricity, which shoots out as bolts of lightning, the plasma cloud also shot bolts of intense currents that became the Birkeland currents we previously discussed.

Most people are unaware of the creative powers of the Birkeland currents. As a matter of fact, the Z pinches of these Birkeland currents are fusion machines that modern scientists are now beginning to understand better and utilize commercially in an attempt to harness the immense energy generated by fusion. On April 7, 2003, Sandia National Laboratories, managed and operated by the National Technology and Engineering Solutions of Sandia, announced that it had created deuterium (heavy water) in its fusion Z machine.

Sandia was also able to create a form of hyperdense hot ice known as "ice vii" (ice 7). On March 15, 2015, Sandia announced that it had created ice in a few nanoseconds. Imagine that! No long process of eons of years. In nanoseconds, the enormous power of the fusion of particles in the Z pinch created ice.

It may be surprising to find out that water abounds in our entire universe. The Genesis record, claiming that water existed before the Earth was formed, was also ridiculed by skeptics as an impossible, unscientific superstition. But modern science has now documented that water abounds in our cosmos to such an extent that few scientists have ever imagined.

Water Everywhere in the Cosmos

In a July 22, 2011, article titled "Astronomers Find Largest, Most Distant Reservoir of Water," written by Whitney Clavin and Alan Bluis of the Jet Propulsion Laboratory in Pasadena, California, and found on the NASA website's Missions News section, this amazing discovery was described:

> *Two teams of astronomers have discovered the largest and farthest reservoir of water ever detected in the universe. The water, equivalent to 140 trillion times all the water in the world's ocean, surrounds a huge, feeding black hole, called a quasar, more than 12 billion light-years away.*
>
> *"The environment around this quasar is very unique in that it's producing this huge mass of water," said Matt Bradford, a scientist at NASA's Jet Propulsion Laboratory in Pasadena, Calif. "It's another demonstration that water is pervasive throughout the universe, even at the very earliest times." Bradford leads one of the teams that made the discovery. His team's research is partially funded by NASA and appears in the Astrophysical Journal Letters.*
>
> *A quasar is powered by an enormous black hole that steadily consumes a surrounding disk of gas and dust. As it eats, the quasar spews out huge amounts of energy. Both groups of astronomers studied a particular quasar called APM 08279+5255, which*

harbors a black hole 20 billion times more massive than the sun and produces as much energy as a thousand trillion suns.

Astronomers expected water vapor to be present even in the early, distant universe, but had not detected it this far away before. There's water vapor in the Milky Way, although the total amount is 4,000 times less than in the quasar, because most of the Milky Way's water is frozen in ice.

Water vapor is an important trace gas that reveals the nature of the quasar. In this particular quasar, the water vapor is distributed around the black hole in a gaseous region spanning hundreds of light-years in size (a light-year is about six trillion miles). Its presence indicates that the quasar is bathing the gas in X-rays and infrared radiation, and that the gas is unusually warm and dense by astronomical standards. Although the gas is at a chilly minus 63 degrees Fahrenheit (minus 53 degrees Celsius) and is 300 trillion times less dense than Earth's atmosphere, it's still five times hotter and 10 to 100 times denser than what's typical in galaxies like the Milky Way.

Measurements of the water vapor and of other molecules, such as carbon monoxide, suggest there is enough gas to feed the black hole until it grows to about six times its size. Whether this will happen is not clear, the astronomers say, since some of the gas may end up condensing into stars or might be ejected from the quasar.

Bradford's team made their observations starting in 2008, using an instrument called "Z-Spec" at the California Institute of Technology's Submillimeter Observatory, a 33-foot (10-meter) telescope near the summit *of Mauna Kea in Hawaii. Follow-up observations were made with the Combined Array for Research in Milli-meter-Wave Astronomy (CARMA), an array of radio dishes in the Inyo Mountains of Southern California.*

The second group, led by Dariusz Lis, senior research associate in physics at Caltech and deputy director of the Caltech Submillimeter Observatory, used the Plateau de Bure Interferometer in the French Alps to find water. In 2010, Lis's team serendipitously detected water in APM 8279+5255, observing one spectral signature. Bradford's team was able to get more information about the water, including its enormous mass, because they detected several spectral signatures of the water (Clavin and Buis).

It may well be that the quasar is producing Birkeland currents in the surrounding plasma accretion disk that is pumping out water in the same manner that our early universe did.

I believe that by the end of the first day of Genesis, all the material that exploded from the Genesis Singularity that was inside the PCC had already exited, and our early plasma universe was subsequently divided into seven concentric circles of differing expansion rates. The outer circles were expanding faster and cooling more quickly. Each concentric circle inward expanded more slowly until they reached the area of the universe that last exited the PCC.

My reason for believing this comes from the empirical evidence amassed first by William Tifft in 1970 and then corroborated by many others such as David C. Koo and R. Krone, who affirmed through red-shift analysis that the galaxies in our universe are grouped into seven concentric shells around our sun-Earth ecliptic at the center.

D. Koo and R. Krone, two University of Chicago scientists, did the same kind of redshift analysis on galaxies. Their results were identical to Napier's and

Guthrie's and even made it to the New York Times. They conclude: ". . . the clusters of galaxies, each containing hundreds of millions of stars, seemed to be concentrated in evenly spaced layers" [i.e., concentric spheres around the Earth]. Incidentally, for those who see symbolic significance in numbers, the number of "evenly spaced layers" discovered by each team of astronomers is seven. There are seven evenly-spaced layers in the north direction, and seven evenly-spaced layers to the south. Koo admits that astronomers are very disturbed at this spacing, obviously because it gives evidence of intelligent design and geocentrism" (Sungenis and Bennett 2014a, 393).

That the God who made the universe in seven days would choose to band the galaxies in seven concentric circles around the Earth is no surprise to me. According to the M-theory, He also created seven invisible dimensions. When He destroyed the First Earth with the Great Flood, He broke the single continent into seven continents and the single ocean into seven oceans. And when the time comes to destroy the Second Earth, it will be done in a seven-year period called the Great Tribulation. It is evidence of His magnificent supersymmetry that He left as His indelible fingerprint at which all can marvel.

The Genesis account then records that water was created on the first day. That means that the material that made up our region of the universe, which was the last to exit the PCC, had to have already exited the primordial critical circumference by that time and that the expansion had cooled the spinning plasma cloud enough to harbor water. I suspect that it was at first in vapor form, and as the concentric circles of plasma moved outward and cooled, the vapor may have cooled down to liquid or ice form. If we study the composition of the Earth, we find that water is contained in areas far too deep to have come from rain drizzling down from above.

If we look at the material that fills our universe, we find first and foremost an overwhelming amount of material still in the plasma state. All stars are composed of mostly hydrogen and helium in the plasma state. By far the most abundant element found in the universe is hydrogen, followed by helium. But we also have large amounts of carbon, oxygen, and nitrogen, which form many organic compounds.

If we consider the creative process of the Birkeland currents, it is possible to create heavier and heavier elements in succession as the compression of the Z pinch merges smaller nuclei into larger ones. Once these elements were produced in sufficient numbers, compounds could be formed through the same process. For example, the union of two hydrogen atoms with one oxygen atom could create the water molecule.

There is an added electromagnetic dimension to the water molecule because of its peculiar position as the two hydrogen atoms bond at one end of the oxygen atom, making the molecule have one pole positive and the opposite negative. It is this property that makes water such a good solvent. This polarity would make the water molecules more susceptible to the electromagnetic force and may account for the ability of water to be shepherded into distinct areas.

This artist's concept illustrates a quasar, or feeding black hole, similar to APM 08279+5255, where astronomers discovered huge amounts of water vapor.

Matt Bradford, a scientist at NASA's Jet Propulsion Laboratory in Pasadena, California, said, "It's another demonstration that water is pervasive throughout the universe, even at the very earliest times" (Clavin and Buis). It should therefore not surprise us to find that water was one of the very first compounds made by the giant Birkeland currents that turned the churning plasma into the most important component of living things.

From the very beginning of God's creative process, the water that would comprise the most important component of all living things was created. That the scriptures documented that long before science could even begin to dream of how this could be true is indeed a testament to the reliability of the Genesis record as a historical space-time account of the birth of our universe. Water was created on the very first day, but it was not separated until the second day, as we shall see.

The Day Light Broke Free of Darkness

First, let me bring your attention to the fact that the book of Genesis tells us that light was separated from darkness. That modern science would document this actual process through the physics of the Big Bang is another incredible attestation of the historical space-time reality of the Genesis record. This is the exact process described by the photons coming through the opaque cloud of plasma. All of this, from God's point of view, took place during day one. The day began in darkness and ended with light. Sound familiar? That is why the Hebrew day begins with darkness and ends with light. And it is the overall silver thread of human history that brings us from the Alpha Point to the Omega Point.

So we see that the Genesis creation account is not in contradiction to science. On the contrary, science has corroborated the story of Genesis. In fact, the prophecies of the scriptures create a scaffolding of truth that gives us a much deeper understanding of creation and God's purpose behind the symmetry of His creative process.

The history written within the scriptures begins with God's perspective and then shifts to human genealogy after the seventh day, the day of rest. The Sabbath was the first day of humankind. From that point forward, time was viewed in scripture as relating to the point of view of human beings. It is an unfolding of the entire future of humanity and the universe we inhabit.

The Sabbath was symbolic of God's desire for humankind. For this reason, He commands us to keep it holy. Humans were meant to be at rest and at peace with the world around them, but humankind fell. Death entered the world, and paradise was lost because God did not want wickedness to become eternal. It is for this reason that there has to be an Omega Point to end all sin.

It is in the garden at Eden that perhaps the second law of thermodynamics kicked into gear. However, although I think it unlikely, it could have been part of our creation from the beginning. If so, then the creation could not have taken place without even more previsioned and specified energy being inputted externally to fine-tune and control the delicate parameters necessary for our universe to have been birthed as it was.

God did not have to undo the second law of thermodynamics. He simply needed to input more external energy into the system to move it from disorder to order. So God hovered over the world and brooded our universe. We see this energy in the form of light that brought order to the chaos on day one. In fact, it is light that allows life to exist even today in both plants and animals, for without photosynthesis, not even animals could survive.

We ought not be surprised at this claim. This is what photosynthesis does. It uses light and water to bring order from disorder and creates glucose to feed the plant. But for that raw energy to become useful, it needs a metabolic motor to channel that energy into useful ordering. In plants, that is done by chlorophyll, which takes the raw energy of the sun and the nutrients and water from the soil to make glucose. The mitochondria perform a similar purpose in the animal kingdom. That motor on day one did not come just from the stuff inside our universe. It was God who guided the parameters of that Big Bang to create the specified expansion rate and formation of the elements of the atom in their appropriate density and temperature to create a universe that can inhabit life.

Hence, it is an open universe. Chlorophyll and the mitochondria are mechanisms designed by God to mimic His creation. It ought to make us stand up and note that these motors have an incredibly sophisticated design that could not have resulted from random ordering, but we will deal with that in Book 3 of this series, *Codes: God's Indelible Fingerprint*.

At this point on day one, there were no sun or stars. The Earth was not revolving around the sun, and its rotation did not necessarily produce a day of 24 hours, as we suppose. That day simply marks a moment that began in disorder and became more ordered, which God called day one.

His light, not starlight, created that day. It is the same light that will be used in the Final Earth, the Fourth Earth, when there will be no stars and when there will be no second law of thermodynamics (Revelation 21–22). In fact, in that day, there will be no time. We, like the electromagnetic waves, will be as timeless as beings of light.

The idea that light could exist before stars or before the existence of our sun was flagrantly ridiculed by many atheists in the past as a tale of superstitious mythology. But our modern understanding of the process of the creation of the atom from radiant energy has once again vindicated the Genesis narrative as true truth. Light did exist before the stars and galaxies.

In fact, all things in this universe came from radiant electromagnetic energy that was bottled up in the Genesis Singularity with infinite density. Sagan was wrong. We are not the children of stars. We are the children of God's radiant energy. We are the children of light, literally.

Stars are but artificial lamps created by God to continue and maintain His creative process and eventually to end it. The dark stars will be the harbingers of the death of our universe. I suspect that the dark stars are symbolized by the supermassive black holes that are growing with time to gargantuan monsters at the edges of our universe where time has sped up.

It is my opinion that those who attempt to reconcile evolutionary claims to the scriptures are trying to fit a square peg into a round hole. The discrepancy between the Genesis record of the creation of humans varies from the evolutionary fable that humans ascended from the apes, and it is irreconcilable. Genesis describes the act of willful and previsioned creation by a loving and personal God rather than some natural random, impersonal evolutionary materialistic mechanism that brought forth our universe and everything it contains, including all living things.

In complete antithesis to all the propaganda constantly buffeting us from those who ardently pontificate their metaphysical choice of the Copernican principle, God specially created the Earth to be inhabited by His most special creation—humans. These naysayers, in the name of humility, deny God's previsioned placement of our solar system and make their puny finite minds the center of reality. By denying the creator, they make themselves gods. The arrogance is theirs, and it is astronomical.

While it is true that the Earth revolves around the sun, and it is also true that our sun is just one of many in our galaxy, the significance of a special creation does not require anything contrary to the present relationship between the cosmic bodies in our solar system. The sun does not have to revolve around the Earth in order for the Earth to have special significance. In fact, the sun's particularly fortunate location in the outer edges of the galaxy is also quite significant. Only in these outlying areas do we have a neighborhood of stars that is quiet enough not to be a danger to life as we know it.

If the sun were to be in the more crowded areas of the universe, the electromagnetic radiation of our neighboring stars would quite likely sterilize our planet. The orbit of our planets could be perturbed, causing massive destruction in our solar system. Our solar system was uniquely designed with the purpose of humans inhabiting Earth. Its central place in the universe, attested by the seven concentric galactic shells discovered by Tifft and the alignment of the anisotropic dipoles and quadrupoles discovered by the three mapping satellites (CORE, WMAP, and Planck), prove without question that we are at the axis of our space-time continuum. That is exactly what the deophobes and their Copernican principle so ardently despise and try to subvert.

Modern evolutionary scientists claim that dying stars are the engines that make the more complex chemicals necessary to form rocky planets. I do not dispute that thermonuclear fusion can create more complex elements, but it is not the only theoretical way. Certainly, God's radiant energy brooding over the surface of the waters can be the very same basic mechanism of this fusion function in stars. In fact, He made the stars to mimic His creation. Stars point to the process of God's creation. They are His fingerprint on our universe, just like the chlorophyll and the mitochondria are His fingerprints of life. In these, He gives us a glimpse of how He did it.

The intense nuclear processes in exploding stars that make our heavier elements is a system set up by God to mimic what He did through immense energy as He brooded over the cosmos. He set forth the system that would continue His creative process. The creative process that God set in motion is continuous. For example, one would not stipulate that Adam had to be recreated many times over to populate the Earth. He made humans to procreate. Neither would we expect Him to do this with the plants and animals. I believe that the process of the formation of suns and galaxies is no different. God set in motion the growth of our universe by establishing a mechanism that would mimic His creative process.

From God's viewpoint, the universe was the canvas upon which He could paint human beings, the pinnacle of His creation. And against the erroneous assumption of the evolutionary Copernican principle, this is not an anthropocentric but rather a deocentric reality. It is at this point that we need to examine the role of the Birkeland currents in the creation of solid matter from plasma. Stars and galaxies were not necessary to create the fusion of the elements. Already from day one, water was created, and the proto planet near the Alpha Point was being molded from a watery chaos.

The Spirit of God Brooded over the Water

The Genesis account is written specifically from God's viewpoint. It is unapologetically a deocentric narrative. And for this reason, we read that the Earth was created before the stars. In fact, the Earth was begun before light broke through the darkness of the opaque plasma. But it was not a planet. It was a formless and chaotic watery entity that God was progressively molding. It was not a defined ball of water but more like a watery cloud. God was moving over—brooding over (*rachaph*)—the surface of the waters.

> *The earth was without form and void, and darkness*
> *was over the face of the deep. And the Spirit of God*
> *was hovering over the face of the waters (Genesis 1:2).*

It is quite possible here to infer that God's radiant energy brooded in fluttering barrages of Birkeland currents that created all the water that now inhabits our universe. Obviously, I am not insinuating that I have some secret knowledge of how God created the universe. I am simply pointing out that there are scientific explanations that dispel the notion promoted by the evolutionists that Jews and Christian, who believe in the literal interpretation of the scriptures, rely on superstition and a magical God.

God uses the natural laws, which He created to accomplish His good will. There is no magic to creating water through the electromagnetic force in the Birkeland currents. In fact, they can also create the higher elements that formed *terra firma* on Earth. And finally, the process of creating the elements can be accomplished almost instantaneously through the Birkeland currents. There is no need for millions of years for the slow gradual process of gravity to create our universe.

Day Two: The Universe of Aquarius

On day two, God continued the process of bringing greater order to the watery chaos that would later be formed into our planet Earth. By the end of day two, from the viewpoint of God, two days had passed. At this point, the brooding or hovering radiant energy of God over the waters caused further change.

> And God said, "Let there be an expanse in the midst of the waters, and let it separate the waters from the waters." And God made the expanse and separated the waters that were under the expanse from the waters that were above the expanse. And it was so. And God called the expanse Heaven. And there was evening and there was morning, the second day (Genesis 1:6–8).

The magnificence of God's words is revealed in the multilayered meanings that they portend. Often, there are primary, secondary, and even tertiary meanings that show an unfolding symmetry to God's design that cannot be the result of human minds or random chance. Our Judeo-Christian cosmological model maintains that this simple passage contains three levels of meanings:

1. The primary meaning of this passage is that God separated the water that would become our oceans from the water that would reside in the cosmos such as comets, cosmic water/ice, clouds, and so on. The narrative in Genesis of our Judeo-

Christian cosmological model claims that water was one of the first compounds made by the giant Birkeland currents, and God separated the waters to mark out the specific region the Earth would hold within the entire cosmos He was creating.

Early skeptics of the Genesis record assumed that the narrative was referring to rain clouds. That could not be further from the facts. In fact, quite the contrary, Genesis expressly claims that there was no rain upon the First Earth until the day of its final judgment when God ended it with a global deluge. We will discuss this more in the second meaning.

Some may think it is strange that the cosmos would contain water. It is not. Modern scientists have known for some time that there is a lot of water/ice in space. Because of the coldness of space, the vast majority of water is in the form of ice clouds. Scientists have long suspected that some water would be in the form of gas but, until recently, lacked the technology to detect it. Water in liquid form does not exist in space due to the pressure and temperature conditions of deep space. The temperature in these deep space regions is -263 degrees Celsius (-441 Fahrenheit). That is just 10 degrees Celsius above absolute zero.

Ground telescopes on Earth can easily detect the ice water. Light traveling through water vapor from distant stars leaves a chemical fingerprint that allows scientists to detect it. But the signature of water vapor is hidden from ground-based telescopes by the water vapor in our atmosphere. Recently, however, using data from the ISO, the European Space Agency's Infrared Space Observatory, Italian astronomer Andrea Moneti and his colleagues found that cold regions have as much total water (ice plus vapor) as warmer regions where stars are actively forming. It seems that 99 percent of these ice clouds are in the form of solid ice, and 1 percent are in the form of gas.

Astronomers estimate that there are millions of cold clouds, or quiescent clouds, in the Milky Way. These cold clouds are areas where stars are not forming and the water vapor condenses on cold dust particles. If we look into the night sky toward the Milky Way, these clouds look like dark patches where there are no stars. But in reality, these clouds contain tiny dust particles that block the starlight behind them. These clouds are even colder than the interstellar medium because of the blocked starlight.

Water seems to be ubiquitous throughout the cosmos. As it turns out, they have found water even in the middle of galaxies. In fact, they have found water on the surface of the sun. A team of researchers led by Oleg Polyansky, a theoretician from the Institute of Applied Physics of the Russian Academy of Sciences; Nizhny Novgorod, who works with co-researcher Jonathan Tennyson, a physicist at University College, London; and University of Waterloo chemistry professor Peter Bernath, an expert in molecular astronomy, have documented that water vapor exists even on the surface of our own sun.

Other members of the team included physicist Serena Viti at University College, London; Nikolai Zobov, a physicist at University College, London; and Lloyd Wallace, an astronomer at the Kitt Peak National Observatory in Tucson, Arizona. In 1995, the team recorded the evidence of water vapor in the dark sunspots on the surface of our sun through infrared spectrum analysis. But this is not a phenomenon of just our own sun. Scientists have now regularly detected that vast amounts of water are ejected in exploding stars.

> *Seven hundred and fifty light-years from Earth, a young, sunlike star has been found with jets that blast epic quantities of water into interstellar space, shooting out droplets that move faster than a speeding bullet (Fazekas).*

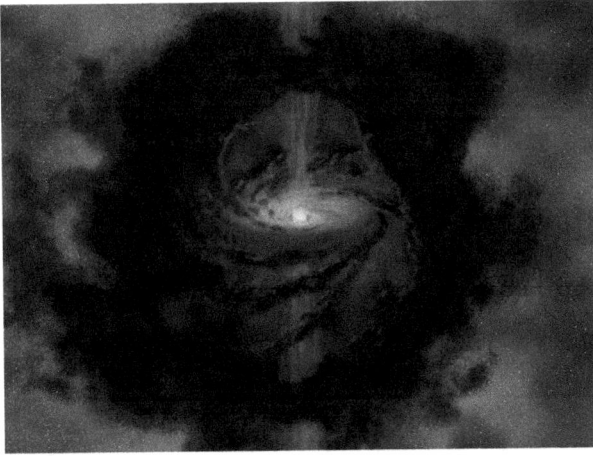

A star is born: Swirling gas and dust fall inward, spurring polar jets, shown in blue in this illustration

Little did the early skeptics of Genesis suspect that water would be found even inside the brightest of stars.

> *Recently, two of the brightest supergiants in the galaxy, Betelgeuse (in the Orion Constellation) and Antares (in the Scorpio constellation), were discovered to actually have water in her photospheres, as well as in her circumstellar material surrounding their photospheres. . . . The structures in photospheres in cool stars is due primarily to the opacity of water, which is one of the most abundant molecules in such stars. The presence of photospheric water in these red supergiants confirms that it is located within the star itself and is not just a component of the dust and gas clouds surrounding the stars. Aging supergiants have been observed to release massive amounts of water as they die (Sungenis and Bennett 2014b, 404).*

Science is just now beginning to understand the true dimension of water's existence in the cosmos. It truly is the domain of Aquarius.

Just within our own solar system we have found worlds of ice. Enceladus, the sixth-largest moon of Saturn, contains a large saltwater ocean. At that distance from the sun, the water should be frozen. But the huge tidal waves created by Saturn's enormous gravity keep the interior ocean warm enough to crack through the surface in huge geysers that pump out 1,000 tons of water every hour into space. In fact, the ice crystals it leaves behind form the F ring of Saturn.

Europa, the smallest of Jupiter's four moons, contains a beautiful white surface of ice that shows the typical cracking and shifting that mimics our own ice poles. NASA's New Horizons space probe has even found ice on Pluto. It appears that Pluto contains hills of ice floating on a frozen ocean of nitrogen. But perhaps one of the most spectacular findings of the New Horizons pictures is a thin blue line around Pluto, which shows that this tiny planetoid has an atmosphere. It is hard to imagine that such a tiny body could keep an atmosphere for the enormous time spans evolutionists claim as the age of our solar system. Nevertheless, the fact is that most of the bodies that form the Oort Cloud beyond Pluto are made of water ice. The cosmos is filled with water. The Genesis record told us this long before man ever knew of a telescope.

The Genesis narrative is once again vindicated by modern science. Water was separated by the firmament. But not all water is the same. There is a distinction between the chemical signature of the water on Earth and that found in comets and deep space. Comets contain a much larger content of heavy water than the oceans in our planet. Nevertheless, we have found water not only on comets, planets, and moons but also in large space clouds hundreds of light years across, located deep in our universe. And so we see that God separated the water from the Earth and the water that inhabits our cosmos on the second day.

It should not surprise us that water was created so early. It is the single most important requirement for life to exist. Human beings are made mostly of water. But just as important, on the second day, God was literally fixing a space-time coordinate for the development of our planet by separating our specific place in the heavens. Our so-called Blue Planet was first an ocean planet, and the significance of its place in the universe cannot be denied.

The Hammering Process

The word translated "expanse" is also translated "firmament" in other translations of the Bible. It is the Hebrew word *raqiya* that is rooted in the word *raqa*, which means to expand by hammering, to pound the earth, to make broad. This leads us to the second level of meaning. Hence, the nuance is that God's light pounded the Earth and made the broad expanse of the sky—our atmosphere. The connotation is that the process was very much like the work of a metalsmith hammering out his project.

2.   The second level of meaning to God's narrative of the second day of creation is that God separated the water from our ocean from a water vapor canopy in the sky. There are good reasons to believe that our early Earth contained a thin water vapor canopy that created a greenhouse effect upon the First Earth. Such a water vapor canopy would explain how the Earth would have a mist that came over the ground throughout the globe as the nightly condensation of the humidity in the air.

There is no doubt that this "hammering" process not only created the vital gases that compose our atmosphere, but also the protective water vapor canopy and the ozone layer that magnificently sheltered the First Earth from cosmic rays. By the end of the second day, we not only had an atmosphere around the surface of the Earth but also a protective water vapor canopy that caused the temperature over the surface of the Earth to be moderated like a veritable greenhouse. This is the second meaning of that verse. We can see this water vapor canopy on the planet Venus, but in a much more exaggerated form than in the form that existed here on Earth.

The very word translated as "heaven" adds weight to the multiple meanings of that verse. The Hebrew word for "heaven" is *shamayim.*

It is interestingly spelled with the plural ending "im." *Shamayim* is plural for the Hebrew word *shameh*, which means to be lofty. This plural ending denotes not a dualism but a cluster, as in *El* and *Elohim*. *El* is the title of God in the Hebrew language, but often it is written *Elohim*, which is the plural form that denotes not two but a unified cluster, as in a cluster of grapes. There is but one God, but He manifests Himself in three forms: the Father, the Son, and the Holy Spirit.

This plural *shamayim*, therefore, is used to mean three things: (1) the space beyond the Earth where the celestial bodies exist (the cosmos); (2) the sky, or our atmosphere; and (3) the abode of God, where His temple resides. Thus far, all of this has been created by the glory of God's light. (For more information on the nature of the First Earth, see the fifth book of this series, *The Death of the First Earth*.)

The first meaning of the word *shamayim*, then, infers that the separation of the waters is the division between the Earth's ocean and the waters in the cosmos. The second meaning of shamayim is the division between the Earth's ocean and the water vapor canopy above our atmosphere.

Hence, the secondary meaning of this passage is that God separated the waters in our oceans from the water in our atmosphere. But the water in our atmosphere was not in the form of the clouds we see today. The Genesis narrative explicitly tells us that the meteorological system of the First Earth was quite drastically different from our Second Earth after the Great Flood. It tells us that it did not rain upon the Earth until the judgment of the Great Flood destroyed the First Earth. Each day, the morning mist watered the land. There were no rain clouds in the First Earth.

> *For the* Lord *God had not caused it to rain on the land, and there was no man to work the ground, and a mist was going up from the land and was watering the whole face of the ground (Genesis 2:5–6).*

Yet the fossil evidence tells us that even without rainclouds, the First Earth was superior to our present Earth in many ways. The fossil evidence tells us that everything grew bigger, and the animals and plants thrived with much greater variety within each species. Land everywhere was full of thick, lush, fertile soil capable of sustaining an enormous variety and quantity of vegetation. Even evolutionists admit this.

The Advantage of the Water Vapor Canopy

This mild, uniform global climate can only be explained by a thin water vapor canopy that shrouded our planet and moderated the temperature from pole to pole, creating a literal global greenhouse. That is one of the reasons for the observed gigantism of that epoch, attested in the fossil record. No doubt, in addition to this, the protection from the deleterious effects of cosmic rays that cause harmful mutations to our genes maintained a more perfect specimen of all the species. That was the reason for what we call the *megafauna*. In reality, that was the normal. Today, we are an inferior shadow of the world that was. We are not normal. We are most definitely subnormal compared to the magnificent First Earth.

> *Huge insects crawled, crept, and flitted across the earth. Two of them were the largest known insects of all time, the centipede Arthropleura, which grew to a length of more than eight feet, and the giant dragonfly Meganeura, which had a wingspan of some two and a half feet. These enormous dimensions were possible because at that time oxygen made up 35 percent of total air volume (rather than our current wimpy 21 percent) (Church and Regis 95).*

Plants were watered daily by the morning dew created by the condensation of the humidity in the air during the cooling of the night. This suggests that a small water vapor canopy enveloped the whole planet in order to provide a literal greenhouse effect that allowed life to exist even in the polar regions, which, in our Second Earth, are so hostile to life.

If the biblical model is correct and that antediluvian world was shielded by a water vapor canopy that moderated temperatures throughout the planet by means of the greenhouse effect, then we would expect the geological record to show that both flora and fauna of temperate creatures would have been able to live in those extreme remote areas of the poles, which are presently under impossible frigid conditions that are incompatible with higher forms of life.

With respect to climate, the fossils show that there was a uniformly mild climate in high and in low altitudes of both the northern and the southern hemisphere. That is, there was a perfectly uniform, non-zonal, mild, and springlike climate in every part of the globe. This does not mean that the climate was of necessity the same in all parts of the earth. There were differences, but not the present extremes. Sir Henry H. Howorth, a noted geologist and competent interpreter of these fossils, says: "The flora and fauna are virtually the only thermometer with which we can test the climate of any past period. Other evidence is always sophisticated by the fact that we may be attributing to climate what is due to other causes. But the biological evidence is unmistakable; cold-blooded reptiles cannot live in icy water; subtropical plants, or plants whose habitat is the temperate zone, cannot ripen their seed and sow themselves under arctic conditions."

Or another outstanding authority, Professor Alfred R. Wallace, says: 'There is but one climate known to the ancient fossil world as revealed by the plants and animals entombed in the rocks, and the climate was a mantle of springlike loveliness which seems to have prevailed continuously over the whole globe. Just how the world could have thus been warmed all over may be a matter of conjecture; that it was so warmed effectively and continuously is a matter of fact." . . .

Or Professor George McCready Price writes: "It would be quite useless to go through the whole fossiliferous series in order, for there is not a single system which does not have coral limestone or other evidence of a mild climate way up north, most systems having such rock in the lands which skirt the very pole itself. The limestone and coal beds of the carboniferous period are the nearest known rocks to the North Pole. They crop out all around the polar basis; and from the dip of these beds, they must underlie the polar sea itself. But it is needless to go through the systems one after another, for they "uniformly testify that a warm climate has in former times prevailed over the whole globe." (Rehwinkel 7–8).

That our world that was our First Earth enjoyed a long period of mild climate throughout the entire globe cannot be disputed from the biological evidence found in the fossils. The overwhelming evidence in every corner of the world is undeniable, even by the staunchest of evolutionary scientists. That our First Earth underwent a violent, extreme meteorological and geological upheaval that suddenly and catastrophically changed it forever, causing the majority of the flora and fauna to become extinct, is now also accepted, even by evolutionists. Most estimates claim that 95 percent of the flora and fauna perished after this catastrophic change in climate.

In effect, the humidity that caused the morning mist was uniformly sustained around the entire planet by the water vapor canopy. It literally caused the entire First Earth to become a lush paradise. Underground springs and streams and the morning mist created by the rich humidity amply watered our primordial planet. It was a superior form of irrigation not causing undo erosion and avoiding the problem of overwatering in low-lying areas and underwatering in higher elevations. Continuous watering with this mist evaded the problems of flash floods or arid areas of washed up soil with rocky, untillable ground, as well as areas where too little rain causes deserts.

The water vapor canopy worked in the same way as a greenhouse. It was a global temperature regulator. The net effect adequately kept the entire planet at a moderate, temperate climate throughout the entire surface of the single landmass. Although there were some minor variations in temperature, there were no frozen poles.

The extreme northern and southern regions of the singular landmass had a temperate climate ideal for grasses, which gave ample food for large Mastodons, rhinoceroses, reindeer, musk oxen, horses, and other grazing animals. The lush vegetation that was initially thriving, throughout the entire landmass provided ample food for an incredible variety of animals.

This water vapor canopy created an idyllic First Earth that was a veritable paradise compared to our present Second Earth. There were no hurricanes; there were no tornados. There were no lightning storms or flash floods. But little did humans appreciate the paradise they lived in until they lost it.

It was a marvelous benefit to our planet, acting in much the same manner as our ozone layer and magnetosphere. That is, it filtered the harmful cosmic radiation battering us from our sun, but with the added benefit that it also moderated the global temperatures and humidity levels. The entire First Earth was basically air-conditioned.

For these reasons, I believe that the secondary meaning of that passage is a description of the water vapor canopy that must have protected the First Earth and provided an ideal climate for life. It seems that God's light containing heat must have created water vapor that rose into the sky to form this canopy.

3. The third meaning of the narrative of the second day is tied to the third meaning of *shamayim*, which is the abode of God. The scriptures tell us that there is a temple of God where He and the angels reside, and from His throne, a river of fire flows to judge the world. The prophet Daniel saw it in the day that God judged the antichrist.

> *"As I looked, thrones were placed, and the Ancient of Days took his seat; his clothing was white as snow, and the hair of his head like pure wool; his throne was fiery flames; its wheels were burning fire. A stream of fire issued and came out from before him; a thousand thousands served him, and ten thousand times ten thousand stood before him; the court sat in judgment, and the books were opened. "I looked then because of the sound of the great words that the horn was speaking. And as I looked, the beast was killed, and its body destroyed and given over to be burned with fire"* (Daniel 7:9–11).

But it is important to understand that this present heaven is also temporary. God says He will destroy it because Lucifer and one-third of all the angels that rebelled against God defiled it. Hence, He shall create a new heaven and a new earth that is undefiled after the Great White Throne Judgment. This new heaven is described not with a river of fire but, instead, with a river of living water flowing from the throne of God in the New Jerusalem.

> *Then the angel showed me the river of the water of life, bright as crystal, flowing from the throne of God and of the Lamb through the middle of the street of the city* (Revelation 22:1–2).

I strongly suspect that the abode of God in this present universe is within those seven invisible dimensions that the string theory has discovered. However, since the meaning here is a bit more esoteric and not in line with the purpose of this book, I will not tarry any more on this point.

What the Genesis narrative does tell us is that Earth went one further step from chaos toward order in God's previsioned purpose to make our world habitable for life. Once more, we went from evening to morning, and it was the second day of God's light.

Day Three

> And God said, "Let the waters under the heavens be gathered together into one place, and let the dry land appear." And it was so. God called the dry land Earth, and the waters that were gathered together he called Seas. And God saw that it was good. And God said, "Let the earth sprout vegetation, plants yielding seed, and fruit trees bearing fruit in which is their seed, each according to its kind, on the earth." And it was so. The earth brought forth vegetation, plants yielding seed according to their own kinds, and trees bearing fruit in which is their seed, each according to its kind. And God saw that it was good. And there was evening and there was morning, the third day (Genesis 1:9–13).

It is not hard to imagine how God's radiant energy could have heated the oceans to create the water vapor that eventually reduced the ocean level to allow the dry land to appear. On the third day, God said twice that it was good after he completed two special tasks. The first was to gather the ocean waters to create dry land. Our first world contained one large landmass of a granitic substrata and a single pristine ocean. This was unknown to modern man until the continental drift theory revealed it in the twentieth century. It was then that modern scientists realized for the first time that at one time, the Earth's continents were united, and we had one single massive ocean, just as the scriptures had insinuated long ago.

The Code of Life

But the second thing He did on that day was create living plants. There was no long period of gradual sophistication in chemical processes that led to the miraculous single cell. Long ago, scientists understood that spontaneous generation could not take place. During Darwin's time, the complexity of the biochemistry of life was utterly unknown. The single cell was seen as an unsophisticated blob of protoplasm. The idea that gradual chemical changes could accumulate and form a single cell was seen as a plausible mechanism.

Today, scientists know that the complexity of the biochemistry in even a single-celled organism is utterly intricate and extremely complicated. There is no such thing as a simple-celled organism. Now more than ever, life is understood in a way that Darwin could not have dreamed of. On this day, the third day, life began on Earth, not by long periods of gradual changes, but by the express design of a master designer.

On this day, DNA was created to provide the seed from which plants could yield other plants after their kind. On this day, chlorophyll was invented to bring food from sunlight. There is no gradual evolution from one kind into another. Each kind reproduces itself. There is no way to merge the evolution of the species with the Genesis text without completely dismantling the meaning of the text. We will discuss this in more detail in Book 3 of this series, *Codes: God's Indelible Fingerprint.*

All living things have a code that determines their kind. Stop for a moment and consider the essence of a code. A code is a set of rules for converting information from one useful form into another. For example, if I were to send you this message, what would you think?

ajckdomudnhengslpalkrshextmkhaxedlpezr
baoywdmvuircnktduossfmuasrmxriwfnaed

Is there any useful information in this set of letters? It is not immediately apparent, but there is, if I were to give you the code: every third letter, left to right. Without a code, it is gibberish.

According to the code, start from the left and choosing every third letter, we see that the first third letter is c, the third letter after that is o, and so on. If you continued, you would get this message: *codesaretheproductofamind.*

Because the reader knows the code of the English language, the first code enables the reader to set the stage for the second code (the completely arbitrary choice of selecting every third letter). There is no chemical reason for this choice. It is a choice that can only be made by a mind. Looking at these letters, we can now deduce that it says, "Codes are the product of a mind."

Perhaps it would be possible to randomly duplicate that sequence of letters, but without a code to translate them, it is gibberish. How does natural selection engineer a double code? Yet that is precisely what we have between the code of the DNA and the proteins in cells.

I could have created any number of codes to send the message. There is no law that requires that I should have chosen the code of every third letter. I could have made another choice. It was an arbitrary choice made by me, the code maker. That choice necessitates a mind to choose. Random ordering is the absence of choice.

The very nature of the code is that it is an arbitrary system that has been selected to decipher what would otherwise be unknowable within a set of asymmetrical data through a permanent template.

Natural processes do not produce codes that pass on information. Only a mind can create a code. In fact, we cannot observe this anywhere in nature outside of living things. How, then, can inanimate matter evolve into living matter without the ability to create a code? Codes are produced only by intelligence, and no empirical evidence exists that can claim otherwise.

The magnificent DNA with its double code could not have accidentally been formed by random chemical processes. Random forces can create simple complexity characterized by repeating patterns such as waves or crystals. For example: ABC, ABC, ABC is a form of simple complexity. Such repetitive forms are capable of carrying very little information. They are symmetrical. Specified complexity is asymmetrical.

The sentence "My name is Henry" is written in the English language. Each of the letters in that sentence has specific information that, when placed together, performs a function. The code allows the reader, who also knows the code, to decipher the meaning.

If the reader understands the code that interprets each letter, he or she is able to understand the sentence. That is viewed as specified complexity because the asymmetrical structure allows the code to carry information-rich messages. The code allows the transfer of functional information.

The numerical sequence 100 1101 101 1001 100 1110 100 0001 100 1101 100 0101 100 1001 101 0011 100 1000 100 0101 100 1110 101 0010 101 1001 also says the same thing. It says "My name is Henry" in the American Standard Code for Information Interchange (ASCII). It is the binary code that my computer is using to store the data I input when I am writing.

Each letter of the English language in which I am speaking is represented by a specified numerical code of the numbers 1 and 0. The exact order of these two numbers is specifically combined to represent a letter that represents a sound that, when placed together, represents a word that represents a thought. It is a multilayered code that cannot be created by random natural processes. It needs a mind or designer.

Asymmetry and complexity alone cannot carry specific information without a code. The sentence "Xt morh br kturs" has the same number of letters as "My name is Henry." It has irregular, unrepeatable, asymmetric complexity, but it has no specific information because it does not have a code.

This exhibits information that scientists call "mere or simple complexity" as opposed to "specified complexity." Codes that exhibit specified complexity (that is, it carries functional information) are not found anywhere in the natural world outside of living things. Specified complexity containing a code cannot appear through random chaotic processes. It must be imputed by a mind.

The DNA, the mitochondria, and the chlorophyll are all complex metabolic motors that are functional only because of the codes that carry specific rich information within them. Random ordering, no matter how many millions of years it is given, could never establish codes. Only minds can create codes, and since these codes preceded humans, it is only rational to conclude that only the mind of God could engineer the many codes found in living systems.

God did not reveal to us how He created these plants. In fact, He did not reveal to us how He took the clay of this Earth and made Adam. What He did reveal is that it only took Him one day to accomplish it. I cannot give you a scientific explanation how that could be done, and no scientist can tell you how it was done. There are no viable materialistic explanations that can explain the transition from non-living to living. What I can tell you is that the evidence of this code in life rules out any accidental random evolutionary process. God designed it so we would know it was no accident. Perhaps as our technology progresses, we will one day begin to get a glimpse of the natural laws He used to create life. But for now, we are looking into a black box.

Here is what we do know: The earth went from the evening of disorder to the morning of order as God plowed through Earth and hammered it into the habitat He had previsioned for humanity to inhabit. There were still no stars and other celestial bodies existing. The very specialness of Earth and the life that thrived in it came not from the sunlight of the churning stars, but from the very glory of God's light. It is that exact same light that will power our future universe when the new heavens and the new earth are created.

Life was first created not from sunlight, but from the pure warm light of the Heavenly Father who, in different intensities, molded and hammered out Earth and brought forth plant life to prepare the way for human beings. Moreover, after the Omega Point, we will live eternally without sunlight, again basking in the peace and warmth of His divine light. The sun is but a substitute, a symbol, an artificial lamp for the light of God.

In that day, when the curse of the second law of thermodynamics is lifted, God will once again be our light. In that day, the living water and the fruit of the tree of life will eternally sustain us. In fact, I imagine that the ancient tree of life no doubt preceded even the stars in the sky. More importantly, it will exist long after the sun and stars no longer exist.

Then the angel showed me the river of the water of life, bright as crystal, flowing from the throne of God and of the Lamb through the middle of the street of the city; also, on either side of the river, the tree of life with its twelve kinds of fruit, yielding its fruit each month. The leaves of the tree were for the healing of the nations. No longer will there be anything accursed, but the throne of God and of the Lamb will be in it, and his servants will worship him. They will see his face, and his name will be on their foreheads. And night will be no more. They will need no light of lamp or sun, for the Lord God will be their light, and they will reign forever and ever (Revelation 22:1–5).

It was that same warm, life-giving light from God that sustained all plant life before the first insect or animal was ever made. Photosynthesis flourished before the first star was ignited. Oxygen flowed freely before the light of stars ever graced the night sky. No evolutionary chronology can be synthesized with the Genesis account. Either one believes that God is omnipotent and capable of doing this, or one does not. There is no middle ground.

But I hope by now you have come to understand that the omnipotent God who was able to control the explosion of the Genesis Singularity does not fit into manmade constructs with finite limitations. The creative power of His logos is unlimited.

The creation of plant life provided the oxygen necessary for our planet to contain other forms of life. Thus the vegetation had to come first. We see in Genesis that it did. So great was this vegetation and so fertile was this soil that the oxygen in our early atmosphere during the First Earth was hovering around 35 percent to 37 percent in volume compared to our measly present 21 percent. It is the marine ecosystem that is mostly responsible for our oxygenated atmosphere. Were it not for the tiny phytoplankton, the oxygen levels in our atmosphere would be greatly reduced. Today, the tiny phytoplankton account for 50 percent of our oxygen and reduce our carbon dioxide levels by absorbing the carbon during photosynthesis.

Too tiny to see with the naked eye, these indispensible creatures form the very backbone not only of our marine ecosystem but of all terrestrial life, even today. When found in large concentrations, they may appear as a green discoloration of the water due to the green chlorophyll. The color may vary with the species of phytoplankton. These form the primary level of the marine ecosystem as they turn light energy into usable energy for living things. They turn light to food. Yes, we are the children of light in more ways than one.

The entire ocean of the First Earth was capable of harboring phytoplankton since there were not vast areas of frozen solid ice in the North and South Poles. That provided a much greater amount of oxygen production. The entire landmass was filled with virgin forests and vast steppes of grass, so the First Earth contained a much richer oxygen supply for animals. It is at that time of high oxygen concentration that God probably produced a much thicker ozone layer than the pitiful layer we have today. It was designed as another added protection against harmful cosmic rays in order to create for humans an ideal habitat. Each step God took formed the First Earth with further order in order to provide human beings with an ideal environment.

From the viewpoint of God, three days had passed. Certainly, the power of God to do all these things is miraculous, whether He did it in one 24-hour period of evening and morning or in any other measurement of time. The time frame is irrelevant to the task performed by the creator who is able to use laws of science and physics in our universe that we have not even begun to imagine. To think that God is limited to the laws that we as finite beings have discovered is arrogance beyond words. How can the condescending clay dictate to the potter what He can or cannot do.

The more technology we possess, the more we are able to use the laws of physics to accomplish what would otherwise seem impossible. For instance, if one had asked a person living in Abraham's time whether or not humans could leave the surface of the Earth and fly to the moon, that person would have thought you were crazy. Even if that person did not know about the lack of atmosphere in deep space, he or she would have known about gravity. We cannot go against gravity. But we did. We used other laws of science such as dealing with aerodynamics and thrusts to counter the law of gravity.

It is supremely arrogant of humanity to consider that our finite knowledge of science can give us the platform to determine what is scientifically possible for a superior intelligence to do. Ironically, modern people can easily accept the idea that aliens could travel through the galaxy at the speed of light, knowing that the speed of light is, in fact, an absolute barrier that cannot be broken, but they balk at the abilities of the author of light.

What we do read in the text is that God spoke forth these events. In other words, the energy of the voice of God is what created these achievements. This is an amazing scientific statement. Matter is the vibration of the voice of God. Matter is vibrating energy. Only in our modern physics have we begun to understand that our reality is composed of vibrating energy. All that we see, feel, and hear is simply vibrating energy in space-time. This is exactly what Genesis described thousands of years ago in language that could be understood by people even now.

At that point in the universe, the celestial bodies had not yet been created. But our Earth was. Here, the scriptures do not harmonize with evolutionary dogma. Life did not rise from the sea. The first living things were plants on both the land and in the sea. Animal life followed after the formation of the heavenly bodies. There is no way to reconcile both timetables and the order that life appeared between Genesis and the atheistic theory of gradual evolution.

Day Four

> *And God said, "Let there be lights in the expanse of the heavens to separate the day from the night. And let them be for signs and for seasons, and for days and years, and let them be lights in the expanse of the heavens to give light upon the earth." And it was so. And God made the two great lights—the greater light to rule the day and the lesser light to rule the night—and the stars. And God set them in the expanse of the heavens to give light on the earth, to rule over the day and over the night, and to separate the light from the darkness. And God saw that it was good. And there was evening and there was morning, the fourth day (Genesis 1:14–19).*

From God's point of view, the celestial bodies began on the fourth day. In the standard cosmological model of evolutionists, the two necessary elements to begin the process of star and galaxy formation are temperature and gravity. The clumping of the disparate contents of the plasma soup had to be cooled down enough in order for this to take place.

One very important consequence of this conclusion is that the differentiation of matter into galaxies and stars could not have begun until the time when the cosmic temperature became low enough for electrons to be captured into atoms. In order for gravitation to produce the clumping of matter into isolated fragments that had been envisioned by Newton, it is necessary for gravitation to overcome the pressure of matter and the associated radiation. The gravitational force within any nascent clump increases with the size of the clump, while the pressure does not depend on the size; hence at any given density and pressure, there is a minimum mass which is susceptible to gravitational clumping (Weinberg 74).

Weinberg goes on to say that the critical temperature for this process to begin is around 3,0000 Kelvin. At this temperature, free electrons are now able to unite with the nucleus of atoms in the plasma soup.

With the disappearance of free electrons, the universe became transparent to radiation; and so the radiation pressure became ineffective. . . .

When the radiation pressure became ineffective the total effective pressure dropped by a factor of about 1,000 million (Weinberg 74–75).

However, the assumption taken by the standard cosmological model is that gravity was the main player in this story. We have already documented that the electromagnetic force through the Birkeland currents can instantaneously produce solid matter by uniting free electrons with the nucleus within a plasma soup. This began on the first day of creation when these giant Birkeland currents were synthesizing large amounts of hydrogen, helium, carbon, nitrogen, and oxygen as well as copious amounts of water. By the time light broke through the plasma on that first day, the temperature of the nascent universe had been greatly reduced by the synthesis of matter and the impact that water had on the overall temperature at the core of the spinning plasma.

It seems to me that God separated the whirling plasma into seven distinct bands that circled around the cooler central region where the Birkeland currents had ejected the material (including water) that God used to make the Earth. How exactly He used these Birkeland currents to create stars and galaxies, I cannot tell you. But I am inclined to think that He used them as a means to shepherd material to one point or another in order to create the stars and galaxies in precisely the place He had previsioned so the constellations, as seen from Earth, would tell the story of the ancient cosmic battle between good and evil.

What I can tell you is this: It was not gravity that created the stars. It was the electromagnetic force that chiefly created the conditions by shepherding hydrogen and helium into specified regions of higher density spinning gas clouds in order for gravity to then do the final work. But even then, the force of gravity alone could not contain these swirling clouds without the help of electromagnetic field lines that were probably created in their spinning cores to hold them together as a unit. The formation of stars did not need millions of years to aggregate slowly by gravity. That entire process was set in motion on a single day, the fourth day of creation.

Of course, this is only speculation on my part, but in this regard, I am in no different of a place than the evolutionists. The truth is that evolutionists have absolutely no idea how galaxies were formed, and they only think they know how stars were formed, but their gravity model is not even close. Gravity alone cannot condense a whirling cloud of plasma, since the hotter it becomes, the faster it spins, and its angular momentum keeps it from condensing into a star. No, it is only the electromagnetic force that can tame plasma. And it can do so instantaneously. No slow, gradual evolutionary process is either possible or necessary.

In spite of the many bold claims made by naïve evolutionists when Immanuel Kant first published his *Universal Natural History and Theory of the Heavens* some 200 years ago, physicists today now know this mechanism to be physically impossible. Kant had proposed that all the galaxies emerged from primal nebulae. A nebula with molecules in constant motion, he theorized, began to aggregate into cores of mass, which eventually began to rotate. In this manner, then, protostars and planets were formed. If evolutionists are honest, they will tell you that they have no inkling as to how the galaxies were formed. Steven Weinberg has the integrity to speak the truth in this regard.

> *This is not to say that we actually understand how galaxies are formed. The theory of the formation of galaxies is one of the great outstanding problems of astrophysics, a problem that today seems far from solution (Weinberg 75).*

On this fourth day, God created the stars and galaxies within seven concentric circles that surround our solar system. Hence, the number four is symbolic of God's overall plan for our creation. That number is repeatedly symbolized throughout scripture as representing the world we inhabit. And the number seven is symbolic of God the Father, who spoke these things into being in seven days.

This First Earth that God created passed away in the Great Flood. We now live in the Second Earth. The Third Earth will begin when earthquakes destroy this Earth and the returning Messiah sets up the Davidic Kingdom. The Fourth Earth will be the final earth, when God re-creates the new heavens and the new earth. Four is therefore the symbolic number of the Earth:

1. There are four seasons divided by the two solstices and the two equinoxes.
2. There are four quadrants: north, south, east, and west.
3. The DNA code uses a four-letter code that allows life to exist.
4. The visible universe has four dimensions: three spatial dimensions and time.
5. Four forces rule our material universe: gravity, electromagnetic, strong nuclear, and weak nuclear.
6. There are four states of matter: solid, liquid, gas, and plasma.
7. Matter particles come in four groups of three. Two of the groups of three are quarks, and two of the groups of three are leptons.

Even the symbol for the Earth is a globe bisected by meridian lines into four quarters. In essence, it is a circle with a cross in the middle. This is no coincidence, for it was on the cross that the Messiah was stretched out to all four quadrants of the Earth when He took on the sin of humankind upon His shoulders. What materialistic mechanism can explain rationally how random events could create such symmetry?

On the fourth day of creation, the day we call Wednesday, the celestial bodies were created. The stars and galaxies were formed, and our sun, moon, and planets were created. Our solar system became complete. It was a new moon. The lunar cycle began that day. On that day, time took on the form of counting by the orbits of the celestial bodies. The year took on the measurement of time that our Earth revolves around the sun. The day took on the form of the time it takes to rotate from the sunlight one revolution. The month was determined by the cycle of the moon around the Earth. The constellation rising on the vernal equinox was Virgo, and the physical cosmic course of our solar system was set toward Leo. The star Vega in the constellation of Lyra was the North Star. On that day, God's creation was perfectly set in motion.

The Lunar and Solar Calendars Were in Synchrony

The purpose for the creation of the celestial bodies is given by God's own words: "And let them be for signs and for seasons, and for days and years (Genesis 1:14).

God made the celestial bodies expressly to provide for humans the signs that would lead them to understand God's design and majesty. These signs, of course, include the story of the zodiac, but I think it also includes the wonder of the magnificent power and design of all the celestial bodies that point directly to God's omnipotence. As I learn more of the structure and properties of galaxies, stars, pulsars, neutron stars, black holes, and all the many exotic celestial bodies, it boggles my mind how extremely complex and powerful they are.

God's second stated reason for creating the celestial bodies was to provide humans a way to discern time, not just the mere counting of days and years. No doubt included in this meaning is also the understanding of the timetable from the Alpha Point to the Omega Point of God's providential plan. So we see in the story of the zodiac that the ancient cosmic battle between good and evil is portrayed, and God's victory is written in the stars even before human beings were made.

It is my suspicion that at the time the celestial bodies were created, the cycle of the months related to the moon was in synchrony with the orbit of the Earth around the sun. In other words, the lunar and solar calendars were in synchrony. This was later altered by the events that brought forth the Great Flood. The intensity of the Earth's seasons is also determined by this motion and axial position of our planet in relation to the sun. This changed at the death of the First Earth when Earth became tilted and was flung further from the sun in the winter months of the Northern Hemisphere by the catastrophic impacts of meteors that cracked our continent into seven and our singular primordial ocean into seven.

I quite confidently suspect that the orbit of our planet was originally more circular than it is today. It probably matched the circular orbit of Venus. Although our orbit is unusually circular compared to most other planets in our solar system as well as those discovered in deep space, it is peculiarly longer at one end of its circuit around the sun. The measure of how much an orbit deviates from a perfect circle is called its eccentricity. A perfect circle has the eccentricity of zero. The eccentricity of Venus is nearly zero at 0.007. Its orbit ranges from 107 million kilometers to 109 million kilometers during its yearly circuit.

The Earth's eccentricity is 0.017 with a slightly more elliptical orbit. It ranges from 147 million kilometers to 152 million kilometers. In its present elliptical orbit, Earth is 5 million miles further from the sun in December than in July at the opposite end or its circuit. This, I suspect, is due to the impact of the meteor strikes that ended our First Earth with a global flood.

The eccentricity of our orbit marks not only the direction from which the meteors came, but also the time during the year in which they struck the Earth. We are told in the Genesis narrative that the fountains of the deep were broken open on the 17th day of the month of Heshvan, which was the second month of the Hebrew lunar calendar. That night, the moon was waning. The first month in the Hebrew calendar, Tishri, marked the autumnal equinox when day and night were exactly equal in duration. That means the meteors struck the Earth during its fall when the Earth was, in fact, moving toward the area of our elliptical orbit around the sun that became more pronounced by 5 million miles following the cataclysmic impacts.

In the six hundredth year of Noah's life, in the second month, on the seventeenth day of the month, on that day all the fountains of the great deep burst forth,

*and the windows of the heavens were opened. And
rain fell upon the earth forty days and forty nights
(Genesis 7:11–12).*

I further suspect that it was the combined powerful force of
the impacts of these meteors that created our 23° tilt of axis. The
consequence was that from that time forward, the seasonal variations
became even more pronounced by tilting the Northern Hemisphere
away from the sun during the winter. Simultaneously, this also
impacted the Southern Hemisphere to likewise lean away from the
sun during the summer season when the Earth is at the opposite end
of its revolution around the sun. The Second Earth was now a much
harsher Earth than God had designed for humans.

By moving the equator of the Earth 23 degrees so it is no
longer aligned with our sun-Earth ecliptic, the constellation rising
on the vernal equinox was also radically changed. I suspect that,
for reasons that are beyond the scope of this book, this largely
bypassed perhaps some of the end of the age of Cancer and, for
certain, all of the age of Gemini, bringing our planet to the age
of Taurus. When Noah disembarked, Taurus was rising on the
vernal equinox.

Not only could these meteor strikes have impacted the Earth's
orbit, but they could also have impacted the lunar orbit around us
to further break the harmony of the lunar and solar calendars. Our
moon is now slowly moving away from us at about one inch per
year. I suspect that the moon's original orbit around us may have
been perfectly timed to coordinate with the Earth's revolutions
around the sun.

It may be that the ejecta from the massive explosive impacts of
the meteors that cracked our crust shot back out into space. These
could have also impacted the moon and changed its rotational
speed by the additional matter added to one side and also by the
force of the impact. In addition, the change in our planet's axis

would have also contributed to the alteration of the moon's orbit around it. Both these events would have easily altered the moon's original course.

I suspect that in a perfectly symmetrical world that would befit God's intended design, the moon would have probably spun twice as fast as today, causing both sides of the moon to be seen every month. The new moon would have been one face, and the full moon, 15 days later, would have been the opposite side of the moon. Now, the rotation of the moon has slowed down so it always keeps one side away from us. I suspect that it was the far side that was facing the First Earth when the meteors struck, because it is the side that is disproportionately cratered. It is as if God said, "This side you have now lost forever because of your rebellion." If that is correct, then on the 17th day, the moon would have been waning, and the far side would have been the face of the full moon during the First Earth. (For more information regarding the moon see the section of "The Evolution of the Moon" in Chapter 19.)

Before that fateful day, the solar year and lunar year were in complete synchrony, and the tilt of the Earth was not inclined to 23°, so the seasons were much milder. That incline is what creates winter in July for those in the Southern Hemisphere who face away from the sun, even being 5 million miles closer from the sun's surface than the Northern Hemisphere at the far end of our ellipse.

It is quite probable that this synchrony was broken when the Earth's surface was fractured at the end of the First Earth. In other words, the meteors that caused the Great Flood may have also changed the trajectories of the Earth and the moon so the solar year no longer was in synchrony with the lunar year. The fact that all the ancient cultures initially relied on the lunar calendar and later had to do many fanciful tricks to match the lunar and solar cycles seems to point to this reality.

The evolutionary myth that the ancients did not learn the solar year until they were more evolved flies in the face of the archaeological evidence that shows they marked the equinoxes and the solstices as far back as we can go and had a much deeper understanding of astronomy than evolutionists concede.

The Witness in the Sky

Prior to that fourth day of creation, time was only measured by God's mind. And so the stars were placed appropriately for signs in the second heavens—the expanse of space specifically for humans to track time. The constellations, which appear in the night sky from our viewpoint on Earth, were propitiously placed by God to provide for humankind signs of the wonders of God and of His previsioned story of humankind. They are the witness in the sky.

They speak of the cosmic battle between good and evil and of the final triumph of the Son of God. All this is declared by the heavens from the very beginning through the witness in the sky, even before humans were made. The scriptures plainly state that the constellations bear a message from God that was known throughout the ancient world.

> *The heavens declare the glory of God, and the sky above proclaims his handiwork. Day to day pours out speech, and night to night reveals knowledge. There is no speech, nor are there words, whose voice is not heard. Their voice goes out through all the earth, and their words to the end of the world (Psalm 19:1–4).*

Philosophers and scientists in every culture and era have asked the same question: Why should so many stars in the universe (estimated at 10^{22}) be created only to illumine such an insignificant speck as the planet Earth? The answer is found in the scriptures. God has written a message to humanity in the constellations formed by the stars. This was the original and true intent of the zodiac.

What is this speech uttered day and night that cannot be heard, that still permeates the planet? Does our human history bear the evidence of such a record of this speech? Yes; it is the zodiac whose story was the Bible of the First Earth. In that day, the star Vega from the constellation Lyra was the North Star. God's plan of the ages was beginning to unfold even before He made the first person. The constellation known as Lyra is depicted as a seven-stringed harp (seven, again, being the number of God), superimposed over a majestic eagle soaring upward. The ancients named this constellation Aquilaria, a composite word meaning "the eagle's harp." In some of the older versions of the zodiac, the constellation is depicted as simply an eagle or a hawk, the natural enemy of the serpent.

The eagle soaring upward is here a symbol of the ascension of the Messiah into the Holy of Holies in the temple of heaven. The harp symbolizes the creation song since the star Vega was the North Star when God created our universe. He is the same eagle that will guard Israel from the antichrist in the Great Succoth for three and a half years (Revelation 15). He will keep them in safe hiding until the Battle of Armageddon. When the eagle, who is the Lion of Judah, returns, the sweet music of the victorious work of Christ and the ensuing new song of the birth of His Millennial Kingdom shall be heard by every creature, great and small.

> And when he had taken the scroll, the four living creatures and the twenty-four elders fell down before the Lamb, each holding a harp, and golden bowls full of incense, which are the prayers of the saints. And they sang a new song, saying,
>> "Worthy are you to take the scroll and to open its seals, for you were slain, and by your blood you ransomed people for God from every tribe and language and people and nation, and you have made them a kingdom and priests to our God, and they shall reign on the earth."

Then I looked, and I heard around the throne and the living creatures and the elders the voice of many angels, numbering myriads of myriads and thousands of thousands, saying with a loud voice,

"Worthy is the Lamb who was slain, to receive power and wealth and wisdom and might and honor and glory and blessing!" [notice that the Lamb was given seven attributes from God the Father] (Revelation 5:8–12).

Then I looked, and behold, on Mount Zion stood the Lamb, and with him 144,000 who had his name and his Father's name written on their foreheads. And I heard a voice from heaven like the roar of many waters and like the sound of loud thunder. *The voice I heard was like the sound of harpists playing on their harps, and they were singing a new song before the throne* (emphasis added) *(Revelation 14:1–3).*

What a marvelous sound it will be when ten thousand times ten thousand and thousands of thousands of angels will join in with the 144,000 Jewish witnesses standing on Mount Zion with the returning Christ to sing the new song. It is the song of the victory of the eagle over the serpent, the establishment of the real king over the universe upon the throne in Jerusalem.

The sound of the heavenly harps will resonate throughout the universe as heaven's doors open wide to announce the beginning of the long-awaited Day of the Lord. The celestial choir, accompanied by the heavenly music from the angelic harps, will harmonize with the glorious melody of this divine new song, extolling the virtues of the new kingdom and the Messiah who has made it possible by the shedding of His blood.

Long has humankind hoped for this day to become a reality. Long has humankind hoped for that sweet melody from on high. I marvel at the power of music to move people's souls. But on this day, the Lord, who is at once the Eagle and the Lion of Judah and Aquarius, will share His music with us. Oh, how marvelous it shall be as our souls are raptured by the divine sound of the new song! May it be soon!

This will mark the precise moment when the Kingdom of God on Earth will begin and the long night of injustice and tyranny that has plagued our world through the greed and avarice of humanity, aided by Satan and his demons, will end. The rich will no longer be able to exploit the poor. The powerful will no longer be able to use their positions to steal and extort from their constituents. Those wishing to desensitize the masses to violence and perversion will no longer control the media.

The strong will not be allowed to prey upon the weak, and the all-powerful monopolies that gouge the unfortunates needing their products will be disbanded eternally. No one will be allowed to discriminate against a fellow human being on the basis of the color of their skin, the language that they speak, their age, the intelligence or education they possess, their outward appearance, their financial status, or any other superficial distinction we have invented between us—the sons of Adam and the daughters of Eve.

The bankers will no longer be able to extract their inflated usury from the poor. The governments of the nations, under the kingship of the Messiah ruling from Jerusalem, will have to answer to Him directly. And in that day, justice will finally prevail. Unjust taxes and evil politicians seeking to line their greedy pockets at the expense of their constituents will no longer exploit the masses. Political corruption and graft will abruptly end. What a glorious day that will be!

In that day, when righteousness and justice rule the nations, at long last we will truly be free. The excessive burden of over-taxation placed on the people by governments seeking to fleece the flock for sordid gain will be prohibited by our righteous and all-knowing king.

The merchants who controlled the distribution of food products and hoarded these goods, keeping them from the poor in order to get the maximum return at the expense of the unfortunate, will be punished severely. Those who pandered in human misery will be wiped out, along with the antichrist and his cronies. Those who instigated and financed the wars that made their coffers rich with the shed blood of the innocents will at long last be rooted out, judged, and banished forever.

Christ will cleanse the Earth with a bloodbath even more severe than the previous judgment of the Great Flood. All of Satan's henchmen will perish at the atoning Battle of Armageddon in order to purify and remove the curse that sin has brought upon Earth so the Earth can be properly cleansed and atoned for His Millennial Temple.

And so we see that from the beginning, from the Alpha, the Omega has been known by God. The spring equinox determines the ruling house of the sun. The sign of Virgo rose on the vernal equinox right before the sunrise. Thus, the great constellation of Virgo from the very beginning foretold that the virgin from the seed of Eve would come and fulfill the prophecy of the dragon doom voiced by God in Genesis 3:15.

During that first age, Gemini rose on the summer solstice. Pisces rose on the autumnal equinox. And Sagittarius rose on the winter

Perhaps my favorite painting is Raphael's depiction of Michael the Archangel defeating Lucifer

solstice of that age. All of these have meanings, but they are not for this book.

It was at this time that the universe began to appear as we know it today. Those seven concentric bands of galaxies continued to expand outward, and the tapestry of the night sky was made evident from the viewpoint of the Earth. For this reason, the Essene Jews believed that this day was Rosh Hashanah, the First of Tishri in the autumnal equinox. The fourth day began the time of the four seasons. Again, the number four is specifically relating to things pertaining to our world. The universe began in the autumnal equinox. Humans appeared two days later when the crescent moon was waxing.

Other Hebrew scholars believe that the year began when Adam and Eve were made on the sixth day. They point out that from this moment forward, time in the scriptures was no longer counted from God's perspective but from people's perspective. Hence, the Genesis account is, in essence, the genealogy of man that sets the stage for the coming seed of Eve who would be the branch of David, as shown in the constellation of Virgo.

I would say that from the point of view of the Earth, time began on the fourth day, but from the point of view of humans, I tend to agree with the latter, for after all, it is people who do the counting, not the celestial bodies. They are counted, but the counters are people. God created it for humankind to count.

However, that statement must be counterbalanced by the fact that our perception of reality is not what gives reality to the thing observed. Truth stands in objective reality quite outside our ability to either perceive it or even understand it. It is real because its intrinsic reality is not dependent on humans, but on God, its creator.

From God's viewpoint, four days had passed. That night, the night sky became the starry wonder we can see today. But if we could have seen it then, it would have been even more spectacular than we can imagine, as all the stars were much closer to Earth than they are today.

Day Five

Once more, disorder was made into order as God prepared our habitat. Once more, evening gave way to the light of day that broke forth in the morning. Once more, God's voice brought further order and molded our reality. Life now could be sustained by the sunlight. Our tropical world was filled to the rim with plant life of every kind that one can imagine and of others we cannot.

These gave off oxygen as a byproduct of their photosynthesis and enriched our atmosphere so it ranged in volume between 33 percent and 37 percent of our air. Today, we have only 21 percent concentration of oxygen in our atmosphere. That is a decrease in volume of 44 percent from the First Earth. Again, we see the number four popping up conspicuously, but it could be mere coincidence. Nevertheless, our First Earth was, indeed, a veritable paradise compared to our diminished Second Earth with almost half the amount of oxygen available to us now.

Returning to our Genesis text, our world was now ready for the next step of God's magnificent creation symphony as God continued plowing our First Earth.

> *And God said, "Let the waters swarm with swarms of living creatures, and let birds fly above the earth across the expanse of the heavens." So God created the great sea creatures and every living creature that moves, with which the waters swarm, according to their kinds, and every winged bird according to its kind. And God saw that it was good. And God blessed them, saying, "Be fruitful and multiply and fill the waters in the seas, and let birds multiply on the earth." And there was evening and there was morning, the fifth day (Genesis 1:20–23).*

On the fifth day, God created all the aquatic creatures and filled the seas with life. He then created all the birds and filled the sky with life. Our sun now shone brightly in the sky and provided the light necessary for life on Earth to exist, both terrestrial and marine. Our ecosystem was balanced. There was no predation.

Peace ruled the planet as all living things lived in harmony. What a magnificent world that must have been! How I wish I could have seen that. The lion ate grass like the ox. This is the paradise for which all people have ever internally yearned. It is our intrinsic collective memory of what we were supposed to be.

Throughout all ages, humanity has dreamed and longed for this enchanted world. It was the way God meant it to be. Not so long ago, I stood upon the peak of a magnificent mountain in Austria and marveled at the beauty that spanned below me as far as the eye could see. In the distance, two brightly colored paragliders looped in slow lazy circles above a pristine blue lake. The snow-capped mountains rimmed the background like a formidable bastion that spanned from horizon to horizon. I took a deep breath of the cool Alpine air with a tinge of the scent of pine. It was a moment of transcendence, and I thought, "If this wilting world is so beautiful even now, how much more wondrous would it have been when the First Earth was in its full glory.

What would it have been like to swim in a crystal clear ocean of pristine sweet water, unspoiled by the emulsified minerals that were churned into it during the tumultuous hydraulics of the Great Flood and the careless pollution of people? How many wondrous and delicious fruits have become extinct? How many beautiful wild flowers with their tantalizing scents, intricate designs, and vibrant colors have been lost?

I took several pictures with my camera, but no picture can capture the peace I felt inside. How sad that selfishness has so marred our planet with so much pollution and violence that Earth has become but a shadow of the world that was. So few are the living things that survived the watery death of the First Earth.

My picture taken in Swiss Alps

Today, the marine ecosystem is divided into trophic levels grouped by the role they play in the food web. The second level after the phytoplankton and seaweed is the herbivorous consumers such as zooplankton and cockles. These support the first level of carnivorous consumers such as crustaceans, sea stars, and small fish, and, in turn, support the larger fish.

Moreover, marine life provides humans as well as many land animals and birds with a primary food source. So the fact that marine life came before terrestrial life is logical. Were it not for this specific order of creation, humankind would not have survived in this world. The chain of life would not have been possible. But I wonder how this chain functioned before the great fall of humankind? I suspect that all animals at that time were herbivores. Since the fall, predation and death has ever increased with time. Cruelty has been inextricably linked with survival. We are a shadow of the world that was. It is not just humans that were impacted by the fall of the first father and first mother. All of nature is now embroiled in ever-increasing violence. But when the Lion of Judah returns, Aquarius will bring harmony back to humankind and to nature.

Day Six

It was not until the morning of the sixth day that God created terrestrial animals.

> *And God said, "Let the earth bring forth living creatures according to their kinds—livestock and creeping things and beasts of the earth according to their kinds." And it was so. And God made the beasts of the earth according to their kinds and the livestock according to their kinds, and everything that creeps on the ground according to its kind. And God saw that it was good (Genesis 1:24–25).*

Here, the scripture once more deviates from the evolutionary lineage, which has the insects evolving long before vertebrates. There is no escalation in anatomical complexity as supposed in the evolutionary tree of life. Insects and vertebrates were made on the same day, on the morning of the sixth day.

In the fifth and the first parts of the sixth day, *nephesh* was created. Animals in the sea, in the air, and on the land were made into living *nephesh*. The Hebrew word *nephesh* means "the soul." All animals are described as having souls. They are conscious living things. Their conscious life is described as a soul. Anyone who has had and loved a pet dog, a cat, or a horse can tell you that animals are truly magnificent friends who have a soul and personality. But they do not have personhood. They are not beings that can look beyond their lives to the future. In contrast, humans, the crown of God's creation, are more than nephesh.

It was not until the evening of the sixth day that God created man and woman.

> *Then God said, "Let us make man in our image, after our likeness. And let them have dominion over the fish of the sea and over the birds of the heavens and over the livestock and over all the earth and over every creeping thing that creeps on the earth."*
>
> *So God created man in his own image, in the image of God he created him; male and female he created them (Genesis 1:26–27).*

We humans are also conscious living things that have a soul (*nephesh*). But the scriptures explain that we are something more than a soul. We also have the breath of God—*nshamah*. This is our spirit. It is who we are, our identity unique among all others. It is our personhood. It is what separates us from all other living things, for we are made in the image of God, and our spirit descended from the very breath of God.

For this reason, only humans contemplate the meaning of death, the beauty of art and music, and the origins of the universe. But more importantly, all of humanity has sought to know God and has developed religions. No animal ever has or ever will do any of these things. We are not gods, but we have a transcendental significance that is greater than all other things on this Earth and in our entire universe.

We did not ascend from apes. We descended from the heavens. We are the children of God. We are the children of light not because we chose it to be so, but because God did. It was not because of some arrogant subconscious motive that wishes to place more importance on our species than all the other species in nature, but it was because God created us to manage the world and the living things He created. He made us unlike all others in His image.

> *Then the LORD God formed the man of dust from the ground and breathed into his nostrils the breath of life, and the man became a living creature [nephesh] (Genesis 2:7).*

No other living thing is associated with *nshamah*. Only humans have the breath of God. For that reason, He says we are created in His image. It is for that reason that humans are endowed with infinite value. It is for that reason that human life is sacrosanct.

> *And God saw everything that he had made, and behold, it was very good. And there was evening and there was morning, the sixth day (Genesis 1:31).*

On the sixth day, the *nshamah* of Elohim made humans into persons who reflected the image of their creator. We alone among all other beings in the world can choose right from wrong. We alone among all other beings in the world can fathom our origins and think of eternity. We alone among all other beings in the world can understand death. We alone among all other beings in the world can begin to understand the mysteries of God's creation from the subatomic world to the far reaches of our universe. We alone among all other beings in the world seek to commune with the mind of God. We alone among all other beings in the world can begin to think the thoughts of God, albeit imperfectly as through a dark glass.

So when God had finished the crown of His creation—humans— He declared a day of rest. That was His intent for humankind, that we should live in paradise and enjoy eternal rest and peace.

> *Thus the heavens and the earth were finished, and all the host of them. And on the seventh day God finished his work that he had done, and he rested on the seventh day from all his work that he had done. So God blessed the seventh day and made it holy, because on it God rested from all his work that he had done in creation (Genesis 2:1–3).*

A curious thing happens now in the Genesis account. There is no recorded evening and morning for the seventh day because there is no movement from disorder to order on the seventh day. God rested on the day after His work was completed on the sixth day. No more plowing was necessary. For this reason, He asks us to rest on the seventh day, for it is a symbol of compliance and association with the creator. It is a time to reflect on the work of our Father. It is a time to reflect on God's desire for us. Unfortunately for us, it is also a time to ponder the loss of that paradise.

The importance of the Sabbath is paramount in the Hebrew Scriptures because it gives testimony to the fact that God was the creator of all things. Christians who try to jam evolution into Genesis are completely missing the entire message that God meant for His people to remember. They are complicit in this rebellion toward the great creator.

The number seven then symbolizes the completion of the work of God, while the number six symbolizes humans outside of the will of God. It describes humans as having rejected paradise. It describes human efforts throughout history to become autonomous from God's grace. It is a rejection of the Sabbath day as a memorial to God's creation of our universe.

Such is the metaphysical proposition of naturalism that posits that the universe is eternal and without a creator. And for this reason, the number 666 is of great importance to the occult, which seeks autonomy from God and wants to establish a global kingdom controlled by the demonic hierarchy that led and instigated the first and second insurrections against the great creator.

Christians and Jews, who naïvely attempt to correlate evolutionary theory with the Genesis space-time historical account of creation, must then allegorize the meaning of the scriptures and counter all the specifics written by simply ignoring them. They fail to understand that their futile attempt to conflate an atheistic proposition with a theistic proposition is an irrational proposition and an exercise in futility. The Genesis record does not oppose science. It opposes an interpretation of science, which is, at its very root, atheistic.

I cannot tell you exactly how God created the animals and human beings. I do not pretend to have all the answers. But I can tell you that life could not have ever been an accident of nature. Such specified complexity tells us that it was designed by a mind.

Evolutionists do not have all the answers, either, no matter how else they pretend. They fail to provide a rational scientific theory that can explain the origin of the universe from nothing. They fail to provide a rational scientific theory that can explain the movement from a chaotic Big Bang to an ordered universe filled with galaxies and stars. They fail to provide a verifiable mechanism that can account for the origin of life and the change from one species to another. Their dysteleological proposition cannot stand the litmus test of reality when our universe is filled with symmetry and elegance and is a fine-tuned cosmic system that can only be logically explained as the work of an all-powerful, intelligent designer.

They must resort to an invisible, repulsive energy that irrationally has more power as the distance increases to explain our accelerating universe. This is the new cosmological constant, or vacuum energy, that is also known as dark energy. The old Enlightenment doctrine of rejecting anything that is not sensible has now been obviously circumvented out of necessity to ignore the evidence that points to a creator. No longer can the invisible be considered mythological superstitions. And yet they continue to argue that because God is insensible, He cannot be considered through the scientific process. That, my friends, can only be considered as nothing more than unabashed schizophrenic reasoning.

CHAPTER 16

● ● ●

THE EXISTENTIALIST PARADIGM

I realized that Bohr's whole philosophy of life was based on an overwhelming need to renounce determinism and detailed space-time descriptions on the atomic and nuclear particle scale in order to achieve a sense of personal freedom, and to accept the mathematical abstraction of quantum mechanics worked out by Heisenberg and Dirac as fulfilling "all demands on rational explanation with regards to consistency and completeness." . . . I was also struck by the fact that Bohr looked down on what he called the "religious belief in harmony" that he felt de Broglie and Einstein maintained.

—Ernest J. Sternglass

Einstein's true genius, in my opinion, is that he was humble enough in spite of his incredible intelligence and accomplishments to admit a mistake and argue that his initial assumptions were incorrect, even when they were the very basis for his fame and notoriety. Almost alone, Einstein stood against the majority of physicists, maintaining that the Copenhagen school of thought (Bohr, Heisenberg, etc.) was an incomplete view of physical reality and that eventually a unified concept that could harmonize all reality would be found.

Sternglass, in one of his encounters with Einstein toward the end of his life, writes of this change in his thinking.

> *"You see the large tree over there," he said. "Now turn your head away. Is it still there?" At first, I was puzzled about what he was driving at, but it soon became clear. He was explaining to me one of the principle aspects of the Copenhagen interpretation of quantum theory that he found particularly unacceptable, according to which an observation or measurement is necessary to bring an object like an electron into definite existence. In this view of atomic phenomena, since one only has a probability distribution for the location or velocity of an electron, given by the de Broglie wave function as interpreted by Heisenberg and Born, it is only when an observer makes a measurement that an object or an event becomes real.*
>
> *It was this mysterious, almost magical aspect of the quantum theory that Einstein told me he abhorred, an approach that he deeply felt to be a great mistake, an opinion that I told him I completely shared. The Copenhagen view of quantum theory was in consonance with the positivistic philosophy [Kierkegaard and the subsequent existentialists that built upon his views] that had become increasingly accepted in the latter part of the nineteenth century, favored in particular by the Austrian physicist and philosopher of science Ernst Mach and the philosophers that shared his ideas in Vienna. . . .*

As Einstein explained his views to me, he said that in his early years he had been strongly influenced by Mach's insistence that theories should be based entirely on empirically derivable or directly observable quantities, but later he concluded that this was a mistake. Particles like electrons and alpha particles exist before they produce a click in a detector or a flash on a fluorescent screen, just like the tree existed before I saw it.

He went on to say that he no longer believed in some of the fundamental ideas of his Special Theory of Relativity that were the result of Mach's early influence on his thinking. It was only subsequently that I understood what Einstein was referring to, namely his assumption at the time he worked on the Special Theory that there was no such thing as an ether, since it could not be directly observed or detected by any physical measurements. The inability to detect the so-called luminiferous ether postulated by Maxwell and others as the medium in which electromagnetic waves are propagated had been discovered in the experiments by Michelson and Morley in the 1880s. These experiments had failed to detect a motion of the Earth through the hypothetical ether. . . .

As Einstein explained in a lecture at the University of Leyden in 1920, "To deny the ether is ultimately to assume that empty space has no physical qualities whatever. The fundamental facts of mechanics do not harmonize with this view." In particular, he cited Newton's conclusion that rotation of an object in empty space, which influences its mechanical properties, has to be taken as something real, so that Newton might well have called his absolute space "ether." . . . However, *as Einstein put it, the ether of general relativity is not a uniform medium as conceived in the wave theory of light. Instead, it is at every place conditioned by the presence of matter at a particular location and in neighboring places. . . .*

"Something completely new has to be found," he said, "something that is somehow based on the ideas of General Relativity" (Sternglass 53–55).

Einstein conceded that our inability to detect through quantitative terms the existence of the matrix of empty space does not imply that something isn't there. If space can be bent, then there is something there. If the motion of an object in space changes the properties of the object, then something in space is interacting with the object to make it so; otherwise, there would be no change. If everywhere in space there is a zero-point energy, then it is not a pure vacuum. As we will see later, the Higgs boson also indicates that Einstein was right.

This, in principle, has been the understanding of the Judeo-Christian worldview and is precisely why it is so vehemently opposed by the scientific paradigm of our times, which automatically precludes anything untestable from the realm of reality. And yet, hypocritically they accept the idea of WIMPs and fourth-dimension membranes outside our universe, which are also presently untestable.

The matrix of absolute space, or the space-time continuum, is a very real entity with certain physical qualities: it can be curved; it is affected by matter around it so that it determines the rate of time; and it contains an intrinsic energy. Technically speaking, there is no such thing as a vacuum of space if space itself has influence on matter around it and on time. When space is stretched, time speeds up. When space is contracted, time slows down.

The objective observer must not ignore this, even though presently we are incapable of observing, as Sternglass said, any "empirically derivable or directly observable quantities" in the fabric of space by our limited technological capabilities (54). Moreover, they have no problem believing that this invisible and untestable dark matter and dark energy also seem to inhabit space. But the idea of the observation of the tree that Einstein was rejecting was the existential idea that is promoted by those who use the Heisenberg Uncertainty Principle to artificially prop up their metaphysical worldview of the macroworld.

The point that Einstein was making to Sternglass in regard to the tree was directed at the existentialist paradigm that permeated the Copenhagen school of thought. This was the idea that later brought forth the anthropic principle, based on the philosophical foundation of Kant's logic, the idea that it is the observer who causes the reality of the thing observed. The idea is that reality is not the thing-in-itself but rather the impression received by the observer. The idea is that each observer creates a separate reality by the very act of observing. The idea is that truth and reality is only in people's minds. That is the sum of existentialist thinking, which abhors the idea of God not only physically existing outside people's minds but outside the very universe.

The Heisenberg Uncertainty Principle has been incorrectly used to promote the ideology of moral relativism through the deceptive use of smoke and mirrors. The fact that we are incapable of measuring with exactitude the position of an electron, because by that very event we disturb it, does not mean that the electron does not have a real position. Our inability to detect it does not mean it is not there. Mortimer Adler sees this discrepancy in logic.

> In the controversy between Einstein and Bohr over quantum theory, Einstein was, in my judgment, philosophically sounder than Bohr. The Heisenberg Principle of indeterminacy has epistemological, not ontological, significance. It should be interpreted as indicating the indeterminacy of our measurements

in subatomic physics, not the indeterminacy of reality in that area. Reality may be indeterminable with certainty, but this does not mean it is certainly indeterminate. *The fact that we cannot assign an equally definite position and velocity to an electron in motion does not mean that the electron really lacks a completely definite position and velocity* (emphasis added) *(Adler 117–118).*

If we wish to measure an object's location, the most direct way is to shine a light on the subject in order to view it. But when we are dealing with small microscopic particles such as an electron, even sending low intensity photons to strike it affects the path of the electron. This is at the heart of the Heisenberg Uncertainty Principle. If we want to measure the velocity of an electron, the best way is to measure it at one initial point and, in a given period of time, measure it at another point. The distance traveled within that period of time gives us the velocity of the electron. But here is the problem: In order to measure at two different points without changing the course of the electron, scientists must use a long wavelength and, therefore, a less energetic electromagnetic wave. But if we want to measure its precise location, the long wavelength is insufficient and gives us a very fuzzy picture. The shorter the wavelength, the more precise and clear the picture, but the shorter the wavelength, the more energetic it is, which impacts the electron.

Hence, we can measure the position accurately with a short wavelength, but we can't measure the velocity because we alter the course of the electron. That means that we cannot determine both the velocity and its exact location at the same time. But the uncertainty created by our inability to measure it is intrinsic to our inadequate, limited powers of observation and not to the intrinsic nature of the electron. That is what Einstein understood. The difficulties of seeing such small particles in no way causes our observations to make the thing viewed real. Our observations do not give rise to reality.

Unfortunately for most of humanity, this marvelous humility possessed by Einstein, which allowed him to change and discard his basic underlying philosophical presupposition (albeit grudgingly at first) when the pure facts dictated so, is quite lacking in many scientists and most people, for that matter. It is quite true that old dogmas die hard.

Most of us are dismayed at the present global conditions. But most of us have not understood the interrelatedness of our adoption of a relativistic ideology with the violence that now permeates our planet. The scientific argument between Einstein and Bohr was a prelude to the social unrest that has pulled our world into a quagmire. In my opinion, Max Planck and Neils Bohr leaned heavily on the relativistic aspects of quantum mechanics to legitimize their existential worldview. Planck, one of the founders of the quantum theory, once said in a brilliant insight, "A new scientific truth does not triumph by convincing its opponents and making them see the light, but rather because its opponents eventually die, and a new generation grows up that is familiar with it" (Planck 33–34).

This is unfortunately all too true and ironically also applies to those who promote the existentialist philosophy based on some aspects of the quantum theory. And for this reason, Einstein stood almost alone after he had changed his mind, in opposition to the existentialist paradigm of his time. But he did not allow his peers to pressure him into abdicating his rational position that an all-encompassing theory must be able to explain the macroworld and the microworld in a unified, comprehensive, and elegant fashion. Einstein now believed in the symmetry of the universe as designed by intelligence. In a randomly ordered reality, no symmetry should be expected. That is also a magnificent testimony to the greatness of this man and the rigorous intellectual honesty he portrayed.

And there was Einstein's recent isolation from the sci-
entific community that largely regarded his almost lone
opposition to the highly successful quantum theory as
unrealistic and anachronistic. Most younger physicists
tended to regard as hopeless his efforts to find a theory
that would unify the forces of nature by relating elec-
tromagnetism to gravity (Sternglass 58–59).

Some, like Sternglass, had the wisdom to continue in the quest for this elusive equation that could unify all the forces of nature. And for this reason, Sternglass astutely recognized the subjective and recalcitrant positioning of Bohr, predicated on his philosophical preconception. Sternglass, in an encounter with Bohr of the Copenhagen school, realized that Bohr was not really interested in knowing the facts if the facts did not agree with his preconceived notion of reality. In a twist of irony, the avowed enemy of determinism allowed his philosophical presuppositions to completely determine his view and interpretation of the facts.

I realized that Bohr's whole philosophy of life was based
on an overwhelming need to renounce determinism
and detailed space-time descriptions on the atomic
and nuclear particle scale in order to achieve a sense
of personal freedom, and to accept the mathematical
abstraction of quantum mechanics worked out by
Heisenberg and Dirac as fulfilling "all demands on
rational explanation with regards to consistency and
completeness." . . .

I was also struck by the fact that Bohr looked
down on what he called the "religious belief in harmo-
ny" that he felt de Broglie and Einstein maintained.
Yet it seemed to me that Bohr had almost the fanati-
cal approach of a fundamentalist preacher, intensely
concerned to save my soul from perdition.

> *Neither de Broglie nor Einstein had felt the need to talk almost continually without making a real attempt to listen, as Bohr did. They were humbler and gentler, much less anxious to defend their views and persuade others of their own ideas with such intensity. . . .*
>
> *Many years later, I read in Lewis Feuer's book* Einstein and the Generations of Science *how deeply Bohr had been influenced by the writings of the Danish existentialist philosopher Søren Kierkegaard. I began to realize where this passionately-felt need to renounce a detailed, deterministic description of phenomena originated. Bohr had apparently been introduced to the ideas of Kirkegaard through his professor of philosophy at the University of Copenhagen, Harald Hoffding.*
>
> *In his lectures, Hoffding had emphasized that "an absolute systemization of our knowledge is not possible," and that "when one thought to extend any analogy based on a part of existence to the whole, no verification is possible" (Sternglass 123–125).*

The denial of the unity and universality of truth is at the very heart of Bohr's fanatical and persistent insistence that the universe cannot have an all-inclusive theory of everything that would harmonize all forces. He had accepted the Kantian view that objective knowledge is impossible and that our observations subjectively change the thing-in-itself observed. Since the existentialist believes that the basic problems of philosophy cannot be solved by an all-inclusive, absolute system, their basic underlying presupposition simply states that we authenticate our "self" by the act of choosing. The idea of finding harmony to the whole of nature is regarded as absurd in their existentialist framework that glorifies randomness and chaos over symmetry.

The argument between Einstein and Bohr is not just a trivial scientific squabble. Relativism as an ideology encompasses every aspect of human existence.

It is therefore no surprise that socialist radicals seeking to overthrow our economic and political system are eager to use Bohr's existential ideology as a cloak of scientific respectability. We can, for example, see this tactic quite plainly in Saul Alinsky's apologetics for a revolution in America.

> Niels Bohr, the great atomic physicist, admirably stated the civilized position on dogmatism: "Every sentence I utter must be understood not as an affirmation, but as a question" (Alinsky 4).

Alinsky's work became a training factory for so-called community organizers who were thoroughly indoctrinated in the Hegelian relativistic ideology. The impact these community organizers have had on our political system cannot be underestimated. In fact, our nation even voted one of his disciples as our president. In explaining this ideology of community organizers, Alinsky wrote:

> To begin with, he does not have a fixed truth—truth to him is relative and changing; everything to him is relative and changing. He is a political relativist (Alinsky 10–11).

This is not merely a scientific matter, not even just a philosophical or moral matter. It is the underlying ideology that undergirds and fuels the Hegelian collectivist ideology that has brought so much death and suffering to countless nations in our world, such as the Soviet Union, China, and Cuba, to name a few. But do not imagine for one moment that this is not an American concern. If you are frustrated at the great polarization that is dividing our nation and want to know its root cause, all you have to do is read the words of the master community organizer.

The organizer must become schizoid, politically, in order not to slip into becoming a true believer. Before men can act an issue must be polarized. Men will act when they are convinced that their cause is 100 per cent on the side of the angels and that the opposition are 100 per cent on the side of the devil. He knows that there can be no action until issues are polarized to this degree (Alinsky 78).

The first step in community organization is community disorganization. The disruption of the present organization is the first step toward community organization. . . .

The organizer dedicated to changing the life of a particular community must first rub raw the resentments of the people of the community; fan the latent hostilities of many of the people to the point of overt expression. He must search out controversy and issues, rather than avoid them, for unless there is controversy people are not concerned enough to act. . . .

Enter the labor organizer or agitator. He begins his "trouble making" by stirring up these angers, frustrations, and resentments, and highlighting specific issues or grievances that heighten controversy. . . .

And so the labor organizer simultaneously breeds conflict and builds a power structure (Alinsky 116–118).

We have seen in 2016 the fruit of this polarization ideology in the streets of our nation as a wave of cold-blooded assassinations of police officers has claimed the lives of countless husbands, fathers, wives, and mothers. This is not some empty intellectual sophistry. It is at the very root an ideology that impacts every aspect of human existence, and we as Jews and Christians must know how to intelligently oppose it. This ideology in essence denies the existence of a good and righteous God and makes man a god, a narcissist whose ego cannot be ignored. If you think I am sensationalizing this, read Alinsky's words.

> *The ego of the organizer is stronger and more monumental than the ego of the leader. The leader is driven by the desire for power, while the organizer is driven by the desire to create. The organizer is in a true sense reaching for the highest level for which man can reach—to create, to be a "great creator," to play God (Alinsky 61).*

The existentialist simply makes a choice, and upon making that choice declares reality, giving the individual the complete freedom to choose as he or she wills. But more importantly, it allows the existentialist the freedom to be consistent with whatever mode of morality he or she chooses as his or her personal reality. The existentialist becomes his or her own god.

And so we see that there is a natural progression that begins first with establishing a relativistic ideology and then fomenting a polarization of our society in order to fuel the violence that can be used as a power base for the revolutions fomented by the socialist/progressivist/communist ideologues who have bathed our planet in blood. If a person is god, then all actions necessary and pragmatic to attain personal power are permissible.

The idea that the very matrix of reality would be completely chaotic (as maintained by the Copenhagen school's interpretation of quantum mechanics) would then be in complete resonance with this philosophical existentialist view that determines all value systems as absolutely relative. How ironic is it that those who hate absolutes claim with absolute fervor that no absolutes exist?

It is quite true that for most of humanity, our morality dictates our philosophy and not the other way around. But it ought not to be so! Hence, this interpretation of the Heisenberg Uncertainty Principle works quite nicely to legitimize this underlying relativistic, existentialist presupposition. In Bohr's view, the quantum or subatomic world mirrors this existentialist system and therefore provides for those who propose this philosophical and political presupposition some semblance of cohesion between material and metaphysical systems. Ironically, they are unaware that in trying to find this cohesion, they are, in effect, evidencing our human intuition that harmony is necessary to determine truth in reality. They are trying to harmonize physical reality with metaphysical choices while insisting that harmony does not exist.

Einstein was right. The tree was still there. There is harmony between gravity and the other forces of nature; in fact, there may even be supersymmetry in certain aspects of our reality. It must be stated that our Judeo-Christian cosmological model recognizes that our universe is not completely symmetrical. The Big Bang was not completely smooth. Some irregularity had to exist to allow for the formation of galaxies and stars. Without asymmetrical specified complexity, there could be no codes rich with information to establish the components of life. But there are reasons for this asymmetry. They are not generated by random events. Instead, they evidence an intelligent choice.

Moreover, the universe is filled with unexplained examples of this willful choice made by the creator. The neutrinos are found to be spinning only left-handed. There are no right-handed spinning neutrinos. Life is composed of only left-handed amino acids. There are no proteins made from right-handed amino acids. What the universe evidences is a finely tuned order that logically points to design and elegance instead of pure random chaotic forces. All forces have a common heritage that points to a universe that is expressly designed for a purpose.

*Although the gravitational force and the strong force
have vastly different properties (recall, for example*

that gravity is far feebler than the strong force and operates over enormously larger distances), they do have a somewhat similar heritage; they are each required in order that the universe embody particular symmetries. Moreover a similar discussion applies to the weak and electromagnetic forces, showing that their existence, too, is bound up with yet other gauge symmetries—the so-called weak and electromagnetic gauge symmetries. And hence, all four forces are directly associated with principles of symmetry (Greene 2003, 126).

In fact, it is now an accepted conclusion that all four forces were once united in one all-encompassing force during the early universe when it was yet in a plasma stage. The four forces broke up and condensed into their present form as the universe cooled down into its respective forms. The specific interaction and precisely designed values that form an exact and delicate balance between them is what makes our universe possible. That harmony cannot be accounted for by a random, chaotic origin.

Our universe is finely tuned to be asymmetrical enough so that matter can exist in different forms and so the forces can exist in different forms; yet they are symmetrical enough that there is a harmonious continuity and purpose in their design so life can exist in our universe. If the universe were too symmetrical, all particles and forces would look the same, and our universe could not inhabit life. If it were too asymmetrical, all particles would not be able to interact the way they do to form a universe that can inhabit life. It is not just Earth that is unique; our universe is also in the so-called Goldilocks zone. It is not just that the Earth is uniquely created, but that our entire universe is uniquely created. That is an incredible miracle of the first order that cannot be explained through random ordering.

According to modern physicists, it is the Higg's field that breaks up the symmetry and gives us the variety of particles that make our universe possible. (We will deal with the Higgs field and the Higgs particle later.) For this reason, many call the Higgs particle the God particle. It is certainly prima fascie evidence that our universe is not an accident of random ordering.

The Ekpyrotic Cosmic Model: Smashing Branes

Lately, a new theory has been proposed in yet another attempt to do away with the great embarrassment for the naturalists. The beginning of our universe poses an insurmountable obstacle for the naturalist's worldview. The Big Bang hypothesis basically states that from no *when* and no *where*, for unexplained reasons, an immense amount of energy magically appeared in a highly concentrated, hot, tiny space-time point. In other words, nothing created something. In fact, to be more precise, nothing created everything. Beyond that, this "everything" occupied a single point in space-time with infinite density. What finite force could overcome the infinite gravitational power of infinite density?

And even beyond that, what force could do so in a controlled and finely tuned explosion that resulted in such an even distribution of matter/energy throughout the entire expanding universe? There are no material forces that could explain any of these things. And yet naturalists insist that all of this took place by random finite forces. Naturally, such an irrational leap is hard to swallow by any rational, objective being with two working brain cells. No matter how you slice it, the Big Bang rationally points to a beginner who created our universe for the express purpose of harboring life.

This, of course, is unacceptable to the naturalist. Since Einstein's theory forced this mathematical conclusion upon naturalists, the religion of scientism has developed arguments that attempt to avert this dilemma. Their religious theological insistence that we live in a closed system automatically disregards any consideration of events occurring outside our universe. Since time was created in the Big Bang, what is before is therefore outside their purview. At least that is the big excuse they have traditionally used to ignore reality and the possibility of an intelligent designer.

However, if there is a beginning, then there can be no closed system, for something outside the universe had to exist before our universe was birthed. That is only logical. The objective mind cannot simply accept by faith that nothing created something. At least one thing stood outside of time and space that created our universe, or we would not be here. Naturalists have been racking their brains to find a way to cancel the beginning and make our universe eternal once again.

Ironically, their rabid distaste for the beginning has caused them to do an about-turn, and now they are trying to reach back before the beginning. A new theory has been posed by Paul Steinhardt (Princeton University) and Neil Turok (Perimeter Institute for Theoretical Physics in Waterloo, Ontario), which attempts to go before the Big Bang and explain how the Big Bang occurred. Ironically, in defiance of the accepted evolutionary catechism, which supposedly scientifically delegitimizes any attempt to delve into the period before the Big Bang, Steinhardt and Turok have done just that.

The name of this theory comes from a Stoic term *ekpyrosis*, meaning "conflagration" or "conversion into fire." Their choice belies their natural philosophical inclination to frame reality on a naturalistic framework. At last, some in the naturalist paradigms are beginning to see that the king is naked. It is absurdly irrational to accept that nothing made something. Hence, Steinhardt and Turok boldly stepped over the red line drawn by the closed system doctrine of the Darwinian catechism and attempted to explain the cause of the Big Bang through a naturalistic mechanism that speculates beyond our universe, or before space-time was created.

According to their model, our three-dimensional universe exists within a giant membrane that is in parallel position to another giant membrane within a fourth dimension called "the bulk." These two undulating branes, or universes, are then embedded in a fourth spatial dimension in which they move, sometimes away from each other and at other times toward each other. Some unknown repulsive force keeps them separated. For some unknown reason, the repulsive energy ultimately decays, and the membranes recoil and bump each other. When they collide, their point of contact becomes a big bang.

This collision, then, is cyclical; that is, it does not require a beginning of time—or so they claim. The two enormous branes would move only a short distance apart and then come together to touch at some point again. This, of course, happens over very long spans of time. Their estimate is that the collisions repeat every trillion years or so. It seems that naturalists cannot do anything in less than billions or trillions of years. I wonder how long it takes them to go to the grocery store.

After the collision, each brane expands until it reaches heat death, and then, through some unexplainable springlike force, they come together again and collide, starting the process all over again. It is stipulated that perhaps the repulsion energy is what evolutionists now call dark energy. When that dark energy decays, it is proposed that the branes eventually return to a homogeneous state. At this point, the branes begin once again to move closer to each other until they once again strike each other and create a new big bang.

It is utterly amazing to me how quickly the naturalist abandoned their previously staunch dogma regarding any scientific inquiry before the Big Bang. Almost overnight, they simply reversed course and smiled like a Cheshire cat as if nothing had changed. It finally dawned on them that their evasion was an irrational and unscientific approach to reality. But, hell-bent on avoiding God, they contrived a quite speculative substitute that requires a great deal of faith in unsubstantiated speculation.

To begin with, there is no such magical force as dark energy that grows in strength with distance. And if they cannot describe the nature or essence or even the source of this dark energy, how can they predict that it would decay?

Second, there is no known empirical data that could contract a universe that is, in fact, speeding up in expansion at the edges of the universe. The empirical data show quite clearly that our universe could not contract back into itself.

Third, their springlike mechanism is as elusive as the ghostly wimps that are supposed to be the essence of dark matter. There are no known springlike forces that could accomplish such a gigantic shift in the momentum of a giant brane the size of our universe.

And last, there is no known mechanism to explain what would happen if two separate universes were to come in contact with one another. Why would the collision of two space-time fabrics explode rather than meld into each other? Once again, they propose a theory based on pixie dust and unicorns, just to avoid the obvious creator.

Those who now champion this theory predict that the big bang caused by these brane collisions would not produce the intense gravitational waves predicted by the Big Bang Rapid In-flationary Model. The gravitational waves would, in effect, leave their mark as ripples in the cosmic microwave background radia-tion (CMBR).

If the European Space Agency satellite Planck Mission is successful, it will detect these subtle ripples in the microwave. If the satellite detects gravitational waves, then the brane collision theory will be disproved. But should the satellite not find these waves, it does not necessarily prove that the theory is correct. It could mean that our technology is not yet up to par, or simply that the Big Bang did not proceed in a rapid inflationary stage, which is what our Judeo-Christian cosmological model would predict.

In the meantime, an experiment in Antarctica is claiming that gravitational waves do exist. In a 2014 article "Gravitational Waves from Big Bang Detected," the author Clara Moskowitz wrote:

The Background Imaging of Cosmic Extragalactic Polarization 2 (BICEP2) experiment at the South Pole found a pattern called primordial B-mode polarization in the light left over from just after the big bang, known as the cosmic microwave background (CMB). This pattern, basically a curling in the polarization, or orientation, of the light, can be created only by gravitational waves produced by inflation. "It looks like a swirly pattern on the sky," says Chao-Lin Kuo, a physicist at Stanford University, who designed the BICEP2 detector. "We've found the smoking gun evidence for inflation and we've also produced the first image of gravitational waves across the sky." Such a groundbreaking finding requires confirmation from other experiments to be truly believed, physicists say. Nevertheless, the result has won praise from many leaders in the field. "There's a chance it could be wrong, but I think it's highly probable that the results stand up," says Alan Guth of the Massachusetts Institute of Technology, who first predicted inflation in 1980. "I think they've done an incredibly good job of analysis." The BICEP2 detectors found a surprisingly strong signal of B-mode polarization, giving them enough data to surpass the "5-sigma" statistical significance threshold for a true discovery. In fact, the researchers were so startled to see such a blaring signal in the data that they held off on publishing it for more than a year, looking for all possible alternative explanations for the pattern they found. Finally, when BICEP2's successor at the same location, the Keck Array, came online and began showing the same result, the scientists felt confident. "That played a major role in convincing us this is something real," Kuo says (Moskowitz).

But as we have already noted, these B-mode polarization patterns that were discovered are too intense to be gravity waves. They are, in fact, evidence of synchrotron radiation created by gigantic Birkeland currents. I suspect that these creative currents were used by God to form solid matter from plasma. All signs point away from any cyclical mechanism for the formation of our universe.

It is amazing to me that this ekpyrotic cosmic model is perfectly acceptable to the naturalist, while the idea that our universe could have been created by an intelligent designer is automatically disregarded because it lies beyond the borders of our universe. It is equally impossible for us to confirm any physical or empirical proof that this fourth dimension outside our universe exists. We cannot test it in any experimental way, nor can we visit it in any physical form. Yet the theory nonetheless commands naturalistic respect, simply because it evades God. They deftly ignore that this has been their standard evasion of the consideration of an intelligent designer that exists beyond our universe.

Furthermore, this speculative springlike force is also unverifiable and quite impossible to explain in any scientific way. It is rather like the punctuated equilibrium theory, a simple philosophic fix of an unknown mechanism without any empirical evidence to back it. What force could repel these two branes that are so very compact and dense, and then, when they have expanded so they are near the density of absolute zero and farthest apart from each other, draw them back together? The forces of dark energy would have to reverse in poles. What can cause such an abrupt change in a force that cannot even be tested?

But perhaps the most glaring logical deficiency of this astronomical speculation is that it answers nothing. What in God's green earth made them think that this theory answers the problem of a beginning? Like the panspermia theory, it just moves the puzzle one step back. Where did these branes come from? How did space and time begin? It gives absolutely no answer to the real questions. It simply moves the chess piece one block back.

If the big bang created by the collision of these membranes creates time inside the membranes, how could the fourth spatial dimension in which the membranes float, called the bulk, exist outside of time? Einstein taught us that there is no such thing as space without time. Therefore, that fourth dimension must also have time. But if it has time, then it must also have a beginning.

What, then, created the fourth dimensional space-time in which these membranes float? And what mechanism inside that empty fourth dimension created the membranes within it? Here, the cause-and-effect linear progression stops abruptly without an antecedent that can be explained in naturalistic terms. They may say that this fourth dimension is timeless and infinite, but that speculation is not based on any material or observable reality. This is essentially nothing more than a metaphysically prompted speculation.

All spatial dimensions in our universe are finite and have a beginning. To propose an infinite dimension is no less speculative than proposing an infinite being. Is it not more logical to accept the fact that whatever existed outside our universe had to be something beyond the finite, inanimate, material reality?

How else can we explain the universal symmetry of the natural world? How else can we explain the fine-tuned parameters so specifically chosen that caused our universe to be habitable for life? Naturalists can readily accept an infinite non-intelligent dimension without time, yet they cannot accept an infinite intelligent designer that exists without time.

But the naturalist obviously does not care to really know that answer. Naturalists are simply content to remove God from the equation so they can be morally autonomous. There is a repugnance to the idea of a creator, which they cannot stomach. Any superficial fix is acceptable as long as God is not in the picture.

We see this intrinsic bias in everything they do. We can see it in their irrational attempts to explain how order evolved from randomness. We see it in their irrational attempts to explain how the DNA code was made through random chemical processes. We see it in their attempt to explain dark matter with these imaginary magical WIMPs. It is a pervasive bias that subjugates and perverts the scientific process to the artificial boundaries of a philosophical dogma, the religion of scientism, which is nothing more than atheistic naturalism. This is not the pursuit of truth through the scientific process. It is the attempt to legitimize an atheistic religious view through rationalizations that use scientific jargon to mask their true intent.

The Space Bubbles Theory: Cloning Space

I don't think I've grown up much. At least that's what my wife tells me, and I think it is true. I still love to make soap bubbles with my grandchildren and chase after them to catch them before the wind blows them away. When I was growing up, there was a small plastic stick with a circle on the end that we stuck in a bottle of soapy water. Today, they have elaborate, giant circles that make huge bubbles. They even have bubble guns, which are awesome.

But the space bubbles in this theory are nothing like the size of even the small bubbles I made as a kid. The bubbles in this theory are in the quantum microworld. Those who promote this idea believe that space can roll into a region of lower energy, which they call a valley in the landscape. There, they propose that a bubble forms that can eventually clone space and create a new universe.

As with bowling balls, so it goes with a patch of space we are following: it will most likely plop into some valley, where it will begin to inflate. A stupendous volume of space will be cloned, all located in the same valley. There are of course lower valleys, but to get to them the universe would have to climb over mountain passes at elevations higher than the starting valley, and it cannot do so because it doesn't have the energy. So it just sits there and inflates forever.

But we've forgotten one thing. The vacuum has the quantum jitters. Just like the thermal jitters of supercooled water, the quantum jitters cause small bubbles to form and disappear. The interior of these bubbles may lie in a neighboring valley, with smaller altitude. This bubbling is constantly going on, but most of the bubbles are too small to grow. The surface tension on the domain walls separating the bubble from the rest of the vacuum squeezes them away. But, as in the supersaturated case, every now and then a bubble forms that is big enough to start growing (Susskind 305).

Scientists such as Alan Guth, Andrei Linde, Paul Steinhardt, Sidney Coleman, Alexander Vilenkin, and Leonard Susskind are convinced that this process can create a menagerie of universes as space is being continuously cloned. The certainty and bravado of their speculative projection is, in my mind, a measure of their desperateness. They must conjure a myriad of universes in order to patch the statistical odds of our universe having developed with such fine-tuned phenomena throughout the breadth and length of all things in this highly specialized universe we live in. It seems to me that their bravado is simply a sign of overcompensation for their real insecurity.

"The bubbling up of an infinity of pocket universes is as certain as the bubbling of an opened bottle of champagne (Susskind 304).

Perhaps Susskind should go easy on the champagne. To begin with, we must consider that the scientific premise in which they build this theory of cloning space bubbles is a phenomenon experienced only in the microworld of quantum mechanics.

The mathematics describing this bubble formation in an inflating universe has been known for many years. In 1977 Sidney Coleman and Frank De Luccia wrote a paper that was to become a classic. In their paper they calculated the rate at which such bubbles would appear in an inflating universe, and although the rate could be very small—very few bubbles per unit volume—it most certainly is not zero. The calculations use only the most trustworthy, well-tested methods of quantum field theory and are considered by modern physicists to be rock solid (Susskind 305).

In other words, relying upon a mathematical potential in the quantum world, whereby there is a probability that "very few bubbles per unit volume" are made and which we have never experimentally proved that they have grown into the macroworld, they pin their hopes in creating a megaverse filled with pocket universes. Never mind that our universe is not slowing down like a bowling ball into lower energy valleys; instead, it is inflating at an exponential rate with higher energy gradients, as evidenced by the increase in the speed of inflation and the high-speed rotation of galaxies at the fringes of our universe. Never mind that this theoretical process is highly improbable. If it explains away God, then it is appealing to the naturalists.

Bubble nucleation, like all other tunneling processes, is rare and improbable. Typically, a very long time will elapse before a bubble large enough to expand accidentally nucleates (Susskind 306).

That quantum mechanics provides for us a window into the reality of the microworld is fair enough, although it is an imperfect view. Since gravity cannot be accounted for in the quantum Jiffyworld, it can hardly be used to predict the events of the macroworld, where gravity is fundamentally important. It is no more logical than trying to use general relativity to deal with subatomic processes within the quantum world. How they can, with all candor and a straight-faced grins, claim that the jiffy jitters in the microworld can be transposed into the macroworld is an improvable extrapolation and nothing more than an elastic speculation.

Furthermore, if this creation of new universes is as "certain as the bubbling of an opened bottle of champagne," then why have we not ever been able to see just one single universe form from this supposed froth of bubbles being continually made? Oh, but we can't, because it is supposed to be a different universe from ours. But then there can be no causal connection to ours, and yet they claim that it came from ours. Well, which is it? It is an oxymoron to state that these bubble universes came from ours and yet propose that they have no connection to ours. Their speculation that surface tension would keep them isolated is quite fantastic and unprovable.

Furthermore, the idea that space, a single component of our universe, could then manufacture all other components of the universe simply by expanding is another extrapolation that has no empirical evidence to sustain it. The law of conservation tells us plainly that matter cannot be created or destroyed. Nothing in this universe can escape our universe to begin a new universe and clone an entire new reality out of nothing.

The very same argument Susskind used to refute Hawking's idea that matter going into the singularity of a black hole disappears from our universe applies to his bathtub universes—the law of conservation is universal and cannot be broken. In order for these space bubbles to become another universe, they would have to contain within them the radiant energy that can form into matter. Space itself cannot produce such energy. Space itself cannot stand alone without its co-inhabitants of our space-time continuum—matter/energy and time.

The space bubble theory that Susskind is so fond of is actually a reworking of an older idea by Thomas Gold, Hermann Bondi, and Fred Hoyle (1948) when they were desperately trying to defend the Steady State claim that our universe had no beginning and was eternal in space and time.

> In 1948 three scientists—Thomas Gold, Herman Bondi, and Fred Hoyle—published a novel theory of the universe that was part steady state and part expanding. It imagined a universe that had always existed and would always exist—but that was constantly stretching. As it enlarged, new material appeared out of nowhere to maintain a constant uniform density. Yes, the authors conceded, their theory violated the law of conservation of energy and matter, but by an amount too small to be detected experimentally (Guillen 88–89).

It always amazes me that the common thread of all naturalistic explanations relies on magical solutions of the creation of matter or space-time from nothing. Naturalisis criticize the Genesis narrative and our Judeo-Christian cosmological model by declaring that we simply believe in the god of the gaps. In other words, when naturalistic science has no explanation, we resort to the miraculous. But it seems to me that it is the naturalistic explanation that believes in the magic of nothingness as their god of the gaps.

Quite the contrary to the evolutionary propaganda, our claim does not come from what we do not know, but from what we do know. We know that the law of conservation is universal. We know that it is irrational to believe in abiogenesis. We know that it is irrational to believe that absolute nothingness could magically become everything.

Regarding these theoretical space bubbles, it must be stated that because something is mathematically possible does not mean that it will inevitably happen. We can, for example, turn to the proton and show that although it is mathematically possible to decay a proton, no proton has ever decayed since the beginning of time.

First, just because quantum field equations tell us that it is possible to nucleate bubbles does not mean that bubbles are nucleated in reality. We have no evidence of such a claim. Mathematical potential is not the same as physical existence. We can use equations of physics to describe the rotation of a golden globe the size of our Earth around a star the size of our sun, but that does not mean that such a golden planet exists.

Second, the idea that space can clone itself is a sheer speculation that cannot be proved. Furthermore, a clone is an exact copy of the initial organism. If it is a clone, then there is no fundamental difference in content that could result in a division of universes between one bubble and the other. They are part of the same universe and are therefore indivisible from one another. The only way they could be separate universes is if they were not interconnected. Thus, the only way they could exist as separate universes is if they were separately created from nothing and completely independent from one another. But if they are brought into existence from nothing, then the law of conservation is broken.

Space can inflate, but the law of conservation tells us that there is no spontaneous generation of new space or matter. Since space requires a zero-point energy to function, and the law of conservation would have to be broken to create that extra energy.

The possibility of space bubbles tunneling into our three-in-one space-time dimensions may be mathematically possible through quantum field theory, but even that does not necessarily mean that this bubble came from nothing. Space tunneling is a phenomenon that is better explained as traveling from one spatial dimension to another within our universe. There are seven invisible spatial dimensions from which these space bubbles can come, according to the M-theory. But every one of those dimensions was created in the Big Bang. Every one of those dimensions is causally connected to our visible spatial dimensions and subject to the same physical laws in our universe.

Ironically, Susskind battled with Hawking over black holes because Susskind believed adamantly in the law of conservation. Hawking believed that the energy in black holes disappeared from our universe. In the end, Hawking admitted that Susskind was right, and the law of conservation prevailed. Now, Susskind seems to contradict the law of conservation by this proposition that cloned bubbles of space can become another universe. Is that universe filled with galaxies and stars? Or is that universe a hollow bubble? If it has matter, where did the matter inside it come from? If it is just an empty bubble, then how does that help the statistical hurdles of evolving a universe that can harbor life that they are trying to overcome with the megaverse theory? There is nothing going on inside an empty bubble that could in any way aid their statistical nightmare. And finally, from where does Susskind think quantum tunneling emanates if the law of conservation is to be obeyed?

Somehow, Susskind envisions his new universes as being spherical bubbles. In fact, he names one of his chapters in *The Cosmic Landscape* "A Bubble Bath Universe" and envisions a bathtub full of bubbles as a megaverse. He goes on to further speculate with great certainty that our universe has already expanded beyond the speed of light, thus, in his opinion, bifurcating our universe into at least two universes. Since, from our side, we can have no communication with the other side of the speed of light, he believes that this automatically bifurcates our universe into two. He further believes that these bubble universes exist in great numbers on the other side. Speaking of the inflating universe, he writes:

At some distance the fluid of space would be rushing away so rapidly that the recessional velocity would become equal to the speed of light. At even farther distances the outgoing points would recede with an even greater velocity! Space in those regions would be flowing away so fast that even light signals, emitted straight toward the observer, would be swept away. Because no signal can travel faster than light, contact with these different regions is completely cut off. The farthest points that can be observed, i.e. the point where the recessional velocity is the speed of light, is called the horizon, or more properly, the event horizon.

The concept of a cosmic event horizon—an ultimate barrier to our observations or a point of no return—is one of the most fascinating consequences of an inflating universe. Like the horizon of the earth, it is by no means an end of space. It is merely the end to what we can see. When an object crosses the horizon, the observer can never have any knowledge of them. But if such objects are permanently beyond the limits of our knowledge, do they matter at all? Is there any reason to include the regions outside the horizon in a scientific theory? Some philosophers would argue that they are metaphysical constructions that have no more business in a scientific theory than the concepts of heaven and hell and purgatory. Their existence is a sign that the theory has unverifiable and, therefore, unscientific elements to it—or so they say.

The trouble with that view is it does not permit us to appeal to a vast and diverse megaverse of pocket universes, an idea that does have explanatory power; most importantly here, the power to explain the anthropic fine-tuning of our region of space. *We will see shortly that all the other pockets are in the mysterious ghostly portions of space out beyond our horizon. Without the idea of a megaverse of pockets, there is no natural way to formulate a sensible Anthropic Principle.* (emphasis added) *(Susskind 300).*

It is plain to see that Susskind's motivation to accept this metaphysical choice is to avoid the conundrum created by our fine-tuned universe for those who insist on a naturalistic genesis. To begin with, I am not so sure that matter caught within the matrix of the space-time continuum could ever cross the speed of light, since Einstein's equations specifically indicate that it would begin to gain mass when approaching the critical speed. The mass would continue to gain into infinity, thus never allowing matter to reach that critical speed. That gain in mass would begin to severely crunch space-time and slow down time.

It is more likely that space-time, in the outskirts of our universe when approaching the speed of light, would revert the Genesis spin acceleration constant and slow down the expansion rate by slowing down time. Thus, it seems quite unlikely that space-time outside the confines of a critical circumference of a singularity could ever travel faster than the speed of light.

More importantly we have no idea what the threshold limit of the stretching of space will be. We do not know how far the zero-point energy will allow space to stretch beyond its capacity to power it. It is much more likely that space will begin to rip. It is much more likely that matter will end up inside super-giant black holes, which may begin a process of swallowing each other until space-time itself is ripped apart and begins to eat our universe backward toward the Alpha Point.

But what about space? Can space travel faster than the speed of light? If only space can travel beyond the speed of light, then that is also speculation because we see that matter is intertwined with space-time through gravity. As we can see in the galaxies at the edge of our universe, matter is speeding up as space is stretched because they are interconnected. One cannot impact one without impacting the other three members of our space-time continuum.

It is ironic that in the same statement in which Susskind admits that no signal can travel faster than the speed of light, he proposes that the matter at the edge of our universe traveled faster than the speed of light. How is that rationally consistent? The imaginary crossing of the speed of light in our universe is not only a metaphysical fix to implant their desperate anthropic principle, but its purpose is clearly to avoid the reality of an intelligent designer because of the enormously improbable statistical chances of all these fine-tuned parameters to occur by blind random ordering within a single universe.

The naturalists completely ignore the explanatory power of an intelligent design model simply because they refuse to accept God as a possible explanation of the origin of our universe. This is not a choice made through the analysis of empirical data, but rather a rationalization accepted because of underlying personal metaphysical choices.

Of course, to naturalists, the idea of an intelligent design is beyond their mental event horizon because they have a priori chosen the metaphysical point of view that there is no God. They have no problem making metaphysical choices as long as those choices aid them in disproving God or, to be more accurate, trying to prove that our universe was created without a God. It seems to me that their souls have become a black hole that does not permit light to escape.

This idea that the universe could expand faster than the speed of light does not take into consideration that the expansion of our universe is, in fact, stretching the fabric of space. That expansion is causing time in these far regions of our universe to accelerate exponentially. This extreme aging is creating extreme entropy and bringing complete disorder and destruction to the space-time continuum and everything it contains.

What we are seeing is that the galaxies are aging as their galactic black holes are consuming everything around them. Contrary to the optimistic speculation that space in this region is cloning, what we are seeing is that the fabric of space-time is ripping into super-duper, super-massive black holes that are burning anything they cannot crunch into their powerful singularities. Given enough time, all that will be left are gigantic black holes gobbling up one another and no more stars and galaxies. All baryonic matter will be consumed into dark singularities. This is the burial zone for our universe, not a nursery for young universes.

No, the outskirts of our universe are not incubation zones for new stars, new galaxies, or new universes. Instead, it is the dark, ominous cemetery of our stars and galaxies. If anything, it is the indication that in due time, this death will work its way back to us as space-time reaches that extreme stretching that rips the fabric—the scaffolding—upon which matter and energy exist.

The Loop Quantum Gravity Theory

Recently, naturalists have endeavored to develop a considerable collection of mathematical procedures for turning the classical aspects of the general theory of relativity into a quantum one. Their basic underlying presupposition prefers a chaotic explanation of reality as described in quantum mechanics rather than a reality with elegant symmetry as described in the theory of relativity. A few physicists have now proposed a new grand unification theory called the loop quantum gravity theory. Abhay Ashtekar of Pennsylvania

State University, Ted Jacobson of the University of Maryland, and Carlo Rovelli of the University of the Mediterranean in Marseille decided to reassess the calculations, changing an initially assumed component of the feature of the geometry of space.

Traditionally, it has been assumed that the geometry of space is continuous and smooth, no matter how minutely examined. But, these scientists began their calculations assuming that the geometry of space-time is not fixed, that it is evolving and dynamic. In contrast, the string theory is not background-independent and is described in a predetermined classical/non-quantum space-time; that is, a smooth, ordered universe.

According to their calculations, they now propose that space is quantified, and many scientists are now endeavoring to substantiate this mathematically. The term *loop*, used in the title of the theory is derived from how some computations involve small loops marked out in space-time. The theory postulates that space-time comes in discreet sets of volume; that is, in distinct pieces (quanta). The possible values are measured in units of quantity called Planck length (10–33 centimeters). The smallest possible non-zero volume predicted would then be a cubic Planck length, or 1099 centimeter cubed. This translates to 1099 atoms of volume in every cubic centimeter of space.

The theory creates a picture of space formed by lines connecting a certain point (node) and radiating out from it. As we draw these lines, we can imagine a space between these lines but in the theory that space does not exist, only the lines. Each graph is defined by the way its pieces connect. But the continuous three-dimensional space that the graph would occupy, according to the theory, would not exist as a separate entity. All that exists are the lines and nodes; these are considered to be space, and the way they connect describes the specific geometry of that space. These are called spin networks.

But it seems to me that in order for their theory to be correct, they must jump out of reality into a world where lines can exist without space between them, a proposition that is, in my mind, quite extraordinary and akin to grasping at straws. It is very illogical to assume that such a spin network could exist, while excluding the natural space that would be between the lines and nodes.

If the road between Miami and Atlanta is 700 miles, then it can be said that there are 700 miles of space between Miami and Atlanta. If there is no space between Miami and Atlanta, then the road is 0 miles, and there is no distance between the two points. To claim that 0 miles exist between Miami and Atlanta and yet place them apart is an irrational conjecture. It seems to me that here is another example of people jumping across the line of despair to an irrational imaginary world in order to remain faithful to their naturalistic paradigm.

There can be no distance without space. There can be no length to a line without space. Again, people are jumping into imaginary time and irrational leaps of faith in ghost worlds in order to evade the obvious because it is philosophically repugnant to their atheistic worldview.

Space is not static; it is dynamic. It can be bent, and it is being stretched, as we can see at the edge of our universe. But our macro-world does not function as quantum mechanical subatomic particles. It follows that to artificially divide space into quanta packets is like trying to stick a square peg into a smooth, round hole. With a big enough hammer, you might get the peg to look like it entered, but it will not fill the hole properly.

This theory provides two juicy carrots at the end of the stick that are quite tempting morsels for evolutionists. The first carrot is that loop quantum cosmology can avoid the beginning. According to their theory, the universe is cyclical; that is, it collapses. But the density of the collapsing cosmos would not squeeze into a singularity with infinite density. Instead, it would squeeze about a trillion solar masses into a volume no larger than a single proton. Then they believe it could bounce back into an expanding universe. Now just what could cause this extremely dense state to bounce is not very clear.

This feature is quite appealing to the naturalists, who otherwise would have to explain how infinite density could be made to expand from a singularity. It seems to me, however, that their proposition is not really much different. It is rather like asking an ant if it would rather bench press the moon or the Earth. Both alternatives are impossible feats.

The second carrot is that these spatial atoms in their theory can stretch and create new spatial atoms out of nothing. That is, they can build new space in quantum spurts out of nothing. Just like a magic wand makes a rabbit come out of a hat, these magical spatial atoms can be constantly creating new space and rolling it out like tracks before a train.

"Say what?" said Foolwinkle. "What about the law of conservation?"

"Never mind," said Rocksy. "Just get the broom."

"What for?" asked Foolwinkle.

"We'll just quietly sweep it under the rug, and no one will notice."

"But what good is new space if there is no new matter to fill it?" asked Foolwinkle while scratching his head.

"Stop asking those silly questions," said Rocksy, wagging his finger at him.

"Why, we'll just pop them out of nothing, too. Heck, we can even pop a new universe."

"We can do that, too?" asked Foolwinkle, still scratching his head.

"Sure," said Rocksy with a grin. "Just ask us."

Like my imaginary friend Foolwinkle, I also have a few questions. If all of space contains an intrinsic zero-point energy, where does the zero-point energy for the new space come from? If space is continuously being created, how can it ever collapse? Moreover, no matter how badly they want to avoid the beginning, the raw data show clearly to any with open eyes that the collapse of the universe is impossible when the universe is expanding at an accelerating pace. As the experimental data from the Hadron Collider begin to favor the M-theory, the followers of the Copenhagen school are being forced to face the mathematical reality that their favored world of chaotic, quantum flux is not as chaotic as they had hoped.

Laws Created Our Universe (a Twist to the Anthropic Principle)

There is a myopic arrogance to people when they begin to see themselves as the lords of creation. It is a sad characteristic of megalomania. If you think I am using terminology that is inflammatory and unrealistic, then please read the following quote from Stephen Hawking's book *The Grand Design*.

> *According to M-theory, ours is not the only universe. Instead, M-theory predicts that a great many universes were created out of nothing. Their creation does not require the intervention of some supernatural being or god. Rather, these multiple universes arise naturally from physical law. They are a prediction of science. Each universe has many possible histories and many possible states at later times, that is, at times like the present, long after their creation. Most of these states will be quite unlike the universe we observe and quite unsuitable for the existence of any form of life. Only a very few would allow creatures like us to exist.* Thus our presence selects out from this vast array only those universes that are compatible with our existence. Although we are puny and insignificant on the scale of the cosmos, this makes us in a sense the lords of creation (emphasis added) (*Hawking and Mlodinow 8–9*).

Hawking believed that somehow humans are the selective creators that choose what kind of universe in which to exist, so their presence today impacts the past creation of the universe. That, my friends, is a leap of unsubstantiated faith based on nothing more than deophobia. The naked truth is here evident—there is no humility involved in the Copernican principle. Quite the contrary, there is an unbridled arrogance to think that finite humans could be lords of the universe.

If Newton's law of universal gravitation was considered illogical and inconsistent with reality because it caused instantaneous effects faster than the known constant of the speed of light, then what do you call the instantaneous effect backward through time to the Big Bang? The sad fact is that those who accept such an anthropic view are lords of creation only in their own minds.

Hawking proposed that the existence of the laws causes the creation of the universe from nothing. In fact, without any physical evidence to support his claim, he insisted that it has happened many times and is happening even as we speak. He calls this science. It is not. It is a metaphysical faith that is unsubstantiated by empirical data.

Laws are simply mathematical descriptions of an external reality independent of the observer. They are descriptions of the behavior of the forces, energy, and matter that comprise our universe. How can the description of the behavior cause anything to appear from nothing? Laws did not cause the Big Bang. There is no raw data that prove such a claim. The behavior of matter and energy is intrinsic to the design of the components and not an entity that stands outside of the thing created. It does not exist apart from or before the thing described. What it evidences is that intelligence designed it with these laws in mind.

The idea that in the instant that the Big Bang occurred all these laws that govern every aspect of the universe developed automatically is rather difficult to swallow. This no doubt is part of the reason Hawking and many others believed that the laws must have existed prior to the Big Bang. Their universal nature points to the fact that the laws must have existed prior to the Big Bang and must have guided the process from the instant the Big Bang singularity popped into existence.

However, laws can only arise from a mind that organizes the various components in a rational form. What naturalistic process can account for the creation of these universal laws? Without a mind, there is no reasoning and no basis for rationality. Hence, the laws must have existed prior to the existence of the universe, but only in the mind of God.

Hawking was convinced that his observation was what made reality exist. He had the cart before the horse. He was proposing that the vector of time runs backward. He could do so with mathematical calculations on a chalkboard, but not in real space-time. Moreover, he was asserting that the very laws that are reflected by all that exists in our universe must have existed outside of a mind and prior to the Big Bang. How can such symmetrical laws in a naturalistic and dysteleological reality appear without a rational mind to have conceived, organized, and engineered them?

Ironically, in their deophobic obsession to deny the supernatural, they propose fantastical improbabilities that could just as well be compared to fairy dust and magic. I believe the proper term for such thinking is fantasy, not science.

Einstein disagreed with this idea that the observer gives rise to reality. He believed that God created the universe with its laws because it was the only way to create a universe that exhibits life as we know it; so did Kepler and Newton. They believed that our universe has an independent existence outside of our minds. They believed that the universe has a true history that could be ascertained by true science. Science, to them, was the pursuit of truth that matched observable reality and not some fantastical trip like Alice took to Wonderland. Einstein was right; the microcosm is not in complete, utter anarchy, and God is still in control, for by Him all things hold together.

The Holographic Principle

The first time I saw a hologram, I was utterly amazed. Almost every year, we took our kids to Disney World in Orlando, Florida, for a few days. I think I loved it as much as they did. It was in Disney World that I first saw a 3-D film and was blown away by the technology that allowed us with those special glasses to see things as if they were standing before us. A ball appeared to be heading right toward my face, and I ducked in a reflex action. Laughing at my reaction, I looked around me and realized that everyone ducked with me. A few years later in one of the exhibits, I saw a hologram. There before me was a three-dimensional ghostly apparition that looked very real. But of course, it was not. It was created by a sheet of film in a circumference with a two-dimensional array of pixels that stored the information of a three-dimensional scene. The light striking at the middle of this circumference then created a three-dimensional image in the center of the filmstrip.

The most amazing thing about this magical light trick is that if one were to see the film, unlike normal two-dimensional pictures, the objects being depicted are not recognizable. The film looks like a patchwork of dark and light patches that seem to form wave patterns. The light that shines on that seemingly random pattern scatters off the film and creates a three-dimensional, free-floating scene that is utterly surreal. No matter the angle from which I watched it, the ghostly apparition looked like a solid form. But of course, it was just an illusion. It was not a solid form but a superficial image created by light. I could, if I wanted to, stick my hand through the head of the person being portrayed.

The hologram, of course, did not contain all the information of the person being projected. It was only a superficial projection of the person's outward appearance. It did not contain the enormous amount of information that a true depiction would entail. It did not contain the internal organs, all the cells, the ribosomes, the DNA, and the thousands of specified proteins of which a human being is composed. And even if the entire information contained in a human body could be projected by a hologram, it would still not be a human being. It would merely be a light depiction of a human being. The depiction is not alive. The depiction cannot make free will choices. It is nothing more than an artificial construct.

Nevertheless, lately a new cosmological theory stipulates that everything we see and touch in our world is nothing more than a hologram. All the living things in our world are not really living; they are simply projections from the edge of our universe. Susskind explains it this way:

> Here, then, is the conclusion that 't Hooft and I had reached: the three-dimensional world of ordinary experience—the universe filled with galaxies, stars, planets, houses, boulders, and people—is a hologram, an image of reality coded on a distant two-dimensional surface. This new law of physics, known as the Holographic Principle, asserts that everything inside a region of space can be described by bits of information restricted to the boundary (Susskind 2008, 298).
>
> It seems that the solid three-dimensional world is an illusion of a sort, the real thing taking place out at the boundaries of space (Susskind 2008, 434).

According to the holographic principle, all the forces, matter, and energy contained inside our universe are simply a holographic illusion—an image reconstructed from the information embedded in the horizon of our universe. The far-out speculative nature of this proposition and the link to the occult worldview is made bare by a footnote in Susskind's book:

> The Holographic Principle raises strange questions—questions of the kind that one might have read about in Amazing Stories or some other pulp science fiction magazine in the 1950s. "Is our world a three-dimensional illusion of some two-dimensional pixel world, perhaps programmed into some cosmic quantum computer?" Even more thrilling, "Will future

hobbyists be able to simulate reality on a screen of quantum pixels and become masters of their own universes?" The answer to both questions is yes (Susskind 2008, 301).

I find it quite illuminating that Susskind finds it "even more thrilling" that people could "become masters of their own universes." Both questions Susskind raised and answered affirmatively are consistent with occult theology. The first is consistent with their dualistic doctrine that states that our physical reality is but an illusion. In occult theology, reality is divided into the phenomenal and the noumenal. In Hindu pantheistic terms, our physical reality is called *maya* (illusion).

The second question lays bare the demonic cosmic quest to achieve godhood. It was the central quest of Lucifer to usurp the throne of God, and it was the same sin with which he tempted Eve: "You will be like God" (Genesis 3:5). For all intents and purposes, the imagined evolutionary ascent championed by naturalism is nothing more than the quest for godhood. It is not only the drive in rebellious humans to become absolutely autonomous, but more pointedly to become "masters of their own universes."

Holograms are an illusion. They are not living entities. They are simply projections created by intelligent minds. But I doubt seriously that most people who can think, act, and choose would accept the notion that they are not real, but simply a projected image from the end of space. Moreover, even if all the information in our universe is imbedded in the horizon of our universe, it does not necessarily follow that we are holograms created by this information.

Furthermore, the holographic principle cannot explain from a naturalistic standpoint how the complexity of a hologram design that is so minutely intricate as to explain the position of every atom in the universe can rise from random ordering. Once again, scientists jump into the irrational to explain away reality without God. To be completely frank with you, this kind of thinking reminds me very much of the imaginative thoughts I had when I was immature and rebellious as a student and smoked those funny cigarettes. What I thought were brilliant insights turned out to be absurdly ridiculous ideas when I was later sober.

The Anthropic Cosmological Principle

> For what can be known about God is plain to them, because God has shown it to them. For his invisible attributes, namely, his eternal power and divine nature, have been clearly perceived, ever since the creation of the world, in the things that have been made. So they are without excuse. . . . Claiming to be wise, they became fools (Romans 1:19–20, 22).

There is yet another recently proposed theory that attempts to diminish the significance of the singularity and the consequence of a beginning to our universe. This new proposition is called the anthropic cosmological principle. It seems that Hawking and Susskind, along with many other prominent scientists, were incorporating this idea into their cosmology. They have come, albeit grudgingly and kicking and screaming all the way, to the point that some naturalists are finally admitting that our universe is so intricately designed that it could not have arisen without a designer.

By sheer necessity, the overwhelming evidence forces them to acquiesce to a need for a designer of our universe. Since the probability that our universe could have developed through mere chance chemical processes is appearing ever so remote, the naturalist must come up with a primal cause that, while being the catalyst of this creation, remains within the materialistic paradigm in a naturalistic closed system. That is how they must have derived this conjecture:

"Man, this is depressing," exclaimed Foolwinkle, with his hands on his face and shaking his head.

"What is?" asked Rocksy.

"The more we discover, it is becoming embarrassingly impossible to continue to insist that the highly ordered universe is not designed. But how can we acquiesce to a designer and still avoid God and His loathsome moral mandates?"

"Oh, oh, oh, I know!" cried out Rocksy, triumphantly pointing his finger in the air. "We will make humans the creator!"

"You mean, humans will be the designers of the universe?" asked Foolwinkle, scratching his head.

"Yup."

"Gnarly, man! But who's gonna buy that?"

"Oh, oh, I know. We will call it by some fancy technical name," replied Rocksy.

"You mean like the Gnarly Man Cosmological Principle?"

"Yup, but that does not sound very scientific." Rocksy tapped his foot as he was thinking and then declared, "How about the Anthropic Cosmological Principle?"

"Dude, that's radical," replied Foolwinkle with a grin.

"Yeah, and to make it more scientific, we will use the Heisenberg Uncertainty Principle as our foundation."

Okay, I admit my sense of humor can sometimes be off-the-wall. Nevertheless, the proposition of the anthropic cosmological principle is just as off-the-wall. The evidence on which they base their cosmology is the sometimes strange behavior of subatomic quantum particles.

A designer must exist. Yet, for whatever reasons, a few astrophysicists suggest that perhaps the designer is not God. But, if the designer is not God, who is? The alternative some suggest is man himself.

The evidence proffered for man as the creator comes from an analogy to delayed-choice experiments in quantum mechanics where it appears that the observer can influence the outcome of quantum mechanical events. With every quantum particle there is an associated wave. This wave represents the probability of finding the particle at a particular point in space. Before the particle is detected there is no specific knowledge of its location—only a probability of where it might be. But, once the particle has been detected, its exact location is known. In this sense the act of observation is said by some to give reality to the particle. . . .

In other words, the universe creates man, but man through his observations of the universe brings the universe into reality. George Greenstein is more direct in positing that "the universe brought forth life in order to exist . . . that the very cosmos does not exist unless observed." Here we find a reflection of the question debated in freshmen philosophy classes across the land:

If a tree falls in the forest, and no one is there to see it or hear it, does it really fall? . . .

It is not that the observer gives "reality" to the entity, but rather the observer chooses what aspect of the reality of the entity he wishes to discern. It is not that the Heisenberg uncertainty principle disproves the principle of causality, but simply that the causality is hidden from human investigation. The cause of the quantum effect is not lacking, nor is it mysteriously linked to the human observation of the effect after the fact (Ross 134).

Those who dogmatically state that a person's observation is what gives rise to reality have severely misinterpreted the evidence of the Heisenberg principle. As Ross so aptly said, "It is not that the observer gives 'reality' to the entity, but rather the observer chooses what aspect of the reality of the entity he wishes to discern. It is not that the Heisenberg Uncertainty Principle disproves the principle of causality, but simply that *the causality is hidden from human investigation*" (emphasis added) (Ross 134).

In fact, what Heisenberg's Uncertainty Principle tells us is that upon the observation of the object, the object is affected and changes; hence, we are uncertain of its true situation. It does not say that the object is uncertain but that our technical powers of perception are inadequate to know its certain position. Michael Guillen, an instructor of physics in Harvard University, explains it like this:

> *Here is one way of seeing it. Suppose we want to know the pressure inside a bicycle tire. To find out, we use a pressure gauge. But the gauge works by sampling the air inside the tire, which alters the pressure ever so slightly. It therefore unavoidably introduces an uncertainty to our assessment (Guillen 106).*

The illegitimate extrapolations of this principle to substantiate a relativistic ideology are not only unsupported by the principle but unwarranted by the scientific methodology that depends on the laws of physics to be empirical truth in order to be able to understand anything about our universe. How ironic that when Christians view the Earth as specially designed for humanity, naturalists derogatorily call it an anachronistic and anthropocentric view. And yet the anthropic cosmological principle is the epitome of anthropocentricism in which humans are not only the center of the universe, as depicted by their existentialist paradigm, but they now are the creators of the universe by the power of their observations of it. Dressed in the clerical robes of religious scientism, postmodernists have moved a step forward toward the occult doctrine of the deity of humanity, whose minds are the creators of the world and the cosmos.

Naturalist cannot seriously question the cause-and-effect system that has given rise to modern science. But since they narrow reality into a closed system, they cannot turn to a cause outside our universe. Hence, humans become the cause of the Big Bang. As irrational as it sounds, the assertion is made that the present is responsible for the past. No matter how you slice it, things that have not yet been are seen to be the cause of the things that were. This is a ludicrous assertion that flies in the face of rational thinking. It completely goes against the vector of time as described in the second law of thermodynamics.

In my mind, this is simply the measure of their desperate philosophical position, which has been made evident by our modern scientific findings. The improbability of this universe evolving without a designer is so mathematically remote that some have finally thrown in the towel rather than embarrassing themselves with such blind faith in the improbable ordering of our universe through random chaos. But they still refuse to move away from their naturalistic presupposition and must therefore come up with a designer that avoids God. Therefore, they have crowned humanity, the observer, as the creator.

Paul Davies, commenting on the anthropic principle, desperately attempts to sidestep this enormous improbability of evolving into life through undirected processes. He therefore resorts to both the multiple universe theory and the anthropic cosmological principle in order to hedge the otherwise impossible odds a bit.

> *All that the anthropic principle can provide is a comment on how lucky it is that we are here. If a vastly greater number of alternative worlds cannot support intelligent life, then they would pass unwitnessed, with no cosmologist to wonder about how improbable they are. We should then regard ourselves as immensely lucky to be alive, and view our existence as an exceedingly improbable accident.*

> *On the other hand, in the Everett-, many - universe interpretation of quantum theory, all the other worlds of superspace are real, each with an equal status of existence. If life is very delicate then most of these worlds are even now devoid of observers. Only ours and those very similar thereto will have spectators. In this case we have, by our very presence, selected the type of world we inhabit from among an infinite variety of possibilities (Davies 45).*

This philosophical assumption, built upon an atomic anomaly, is the very shaky foundation Davies uses to claim that we have given rise to the macroworld: "We have, by our very presence, selected the type of world we inhabit from among an infinite variety of possibilities."

Davies further assumes that there are multiple real and independent universes that exist. He imagines that this quantum suprastructure can hold independent parallel universes. But, as stated previously, the concept of parallel universes is an oxymoron. If they are associated within a superstructure, then they are interrelated and testable or observable. They are part of the same reality or universe that humans inhabit. Then our specific universe is an open system. If our universe is a closed system, then there is no interplay between them. They are mutually exclusive and a moot point since their relationships are of no consequence to the odds of developing life within each of the other individual universes.

Hence, the addition of these multiple parallel universes, in order to lessen the peculiarity of the odds of the universe developing in such a fashion, is simply wish projection. They serve no purpose in lessening the odds within each separate universe if they are not interrelated and are truly separate universes. Hawking correctly objected to this absurdity:

> *There are a number of objections that one can raise to the strong anthropic principle as an explanation of*

the observed state of the universe. First, in what sense can all these different universes be said to exist? If they are really separate from each other, what happens in another universe can have no observable consequences in our own universe. We should therefore use the principle of economy and cut them out of the theory. If on the other hand, they are just different region of a single universe, the laws of science would have to be the same in each region, because otherwise one could not move continuously from one region to another. In this case the only difference between the regions would be their initial configurations and so the strong anthropic principle would reduce to the weak.

A second objection to the strong anthropic principle is that it runs against the tide of the whole history of science. We have developed from the geocentric cosmologies of Ptolemy and his forbears, through the heliocentric cosmology of Copernicus and Galileo, to the modern picture in which the earth is a medium-sized planet orbiting around an average star in the outer suburbs of an ordinary spiral galaxy, which is itself only one of about a million million galaxies in the observable universe. Yet the strong anthropic principle would claim that this whole vast construction exists simply for our sake. This is very hard to believe. Our solar system is certainly a prerequisite of our existence, and one might extend this to the whole of our galaxy to allow for an earlier generation of stars that created the heavier elements. But there does not seem to be any need for all those other galaxies, not for the universe to be so uniform and similar in every direction on the large scale.

One would feel happier about the anthropic principle, at least in its weak version, if one could show *that quite a number of different initial configurations for the universe would have evolved to produce a universe like the one we observe. If this is the case, a universe that developed from some sort of random initial conditions should contain a number of regions that are smooth and uniform and are suitable for intelligent life. On the other hand, if the initial state of the universe had to be chosen extremely carefully to lead to something like what we see around us, the universe would be unlikely to contain any region in which life would appear. In the hot big bang model described above, there was not enough time in the early universe for heat to have flowed from one region to another. This means that the initial state of the universe would have to have had exactly the same temperature everywhere in order to account for the fact that the microwave background has the same temperature in every direction we look. The initial rate of expansion also would have had to be chosen very precisely for the rate of expansion still to be so close to the critical rate needed to avoid recollapse. This means that the initial state of the universe must have been very carefully chosen indeed if the hot big bang model was correct right back to the beginning of time. It would be very difficult to explain why the universe should have begun in just this way, except as the act of a God who intended to create beings like us (Hawking 130–131).*

Hawking explains that the design of our universe is so smooth (the background radiation proves it) that it is quite impossible to imagine that all of the parameters necessary for carbon-based life to exist on our planet could have come by trial and error in random chance processes. The fact that our expansion rate is at the razor's edge between recollapse and complete disintegration through expansion is a parameter that cannot be logically assumed as accidental. It shows that there seems to be a design to the universe in order for it to inhabit life.

The design of the matrix of reality is, from the very beginning of time, uniformly pervasive throughout the entire span and age of the universe. And its early development seems to have been controlled in such a way as to produce what we see around us today. We live in a universe that exhibits a very narrow window or parameter of possibilities without which life could not exist.

The plain, scientific data show that in an enormous number of situations (each of which would have had an infinite number of possibilities, were they to be produced purely by random chance processes), very minute, specific parameters were chosen so life could be harbored. The problem for the naturalist is to answer how this design could exist without a designer. The naturalist is forced to concede a designer. But naturalists leap across the line of despair into the irrational by positing that humanity is the designer.

But, as Hawking pointed out, if the cause of the universe is humans, then we would not have needed the rest of the universe in order to create a habitat for ourselves. Perhaps we would only need our solar system or, at best, our galaxy. The existence of the rest of the universe is completely superfluous to the anthropic need. For what purpose is the rest of the universe if not to exclaim the infinite magnitude of the power of God? Therefore, the anthropic principle cannot really account for the existence of our vast universe. But naturalists' failed logic does not end there.

First, they claim that humans, the observers, by the mere fact that they are observing, choose the reality they observe. Hence, in a sense, they really are the creator. Naturalists finally come kicking and scratching to the point in their observations of the universe that they must concede that the design is too intricate and complex to not have had a designer. But they cannot bring themselves to admit that the creator is God. Therefore, they posit that humans, the only thinking minds they know (in a naturalistic closed system) who are capable of understanding this creation at least superficially, must be the creator. For who else is there? In their view, physical reality does not exist outside the observer.

> The observer and the world were so inextricably connected that "an independent reality in the ordinary physical sense can neither be ascribed to the phenomena nor to the agencies of observations." In other words, many physical properties of atomic particles did not even exist before the act of observation; the act of observation was necessary to bring these properties to existence (Barrow and Tipler 83).

They propose, in essence, that the act of observation is "necessary to bring these properties to existence." They propose that humans are the conjurers of their reality, and they are now not only gods, but the conjurers of worlds.

First, there is in this observer is creator logic an irrational foundation of circular reasoning. We are here because we observe that the conditions in our universe can support life; therefore, the conditions in the universe can support life because we observe that they are here. This logic is preposterous! And if scientists who have traditionally, supposedly, refrained from the speculative are now turning to this, then it only serves to underscore their desperation with their failed naturalistic paradigm. It makes no more sense than to say that we are responsible to give reality to a murder reported on the news on television simply because we observe it in the comfort of our homes.

Second, in order for this postulate to be true, the vector of time must be other than forward. To say that the observer may choose to validate the existence of the thing observed may be, to a limited degree, acceptable. But to say that the observer chooses the existence by mere observation is to give the observer the ability to cause to happen what occurred before the observer's own existence—a magical feat that runs in direct contradiction to the universal second law of thermodynamics. No matter how we deny the fact that we are not negating cause and effect by saying that it is hidden from us, it completely undermines the entire scientific process of cause and effect within the framework of time. It divorces time from space. It denies the reality shown to us by Einstein that space-time is interconnected in our universe. Space is trapped in time, and it cannot be extricated by any known scientific law.

The old philosophy 101 question posed to all incoming freshmen is considered here: If a tree falls and you are not there to hear it fall, did it really fall?

The answer is this: Ask the chipmunk that got squished by it!

This line of thinking is the inevitable and the natural offspring of Kant's view of reason as delineated in his *Critique of Pure Reason*. This was later developed further by Josiah Royce to propose the anthropocentric concept that the viewer and the object are intertwined in such a way that reality is inseparable from the viewer.

Kant argued that we could not come to an objective reality or knowledge of an object. Rather, our cognitive powers are such that they skew the real object, or what he calls the "thing-in-itself," through a grid of sensory experience, which creates in us a pattern, concept, or interpretation of the thing-in-itself. Hence, the object is not objectively known, and reality is intertwined with a person's interpretation or thought. But Einstein disagreed.

> *Einstein argued that it was not true that there were certain specific concepts, without which thought was impossible, as Kant had maintained. Rather it was the case that without concepts in general, thinking was impossible. But, exactly what the nature of these concepts were could not be known a priori (Sternglass 78).*

While it is true that the limit of our perceptive powers may mar the logical and true interpretation of the object observed, the reality of the object observed is independent of the observer. We are amused and often deceived by the story of the three blind men interpreting an elephant. And so the story goes that while feeling his huge leg, the first blind man erroneously concluded that he was a tree. The second one felt his tail and thought he was a rope. The third one felt his trunk and thought he was a snake. But the fact is that the elephant does not become a tree, a rope, or a snake. It is the fact that the observers are blind and their perceptions are incomplete that causes them to come to invalid conclusions. The elephant is still the elephant. All this illustration proves is that logic is valid, and misimpressions can be corrected by further information that eventually will bring the observer in line with the reality of the object observed.

Here is where the existentialist philosophical position becomes the enemy of true science. For if one accepts this underlying view, then there is no impetus to attempt to learn the unknown, because the unknown is unknowable. In this sense, then, we can say that true science is sequential and orderly. That is, as evidence mounts, our logic leads us more deeply into the truth of the reality observed. But the reality observed is not changed one iota by our perception or misperception of the object observed. The perception is merely corrected to be in congruence with the reality of the thing-in-itself through accumulated knowledge and the passage of time. The sun did not revolve around our planet, no matter how often we observed it.

Crucial evidence against the naturalists' argument is the fact that the time vector cannot be backward. This is true reality, and any who say otherwise are simply jumping into the irrational without any scientific data to support their ethically motivated presuppositions. Ironically, the anthropic principle is nothing more than the denial of observable reality that the observer wishes not to concede; that is, that the universe does evidence a master designer.

Now, correction is not the same as contradiction. When we correct something with more evidence, we do not necessarily throw out every bit of evidence we gathered before. We simply discard the evidence found to be erroneous and add the corrected information to the previous evidence that remains rational and logical, thus shedding more light on the matter and bringing the observer closer to the reality of the object observed.

It is for this reason that modern existentialist thought is, in fact at odds with true science. Only the Judeo-Christian concept of reality could have given birth to the sequential, orderly scientific process of inquiry. It is because scientists in the past understood that the world was created by a rational God who made humans capable of reasoning that they believed the universe was intelligible. That is specifically the product of the Judeo-Christian worldview, and it is what motivated people like Copernicus, Kelvin, Ray, Linnaeus, Cuvier, Agassiz, Boyle, Newton, Faraday, Rutherford, and many more to understand the mind of God in their pursuits of understanding His creation. No other culture in the world developed the scientific revolution birthed by the Judeo-Christian worldview.

That is not to say that there were no scientific discoveries in other parts of the world such as China and India, for example. There were very valid scientific discoveries in those places. But there was not the wholesale burgeoning of the scientific process because they lacked the philosophical foundation for such a view.

I do not agree with Richard Tarnas's Platonic-Pythagorean-Gnostic convictions, but I do believe his observation regarding Copernicus to be accurate.

> *The early scientific revolutionaries perceived their breakthroughs as divine illuminations, spiritual awakenings to the true structural grandeur and intellectual beauty of the cosmic order. These were not merely abstract conceptual innovations or empirical find-*

ings of purely theoretical interest. They were not, as had been true of astronomy since classical antiquity, merely instrumentalist mathematical constructs, epicyclic elaborations ingeniously devised for the purpose of marginally increasing predictive accuracy. The new discoveries were triumphant fulfillments of a scared quest. For thousands of years, the celestial and terrestrial realms had been regarded as unalterably separate realities, as incommensurable as the divine was to human. Because of their extreme complexity, the true nature of the planetary motions had come to be seen as fundamentally beyond the capacity of the human intellect to understand. Concerning heavenly and divine matters, it seemed, only the Bible could reveal the truth; human astronomy could produce nothing but artificial constructions, as through a glass darkly. But now the true reality of the divine ordered cosmos had finally been revealed. The deep mysteries of the universe were suddenly unfolding within the awestruck minds of the new scientists through the grace of the sovereign Deity whose glory was now dramatically unveiled. The stunning mathematical harmonies and aesthetic perfection of the new cosmos disclosed the workings of a transcendent intelligence of unimaginable power and splendor. In that very epiphany, the human intelligence that could grasp such workings was itself profoundly elevated and empowered.

The heliocentric discovery thus became the source and impetus for a tremendously magnified confidence in human reason. It revealed the human being's divinely graced capacity for direct, accurate knowledge of the world at the most encompassing macroscopic level, something never before known in the entire history of Western astronomy. . . .

Moreover, contrary to the human-decentering consequences later drawn from the Copernican shift, all of the great Copernicans from Copernicus through Newton were deeply convinced that the cosmic order was expressly created to be known and admired by the human intelligence. Here and now, after millennia of dark ignorance in an exile that had been as much spiritual as intellectual, the human mind had finally achieved direct contact with the true cosmic order as the divine mind had long intended. Only thus can we understand the full exaltation of Kepler, the pivotal figure of the Copernican revolution, as he announced his discovery of the third law of planetary motion, which completed the early mathematical foundation of the heliocentric theory:

"Now, since the dawn of eighteen months ago, since the broad daylight three months ago, and since a few days ago, when the full Sun illuminated my wonderful speculations, nothing holds me back. I yield freely to the sacred frenzy; I dare frankly to confess that I have stolen the golden vessels of the Egyptians to build a tabernacle for my God far from the bounds of Egypt. If you pardon me, I shall rejoice; if you reproach me, I shall endure. The die is cast, and I am writing the book—to be read either now or by posterity, it matters not. It can wait a century for a reader, as God himself has waited six thousand years for a witness" [Kepler] (Tarnas 5–6).

Indeed, a tabernacle to God both Copernicus and Kepler made; not with brick and mortar, but with the revealing of the intelligently structured and ordered universe. Both Copernicus and Kepler believed that the sun, whose centrality in our solar system they perceived, was the symbol of the Godhead. And so it was in ancient history throughout the planet in every civilization from the Americas to China. The sun has from antiquity been the symbol of God and the full moon the symbol of the Messiah (see my upcoming book, *The Secret of the Lost Knowledge*).

It is hard for us to fathom the hardships and ridicule they faced from the secular as well as the religious establishment. So counterintuitive was this insight that decades went by with only a handful of believers writing and encouraging one another from various countries to remain steadfast in their lonely vigil of truth.

Steadfast they remained because of their inner conviction that God had made human beings to reason and discover His majesty through understanding the intricately harmonious marvel of His creation—a creation filled with coherence, harmony, and elegance in the structure of its components, which embodied the intellectual and artistic mind of the creator. It mattered not if all the intellectuals of the day believed otherwise. It mattered not if the religious institutions ridiculed and harassed them. Their faith in God and in our ability as human beings to reason and discover truth held them steady.

It took generations before the scientific paradigm was finally toppled. In fact, it took the death of those philosophers, intellectuals, and religious leaders of the day before the light of the sun of reason shone in the minds of the people. It did not come through intellectual persuasion of those entrenched in their failed paradigm. It did not come through epiphanies. It came through the death of those steadfast minds and the illumination of the next generations.

Deceitfully, men such as Copernicus, Kepler, and Galileo are today used by postmodernists and Darwinists as champions of their cause against the Judeo-Christian worldview. Nothing could be further from the truth. They stood against the tyrannical religious establishment but not against the Judeo-Christian faith. It was their deep-felt faith, like Newton's, that illuminated their minds to reason above and beyond the blinders of the paradigm of the day. We follow humbly in their shadow.

The modern religious existentialists who champion Kant's idea that humans cannot through reason ever know God, have failed to understand that without God, we cannot understand reason. Without a standard for that which is reasonable, all is simply static noise— chaotic, undirected energy. But because there is a God, then there is such a thing as reason. And if that God has created the universe, then there is the possibility of a rational approach to understanding His creation. And more importantly, we can also begin to know the mind of God that created it. This is precisely what the scriptures say, in direct contraposition to Kant:

> For what can be known about God is plain to them, because God has shown it to them. For his invisible attributes, namely, his eternal power and divine nature, have been clearly perceived, ever since the creation of the world, in the things that have been made. So they are without excuse (Romans 1:19–20).

Returning to the elephant in the room, we must therefore also conclude that reason and logic are, in their pure forms, objective. That is, false reasoning may be corrected by further reasoning, which then aligns our perception with the reality of the object, the "thing-in-itself," even if imperfectly. Therefore, the very existence of reason outside the person who is reasoning proves to us that the world is not anthropocentric, but rather deocentric.

For in a universe governed by blind chance and random chemical reactions, reason and logic are without foundation. How could pure randomness result in the structures of reason and logic? If reason exists at all, it does so because God is rational. And if we accept the postmodernist paradigm, then reason is but an illusion in the minds of humans. There is no basis here for any real science. The entire scientific enterprise could not ever have risen from the naturalistic worldview. Had it not been for the Judeo-Christian worldview, science (the rational study of an ordered and lawful universe) would have been impossible.

Let me be more specific. To declare something as rational and logical, one must begin with a standard or structure of what is rational or logical. But how do we arrive at such a structure by random processes? How does a relativistic metaphysical structure arrive at any absolutes in the physical or any other realm? How do we agree on the meaning of logic without a paragon? The very existence of rationality implies a rational creator, separate from our material existence. There is no way to escape this. Reason is, because God is. In fact, He is the reason things are.

But the conflict that rages between naturalists and the Judeo-Christian worldview has nothing to do with reason and scientific facts. The conflict has nothing to do with blind faith, scientific inquiry, and causal reasoning. The conflict we face in our postmodern culture is all about the undisputable reality that people simply do not like the idea of being morally accountable to a supreme being. The naturalist credo stems foundationally from a metaphysical choice, dressed in the vestiges of science. Naturalists have subjectively made the a priori choice to proclaim any consideration of God as unscientific.

We can see this, for example, in Susskind's analysis of the anthropic principle. He precedes the discussion by stating that the anthropic principle cannot be ignored on mere philosophical objections.

> *As for rigid philosophical rules, it would be the height of stupidity to dismiss a possibility just because it breaks some philosopher's dictum about falsifiability. What if it happens to be the right answer? I think the only thing to be said is that we do our best to find*

explanations of the regularities we see in the world.
Time will shake out the good idea from the bad ideas
and they will become part of science. The bad get
added to the junk heap. As Weinberg emphasized, we
have no explanation for the cosmological constant
other than some kind of anthropic reasoning
(Susskind 2006, 196).

Susskind's entire rationale for accepting an anthropic principle is quite simple; there is nothing else anywhere in the materialistic horizon that can even begin to explain the fine-tuned parameters that guide our universe, and chance random action cannot rationally accomplish the crossing of such unfathomably large statistical hurdles. Hence, by default, since the consideration of a creator is unacceptable, the only thing left is to believe that humans caused the creation. He goes on to give an example of anthropic reasoning by using the human brain as the subject in question.

I call it the Cerebrothropic Principle. The Cerebro-
thropic Principle is intended to answer the question,
"How did it happen that we developed such a big,
powerful brain?" This is what the principle says: "The
laws of biology require the existence of a creature with
an extraordinarily unusual brain of about fourteen
hundred cubic centimeters because without such a
brain there would be no one to even ask what the laws
of biology are."

That is extremely silly even though true. But
the Cerebrothropic Principle is really shorthand for
a longer, much more interesting, story. In fact two
stories are possible. The first is creationist: God made
man with some purpose that involved man's ability to
appreciate and worship God. Let's forget that story.

The whole point of science is to avoid such stories. The other story is far more complex and, I think, interesting. It involves several features. First of all it says that the Laws of Physics and chemistry allow for the possible existence of computer-like systems of neurons that can exhibit intelligence. In other words the Landscape of biological designs includes a small number of very special designs that have what we call intelligence. That's not trivial. But the story requires more—a mechanism to turn these blueprint designs into actual working models. . . . "Why did I wake up this morning with a big brain?" is exactly answered by the Cerebrothropic Principle. Only a big brain can ask the question (Susskind 2006, 195–196).

First, he adamantly rebukes those who through some philosophical dictum would automatically refuse to consider a scientific possibility: "As for rigid philosophical rules, it would be the height of stupidity to dismiss a possibility just because it breaks some philosopher's dictum about falsifiability. What if it happens to be the right answer?" (Susskind 2006, 196).

Then he completely ignores his own advice and does exactly what he said would be the height of stupidity: "In fact two stories are possible. The first is creationist: God made man with some purpose that involved man's ability to appreciate and worship God. *Let's forget that story. The whole point of science is to avoid such stories*" (emphasis added) (Susskind 2006, 195).

With all due respect, Susskind is being schizophrenic in his logic. Since when is it the role of science to avoid the scientific inquiry of the possibility that our universe bears signs of having been intelligently designed? The whole point of science is not to avoid any potentiality. What he is declaring, in effect, is that he wishes the whole point of science to be the ignoring of God.

Science is the pursuit of truth to explain reality in a way that is rational and consistent with the empirical data. If the data point to God, then the avoidance of that possible avenue of inquiry can only be termed, as Susskind so aptly said, "the height of stupidity to dismiss a possibility just because it breaks some philosopher's dictum about falsifiability. What if it happens to be the right answer?" (Susskind 2006, 196). There is stark evidence here of a conscious metaphysical choice to avoid considering the possibility that God is real.

There is a predetermined choice to avoid such questions due to an obvious philosophical and metaphysical bias. In fact, he contradicts himself by glibly stating that "the whole point of science is to avoid such stories" (Susskind 2006, 195). Hence, the only option left for them in light of the obvious elegantly ordered fine-tuned parameters pervasive throughout all reality is to make human being, the only other minds outside of God, the creator. Moreover, if we continued with this preposterous anthropic logic, we could equally conclude that computers, because they can understand the computer language, must have been responsible for creating the binary code.

What I find absolutely intriguing is that Susskind is acutely aware of the irrational inhibitions that accepted paradigms pose against the application of true, unfettered scientific inquiry. Unlike most scientists who simply accept the ruling paradigm as absolute, Susskind felt deeply that ideologues in the scientific community should never constrain the scientific process. His battle against the proponents of the scattering matrix theory regarding hadrons is a perfect example. Hadrons are the nuclear particles that are made of quarks, antiquarks, and gluons. These are the particles that the Large Hadron Collider in Switzerland attempts to break open in order to investigate their constituents.

> Sometime in the early sixties, while I was a graduate student, some very influential theoretical physicists, centered in Berkeley, decided that physics had no busi-

ness trying to explain the inner workings of hadrons. Instead they should think of the Laws of Physics as a black box—a black box of the Scattering Matrix, or S-matrix for short. Like the behaviorists the S-matrix advocates wanted theoretical physics to stay close to experimental data and not wander off into specula- tion about unobservable events taking place inside the (what was then considered) absurdly small dimensions characteristic of particles like the proton. . . .

The Berkley dogma forbade looking into the box to unravel the underlying mechanisms. The initial and final particles are everything. . . .

The S-matrix is basically a table of quantum- mechanical probabilities. You plug in the input, and the S-matrix tells you the probability for a given output. The table of probabilities depends on the direction and energy of both the incoming and outgoing particles, and according to the prevailing ideology of the mid- 1960s, the theory of elementary particles should be confined to studying the way the S-matrix depends on these variables. Everything else was forbidden. The ideologues had decided that they knew what constituted good science and became the guardians of scientific purity. S-matrix theory was a healthy reminder that physics is an empirical subject, but like behaviorism, the S-matrix philosophy went too far. For me it turned all of the wonder of the world into a gray sterility of an accountant's actuarial tables. I was a rebel, but a rebel without a theory (Susskind 2006, 203–204).

It was his indomitable curiosity for understanding reality that gave this brilliant man the impetus to come up with the string theory that completely revolutionized our understanding of the true elementary particles. His intuition that true science cannot be constrained by preconceived bias provided the groundwork for the Grand Unified Theory that is able to unite all four forces of nature into one elegant equation. We will recount this story later on, but for now, my point is that of all scientists, Susskind should know that the goal of science is to pursue all possible alternatives. It is my prayer that one day this brilliant mind will come to understand that true science cannot be subjectively limited to exclude the possibility that our universe was intelligently designed.

Susskind has inadvertently fallen into the same trap that his fellow physicists plunged into during the time in which the S-matrix paradigm reigned supreme. And so humans, in order to be free from these moral constraints inevitable if there is a God, suppress the truth and blindly walk into the trap of the enemy of man who deceptively whispers in our ears and convinces us that joy can only be found in unrestrained souls. Our proclivity to desire to be our own gods subliminally causes us to suppress the truth. Reason is then profaned into rationalization. Science is constrained by fiat parameters artificially imposed to negate the possibility of God. What is worse is that they ridicule and label as crackpots any who endeavor to follow the logical conclusion that our universe had to be intelligently designed.

> For the wrath of God is revealed from heaven against
> all ungodliness and unrighteousness of men, who by
> their unrighteousness suppress the truth (Romans 1:18).

It is our human, rebellious desire to usurp the authority of God that brings us to suppress the truth in our hearts; some go beyond that to suppress the truth for others. But the problem is not that some people do not believe in God; it is that they just don't like Him. They abhor authority and erroneously think they can be free without restraints.

Anarchy never made anyone free. On the contrary, many have been robbed of their freedoms because of anarchy. For each step we take to suppress truth, we walk one step closer to the true suppression of our real freedoms. The Kantian view reflected in the anthropic principle is just another rung in the ladder of the master deceiver's plan toward a more sinister ploy.

I believe that this type of thinking that intertwines the reality of the object to the observer is being promoted as a stepping stone toward a more sinister conclusion. The natural conclusion of this line of thought is that the universe is an organic whole, since the thing-in-itself and the observers are being proposed, as inherently intertwined. This, not coincidentally, is an occult and pantheistic doctrine. And it is not surprising that this occult view was also held by Erasmus Darwin, the father of Charles Darwin, and which was also catapulted into a wide social and scientific acceptance through the work of such scientists as Pierre Tielhard de Chardin and Josiah Royce.

> In Royce's view, Nature arises from a sort of mutual interaction between the knower and the known: "Reality is not the world apart from the activity of knowing beings, it is the world of the fact and the knowledge in one organic whole" (Barrow and Tipler 158).

Naturalists are finding it more and more difficult to stay in that bleak impersonal materialistic universe that they have contrived in their minds by negating the creator. The anthropic principle provides a way to personalize that universe. Humankind is the center that has caused it all to happen. The incredible evidence amassed by modern science has, without reservation, pointed to an ordered and designed universe, which uncontrovertibly necessitates a designer as the primal cause.

But since postmodernists are subjectively unwilling to concede the existence of a personal creator who would then hold them accountable morally, the only choice left is to make humans the designer. How improbable the idea that we could cause to happen what has transpired before our existence. Such is the blind leap of faith necessary to avoid confronting the reality of the existence of God.

Thus, postmodernists take the blind leap of faith into the irrational in order to escape God, and in so doing, they must escape reason. This view is then the bedrock of the anthropic principle. But it is only plausible if, in fact, the universe is not deocentric. A deocentric universe measures reality on the basis of the existence of the creator, and not humans. The tree is there, whether you look at it or not. If God exists, that is, if the evidence of creation is such that it necessitates a creator, a designer, and a maintainer, then true reality is deocentric and stands completely outside humanity's ability to either perceive it or understand it. It is not dependent on their existence or knowing, although that has been the longing of God from the beginning. The splendor of the universe was created in order for human beings to grasp the grandeur and majesty of the creator—not for the service of human beings—not for us to exploit and abuse.

Therefore, not only reason but also knowing is impossible without God. As a matter of fact, humans knowing would be complete absurdity without the existence of a creator to anchor epistemology. But the problem with the anthropic principle is, as J. G. Fichte pointed out in 1797, that a finite rational being could never come to the absolute knowledge of anything. Hence, in a naturalistic universe, the very anthropic principle is unknowable.

> *A finite rational being has nothing beyond experience; it is this that comprises the entire staple of his thought. The philosopher is necessarily in the same position; it seems, therefore, incomprehensible how he could raise himself above experience. . . .*

> *The thing-in-itself is a pure invention and has*
> *no reality whatever. It does not occur in experience:*
> *for the system of experience is nothing other than*
> *thinking (Fichte 8, 10).*

There is therefore a sort of super-importance that humans place on themselves as observers. A sort of inflated ego, a form of narcissistic scientism has evolved from this thinking, for postmodernists suppose that their observance of the falling tree is what makes the tree real. This in every sense is the essence of the anthropocentric view.

But in a naturalistic system, individuals' experience of observation is nothing other than an electrical stimulus across the nerve synapses in the brain. Even if reality were truly in a closed system and the naturalistic paradigm were correct, a finite individual observer with limited powers of observation could not ascertain true reality and could not know if it is true reality. This could only be ascertained if the observer were the final resting point for all observations.

Consequently, this places the human observer out of the loop. For unless observers are truly God and can observe all things in all places throughout all time, they cannot give rise to all reality in any concrete way. Humans are therefore left in a diminutive world of unknowable substance with no concrete way of knowing if their perceptions are even true to the thing observed. They are alienated from the world. Reality becomes only illusions in their minds. Reality is merely an electrical hologram.

Here is where this existentialist foundation of the anthropic principle is also mirrored in the field of psychology. Those like Maslow and Rogers who pose phenomenological psychology, which states that the only true reality is one's inner experience, herald the psychological side of this anthropic argument. This, however, is at its very best only an approximation of reality due to humans' limited instrument of perception. In this existentialist frame, knowledge is unknowable in an absolute sense.

The fact that our powers of observation are limited and subjective and can, on occasion, be very deceiving creates for naturalists an insurmountable obstacle. Their subjective impressions then become the only reality possible in this system. It is precisely this situation that would render the scientific process null and void. And as a consequence, it negates all hopes of really knowing anything in an absolute manner. In fact, in this system, there is really no concrete way to prove either that our observations really influence reality in the macroworld or that reality is really in accord with our observations. Reality, then, is nothing more than a hologram in our minds.

Bohr could not state with such certainty and fanatical zeal that his view was correct. Maslow and Rogers could not promote their form of psychology as applicable to any other human beings than themselves. In such a world, the real is only the perceived within our brains. What Tartufery to think that their finite psychological constructs could ever be a universal!

Yet the reality of a car driving 30 miles per hour at a psychotic, deluded individual standing in the middle of the street with his hand stretched out, thinking he is Superman, is not in any way affected by the deluded interpretation of his observation. Take my word for it—the car will win. I have seen it with my own two eyes. If you don't believe me, try it.

This "observer is creator" nonsense is nothing more than clever semantics and a sleight of hand. But I am afraid the only ones deceived are the ones who believe it. The foundational assertion of the anthropic principle is clearly an impossible task if the vector of time is forward, and it has never been anything but forward in the history of the universe. Einstein saw this, and thus came the split between his ideas and the school of Copenhagen, which champions Bohr's existentialist ideology.

No one disputes the intelligence of these naturalists. Intelligence, however, is not the father of reason, but rather its servant. Reason uses intellect, but possessing intellect does not necessarily require the person to use it rationally. That choice is made by the free will of the person; that free will is, however, influenced by underlying subjective motives that may impede a truly objective and rational approach to any inquiry. If there are underlying metaphysical and emotional desires that influence the free will of the person, the rational process may be compromised. That leads us to contrived rationalizations rather than objective reasoning.

Rationalizations are, in fact, profaned forms of reason that serve not the search for truth in order to properly vet it but rather to legitimize and falsely prop up our underlying metaphysical and emotional choices. Because of this, reason has often been a stranger to those persons filled with arrogance. Sadly, many who are quite intelligent become quite arrogant, and arrogance is the enemy of objectivity. It seems to me that no higher arrogance can be claimed than to make humans the creator of our universe. But such is the twisted rationalization proposed by the anthropic principle.

Crucial to the spontaneous generation of the Genesis Singularity, as championed by those holding to the anthropic principle, is the phenomenon known as quantum tunneling. As stated earlier, Davies notes that virtual particles can pop into our space-time reality from seeming nothingness, through quantum tunneling. Quantum tunneling is described as the process in which particles penetrate barriers that are insurmountable or impenetrable by classical objects. Davies then postulates that whole universes could similarly pop into existence from nothingness. Thus, naturalists myopically posit that our universe could have been born through this mechanism.

To begin with, virtual particles don't exist in nothingness. They are there, although in a different dimension from our three-in-one space-time dimension. We know they are there because we can measure their effect on particles around them, which exist in our visible spatial dimensions. They are in an extension of our universe, which we cannot presently perceive with our natural or enhanced senses due to our yet rudimentary technology. But they are still in our universe.

What we call a separate dimension is not a separate universe. It is in correlation and in interplay with our three-in-one visible dimensions because we can measure its effect on our dimensions. The problem for the naturalist is twofold:

1. Naturalists are subjectively antagonistic to the existence of an unseen dimension because it smacks of the biblical doctrine of a spiritual dimension.

2. Naturalists would much rather have a separate and different universe in order to give evolution a greater chance to succeed in the statistically almost impossible task of the chance recombination of chemicals needed to create life. The more universes that exist, they reason, the more time there is available for the improbable to become statistically less improbable by its multiplicity.

This invisible dimension from which these particles seem to pop up is not other universes, but rather an extension of our visible universe. If it were another universe, it would not be interconnected. It is simply a presently undetectable continuum of our physical and observable dimensions in our universe. Davies' assumption that these particles originate from another universe is fundamentally flawed since the very fact that they appear in our observable reality means that the place where they exist is connected to our reality and therefore part of the same universe. Otherwise, there would be no possible interconnection, and they would be mutually exclusive.

And if they are not connected, then they have no bearing on any statistical quantification that we could measure regarding our universe. Hence, Davies and Susskind have not helped their statistical nightmare with the multiuniverse or megauniverse wish projection.

Quantum tunneling, therefore, could not account for creation, for it is not the popping up of something from nothing. It is simply the interchange between at least two dimensions that are part of the same universe, which needed to be created together in order to have any interconnection at all. In other words, the invisible dimensions in which these virtual particles exist were equally created at the time of the creation of the visible universe as part of our cosmos.

Even so, using Davies' own arguments of the existence of multiple universes, if such redundancy did exist, the very fact that such redundancy supposedly exists in order to support our visible reality is evidence that it is merely an invisible extension of our visible reality. And in spite of his objection, this, of course, is also evidence of a much more intricate design and purpose to the superstructure of our reality than initially understood. It would mean that all of it was designed for the purpose of ensuring our existence. It is evidence of a much more extensive and intricate design to this marvelous universe, which makes our existence even more special and unique.

How is it that a completely random process has built into it the redundancy that Davies proposes, observed in this multiplicity of invisible universes from which he thinks these particles mysteriously pop up? What selective pressure could engineer such a feat across universes that are not connected? Are we to believe that Darwin's hypothesis is some mysterious overarching force that connects all universes? Is not the acceptance of such an overarching Darwinist principle behind, above, beyond, and around all material universes not in any way substantially different than belief in God?

CHAPTER 17

● ● ●

DEOCENTRIC COSMOLOGICAL TELEOLOGY

For the word of the cross is folly to those who are perishing, but to us who are being saved it is the power of God. For it is written, "I will destroy the wisdom of the wise, and the discernment of the discerning I will thwart." Where is the one who is wise? Where is the scribe? Where is the debater of this age? Has not God made foolish the wisdom of the world?

—1 Corinthians 1:18–20

History records for us that the tendency for humans to find purpose for their existence has ever been universal. Every civilization has carried in its history a creation story, which attempts to explain their existence in relation to their god and the world they live in. The modern evolutionary paradigm is no different.

Modern scientists have incorrectly dubbed the Judeo-Christian worldview an anthropocentric view of humankind. And those who truly understand the implications of an anthropocentric concept of reality in a closed system see it as an ironic smokescreen set up by deophobes to hide their real anthropocentrism. It is naturalists who hold to an anthropocentric view of reality, making their minds the center of the universe. Our Judeo-Christian cosmological model does not make God after a human's image. It is God who made humans after His image.

After careful inspection, what they have dubbed as anthropocentric in the Judeo-Christian worldview should more accurately be labeled deocentric. Specifically in the Genesis narrative, but also in many of the mythologies throughout the world, the center is not humans, but rather the gods or God of their cosmogony. The universal nature of these traditions that uniformly place humans in an elevated position in relation to the rest of creation does not necessarily make it an anthropocentric worldview. They are simply retelling, albeit in varied corrupted forms, the story initially handed down by the descendants of Noah.

That was the choice made by God, not humans. And in the final analysis, God is at the center of all reality. It is God who recognizes every human being as special, unique, and endowed with infinite worth and value, a concept that is considered pure imagination by those who have accepted a naturalistic presupposition. And yet, even for the staunchest and most vitriolic atheist, the inescapable human need for transcendence betrays this intrinsic truth of divine genesis. It is evidenced in the soul of every person. It is an inescapable sign of our divine origins. It is the *nshamah* of God that breathed life into us and made us into living beings connected to Him by His very breath. The inescapable truth is that we are hardwired by God to long for transcendental meaning, love, justice and a feeling of belonging to a family. The random ordering of impersonal chemicals can create none of these things.

It is because they outright reject the existence of God that naturalists then conclude that all of this is simply humans making themselves the center of the universe in the order of importance. Yet in reality, it is naturalists who place humans at the center of the universe, thinking that their brains are the final arbiters of all truth and reality, the conjurer of worlds. Therefore, it is the naturalistic/existentialist view that may more correctly be termed anthropocentric.

The search for some form of transcendental significance is a universal human trait. Goethe, like Francis Bacon, also recognized this universal intrinsic characteristic in human beings. But in trying to be true to his naturalistic presupposition, he naïvely rejects the human disposition to make themselves the center and end of creation, and then, by denying the creator, ironically establishes a truly anthropocentric reality through a naturalist worldview.

> *Man is naturally disposed to consider himself as the center and end of creation, and to regard all the beings that surround him as bound to subserve his personal profit (Barrow and Tipler 83).*

Goethe was right to point out that all people are inclined to have a transcendent need. It is the evidence of our divine genesis. But in order to be consistent with the naturalistic presupposition, one must come to understand that there can be no inherent worth or intrinsic value in humans that they should regard themselves as superior in any concrete way to anything that has evolved alongside of them in the chance-directed natural selection process of evolution. Bertrand Russell understood this well:

> *But even more purposeless, more void of meaning [than a world in which God is malevolent] is the world which science presents for our belief. Amid such a world, if anywhere our ideals henceforward must find a home. That man is the product of causes which*

had no prevision of the end they were achieving; that his origin, his growth, his hopes and fears, his loves and his beliefs, are but the outcome of accidental collocation of atoms; that no fire, no heroism, no intensity of thought and feeling, can preserve an individual life beyond the grave; that all the labors of the ages, all the devotion, all the inspiration, all the noonday brightness of human genius, are destined to extinction in the vast death of the solar system, and that the whole temple of man's achievement must inevitably be buried beneath the debris of a universe in ruins—all these things, if not quite beyond dispute, are yet so nearly certain that no philosophy which rejects them can hope to stand. Only within the scaffolding of these truths, only on the firm foundation of unyielding despair, can the soul's habitation henceforth be safely built (Russell 106–107).

In the same way that Heidegger came to understand that angst is the only real and central validating factor in a naturalistic framework, Russell rejected any teleological explanation of reality and realized that the matrix of naturalism is "the firm foundation of unyielding despair." There can be no transcendental value to people in a universe that has evolved without prevision and guided by purely chance chemical reactions. And yet this intrinsic transcendent need in people remains a universal characteristic of the human soul. How can a naturalistic system account for this?

The naturalistic ideology can only lead to the "firm foundation of unyielding despair." The advent of Einstein's general theory of relativity and the discovery of the universality of the second law of thermodynamics brought to light the very reality that our universe will eventually die. The concept of heat death then began to preoccupy people's minds, and the intensity of the despair that is central and unavoidable to the naturalistic view of reality began to deeply impact and negatively color their philosophical worldviews. Optimistic existentialism gave way to nihilism.

The vibrantly clear message of the cosmos, as understood by modern physics and mathematics, is that "earth-man" does not have a permanent home in the cosmos. The expanding universe may perhaps one day grind to a halt and burn up in the death grip of supermassive black holes or expand until it dissipates altogether. But long before that, if space-time does not rip from these supermassive black holes, our sun will grow cold. And long before that, our planet's resources will have been exhausted by the expanding nature of humanity. And long before that, we will most likely have poisoned our oceans and waters, making our biosphere a dead planet. That is, if we somehow manage not to first kill each other off by war.

But people cannot stay long in the bleak, sterile, meaningless matrix of this coherent and faithful actualization of the inevitable implications of the naturalistic presupposition. Sadly, instead of admitting the failure of the naturalistic presupposition to account for the way people are, modern humans are slowly crossing over the line of despair into the mystical and irrational field of knowledge to avoid the pain of the unavoidable angst, which is the only natural and inevitable conclusion of this worldview. Our transcendent need cannot be long avoided or ignored. Humans are internally driven to find some sort of significance to assuage the cold sterility of the machine universe they have constructed in their minds to reject the creator.

Naturalists cannot long remain in the "firm foundation of unyielding despair" created by the meaningless relativistic mist of nothingness. So they must seek some higher meaning or "ultimate truth."

In recent years a number of philosophers of science have attempted to describe the "progressive" teleological development of science in terms of Darwinian

> evolutionary concepts. However Stephen Toulmin has
> emphasized, most of these philosophers have depicted
> the teleology as acting in the large to cause an inevi-
> table development of science towards ultimate truth.
> Both Toulmin and Thomas Kuhn have attempted to
> argue that teleology is local just as in evolutionary
> biology; . . . but there is evidence that the historical se-
> quence of physical theories is approaching some limit,
> which could be termed "Ultimate Truth" (Barrow and
> Tipler 142).

But why search for this ultimate truth in a naturalistic reality? In a naturalistic universe, all truths are relative. That longing for an ultimate truth is branded in our hearts because humans were designed by the ultimate truth. When God breathed the *nshamah* spirit into us, that connection was sealed. And for this reason, each of us intuitively knows that we have an intrinsic sense of worth and value. This runs counter to the implications of the naturalistic paradigm of our day. There can be no reconciliation between the Darwinist model and our Judeo-Christian cosmological model. They are antithetical constructs. A human being is not just a biological machine. Humans are beings created in the image of the Being, who brought all things to be by His spoken word.

It is revealing that those who claim that all truths are relative now feel a need for science to come to some ultimate truth. What will this ultimate truth be? I am not a betting man, but I will wager all that I own on the premise that science will eventually accept as rational and scientific a pantheistic-occult concept of reality. The universe will be looked upon as an organic entity in the whole. People will be looked upon as intertwined with the cosmos in such a way as Tielhard de Chardin has proposed.

The vibrating membranes that comprise all matter will be endowed with some "life force." All matter will be considered part of the Akasha spirit. God will be presented as an evolving god, not in the Judeo-Christian concept of God as an individual entity with personhood and choice that stands outside of space-time.

Instead, God will be, in the pantheistic sense, an intrinsic vitality in matter, an impersonal energy, perhaps the very energy of the vibrating strands of the string theory. The universe, in other words, will possess some form of shadowy ghost interconnected with humans, which will be the equivalent of the occult Akasha.

By taking this leap of faith into the mystical, people will attempt to remain firmly rooted in the naturalistic presupposition while trying to assuage their transcendental need. In this way, they can still evade accountability to the creator. Trapped in the sterility that their naturalistic worldview has wrought, they will attempt to alleviate their loneliness by interjecting a ghost into the machine. Naturalists will evolve from ghost worlds to the idea of a unifying ghost intrinsic to all matter. Their utterly vain hope is to assuage the despair of the naked, cold, and sterile machine. But it will not be enough. It will not provide for humans the foundation for human rights and infinite worth. The single cell is expandable for the practical benefit of the whole. Individual rights cannot arise from such a worldview.

Humans are not machines. We are people, and there is a palpable "humanness" inherent in all humanity that screams out for meaning, purpose, justice and transcendence. We are not simply the sum total of a bucket of organic nuts and bolts, as the reductionists so adamantly proclaim. The self is not part of some electrical synaptic delusion or illusion created by the brain. The self is not the mere sum of computational algorythms performed in the synapses of our brains.

> "Computation is computation because it's never about anything. It is non-intentional. The mind is the mind because it's always about something. It's intentional. Computation is the opposite of the mind.

If it is computation, it is not mental. If it is mental, it is not computation. The Venn diagrams never cross.

Note what this means for "artificial intelligence." A computer can't be conscious, because computation is the antithesis of consciousness. Computation is a mechanical process of mapping without reference to the content of the map. Mentation is a mental process of reference to an object—to content—other than itself.

The Times (paraphrasing neuroscientist Graziano):

"... consciousness is a kind of con game the brain plays with itself."

Gibberish. Brains don't play con games. Brains play no games at all. The brain is an organ. It generates action potentials, secretes neurotransmitters, floats around in spinal fluid inside the skull, etc. Only people play con games. Organs play no games at all. The idiotic claim that "brains play con games" is the mereological fallacy—the error of attributing to parts that which can only be attributed to the whole. Brains no more play con games than feet run marathons or hands play piano. People play con games and run marathons and play piano, using their brains and feet and hands. The mereological fallacy is perhaps the most common fallacy in neuroscience (a discipline beset with fallacies). Only a human being thinks or has emotions or has perceptions. Brains don't think or emote or perceive. Brains do organ things. People do people things (Egnor).

We are not our bodies. We are not our brains. The brain is an organic computer—the hardware. But our spirit is the personhood behind the brain. We are self-conscious and choose instantaneously which software to run on an existential basis. The true delusion belongs to those reductionists who claim the self is but the sum total of electrical impulses in the brain—an illusion. The great appeal to the reductionist/naturalist is that if the self is nothing more than an illusion, then our choices have no spiritual consequences. It is this intrinsic rebellion toward accountability for our moral choices that deludes the naturalist into rationalizing instead of reasoning.

The spirit of a human is the person behind the brain. And to declare anything less is counter to our intrinsic knowledge of who we are. It runs contrary to our internal instinct that tells us that there is something of transcendent value in each of us.

Richard Tarnas grasps the predicament created by the naturalistic paradigm. He eloquently addresses the bleak dead end at which postmodernists have arrived.

> By contrast, the modern mind experiences a funda-
> mental division between subjective human self and
> an objective external world. Apart from the human
> being, the cosmos is seen as entirely impersonal and
> unconscious. Whatever beauty and value that human
> beings may perceive in the universe, that universe is
> in itself mere matter in motion, mechanistic and pur-
> poseless, ruled by chance and necessity. It is altogether
> indifferent to human consciousness and values. The
> world outside the human being lacks conscious in-
> telligence, it lacks interiority, and it lacks intrinsic
> meaning and purpose. For these are human realities,
> and the modern mind believes that to project what is
> human unto the nonhuman is basic epistemological
> fallacy. The world is devoid of any meaning that does
> not derive ultimately from human consciousness. . . .
>
> The systematic recognition that the exclusive
> source of meaning and purpose in the world is the

*human mind, and that it is a fundamental fallacy
to project what is human onto the nonhuman, is
one of the most basic presuppositions—perhaps
the basic presupposition—of the modern scientific
method. Modern science seeks with obsessive rigor
to "de-anthropomorphize" cognition. Facts are out
there, meanings comes from in here. The factual is
as regarded plain, stark, objective, unembellished by
the human and subjective, undistorted by values and
aspirations. We see this impulse clearly evident in
the emergence of the modern mind from the time of
Bacon and Descartes onward (Tarnas 17, 19).*

Postmodernism is a dualistic reality created by the Neo-Platonist worldview. It separates the factual and knowable (science), where absolutes are possible in a completely segregated realm, from meanings and the metaphysical. In the lower story, science and math can be spoken of as facts. But in the upper story, which is completely disconnected and unbridgeable to the lower story, they place the metaphysical or meaning questions of life. These are seen as simply subjective, relativistic choices. They are seen exclusively as projections from our human minds.

Naturalists in the lower story live in a barren universe devoid of meaning, beauty, and purpose. They cannot long dwell where their souls are unrecognized and rationalized as a figment or projection of their minds. Human beings, made in the image of God, cannot dwell in such a stark universe. It is in complete antithesis to the way people see themselves. They must opt for something other than what could give them purpose and meaning. But they refuse to bow to the creator. What option is left if they wish to remain in a closed system?

The time will quickly come when the line of demarcation between machine and humanity will become fuzzy. Machines will be intertwined with living things, and lifeless matter will be deemed as living. Modern thinkers view this transhumanist stage as the next step in human evolution. It is the logical terminus for the Darwinian religion.

In an article in the well-respected journal *Scientific American*, Bart P. Wakker and Philip Richter wrote on the dynamics of galaxies in their article "Our Growing, Breathing Galaxy." A prime example of this trend is dramatized for us, which in years past would have been unheard of. The article begins with these words:

> *"Long assumed to be a relic of the distant past, the Milky Way turns out to be a dynamic, living object (Wakker and Richter 38).*

These are not just word games. They reflect a fundamental shift in thinking that is rising in the near horizon as postmodernists become disillusioned with the stark, meaningless void created by their worldview. When our space scientists discover a planet or moon that is geologically dynamic, they call it a "living" planet or moon.

The answer to the atheist will become the occult theology that dresses the closed system with a ghost. Their answer is god made in the image of humans. Tarnas recognizes the division created between objective knowledge in the lower story and subjective knowledge in the upper story, but he offers a way to unify the two through the occult pantheistic worldview, which he calls "primal man" or the "primal perspective." The universe is now declared to be living and filled with meaning and purpose.

> *The primal world is ensouled. It communicates and has purposes. It is pregnant with sign and symbols, implications and intentions. The world is animated by the same psychologically resonant realities that human beings experience within themselves. A continuity extends from the interior world of the human to the world outside. In the primal experience, what we*

would call the "outer" world possesses an interior aspect that is continuous with human subjectivity. Creative and responsive intelligence, spirit and soul, meaning and purpose are everywhere. The human being is a microcosm within the macrocosm of the world, participating in its interior reality and united with the whole in ways that are both tangible and invisible.

Primal experience takes place, as it were, within a world soul, an anima mundi, a living matrix of embodied meaning. The human psyche is embedded within a world psyche in which it complexly participates and by which it is continuously defined. The workings of that anima mundi, in all its flux and diversity, are articulated through a language that is mythic and numinous. Because the world is understood as speaking a symbolic language, direct communication of meaning and purpose from world to human can occur (Tarnas 16–17).

Hear my warning—atheism is but a bridge to mystical pantheism, which in turn is a bridge to the occult. The natural terminus of atheism is the occult. Those of us who begin with the Judeo-Christian concept of reality do not build our hopes and dreams on angst or the firm foundation of unyielding despair, because our cosmological teleology is deocentric. The initial cause is not becoming the final cause, as Tielhard de Chardin would say. God is the initial Alpha and Omega all at once, for He is not trapped in the space-time continuum of our existence. He is not evolving!

God stands outside our universe, having caused it out of the infinite energy of His being. And the vibrating strings that form the matrix of reality are in resonance with that initial symphony of His vibrating voice spoken at the Alpha Point. Furthermore, He is sustaining His creation with the same effortless causality. There will be no heat death in the naturalistic sense, but there will, however, be an end.

There is an end, with prevision and design, foretold by the stars and the prophets of old, in which the almighty creator will, out of His infinite energy, create a new universe in which the second law of thermodynamics will not be operational. This is the design of the universe that, against the natural order caused by the fall, will, at the Omega Point, be an eternal reality of a physical nature in which God will reorder or regenerate the universe into an eternal, incorruptible reality.

It will be a reality in which time will no longer be a factor in the new matter-space continuum—not in a static form, but in a dynamic, ever-creative, growing reality in which people will dwell in the utopia that their inner beings have always longed for and for which they were originally intended. It is why all people wish for a "Camelot" in the deepest recesses of their souls.

The deocentric cosmological teleology as the foundation of our Judeo-Christian cosmological model is the only teleological model that is true to the empirical data ascertained from our physical reality, which can provide a rational explanation for the overwhelming order and design of every facet of our universe from the microcosm to the macrocosm and can account for the way humans are.

Our Judeo-Christian cosmological model provides the only rational explanation that can account for the basic nature of humankind. It provides for an understanding of our universal transcendent need since we have been created in God's image. It explains the interplay between freedom and form and between free will and the sovereignty of God as reflected by the interplay between the microcosm and the macrocosm. It provides for an explanation of evil and for eternal justice. It provides for the redemption of humankind through the loving sacrifice of God. It provides for the fencing in of evil for eternity. Hence, in conclusion, reality is not anthropocentric, but rather deocentric.

Connecting the Dots

Modern science has opened our eyes to a reality that is quite distinct from the Enlightenment understanding of our material existence that gave birth to the materialistic Darwinian revolution. Matter is not the solid, sensible object once thought. It is comprised of mostly space and composed of components that can hardly be understood as anything more than energy in motion. Material reality is nothing like the Enlightenment thinkers wrongfully and simplistically believed. The microworld has been found to be full of surprising features never before imagined by humans. The false idea that the sensible world that we can see, touch, hear, taste, and feel is all that is real has been shown to be an ignorant and myopic view of reality. The dogmatic classification that bifurcated the physical from the spiritual by Enlightenment thinkers has been shown to be an ignorant line of demarcation.

Today, testimonials of prominent physics researchers from institutions such as Cambridge University, Princeton University, and the Max Planck Institute for Physics in Munich claim that quantum mechanics predicts some version of life after death. They assert that a person may possess a body-soul duality that is an extension of the wave-particle duality of subatomic particles.

Wave-particle duality, a fundamental concept of quantum mechanics, proposes that elementary particles such as photons and electrons possess the properties of both particles and waves. These physicists claim that they can possibly extend this theory to the soul-body dichotomy. If there is a quantum code for all things, living and dead, then there is an existence after death (speaking in purely physical terms). Hans-Peter Dürr, former head of the Max Planck Institute for Physics in Munich, posits that just as a particle "writes" all of its information on its wave function, the brain is the tangible "floppy disk" on which we save our data, and the data are then "uploaded" into the spiritual quantum field. Continuing with this analogy, when we die, the body, or the physical floppy disk, is gone, but our consciousness, or the data on the computer, lives on.

"What we consider the here and now, this world, it is actually just the material level that is comprehensible. The beyond is an infinite reality that is much bigger. Which this world is rooted in. In this way, our lives in this plane of existence are encompassed, surrounded, by the afterworld already. . . . The body dies but the spiritual quantum field continues. In this way, I am immortal," says Dürr.

Dr. Christian Hellwig of the Max Planck Institute for Biophysical Chemistry in Göttingen, found evidence that information in our central nervous system is phase encoded, a type of coding that allows multiple pieces of data to occupy the same time. He said, "Our thoughts, our will, our consciousness and our feelings show properties that could be referred to as spiritual properties...No direct interaction with the known fundamental forces of natural science, such as gravitation, electromagnetic forces, etc. can be detected in the spiritual. On the other hand, however, these spiritual properties correspond exactly to the characteristics that distinguish the extremely puzzling and wondrous phenomena in the quantum world."

Physicist Professor Robert Jahn of Princeton University concluded that if consciousness can exchange information in both directions with the physical environment, then it can be attributed with the same "molecular binding potential" as physical objects, meaning that it must also follow the tenets of quantum mechanics. Quantum physicist David Bohm, a student and friend of Albert Einstein, was of a similar opinion. He stated, "The results of modern natural sciences only make sense if we assume an inner, uniform, transcendent reality that is based on *all external data and facts. The very depth of human consciousness is one of them."*

Although there is no definitive concrete evidence for this theory, one could arguably afford some weight to these claims if some of the most brilliant minds in quantum mechanics believe that it is consistent with the general patterns and trends of modern science. If proven, this theory could have monumental implications; if humans do "download" their consciousness into a thus far unobservable field, then a person's consciousness could, in Dürr's words, truly be immortal (Fröbose 90–91).

Since the time of the Enlightenment, due to the atheistic/materialistic bias of the naturalist presupposition, the idea of invisible dimensions has been utterly ridiculed as mythological superstitions by Darwinian scientists. Only that which was sensible to our human senses and measurable quantitatively was considered real by the Enlightenment thinkers. Their rejection of the Judeo-Christian concept of a real heaven or spiritual dimension was utterly ridiculed as mere superstition.

Perhaps it was this feature that caused many physicists to initially ignore the string theory. We may be surprised to find out that there are actually other spatial dimensions distinct from our three-in-one visible space-time continuum, which happens to be invisible. This claim is not a mystical concept based on superstition but the result of empirical scientific mathematical equations.

The existence of seven other invisible spatial dimensions is actually predicted by the M-theory. If these invisible dimensions are part of the reality of our universe, then on what grounds can naturalists disregard the possibility that the spirit world may be real? How can they continue to insist that an invisible spiritual realm is nothing more than a fairy tale, the mythological fabrication of humans?

The beauty of the M-theory is that it shows quite plainly that there is a symmetrical continuity between the microworld and the macroworld. It took the M-theory, which is a reworking of the string theory, to connect the dots between the two seemingly incongruous worlds. There is an obvious tension between the descriptions of Jiffyland and the macroworld, but it is not oil and water. The M-theory shows that connection.

Curiously, it reflects the tension between the human free will and the sovereignty of God. In the very essence of reality, God has given us a picture of His continuous interplay with matter, which is symbolic of His interplay with humanity, soundly refuting the claims of the deists, for He is in continuous interplay with matter and humans. He is absolutely sovereign and holds the universe together, and yet we have free will.

Einstein's initial reservation to believe in a personal God was because of his inability to reconcile free will and sovereignty. But he failed to notice that it existed in balance within the most basic level of reality, which ironically he helped to discover. His insistence that a unified theory would eventually explain the two aspects of reality should have been mirrored in his acceptance of God as the overarching medium that connects free will and sovereignty in elegant harmony.

The fact that we have an inadequate understanding of the interplay between the two does not imply that there is no interplay at all. We, like the subatomic world, live in a state of seeming anarchy in which we have true free will to act as we choose. And yet our actions do not ever restrict the sovereignty of God. There is control exerted over this anarchy that maintains order in spite of our free will. As we stated earlier, chaos in our universe is not absolute randomness. It is deterministic.

In an objective analysis of Einstein's initial argument during his early years against the existence of God, one must conclude that his reservation was not predicated on the basis of the denial of God's existence, but on the basis of his rejection of God's morals. Einstein felt that God could not be good if He allowed humans to do evil (free will). If people are free to choose and can create evil, then God, who allows it, must not be good.

The problem with this reasoning is that we place ourselves with a finite mind over the infinite as court and jury. The argument here speaks not about the existence of God but rather about His character. It is people's disapproval of God's choice to allow free will that is reflected by this position. That is, it questions God's morality, not His existence.

The very existence of evil has been a major problem for many thinking people. I have often heard people say that if God is the creator, then He created evil, and they cannot accept the existence of a God that created evil. But the problem is that this argument does not penetrate the surface. These people have not thought through this deeply enough. Our human ability to perceive is finite, and for this reason, much of reality escapes our notice. Even what we do notice is often improperly perceived until our knowledge of the subject deepens.

For instance, we say that sometimes a given substance is cold. But cold is only a perception. God did not create cold. Cold is the absence of heat. Heat is the measure of the intensity of molecular vibration in a given substance. The more the molecules vibrate, the hotter the substance is. The less the molecules vibrate, the less heat the substance has. And that reduction in heat we call cold. Nevertheless, cold is but a perception; what we are measuring is still heat. When the molecular vibration ceases altogether, then we call that absolute zero on the Kelvin scale. It is still a measure of heat.

Likewise, darkness is simply the absence of light. But our ability to perceive electromagnetic waves is quite limited. We cannot sense the long radio waves, for example. We cannot even sense minute quantum packets of light. In fact, as we age, it diminishes radically. By the time we are 60, our eyes can perceive 60 percent less light than when we were young. We say that there is darkness in outer space. But everywhere we can go in our entire universe, there is still light. There are photons in every square inch of space.

If you can see a star, then there is light. Even if we were to go into the deepest recesses of the Earth, tiny neutrinos are passing through, and there is still ambient energy all around us. Molecules are vibrating around us, and there is no place we can go where the energy that God created is absent altogether. The Higgs field is everywhere a non-zero resting point energy.

Similarly, evil is but the measure of the absence of God. God did not create evil. The very fact that we can perceive evil is direct evidence of the existence of God. For if God did not exist, then there would be no such thing as evil. What is, would then just be. And there can be no comparative quantification of that which we call evil without the existence of a paragon of goodness on which to anchor our measurement of evil.

It is the naturalist who is hard put to explain the existence of evil. We are all born with a natural instinct that is inherent to our souls, as created in the image of God, and therefore, we naturally know that there is such a thing as justice. We hurt when an injustice is committed, and we instinctively know that it is evil. But if, as naturalists say, there is no God, then they must accept this to be nothing more than an illusion in the human mind. To say that we cannot believe that God exists because evil exists is therefore an oxymoron. The true reality is that we cannot assert that evil exists without there being a God. Furthermore, without God, human life has no transcendence. It is merely the accident of chemicals in an impersonal chaotic reality that is void of any absolute, transcendent value.

Some, blinded by their subconscious desire to be free of any moral constraints, attempt to convince themselves that good and evil are but illusions—evolved social constructs that have no absolute framework. But they cannot live true to their philosophy in their everyday lives. They cannot escape the reality that they have been created in God's image. They live in a dichotomy, for they cannot truly practice what they believe. Their God-created soul instinctively knows that their loved ones are precious and not just meaningless blobs of protoplasm. They instinctively revolt against abuse and injustice. But if there is no God, there is only the survival of the fittest and the consequent and natural pragmatic abuse of the weak to benefit the powerful. Exploitation would be the only true matrix to reality.

Each of us has been equipped with the God-given design that allows us to channel God's Spirit through us to do good. We are all potential conductors of this energy. But each of us has also been given an electrical capacitor, if you would permit the metaphor. The capacitor has the ability to withhold the electricity that passes through a conductor. This capacitor is our will. Those who seek the face of God allow this current to flow through them, and many good, selfless deeds can then be done. Those who do not seek the face of God short-circuit the current, and the absence of the current is what we perceive as evil or selfish deeds.

The physical world is filled with examples of this interplay between good and evil. For instance, it is a fact that particles with electric charge always come in pairs. One particle has a positive charge, and the other particle has a negative charge. To get a charged particle, you need two fields that can rotate into each other under the symmetry of the electromagnetic force. A single field has no electric charge since there is nothing upon which the gauge symmetry can act. If evil did not exist, we would not know goodness.

If we are to believe that goodness exists, then evil is but the absence of goodness. If you admit that evil exists, then there must be good in order to differentiate the two. Neither evil nor good can exist in a universe where there is no God, who alone differentiates between the two. Without God, what is, just is. It is a neutered sameness void of any value quantification or description.

There must be a paragon of good in order to even define what is good. But there can be no goodness without a supreme God who is good, holy, and righteous. He alone can be the absolute measure of goodness. If God exists—and the thrust of this book is to hopefully lay out a lucid and logical framework for that—from which this decision can be made rationally, then it follows that we must subjugate our standard of morality to His. Ah, but this is the sticky point for many who abhor the idea that their will cannot be supreme and the god of their reality.

Limited Autonomy

God chose to create beings that have free will. Yet in His sovereignty, He still holds the macrocosm in balance, and His will is accomplished, nonetheless. Such is the nature of this awesome God whose will cannot be thwarted by even our most evil designs. This is called limited autonomy as opposed to absolute autonomy or anarchy.

At the other extreme of anarchy, we find tyranny. He could have made us into robots that were programmed to do no wrong; that is, to function in a completely mechanistic and Newtonian fashion. But then we would not have the things we treasure most in our lives: our individuality; our freedom to create, discover, and design; and most of all, our ability to choose. The choice to love or hate, to give or take, to reflect who we wish to be—that is the very definition of personhood.

Like the Jiffyworld, we have limited autonomy, but not absolute autonomy. We have true freedom to choose, but in the end, God's will cannot be overridden. The macroworld functions in order. We are not predetermined, computerized automatons. We are persons with real freedom to choose. But those choices have consequences, because God runs the macroworld in an ordered and just fashion.

I surely would not have enjoyed being a computerized, mechanically controlled, and completely deterministic entity. I would not have liked it very much if I would have been a Newtonian-functioning, organic automaton. The beauty of God's creation is that He made us into beings like Him. We have the free will capacity of the subatomic particles in the microcosm, and yet it is not total anarchy; God maintains the macrocosm nonetheless.

Hence, Einstein's argument was not about the existence of a creator; rather, he questioned the appropriateness of God's will in giving humans a free will. Einstein initially was reticent to accept God on His terms and placed his level of understanding above the creator's. This was a tragic mistake, albeit understandable considering the horrors he saw perpetrated on his people by the Nazis in World War II. But it is a testimony to his greatness that in spite of his initial hesitance, he eventually reconciled, through his great intellect and humility, to the existence of a personal God. Ironically, it was his understanding of the universe as depicted by his theories that brought him to the light.

I am glad that we are not created in a Newtonian, deterministic, robotic fashion. I like being a being with choice, possessing creativity and limited autonomy. It is ironic that those who use this argument to deny the existence of God do not realize that if God had behaved as they wish, He would have given us no autonomy whatsoever. Those who champion absolute autonomy discredit God because human beings have limited autonomy. I fail to find the rational component in this argument.

We do have some form of limited autonomy, because we have been created in His image, which is what separates us from the rest of creation. While it is true that as free beings we have the ability to do evil and choose against God, in the end, we will see that this is not absolute anarchy, either. For God is still sovereign, and His will maintains the chaos in our human hearts by overriding the order of the supraworld He controls. The Alpha shall bring forth the Omega in His previsioned time.

Einstein's initial rejection of the Judeo-Christian concept of God was missing the very point that God has tried to show us throughout our rebellious existence; namely, His unfathomable grace toward us, in spite of our rebellion towards Him. This is the heart of the matter. It is rather like a child rejecting parental authority or, better yet, denying the existence of the parents on the grounds that his brothers don't behave very well.

There are those who, when faced with the need to recognize that the world had a beginning, refuse to accept the notion of a personal God because it then comes into tension with the moral choices they want to make for their lives. To this extent, it is a certainty that our morality dictates our theology and, by extension, our philosophy of science.

But to be fair, I am not convinced that this was the overriding motivation behind Einstein's hesitance, although it may have been a component. The injustice that his people experienced during Adolf Hitler's regime and in his personal childhood created in him a strong personal sense of justice, and he found it difficult to reconcile such a cruel, violent reality with a personal, loving God.

But Einstein's rejection of God's sense of justice due to his individual sense of justice makes justice meaningless in the whole. If every person is the final arbiter of justice, then there is no absolute justice. Then why did Einstein think that justice was important? What we can say for sure is that if there is no justice beyond this world, then there is no justice at all.

Ironically, the microcosm of Jiffyland is an exact parallel to our free will as humans. At any given point we are free to choose, yet somehow the fabric of the macrocosm does not rip apart, and existence is, in fact, stable. The will of God will be accomplished, and justice will prevail, albeit not always in our timetable. There is no better refutation of the deist doctrine. God maintains control of the universe continually in a dynamic relationship with matter and humans.

Einstein's brilliance caused him to see through the superficiality of religion, and he correctly understood that the authoritarian rule of religious leaders pandering religious dogma without a basis of reason, simply predicated on blind faith, was wrong. But he could not avoid his conclusion that there must be a God, for the universe bears the direct evidence through its elegance and obvious design. Thus, he wrote:

> *"Science without religion is lame, religion without science is blind. . . . If something is in me which can be called religious, then it is the unbounded admiration for the structure of the world so far as science can reveal it. . . . I'm not an atheist and I don't think I can call myself a pantheist. We are in the position of a little child entering a huge library filled with books in many different languages. The child knows someone must have written those books. It does not know how. It does not understand the languages in which they are written. The child simply suspects a mysterious order in the arrangements of the books but doesn't know what it is. That, it seems to me, is the attitude of even the most intelligent human being toward God. We see a universe marvelously arranged and obeying certain laws, but only dimly understand these laws. Our limited minds cannot grasp the mysterious force that moves the constellations"* (Kaku 128–129).

Today more than ever, science has shown through the non-emotional field of mathematics the true interconnection between the microcosm and the macrocosm. And the interconnection is masterfully designed to be beautiful, eloquent, rational, and grand.

CHAPTER 18

• • •

THE GRAND UNIFICATION THEORY (THE STRING "M" THEORY)

God does not play dice with the universe.
—Albert Einstein

The search for a theory that could encompass all known forces and provide for humankind a cohesive "theory of everything" has been the Holy Grail of modern nuclear physicists. Scientists believe that in the first second of the Big Bang the one grand unified force present at the moment of creation separated into the four fundamental forces we currently observe in our universe.

We do not know how it came to be, nor can we explain why that original singular force appeared from nothing, except through a divine act. But we do know that at the very beginning, when the universe was about a thousandth of a trillionth of a trillionth of a trillionth (10^{-39}) of a second old, its temperature was on the order of 1028 Kelvin, and all the forces in nature were united in one single force. At that time, the universe was hotter than anything has ever been or will ever be.

Nevertheless, as the universe expanded and cooled, that single force broke symmetry and coalesced into our four forces that presently govern the very nature of our universe. These four known fundamental forces are gravitational force, electromagnetic force (responsible for the many modern marvels of society through the use of electricity), a strong nuclear force (binds the protons and neutrons within the nucleus of the atom from which comes the enormous nuclear energy released in atomic bombs), and the weak nuclear force (responsible for the effects of radioactive decay).

When Einstein discovered the general theory of relativity, he provided for us a way to understand an aspect of our universe; as he said, a way to begin to understand the mind of God. But as revealing and enlightening as his groundbreaking theory was, it was still an incomplete aspect of the entire picture of reality. It allowed us to accurately perceive and predict things, only in the macroworld. That is, the mathematical equations derived from it are able to explain the larger cosmos of our universe, or the supraworld, but fail to explain the microworld.

Through the mathematics of general relativity, we can understand the inner workings of the stars and galaxies. However, it explained our world through a single lens that understood the force of gravity, but it could not explain reality incorporating the other three forces of nature.

Later was developed the theory of quantum mechanics, which were founded on Einstein's theories. It seemed to provide another slice of the pie and allowed us to understand and perceive things in the microworld. Quantum mechanics explained our reality through three lenses that incorporated the electromagnetic force, the strong nuclear force, and the weak nuclear force. However, it could not explain gravity.

Neither the theory of relativity nor quantum mechanics could provide a comprehensive understanding of all reality with a singular equation that could intertwine all four forces of nature. This has produced no small measure of consternation among modern physicists, for the two foundational pillars of modern physics are at their core fundamentally incompatible. Both these equations have been shown beyond question to be accurate in their descriptions and predictions of the way nature behaves, and yet they are somehow incomplete by themselves.

For years, physicists attempted to mathematically unite the two equations in a way that would form a single unified theory of everything. Some believed that a comprehensive theory that could explain all known forces could never be possible. But Einstein did believe, and to this end he labored furiously until the day he died. The vast majority of scientists thought he was on a fool's errand.

The major quarrel between the theory of relativity and quantum mechanics lies in that the two cannot coexist peacefully, for they cannot be unified mathematically. In addition to their inability to unify all the forces, the two theories render a completely different description of the universe observed. Each seems to paint a picture of reality that is diametrically opposed.

The general theory of relativity gives us a picture of a smooth, orderly universe. Quantum mechanics, on the other hand, gives us a radically different picture. Here, the microcosm of Jiffyland seems to function in an irregular, chaotic fashion, a feature that is much adored by those with an existentialist or naturalist presupposition. Instead of a smooth, ordered universe, it describes a seething, frothy texture that is constantly fluctuating. The paradox is that both cannot be true, and yet both have been substantiated by experimental evidence that is irrefutable.

Somehow, these equations must be pointing to a larger, more elusive truth behind them both. But the difficulty of mathematically uniting the forces under one theory has proved to be a daunting obstacle for theoretical physicists. Unlike Einstein, many physicists since the development of these two equations have traditionally simply chosen to ignore this reality. In the past, most have simply chosen to stay segregated within this unresolved schizophrenic world of physics whereby the reality of the quantum world is fundamentally at complete odds with the reality of the macroworld.

Adding to the difficulty in uniting the equations that explain these forces is the disparity between the magnitudes of these forces. There is a seemingly huge difference between the strength of the force of gravity and the other forces. We may think of gravity as being a strong force; after all, if we fall from a building, the impact on the ground is quite undesirable and forceful. And we know that objects accelerate at the rate of 32 feet per second squared when acted upon by gravity in our world. But compared to the other forces, it seems, in fact, quite weak.

This disparity can be illustrated by the observation that in spite of the gravity of the mass of the entire Earth pulling us down toward its center, we are still able to overcome it and walk on its surface. Not only do we not fall through to the core, but gravity is incapable of even pinning us down. Our rather small muscles can literally overcome the gravity of the entire mass of the Earth.

A simple, small magnet can pick up a paper clip against the gravity of the entire Earth. It is the electromagnetic force in the atoms of the Earth beneath our feet and in our bodies that keep us from sinking into the core of the Earth. It is the strength of the electromagnetic bonds in atoms and molecules that give integrity to matter. Remember, matter is mostly space. Were it not for the electromagnetic forces in atoms, gravity could pull us down into the ground and meld our bodies with the rocks below our feet.

As a result, the relatively minute forces of our muscles are able to counter the force of gravity and allow us locomotion. The strength of the electromagnetic force appears to be stronger than gravity by an overwhelming 1,000,000,000,000,000,000,000,000,000,000,000,000, 000 times; that is a 1 with 39 zeros (10^{39}).

> The only reason the electromagnetic force does not completely overwhelm gravity in the world around us

is that most things are composed of an equal amount of positive and negative electric charges whose forces cancel each other's out. *On the other hand, since gravity is always attractive, there are no analogous cancellations—more stuff means greater gravitational force. But fundamentally speaking, gravity is an extremely feeble force* (emphasis added) *(Greene 12).*

And yet gravity has a property that it interacts with even massless particles such as photons. We have seen this through the gravitational lensing created by the sun as starlight bends by the massive gravity of the sun. It was this prediction that helped prove Einstein's controversial theory. Gravity shapes the space around it and that impacts time.

What are the odds that random chemical processes, devoid of divine prevision and design, could have created a world in which the electromagnetic force is so carefully balanced in order to not overwhelm the gravitational force, and as a consequence, the carefully balanced and negated charges could allow us to function in this world?

The strong nuclear force is also immense in relation to the force of gravity. It is the energy released when an atom bomb explodes. For within that tiny nucleus, the protons of positive charge are pushing against each other with incredible force since they are of the same polarity (like poles repel; opposite poles attract). Yet somehow they are kept within that tightly packed area of the nucleus by the enormous force known as the strong nuclear force. When the atom is smashed, the enormous force that holds these particles together is released.

Physicists have calculated that the strong nuclear force is about one hundred times as strong as the electromagnetic force and a staggering 100,000 times as strong as the weak force. But why or how could random chemical processes engineer the evolution of matter from the absolute moment of inception of the Big Bang with the results that are featured by the exact balance in the differences in these forces that allow our universe to exist in such a way that it could inhabit life? The choice is quite unique, for without the exact ratios of these forces, our world could not exist as we know it, and life could not be harbored.

> For example, the existence of the stable nuclei forming the hundred or so elements of the periodic table hinges delicately on the ratio between the strengths of the strong and electromagnetic forces. The protons crammed together in atomic nuclei all repel one another electromagnetically; the strong force acting among their constituent quarks, thankfully, overcomes this repulsion and tethers the protons tightly together. But a rather small change in the relative strengths of these two forces would easily disrupt the balance between them, and would cause most atomic nuclei to disintegrate. Furthermore, were the mass of the electrons a few times greater than it is, electrons and protons would tend to combine and form neutrons, gobbling up the nuclei of hydrogen (the simplest element in the cosmos, with a nucleus containing a single proton) and, again, disrupting the production of more complex elements. Stars rely upon fusion between stable nuclei and would not form with such alterations to fundamental physics. The strength of the gravitational force also plays a formative role. The crushing density of matter in a star's central core powers its nuclear furnace and underlies the resulting blaze of starlight. If the strength of the gravitational force were increased, the stellar clump would bind more strongly, causing a significant increase in the

rate of nuclear reactions. But just as a brilliant flare exhausts its fuel much faster than a slow-burning candle, an increase in the nuclear reaction rate would cause stars like the sun to burn out far more quickly, having a devastating effect on the formation of life as we know it. On the other hand, were the strength of the gravitational force significantly decreased, matter would not clump together at all, thereby preventing the formation of stars and galaxies (Greene 13).

It's not that a different world would exist, but that no world would exist. All the stars and planets and galaxies in the universe could not have formed altogether. If the mass of the electrons were just a few times greater than it is, electrons and protons would tend to combine and form neutrons, eating up the nuclei of hydrogen. In other words, if the electron and the proton combined much more readily, there would not even be hydrogen atoms.

Hydrogen is the simplest element in the cosmos, with a single electron and a nucleus containing a single proton. Without it there would be no stars. It is hydrogen that fuels the stars. Without it there would be no organic elements. Hydrogen bonds are essential in the formulation of complex organic chemicals. Life could not exist without it. In fact, none of the more complex elements beyond hydrogen could form for the very same reasons. Our universe would be a monolithic soup of subatomic particles.

If the strong nuclear force were just a bit less powerful than its present balance with the electromagnetic force, then the present symmetry that allows for atoms to combine would not exist, and the formation of compounds could not proceed. A deuteron (the second nucleus in the periodic table of the elements) is also able to exist because the attractive strong force acts between neutrons and protons. From the point of view of the strong force, neutrons and protons behave identically, and yet the neutron has no charge and is not being repelled by the proton. On the other hand, two protons of like positive charge do have an enormous repulsive electromagnetic reaction to one another, which is sufficient to keep them from binding to one another if it were not for the enormous attraction of the strong nuclear force. That finely tuned, precise balance between the electromagnetic force and the strong nuclear force is what allows the nucleus of atoms to build with more protons and create all the different elements in our universe.

If the strong force were only a little bit stronger, the protons would bind with one another and overpower the repulsive electromagnetic force of their charges. Then the nuclear furnace of the sun would proceed at a completely different rate, burning its fuel quickly, and life as we know it would be impossible. If the balance went the other way, causing the repelling force of electromagnetism to overcome the strong nuclear force, then the nucleus would disintegrate, and elements could not form at all.

Yet these forces do exist in the seemingly prescribed ratios that are so vastly different from each other and yet obviously originating from the same source by the harmony created from their design. But herein lies the problem of finding a unified theory that could explain the unity and yet diversity of these fundamental forces in the universe. The problem that Einstein and many other physicists since have encountered is that they could not reconcile these forces mathematically. Their forces were irreconcilable, or so we thought.

The two foundational theories that shaped the very core of our understanding of reality in the twentieth century can each only answer a slice of reality. The theory of general relativity allows us to understand gravity, but not the other three forces. On the other hand, quantum mechanics and, more recently, particle theory as an outgrowth of quantum mechanics, could explain the other three forces. All failed equally to reconcile all known forces in our universe.

It is unquestionable that quantum mechanics is able to make accurate predictions within the microcosm of our universe at the molecular level and that general relativity performs equally perfectly in the macroworld, and yet they are in complete contradiction.

> *The gently curving geometrical form of space emerging from general relativity is at loggerheads with the frantic, rolling, microscopic behavior of the universe implied by quantum mechanics. . . . this conflict is rightly called the central problem of modern physics (Greene 6).*

It is as if each looked out into the world through a different window and caught a glimpse of the landscape of reality that was completely different from the other. But the complete picture evaded both. Each was incomplete in explaining the total reality of the universe, and both could not be exclusively right, in and of themselves. Einstein's world was ordered and predictable, but the world of quantum mechanics was chaotic and unpredictable. Both could not be right, and yet both were right from their particular reference points.

Einstein knew that general relativity was only a partial view but also understood that quantum mechanics was equally deficient to explain the whole. And for this reason, he dedicated the latter part of his life to the search for the mathematical equation that could reconcile gravity with the electromagnetic force. It was this elusive dream that Einstein so furiously labored to find before his death. He searched for an equation that could mathematically bind all the forces together in a grand unification theory.

> *This quixotic quest isolated Einstein from the mainstream of physics, which, understandably, was far more excited about delving into the newly emerging framework of quantum mechanics. He wrote to a friend in the early 1940s, "I have become a lonely old chap who is mainly known because he doesn't wear socks and who is exhibited as a curiosity on special occasions."*

Einstein was simply ahead of his time. More than half a century later, his dream of a unified field theory has become the Holy Grail of modern physics (Greene 15).

This unfinished unified field theory and his thirty-year search for a "theory of everything" was by no means a failure—although this has been recognized only recently. His contemporaries saw it as a fool's chase (Kaku 14).

Einstein's belief in the creator caused him to instinctively know that there would be a theory that could unify the forces in a symmetrical and orderly design. His repudiation of the idea, resulting from the implications of quantum mechanics (championed by Bohr and which stipulated that the matrix of the universe is absolute chaos) was founded on this intuitive belief in the elegance and beauty of a symmetrical universe created in an orderly fashion by a supreme God.

Bohr, on the other hand, viewed the universe as controlled by the unpredictability of throwing dice, which dovetailed with his underlying existential metaphysical worldview. Thus, Einstein's famous statement, "God does not play dice with the universe."

Symmetry is a reflection of design and is therefore the very antithesis of randomness, which is at the heart of the quantum theory. Einstein instinctively knew that there was an elegant symmetry at the core of the universe. So he sought for an equation that could incorporate all forces and allow us to view reality through all the windows at once—a new reference point, a theory of everything.

The more beautiful an equation is, the more symmetry it possesses, and the more phenomena it can explain in the shortest amount of space (Kaku 76).

Einstein was convinced that the seeming chaos turning up in our observations of fundamental particle physics was due to some basic incompleteness in our understanding (like the subtle reasons that cause a roulette wheel to seem chaotic in its result), only because our powers of perception cannot pick up the subtle differences in the forces involved.

On the other hand, those who were drawn to the existentialist camp were more than happy to look upon the universe through the partial window of quantum mechanics. They did this in order to substantiate their philosophical existentialist view that presupposes that the universe is purely the result of random chemical reactions. To the deophobes, quantum mechanics was their catechism.

Their intrinsic antipathy to the notion of a creator/designer compels them to lean toward a view of reality that is not symmetrical and ordered. Randomness is a key, fundamental element to the evolutionary model, and therefore, Einstein's refusal to accept this randomness as the matrix of reality was at complete loggerheads with Bohr's existentialist presupposition.

Now it must be stated that the theory of quantum mechanics and the development of quantum electrodynamics as a result of it have been experimentally verified to such an extent that it cannot be logically refuted. The view through that window was correct, at least through the reference point, in that limited part of the landscape of reality.

Quantum electrodynamics is arguably the most precise theory of natural phenomena ever advanced. An illustration of its precision can be found in the work of Toichiro Kinoshita, a particle physicist from Cornell University, who has, over the last 30 years, painstakingly used quantum electrodynamics to calculate certain detailed properties of electrons. Kinoshita's calculations fill thousands of pages and have ultimately required the most powerful computers in the world to complete. But the effort has been well worth it: the calculations yield predictions about electrons that have been experimentally verified to an accuracy of better than one part in a billion. This is an absolutely astonishing agreement between abstract theoretical calculation and the real world. Through quantum electrodynamics, physicists have been able to solidify the role of photons as the "smallest possible bundles of light" and to reveal their interactions with electrically charged particles such as electrons, in a mathematically complete, predictive, and convincing framework (Greene 121–122).

How, then, can we make sense of this paradox if each equation has been substantiated beyond a reasonable doubt and each presents a conflicting answer to the matrix of reality? Can they both be true?

In 1968, a young Italian physicist Gabriele Veneziano who was attempting to understand the workings of the electromagnetic force happened upon an old mathematics book written 200 years earlier by Swiss mathematician Leonhard Euler. In this dusty old book, he discovered an equation that seemed to provide the kernel to an explanation of the strong nuclear force.

After a year's effort, Veneziano was able to come up with a mathematical equation that seemed to explain this phenomenon. No one paid much attention at the time. But another young physicist, Leonard Susskind in the United States, began to ponder on this new, fascinating equation. In 1968, Susskind was visited by a friend from Israel.

But just at that time a friend from Israel visited me in New York. The friend, Hector Rubinstein, was extremely excited about Veneziano's work. At first I was not very interested. Hadrons were exactly what I wanted to forget about. Mainly out of politeness I decided to hear Hector out.

Hector became so excited while explaining the Italian's idea that I really couldn't follow the details. As far as I could make out, Veneziano had worked out a formula for describing what happens when two hadrons collide. He finally wrote down Veneziano's formula on the blackboard in my office. It immediately struck a chord. It was extremely simple, and the features of the formula looked familiar to me. I recall asking Hector, "Does this formula represent some kind of simple quantum mechanic system? It looks like it has something to do with harmonic oscillators." Hector didn't know of a physical picture that went with the formula, so I wrote it down on a sheet of paper to remember.

I was intrigued enough to postpone thinking about quantum gravity and give hadrons another chance. As it turned out I didn't seriously think about gravity again for more than a decade. I pondered the formula for several months before I began to see what it really represented (Susskind 2006, 204–205).

Secluded in the attic of his house, Susskind reworked the equations for some months until he realized that they were describing strings of vibrating strands of energy that could recoil and stretch at different vibrations.

The idea was born that the elemental particles of all matter were not, in fact, point particles, as the particle theory suggested, but instead, vibrating strings. But the scientific establishment was not yet impressed. Moreover, few theorists were willing to abandon the existentialist presupposition of the Copenhagen school to look for an all-encompassing theory that would again picture the universe through a more orderly, symmetrical grid.

Susskind came to the conclusion that the elemental parts of the atoms were not the quarks, which composed the protons and neutrons of the atom, but instead, these subatomic particles were themselves composed of tiny vibrating strands of energy. These vibrating strands were so small that if the atom were the size of our entire universe, the strings would only be the size of an average tree. That is an amazingly small string, equivalent to the Planck length.

The properties of the subatomic particles observed were, in fact, now theorized to be the reflection of the various ways in which the string could vibrate. That is, the mass and force charge of a particle is determined by the string's oscillatory pattern. Excited at the possibilities of this brilliant deduction, Susskind attempted to publish his work. But as can be expected, he was met with resistance from the scientific community, which looked upon this string theory as unworkable.

There were some technical problems with the theory, for it predicted the existence of a mysterious massless particle, which scientists had difficulty accepting at that time. Moreover, it also had some fatal mathematical anomalies that rendered the formulas inconsistent. Only consistent equations have solutions. Solutions that produce infinite numbers are, therefore, considered inconsistent mathematical equations. In addition, one of the more curious aspects of this theory was that it predicted the existence of six more dimensions than the standard spatial dimensions accepted at that time—an idea that was also held as unacceptable by most physicists.

Physicists who were strongly entrenched in the naturalistic paradigm of our age considered scientists who had previously theorized the possibility of other dimensions as crackpots. This new string theory was challenging our knowledge of the three-in-one spatial dimensions in our universe, and few gave it any credence. To complicate matters further, the strings are so small that with the available technology, there was no way to document that they were really there. Some scientists, therefore, ridiculed the theory as a philosophy rather than a scientific theory since they were convinced that it was not testable by our present technology.

Those who favored the chaotic aspect of reality, which seemed to be supported by the quantum theory, found this unification theory of everything quite repulsive. The effect of the theory is that it tames down the chaotic aspect of Jiffyland and mathematically ties all forces under a single theory of everything. Therefore, it describes a universe with an intrinsic order and elegance that is cohesive and symmetrical in its totality—a philosophically repugnant feature for those of the Copenhagen school who favor an existentialist worldview.

> *String theory offers a novel and profound modification to our theoretical description of the ultramicroscopic properties of the universe—a modification that, physicists slowly realized, alters Einstein's general relativity in just the right way to make it fully compatible with the laws of quantum mechanics. According to the string theory, the elementary ingredients of the universe are not point particles. Rather, they are tiny, one-dimensional filaments somewhat like infinitely thin rubber bands, vibrating to and fro. But don't let the name fool you: Unlike an ordinary piece of string, which is itself composed of molecules and atoms, the strings of string theory are purported to lie deep within the heart of matter. The theory proposes that they are ultramicroscopic ingredients making up the particles out of which atoms themselves are made. The strings of string theory are so small—on average they are about as long as the Planck length—that they appear pointlike even when examined with our most powerful equipment.*

Yet the simple replacement of point particles with strands of string as the fundamental ingredients of everything has far-reaching consequences. First and foremost, string theory appears to resolve the conflict between general relativity and quantum mechanics. As we shall see, the spatially extended nature of a string is the crucial new element allowing for a single harmonious framework incorporating both theories. Second, string theory provides a fully unified theory, since all matter and all forces are proposed to arise from one basic ingredient: oscillating strings. Finally ... beyond these remarkable achievements, string theory once again radically changes our understanding of spacetime (Greene 136).

The mathematical anomalies and the intense antipathy toward a symmetrical and orderly universe caused most scientists to reject the string theory outright. In 1984, another brilliant physicist, Michael B. Green, was able to work out the equations in such a way that he effectively removed the mathematical anomalies that had plagued the string theory and had consequently previously discredited it. String theory became a bona fide consistent equation. Almost overnight, many of the young theorists of the scientific community became interested in the possibility of finding a grand unification theory.

Work on the string theory immediately commenced everywhere. Some have theorized that the mysterious massless particle predicted was in fact the graviton, which is described as the substance that creates our perception of the force of gravity. Before long, the combined efforts of these scientists brought us five separate string theories, which as a consequence seemed to again bring the idea to a grinding halt in the scientific community. For if the string theory was to be the grand unification theory, then how could there be five distinct and separate ununifiable equations that each seemed to explain a different piece of the puzzle? Scientists were once again looking at reality through five different windows. And all five could not be right—or could they?

In 1995, perhaps the most brilliant physicist of our time, Ed Witten (considered by some to be the Einstein of our age), announced that he had discovered the solution to the problem that had brought the string theory into its present quagmire. In his equations, he was able to show that the five separate string theories could be unified by looking at them from a different point of reference, by adding another dimension—a seventh invisible spatial dimension to the other six predicted previously.

This brought the total number of spatial dimensions from the visible three and one, which we know as height, width, and length in time, and added seven more invisible spatial dimensions. That means that if the theory is correct, our universe has 10 spatial dimensions in time (3 visible + 7 invisible = 10 spatial dimensions). It should be noted that some scientists refer to time as a separate dimension, making it 11 dimensions.

Scientists, before the advent of the string theory, believed that there were four dimensions. They considered time to be a separate dimension from height, width, and length. But as I understand it, time is not a separate dimension; it is an integral component of all spatial dimensions since none can exist without time and are inextricably intertwined with time, and thus the idea of space-time that we learned from Einstein. All spatial dimensions are intertwined in time and cannot be separated from time; thus the term *space-time*.

The three fundamental visible spatial dimensions can be better described as three in one and not four dimensions; symmetry with the Godhead is here reflected in the very nature of our space-time universe. The three visible spatial dimensions correspond to the visible part of the Godhead: Christ (three is the number of the Messiah). The addition of the seven invisible dimensions (seven is the number that symbolizes God the Father, the invisible) then brings us to a total of 10 spatial dimensions intertwined in time.

The reworking of the equation then showed that the strings could be vibrating membranes, and Witten named his reconstructed grand unification theory the M-theory (membrane); the "M," according to Witten, could also represent the magic or matrix of reality.

Some of these strings of vibrating energy were now seen to be open-ended (strings or membranes) and attached at one end to the membrane of our three-in-one dimensional space-time continuum. But others were predicted to be closed-ended, in the form of a doughnut or loop, and therefore not attached to our visible space-time membrane. The upshot of that is that these closed-ended strings could theoretically move out of our visible space-time membrane into the other existing dimensions, which are theorized to exist alongside of it. That is so because the closed-ended strings were not tied down or anchored on one end to our three primary and visible spatial dimensions.

This feature may explain the seeming weakness in the force of gravity. If the graviton is found to be composed of closed-ended strings, then the force may seem weak to us, only because its energy escapes from one dimension into a parallel invisible dimension. This could theoretically dissipate the energy felt within our perceived or visible three-in-one dimensional universe.

In addition, the theory predicts that these closed-ended strings could expand and flatten into a membrane so these branes, as they are now called, could stretch to enormous sizes. Our universe is now seen, according to the mathematics of this theory, to exist in a membrane within a higher dimensional space outside of it; that is, within a very real hyperspace outside of our universe. These membranes are perceived to be undulating or wavelike membranes.

If this speculation is true, then by definition, scientists no longer believe in a closed system. Something exists beyond our universe. Yet somehow that issue is absolutely ignored. The Tartufery of their dogmatic claim to an absolutely closed system is now made evident for those with eyes to see.

Further speculation has suggested that perhaps the crashing of two membranes together caused the singularity of the Big Bang. There are two important deductions that we can make from that conclusion:

1. The M-theory then insinuates that the Big Bang did not originate from nothingness, but that there may be a higher dimensional existence outside of our universe from which came the Big Bang. The idea championed by naturalists that we live in a closed system and that science cannot delve in speculations outside our space-time continuum is now categorically shown to be a false, myopic assumption arising from our ignorance of the true nature of reality. It must be recognized here that the subjective naturalistic imposition of a closed system to our reality is, in fact, a matter of faith—and not science. The naturalists want to have their cake and eat it, too. But of course, they do so for subjective reasons in order to sidestep the obvious inference to a creator. In the same token, the idea that the Big Bang could come from two colliding membranes is also a speculation that cannot be proved mathematically. Nevertheless, their readiness to accept this proves that their previous dogma is nothing more than subjective bias predicated on theological preferences.

2. This idea of membranes colliding to make the Big Bang is quite appealing to naturalists who desperately want to avoid any beginning from nothingness since it smacks of a special creation by a creator. But the question of origins is not answered here. They have only moved it one step backward. The origin of these membranes still leaves us at the exact point they tried to avoid. In their quest to explain away the enormous statistical hurdles they must leap over in order to have our complicated, intricate forces of nature, which

coexist in perfect balance, as well as the impossible odds of evolving life through chance chemical processes, they lean toward enhancing these chances by imagining multiple universes in which many renditions failed, and we just happen to be in the one that, through chemical serendipity, turned out just right.

The Large Hadron Collider, Sparticles, and the Higgs Boson

Although experimental evidence to substantiate the M-theory has not yet been developed, at this moment in history, the Europeans have finished a giant particle accelerator on the border of France and Switzerland. The Large Hadron Collider (LHC) is seven times stronger than Fermilab in the United States, the most powerful one up to that point.

It is hoped that the more powerful accelerators will yield experimental results heretofore not technically possible. In these giant particle accelerators, hydrogen atoms are exposed to a huge amount of electricity and stripped of their electrons, leaving only the protons. These protons are then sent speeding through the accelerator. When they approach the speed of light, they are met with protons shot the opposite way. In most of the collisions, the particles merely glance off of each other, but on the occasions when they hit head-on, they explode into a shower of subatomic particles from the impact, which are then captured at that precise moment by detectors that allow us to study those subatomic entities.

Since we know from Einstein that energy can be turned into matter, by increasing the energy of these collisions, we can produce larger particles. At the moment of collision, the enormous energy of both particles is turned into mass for a brief moment. These large particles are unstable and decay quickly, but by studying the particles into which they decay, we can deduce the type of particle that was created. That is precisely the goal of the Large Hadron Collider.

It is hoped that eventually the detectors will be able to capture the moment a graviton escapes from our three-in-one space-time dimension, giving scientists the circumstantial evidence that the string theory may be correct. Perhaps our technology will never reach the point of being able to directly see objects of that minute size, much smaller than the length of a light wave or even an x-ray. But we may be able to see their impressions and other particles that the theory predicts exist.

The theory also predicts the existence of sparticles, which are in symmetry to the particles we now know. These sparticles of supersymmetry are predicted to be incredibly heavy and may perhaps also be documented by the new generation of atom smashers being built by the Europeans.

A recent experiment at Fermi National Accelerator Laboratory (Fermilab) has shown that the estimated mass of a top quark is about four billion electron volts more than previously estimated. Instead of the mass being some 174 billion electron volts (mass is expressed in terms of energy; remember that they are interchangeable), the newly estimated mass of the top quark is now estimated to be about 178 billion electron volts. Although this adjustment may seem trivial, it influences our understanding of the other particles because of their interrelationship. As a result, scientists obtained a better idea of where to look for the Higgs boson, which is believed to endow all particles with mass.

Based on the new calculations, scientists computed the Higgs boson to be even weightier than previously expected, perhaps around 117 billion–127 billion electron volts. That explains why experiments have previously failed to isolate it since they did not have the power necessary to generate such force. Since the Large Hadron Collider was completed, scientists have been working feverishly to close in on these elusive subatomic entities.

The difficulty is that the mass of the Higgs boson is too large and therefore quickly decays before we can detect it. The lifetime of a

Higgs boson is estimated at around 10-21 seconds, which is less than a zeptosecond. This means that it can only travel less than a billionth of an inch before decaying. We cannot, with our present technology, hope to catch a glimpse of this elusive particle. But we can detect the particles into which it decays. This is a bit tricky, for sometimes other particles can produce similar results. Hence, the last step that physicists must take is to convince themselves that the decay particles really came from the Higgs boson and not something else.

In 2012, the Large Hadron Collider had been slowly upping its energy levels and was now functioning at a whopping 8 TeV—higher than anything before. On top of that, they were functioning at a higher luminosity, which means that they were getting more events per second, thus increasing their chances of finding a significant event.

This, of course, meant that more collisions were happening in the detector at the same time, making it much harder to separate the pileup. The LHC has two testing programs running side-by-side; one is the Atlas, and the other is the CMS. In July 2012, the Atlas found a Higgs mass of 126.5 GeV in several events, including two photon events and four charged leptons that resulted from the Higgs decaying to two Z bosons. These had a combined 5.0 sigma.

The sigma number is a measure of deviation that indicates something is going on. A 3 sigma indicates something that only happens 0.3 percent of the time. Within particle physics, an informal standard has risen that a 5 sigma is needed to prove the discovery of something new.

CMS detected a particle made by this collision at 125.3 GeV and also calculated a 5.0 sigma for the same events. However, they analyzed more channels in addition to the two photons and four leptons, and as a result, their combined significance ended up dropping to 4.9 sigma. Nevertheless, most physicists are convinced that they have found a new particle. The scientific community poured over the data collected to analyze if they were definitively the Higgs boson.

In October 2013, Peter Higgs and François Englert shared the Nobel Prize in Physics for "the theoretical discovery of a mechanism that contributes to our understanding of the origin of mass of subatomic particles, and which recently was confirmed through the discovery of the predicted fundamental particle, by the ATLAS and CMS experiments at CERN's Large Hadron Collider" (Nobel Foundation).

By 2014, CERN's computing capacity had reached an amazing 100 petabytes of stored energy. This has now been significantly expanded with an extension to the data center as well as the establishment of a new data center in Budapest, Hungary. After some maintenance and consolidation, the physicists again increased the power in 2015. The LHC plans to gradually up its strength to a whopping 14 TeV. That is almost double the strength that has ever been used in such experiments. If we find the graviton and the sparticles predicted by the string theory, perhaps then we will find the indirect evidence of the grand unification theory.

Because of the difficulty in seeing such small increments as the oscillating strings, some have criticized the M-theory as a philosophy of science and not a scientific theory. The LHC may provide proof that the theory is right. Time will tell.

The Explanatory Power of String Theory

But there is another way to look at this grand unification theory that I believe to be of utmost importance. There are three universal constants in nature. The grand unification theory must be able to explain nature through these three constants at every level of reality.

There are only three universal constants in nature common to all aspects of nature, all interactions, and all particles. They are Planck's constant h; the speed of light (denoted by c), which is constant under all conditions; and Newton's constant G that measures *the strength of the gravitational force. Since Einstein proved that energy and mass are convertible into one another, and gravitation is a force proportional to the amount of energy a system has, everything in the universe feels the gravitational force. In fact, using these three quantities—h, c, G—we can construct combinations that have the units of length, time, and energy. We expect all the quantities that enter into the final theory or are solutions of the equations of the final theory to be expressible in terms of the units constructed from h, c, G. [For the interested reader, the result for the Planck length is $(Gh/c^3)^{1/2}$, which as noted earlier, is about equal to 10^{-35} meters. For completeness, the Planck time is $(hG/c^5)^{1/2}$, which is about 10^{-44} seconds, and the Planck mass is $(hc/G)^{1/2}$, which is about 10^{-8} kilograms.] (Kane 55–56).*

Planck's constant is the fundamental, universal constant of the quantum theory. The Planck length is the smallest measurable unit of reality. Yet quantum theory is incapable of describing nature at that level. On the other hand, the M-theory does explain things at the fundamental level of the Planck scale. Hence, what some call a deficiency is, in fact, an incredibly important positive property of the equation.

The Standard Model by itself has a very serious conceptual problem. We saw in chapter 3 that the natural scale for the final theory was the Planck scale, about 10^{-35} meters. The Standard Model is a description of quarks and leptons and their interactions at a scale about 10^{-17} meters (Kane 64).

From the scale of the universe to the Planck length, there is a difference of 62 powers of ten. From the scale of the universe to the quantum scale of quarks and leptons, there is a difference of 46 powers of ten. From the quantum scale to the Planck length, there are 16 more powers of ten. Hence, the Standard Model can explain only about three-fourths of the length scale domains of reality, while the M-theory covers the whole spectrum. That is a powerful argument for the legitimacy of the M-theory. It alone can explain all of reality.

The elegance of the string theory, and perhaps the reason why some so ridicule it is that it describes a universe that is cohesively designed to incorporate all matter and forces. In doing so, it describes our subatomic world in a more ordered and less chaotic fashion. The seeming paradox between the subatomic world and the supraworld is now resolved, completely debunking the idea that the matrix of our universe is absolute chaos. Instead of the unpredictable quantum flux so loved by Bohr, what we find is that the universe as described by the M-theory is supersymmetric. Thus the title of this book, *Supersymmetry or Chaos*.

> *In physics as in art, symmetry is a key part of aesthetics. But unlike the case in art, symmetry in physics has a very concrete and precise meaning. In fact, by diligently following this precise notion of symmetry to its mathematical conclusion, physicists during the last few decades have found theories in which matter particles and messenger particles are far more closely intertwined than anyone previously thought possible. Such theories, which unite not only the forces of nature but also the material constituents, have the greatest possible symmetry and for this reason have been called supersymmetric. Superstring theory, as we shall see, is both the progenitor and the pinnacle example of a supersymmetric framework (Greene 167).*

The seeker of truth must ask, "Is our universe interconnected with supersymmetry, or is it fundamentally chaotic and randomly disordered?" The answer to this question reveals whether our universe was purposefully designed or the accidental serendipity of randomness and chaos.

If the Large Hadron Collider is able to find these mysterious sparticles as predicted, the M-theory would be vindicated. That, however, does not in any way prove that the Big Bang came from the collision of two membranes. It simply proves that our universe is symmetrical and ordered, a reality that is difficult to correlate with a chaotic, randomly guided universe. It is an indisputable indication that our universe was indeed purposefully designed for life.

The idea that God created the universe out of His spoken word is now understandable from a scientific perspective. For there can be no better way to explain the matrix of the universe than resulting from vibrations, which emanated from God—His voice in the vacuum of nothingness. The very fabric of our universe is continuously resonating as the symphony spoken by the creator at the Alpha Point.

As previously stated, sounds are, in essence, vibrations in the air, which are picked up by the diaphragm in the ear and interpreted by our brain. In space, there is no atmosphere to vibrate, but the concept remains the same—the vibrations created out of the energy of God are distributed in tiny packets of vibrating strands or membranes of energy that compose our universe and everything in it, including the space. His voice created the space-time continuum that we inhabit and everything in it. The matrix of all reality is then the very energy that has resulted from the voice of the creator in the form of vibrating strands of energy described by the M-theory.

> By the word of the LORD the heavens were made, and by the breath of his mouth all their host. . . . Let all the earth fear the LORD; let all the inhabitants of the world stand in awe of him! For he spoke, and it came to be; he commanded, and it stood firm (emphasis added) (Psalm 33:6, 8–9).

The biblical creation story in Genesis clearly states that Jehovah created the universe by the spoken word:

And God said, "Let there be . . ." (Genesis 1:3).

This spoken word is one and the same as the logos (word) of the New Testament:

In the beginning was the Word, and the Word was with God, and the Word was God (John 1:1).

He has delivered us from the domain of darkness and transferred us to the kingdom of his beloved Son, in whom we have redemption, the forgiveness of sins. He is the image of the invisible God, the firstborn of all creation. For by him all things were created, in heaven and on earth, visible and invisible, whether thrones or dominions or rulers or authorities—all things were created through him and for him. And he is before all things, and in him all things hold together. . . . For in him all the fullness of God was pleased to dwell (Colossians 1:13–17, 19).

The idea that God's creative voice brought forth all the matter and energy in the universe is corroborated by the M-theory. Moreover, it is His continued voice that holds all things together. The old deistic idea that God had set the clock running and walked away is here refuted with the loudest voice. He is not a static God. He is a dynamic God whose voice continues to make our universe dance in His good pleasure. It is He who made the Morning Star sing at the wonder of His creation. Each and every one of us is a very special, individual song of God.

But the idea that the universe was created by the spoken word of God is not peculiar to the Hebrew Scriptures, for it has been reflected in most ancient civilizations. The ancient Ohlmec and Mayan cultures of Central America stated that the universe was created by the spoken word of the supreme God Zamna/Tonatiuh.

The God of the Olmec people who inhabited Central America was Zamna. He was one and the same as depicted by the later Mayans as Tonatiuh. Both of them were depicted in the act of creation with their mouths open, speaking the universe into existence. The ancient Egyptians believed that Ptah created the universe by His spoken word. The Hindu traditions also speak of a creation from the spoken word of God, the Great Omm.

But this universal commonality is not the only one observed throughout the ancient cultures. The motif of the creator fashioning the first man out of clay is also widely depicted throughout all the ancient cultures of the world. And so are the stories of the global deluge, the confounding of the languages, the idea of heaven and hell, the existence of angels and demons, the tree of life, the tree of the knowledge of good and evil, the concept of the resurrection, and I could go on and on. You get the idea.

The biblical narrative clearly states that God brought forth the universe by His spoken word. He spoke, and where there was nothing, the universe and all that is in it came to be as He turned energy into matter. After the Earth was created, He then took clay and fashioned a man, and from the rib of the man, He fashioned a woman—both of them created in His image.

Naturalists would automatically scoff at such a notion, imagining God as a potter molding a clay figurine, waving a magic wand, and turning him into "Pinocchio." But what is clay? Clay is formed from water and minerals. And what is our body made of? The answer is about 99 percent water and 1 percent minerals. If there were a God, would it be too difficult for the creator of the universe to make a body out of water and minerals?

The ancient Egyptians were in complete accord with this account of creation as reflected in their ancient myths.

In some myths life is brought forth when the god
merely utters a word; in other, man is molded by the
deity out of the clay (Mercatante 21).

This commonality, reflected in the major ancient cultures throughout our world, brings us to an interesting enigma. How could this idea be held in common by the major ancient civilizations, separated from each other by vast distances and oceans, if they were all independently evolving civilizations? The obvious inference is that these accounts emanated from a common source. Perhaps the funnel created by the biblical account of the Great Flood explains this enigma. But this is a matter that I will deal with in my upcoming book *The Secret of the Lost Knowledge*.

For now, we can see that our universe is not a closed system, as the naturalists have dogmatically insisted for so long. It did not begin in the Genesis Singularity of the Big Bang from utter nothingness but from an existing higher dimensional reality, where God exists.

Our space-time reality is composed of three visible and physical spatial dimensions intertwined in time. Thus the reality in our visible dimensional universe is described as three in one. What a coincidence that this reflects the triune Godhead. Hmm!

It just so happens that the number three is the number of the Hebrew Messiah, who died in his 33rd year, ministered for three years, rose from the dead on the third day to ascend into heaven, and sits as the third member of the triune godhead with God the Father on the throne.

But our M-theory now tells us that coexisting in our universe are seven invisible spatial dimensions, also intertwined with time. It just so happens that the number for the invisible Father in the triune Godhead is the number seven. He built the universe in seven days, resting on the Sabbath. He will complete the history of Israel on the 70th week of Daniel. He freed the Jews from the slavery of Egypt in the seventh month that then became the first month.

He will free the Jews in the end of days on the seventh month in Yom Kippur, which used to be the first month. He destroyed the City of Jericho on the seventh day and on the seventh circumambulation and at precisely the seventh trumpet (notice that there are three sevens). He will end our Second Earth with a judgment of seven seals. On the seventh seal, there shall be seven trumpets of judgment. On the seventh trumpet, there shall be seven bowls of judgment. The seventh bowl will cause a global earthquake that will bring every city in the world to ruins (notice that, again, there are three sevens). The earthquake will build in intensity until all who walk the Earth will know that God is on the march against the terrorism and wickedness that spread throughout our world.

Our three-in-one visible spatial dimension is now seen to exist alongside another seven invisible spatial dimensions—the number that just so happens to correspond to the invisible Father. These are unseen and coexisting concurrently around us and all within the same time continuum. Is it a coincidence that Hebrew tradition tells us that there are seven dimensions of heaven beyond our visible spatial dimensions? According to the Talmud, there are seven heavens (shamayim). Is it coincidence that the Hindu religion also claims there are seven heavens? According to the Puranas, there are seven heavens called vyahritis. Even the Muslims believe in seven heavens. The Qur'an speaks of seven heavens called samaawat.

The idea that there are seven invisible dimensions smacks too much of the biblical concept of the spiritual world and the Hebrew tradition that there are seven levels of heaven. Since it is equally possible to mathematically propose that these invisible dimensions are not large but rather small, as in the Planck scale, the evolutionists naturally prefer this diminutive interpretation for subjective metaphysical reasons.

Some theorists have considered the possibility that some or all of the extra dimensions could be larger than that, which is logically possible. In this book we focus only on the approach that the extra dimensions are typically near the Planck-scale size (Kane 120–121).

In this multidimensional universe, therefore, beings perhaps composed of closed-ended strings, could then be capable of entering into our three-in-one visible space-time membrane, while not being tied down to it. What justification then can the arrogance of the naturalist have to claim that the spirit dimension is pure mythology and unsubstantiated by science? But I must again clarify that these dimensions are not parallel universes, as some try to suggest, but part of the very same reality of our universe. They are interconnected within the presently all-encompassing framework of time.

The idea that gravitons could travel into them, escaping our visible, three-in-one dimensional space-time continuum, opens the possibility of beings of this type being able to travel into ours. It is therefore quite evocative to note that Hebrew, Christian, Hindu, and Islamic traditions state that heaven is in seven spheres; perhaps these very seven dimensions. The elegance of this theory, therefore, surpasses even that which the scientists are able to discern. For it harmonizes with the numerical design that God has imprinted throughout creation as His fingerprint—for those with eyes to see and minds open enough to grasp His majesty.

The three visible dimensions correspond to Christ, the visible form of the invisible God, and the seven invisible spatial dimensions correspond to the invisible creator. The more we learn about our magnificent universe, the more we understand about the intricacy of its design and the more it corroborates with the truths written in the scriptures.

The supersymmetry in our universe is much deeper than most physicists realize. For this reason, God designed our universe with three universal constants (h, c, G), three particles in the atom (proton, neutron, electron), and three types of quarks with three distinct families. Gluons come in three colors—red, green, and blue. These color labels are arbitrary terms used by physicists to describe three types of gluons. At this minute level, shorter than the electromagnetic waves visible to humans, there are no colors.

We see that the electromagnetic spectrum is divided into three groups of three. On the short wave length, we have three invisible electromagnetic waves. In the middle, we have three visible electromagnetic waves. And in the long wavelength, we have three more invisible electromagnetic waves. Time is divided into three (past, present, and future).

So we see that the basic structure of matter is based on a three-quark code, and strangely enough, the basic building blocks of life are based on three families of macromolecules: the nucleotides of the DNA, the proteins, and fatty acids.

Life is described by three characteristics: respiration, metabolism, and replication. Is it also just serendipity that all ancient cultures divided the universe into three parts: heaven, earth, and hell? Humans are divided into body, soul, and spirit. Of course, we can speculate that all of this symmetry is purely accidental. But that would require a great deal of faith in improbable statistics.

The universe bears the indelible fingerprint of an intelligent designer. There is no artificial division to truth. There is no contradiction between true science and true faith. The unity of truth applies to quantum mechanics and the theory of relativity. It was ignorance that kept us from seeing the unity between these mutually inclusive aspects of our reality.

Most scientists who heretofore have universally acted as if the king was dressed, when all along he was naked, have finally recognized the schizophrenia in the field of physics created by the incompatible theories of quantum mechanics and the general theory of relativity that has plagued scientists in the twentieth century. The natural conclusion that for physics to be logically consistent there has to be a theory that could somehow unite quantum mechanics and general relativity has finally become accepted, even by many mainstream scientists.

But those with a dogmatic existentialist mindset have little regard for a theory that tames down the seeming chaos of the microcosm. Hence, the string theory is, for them, an unacceptable grand unification theory, as it destroys the very substratum of their metaphysical, existentialist paradigm that they prop up with the incomplete view of the total reality afforded by quantum mechanics.

CHAPTER 19

● ● ●

THE EVOLUTION OF OUR PLANET

The failure of the naturalistic paradigm is not, however, limited to the realm of metaphysics. To date, there are no adequate naturalistic theories on the formation of our solar system that can explain the unique features of all the planets and their moons. The standard naturalistic model has for some time now speculated that our sun and the planets condensed from an enormous swirling cloud of dust and gases, some 4.6 billion years ago. According to their standard model, orbiting particles began to collide with one another, producing heat as gravity pulled them to the center. When the pressure and density reached a certain point, the internal heat generated caused the sun to begin its thermonuclear reaction, and our sun was birthed.

In somewhat similar action, they claim that the planets began to accrete, scooping up most of the debris that had condensed from the primordial planetesimal cloud. They reason that the density in these smaller objects was not enough to ignite the thermonuclear process in the sun, so they developed into planets. At first, as the theory goes, planets crashed and jostled one another, until the last surviving planets came to their stable and present orbits. Whatever debris the planet did not use, they theorize, was eventually sucked into the sun by gravity or flung deep into outer space.

There are a few serious problems with this standard gravity accretion model that render the hypothesis all but obsolete. To begin with, it is highly doubtful that the mutual gravitational attraction of particles could lead to the accretion of a large planet. Collisions of small particles with other small particles would lead to dispersion rather than accretion, in accord with the second law of thermodynamics. Particles in collision, even gentle collisions, are much more likely to promote further chaos in an exponential fashion rather than harmony. This is, in fact, the principle that creates atomic fission in an atomic reactor, and it is clearly illustrated by the Kessler effect, which space scientists have recently become aware of due to the danger it creates for satellites and manned missions in space.

In 1978, NASA scientist Donald J. Kessler warned of a coming cataclysm in space because of the growing density of space junk and the cascading effect of their collisions. It is presently estimated that there are more than 600,000 pieces of space junk orbiting the Earth. These range from a centimeter to 10 centimeters in length. There are actually more space junk objects in orbit, but the larger pieces are observable to us, and therefore we can prepare for them by moving satellites out of their way ahead of time. The problem is with the objects smaller than 10 centimeters is that we cannot prepare for them because we cannot see them. A single screw traveling at orbiting speeds can have the impact of a Volkswagen traveling at 70 mph, which would devastate any satellite and kill any astronauts unlucky enough to be in their crosshairs. We are, in fact, on average losing one satellite per year because of these lethal, unseen flying bullets.

The cascading Kessler effect illustrates that striking debris fractionalizes exponentially into much smaller fragments and creates an ever more fine structure rather than accruing into larger bodies. In fact, they now realize that the Kessler syndrome is exponential in its cascading effect. Unless we are able to stop this cascading effect, the time will come when it will be impossible to send manned missions into space because of the danger these flying bullets of space trash will pose.

In the very smallest dust particles where electrostatic charges may outweigh the angular momentum of the striking particles, some limited accretion may grow. Many of us have seen the experiment in the Space Station where sugar inside a cellophane bag was shaken, and in the low gravity orbit, it seemed to clump into small conglomerates. Gravity has nothing to do with this. It is caused by the electrostatic charges of the sugar molecules. But these electric charges will not hold them together once they go beyond a miniscule scale. That entire argument is a complete, fabricated misrepresentation of the true scientific facts.

Neither will the small particles of sugar have the power to accrete if the speed of the interaction between the particles is great and the angular momentum differs in even a minute way. It is one thing to accrete inside a cellophane bag where all the particles' orbital speeds are equivalent and their angular momentum the same, and quite another when they are independent bodies coming from every angle at varying speeds. The only possible way for this material to coagulate properly would be to continually and periodically radiate them by lethal beams of gamma rays that could melt them once they struck. One way that could happen would be in places where gamma ray bursts exist. But it seems quite unlikely that every planet ever created did so only next to pulsars or gamma ray bursts.

There is another possibility whereby lightning created by the static charges produced in dust clouds by friction could take part in melting these conglomerated "dust bunnies." But such electrical discharges would have the tendency to explode those particles outward and disperse the cloud rather than conglomerate it.

In order for large particles in motion to accrete, they would have to be moving in the same direction and with almost identical speeds so they could gently arrive at each other's sides and allow gravity to hold them together. The statistics that this could happen in a chaotic environment through random ordering is quite astronomically small compared to all the other possible alternatives that would simply either sideswipe each other or, if one is traveling at a faster speed, collide and break off into smaller pieces.

How strange that we have not seen space trash in orbit around the Earth accrete to become space trash planetoids! In fact, what we witnessed on January 29, 2010, when the Hubble Telescope took a picture of colliding asteroids is the exact opposite of their gravity accretion models.

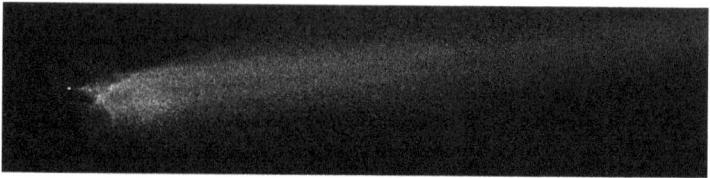

Hubble WFC3 Image of P/2010 A2 (January 29, 2010).

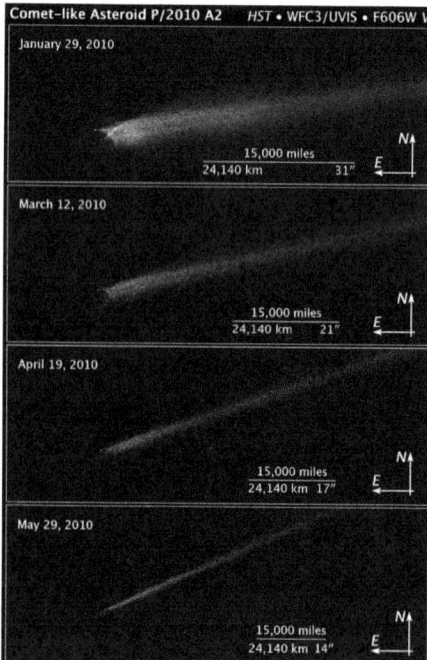

Comet–like Asteroid P/2010 A2   *HST • WFC3/UVIS • F606W V*

January 29, 2010

15,000 miles
24,140 km   31″

March 12, 2010

15,000 miles
24,140 km   21″

April 19, 2010

15,000 miles
24,140 km   17″

May 29, 2010

15,000 miles
24,140 km   14″

These four Hubble Space Telescope images, taken over a five-month period from January to May 2010 with Wide Field Camera 3, show the odd-shaped debris that likely came from a collision between two asteroids.
Illustration Credit: *NASA, ESA, and Z. Levay (STScI) Science*

NASA's Hubble Space Telescope has captured the first snapshots of a suspected asteroid collision. The images show a bizarre X-shaped object at the head of a comet-like trail of material.

In January, astronomers began using Hubble to track the object for five months. They thought they had witnessed a fresh asteroid collision, but were surprised to learn the collision occurred in early 2009.

"We expected the debris field to expand dramatically, like shrapnel flying from a hand grenade," said astronomer David Jewitt of the University of California in Los Angeles, who is a leader of the Hubble observations. "But what happened was quite the opposite. We found that the object is expanding very, very slowly."

The peculiar object, dubbed P/2010 A2, was found cruising around the asteroid belt, a reservoir of millions of rocky bodies between the orbits of Mars and Jupiter" (NASA).

The photographs say it all. But the comment made by NASA seems to ameliorate the embarrassment by claiming that the debris is expanding more slowly than they expected. What is obvious to any with eyes to see is that there was no gravity accretion. IT is the angle of the trajectories of the two striking asteroids that determines how wide the dispersion cloud becomes. These two were obviously in similar enough trajectories that the shattered particles continued in that direction.

We can point to the asteroid belt between Mars and Jupiter and find no evidence that these space rocks are amalgamating into a planet. When we see collisions, the opposite is observed. The rocks are broken into smaller pieces and often sent into different trajectories. Even when rather small rocks strike larger rocks, the gravitational pull is so slight that they either bounce off or send debris flying outward, depending on the energy and the angle of their strike. It is for this reason that the satellites sent to land softly on these asteroids are equipped with harpoons to anchor them so that they do not bounce.

Because gravity is relatively weak compared to all the other forces, the conglomeration of particles could not cement together into a solid piece of rock in small asteroids. They all would then be composed of loosely held rubble piles. That is not what we commonly find in asteroids. Many of the asteroids are quite solid. These few with loosely held rocks may more properly be categorized as fossil comets of frozen water that had rocks trapped inside them, which was evaporated by the cosmic wind and therefore simply left behind a rubble pile. It is rather doubtful that independent asteroids clumping together created these.

Moreover, we witness this phenomenon in the giant debris rings of the giant planets. When meteors strike their moons, debris is flung out, and the gravity of the planet keeps the fragments from accreting. The sun would have acted in similar fashion to the supposedly evolving planets, a problem that few scientists readily admit because they have no other way to explain the origin of our planets in an evolutionary random process.

In fact, the massive gravitational pull of the giant planets can be seen to massage the smaller moons to the point that their bodies are elongated and stretched, causing severe warping of these celestial bodies. If these moons are drawn nearer to the surface of the giant planets, they will instead be broken apart. The effect of gravity seems to work contrary to their wishful thinking. Some have the honesty to almost admit this.

> It turns out to be surprisingly difficult for planetesimals to accrete mass during even the most gentle collisions (Asphaug 54).

I beg to differ. It is only surprising to those who ignore the second law of thermodynamics. In fact, it is not just surprisingly difficult, it is almost impossible. From a statistical perspective, the number of possibilities of trajectory strikes, which will cause dispersion, is enormously greater than the few miniscule possibilities that could end up in a rubble pile.

As we peer into space, we can observe giant gas and dust clouds called nebulae. These clouds are seen to be mushrooming in giant pillars as asymmetrical structures. This is what would be predicted through Brownian motion. They are not in uniformed swirling rings as imagined by the evolutionists.

The famous three pillars of the Eagle Nebulae (*Photo credit NASA*)

Second, if gravity were to cause matter in the space cloud to condense, the condensing cloud would increase in friction as it would increase in density. Subsequently, this would cause the gases to expand outward, opposing the condensation or accretion process. Our observation of gases in space shows that their tendency is to dissipate, not to coalesce or accrete. That is why Mars has lost its atmosphere (in spite of the gravity of the planet) and not the other way around.

Furthermore, even if one were to miraculously accrete, the condensing clouds could simply not have developed such spinning speeds we observe in the planets from the physical process of accretion or condensing.

"We came to the conclusion," says Lissauer, "that if you accrete planets from a uniform disk of planetesimals, [the observed] prograde rotation just can't be explained." The simulated bombardment leaves a growing planet spinning once a week at most, not once a day (emphasis added) *(Kerr 548).*

The fastest spinning planet in our solar system is the giant Saturn, which was measured by the space probe Voyager in 1980 to be spinning at an astounding rate of once every 10 hours, 39 minutes, and 22.4 seconds. That is more than twice as fast as the much smaller planet Earth and impossible to correlate with the standard accretion model for the origin of planets.

The tremendous spin rate of Saturn has baffled evolutionary scientists for years. More recently, the Cassini space probe measured once again the spin rate of Saturn, and the results have sent the evolutionary scientists spinning. Cassini recorded the spin of Saturn to have slowed down by eight minutes since 1980 to 10 hours and 47 minutes.

Of course, evolutionary scientists exclaim that nobody in their right minds would think the planet has slowed down by eight minutes; it must be a fluctuation of some kind. Their obvious consternation is caused by the conclusion that if, in fact, the planet is slowing down at that rate, then it is proof that the planets are young, for they would have slowed down to a snail's pace had they existed for several billion years, as claimed by the evolutionary model.

In addition, the giant planets could not have formed so far from the sun in such a swirling cloud, as they envisioned. There just would not have been enough gas and material that far from the sun in the thinning tip of the swirling gas cloud for them to form in such a fashion. This causes a problem for the evolutionists.

Therefore, in order to force-fit their model, they must begin with the four giant planets forming closer to the sun and then somehow move them all to the periphery, against the gravity of the sun. That one of them might have, by chance, been perturbed from its original orbit and be thrown out further than its original orbit might be more palatable. But for all four of them to be fortuitously moved to the outer regions together is quite a coincidence; the precise logistics for this to have happened, without destroying the interior planets altogether, is quite miraculous, not to mention the destruction that these giant planets would have caused to one another. The strength of their gravitational pull against one another would have been disastrous. Furthermore, such perturbations would quite likely move them away from the ecliptic in which they are found.

Most gravity-based evolutionary models, whether through gravitational accretion (the collision of meteors) or the contracting of dust clouds, maintain that at one time, the Earth was a molten planet. The heat released by the impact of falling meteors in an accretion process would have, indeed, melted the planet. But if our planet had ever been completely molten, then we would expect that those elements, which are heaviest, would have stratified according to their density through the constant force of gravity during the long expanse of times they imagine as the Earth cooled off.

Such dense, unreactive elements as gold would have plummeted to the center of Earth's core. How, then, is it possible to find gold, which is twice as dense as lead, on the surface of the Earth? It would be more understandable if the heavy metals were only found within the vicinity of volcanic areas, but this is not always the case. It therefore seems highly improbable that the Earth could have been molten at any time in its past history.

Water, Water Everywhere: The Universe Ruled by Aquarius

The Genesis record is clear that God first created a universe filled with water before creating the Earth and stars. "Let there be an expanse in the midst of the waters, and let it separate the waters from the waters" (Genesis 1:6).

In every part of our universe, we find water. We now also know that beneath the crust of the Earth, there is more water than on the surface of the Earth. Had the planet been molten, the water being much lighter than rock would have come to the surface, bubbling up as steam through the molten crust. The vast amount of water found so deep in the magma could not have been the work of tectonic overthrusting, as evolutionists try to explain. It is rather difficult to imagine that two continental plates under such enormous pressure from their gargantuan weight of the rock could have allowed much water to seep between the two grinding plates. And even if some water seeped into the upper mantle that way, it could not explain the volume of water found almost 700 kilometers deep in the transition zone between the upper and lower mantles.

An enormous volume of water greater than our oceans has been found in this transition layer from 410 km to 660 km deep. The breakthrough study by Schmandt and Jacobsen, using seismic waves, detected enormous reservoirs of water dissolved in high-pressure rock beginning near the top of the lower mantle at about 700 kilometers deep. This is the bottom end of the transition zone. There is no physical process that could shepherd that much water from the surface to a depth of 700 kilometers below the surface of the Earth. That enormous volume of water could only have been placed there during the creation process of the Earth.

The gravity accretion process just cannot explain the empirical data we now know. It is more likely that the spin of planets and the coalescing of the material that composes these heavenly bodies were the result of the work of Birkeland currents in our early plasma universe. The angular momentum of the different planets points to several distinct steps. Not all the planets were made together. The planets seem to line up in pairs regarding their angular momentum. That leads me to believe that several huge Birkeland currents may have produced the material for these twin systems, and the electromagnetic force played a much greater role in the creation of our solar system then evolutionists imagine. And as we have previously discussed, these Birkeland currents were responsible for creating a universe filled with water.

There is yet another enigma for evolutionists in their attempt to explain our origins through random processes. How did our planet get all of this water to begin with in a gravity-based scenario? The Genesis record begins with the creation of water in massive amounts. The abundance of liquid water on our planet is a remarkable asset, which, in its liquid form is capable of harboring life. If the Genesis record is correct, one would expect to find vast amounts of water, not only throughout the universe but also in our solar system. We have already spoken of the water in outer space throughout the universe, but our solar system also indicates that water was everywhere when the planets and moons were first created.

Some smaller moons in our solar system contain vast volumes of water, and some of it might be in liquid form. In 2005, the Cassini spacecraft took pictures of Enceladus, one of Saturn's moons, which surprisingly showed geysers of water shooting from the surface in its southern pole. At first, scientists believed that an ocean may be localized in this southern region, but gravimetric data between 2010 and 2011 confirmed that the geysers came from a subsurface ocean that is now believed to be global. It is the massive gravitational pull of Saturn that provides the energy through tidal flexing to liquefy the subsurface ocean of Enceladus in a region of space that is so far away from the sun and so cold that it should be frozen solid.

On Saturn's moon Titan, scientists hope there may be water underground. On the surface, there are lakes of methane, which, at that freezing temperature so far from the sun, are in a liquid state. Our probes have found lakes, rivers, and seas of liquid methane, whose contours resemble the Earth's system. Rhea may also have a surface partially composed of water ice.

The same geyser phenomenon observed in Enceladus has also been observed on Jupiter's moon Europa. Slightly smaller than Earth's moon, Europa is thought to be composed of a silicate rock core with a water-ice crust that envelops it. It is visually remarkable because of its smooth surface whose features highly resemble our polar oceans when frozen. The cracks on the surface are ever changing by the heat of tidal flexing created by Jupiter's massive gravitational force. Scientists estimate that this water-ice crust is somewhere between 10 kilometers and 30 kilometers deep. If that is correct, it means that the enormous ocean in Europa is slightly more than twice the volume of Earth's surface oceans.

Ganymede, another moon of Jupiter, is also believed to have a subsurface ocean that scientists believe to be around 100 kilometers deep below a crust of ice above it some 150 kilometers high. Callista, also a moon of Jupiter, may also contain trace amounts of water in the form of ice.

We now know that water in the form of rivers and oceans once existed on Mars and simply evaporated due to its small atmosphere and the constant bombardment of cosmic rays because it has no magnetosphere to protect it. It is thought that water-ice still exists in the polar regions. On Pluto, we have discovered mountains of ice floating on an ocean of frozen nitrogen. Almost all the planets and moons in our solar system have shown that at one time, water existed in greater proportions.

Uranus is thought to be composed of methane and ammonia and may contain an internal layer of water. It is believed that a middle layer in this gas giant may be composed of a vast water and ammonia ocean. The heat created by the intense pressure may cause this middle layer to actually be liquid. But unfortunately for the evolutionists, even if the middle layer contained liquid water, the ammonia liquid mix could hardly be considered a habitable place for life. This is especially so when considering the lack of sunlight at that depth and the enormous buffeting winds created by the incredibly rapid rotational speed of the planet. By the way, two of the moons of Uranus , Titania and Oberon, may also contain surfaces partially composed of water-ice.

What we can conclude is that the empirical evidence we have gathered from our solar system as well as deep space has substantiated the Genesis record; our universe was from the beginning a water-rich universe.

Our Earth is often referred to as the Water Planet or the Blue Planet because deep ocean waters cover 71 percent of the surface of our planet. Earth's oceans contain 10 times more volume of water than dry land above sea level. It is the indispensable necessity of life. Our bodies are composed of almost all water and a few trace elements, and without it, life could simply not exist. It is obvious that a universe designed for life had to be a water-filled universe.

It also so happens that the composition of comets is mostly frozen water, and some frozen gases along with dust. It is as though comets are nothing more than clumps of frozen, muddy snowballs circling through space. Through the study of the light spectrum given off by heavenly bodies, we have come to catalog the specific colors, intrinsic to each type of element they contain. We are able to do this because molecules of each element absorb certain colors of light and deflect others in a constant pattern. Thus, through the observed color spectrum, we can ascertain what elements are contained within the object of our inquiry.

As the light of the sun reflects off of the dust particles in the tails of the disintegrating comets, we can determine their chemical composition. Such studies have revealed that comets are indeed composed of dust (approximately 15 percent) and mostly water (approximately 85 percent), a characteristic that is an anomaly to most other known heavenly bodies except for Earth. Remarkably, even some asteroids have been reported to contain salt crystals. The similarity with the vast majority of comets and Earth, however, ends there. The actual composition of their water evidences that the vast majority of comets have been made in deep space and do not contain the same chemical signature of water as our planet.

The problem with the evolutionary claim that comets seeded our planet with life and the water that filled our oceans is what the study of the composition of water in comets reveals is highly improbable, if not impossible. Not all water is created equal. Two hydrogen atoms and a single oxygen atom combine into the molecule we recognize as H_2O—the normal water molecule. The normal hydrogen atom is composed of one proton and one electron.

However, a hydrogen atom is sometimes composed of a proton and a neutron in the nucleus, plus the electron revolving around the nucleus. This configuration is called a deuterium atom, which has more mass because of the neutron that is added to the single proton in the nucleus of the hydrogen atom. When water is made of deuterium and oxygen, it is therefore called heavy water. On Earth, every 10,000 molecules of water have three molecules of heavy water created by the heavier deuterium in place of the normal hydrogen atom. It turns out that the extra mass of the deuterium causes water molecules to be 10 percent heavier than normal water molecules made of the single proton normal hydrogen atom.

If, in fact, comets were the source of our oceans, then the chemical signature of the water molecules in comets would be equal to those of the oceans on Earth. Unfortunately for the evolutionists, the empirical data says otherwise. A Cal Tech study of the Hale-Bopp comet showed that the water in the comet contained a significantly greater percentage of deuterium, or heavy water, molecules than what is found on Earth. Similar studies of the Comet Hyakutake and the famous Halley's Comet showed conclusively that they contained from two to three times the amount of heavy water.

The famous Rosetta mission that was the first to send a spacecraft to orbit a comet (67P/Churyumov–Gerasimenko) found that it contained double the amount of heavy water. Out of the 11 comets thus far surveyed, only one had a ratio that was even close to Earth's water composition. This was comet 103P/Hartley 2 from the Kuiper belt, which contains hundreds of thousands of icy bodies up to 100 kilometers in diameter. These contain what astronomers call short-period comets, which circle the sun within 200 years or less. The largest Kuiper objects are Pluto, Quadar, Makemake, Haumea, Ixion, and Varuna.

There are two major regions in our solar system that contain these icy comets. The Kuiper belt spans from around the area of the orbit of Neptune, which is 30 astronomical units from the sun (1 AU is the distance from the sun to the Earth, which is about 93 million miles). It extends outward past Pluto to about 55 AUs. The other region is called the Oort Cloud, which begins way past Pluto at about 5,000 AU and extends to about 100,000 AU at its outer edge.

Using our previous scale of one inch per every AU, our sun would be a single pencil dot on a paper. Earth would be one inch away. Neptune would be two and a half feet away from the pencil dot of the sun. That represents the interior region of the Kuiper belt. Three and a half feet away from the pencil dot would represent the outer region of the Kuiper belt.

Traveling outward from that pencil dot to 416½ feet, we would find the inner region of the Oort Cloud. In that area, we find the planetoid Sedna, discovered in 2003, which is larger than Pluto. Using our diminished scale of 1 AU being equal to 1 inch in length, that Oort Cloud is thought to extend outward from that pencil dot a whopping 7,916½ feet. From this region come the long-period comets, marking the very end of our solar system. These objects in the Oort Cloud are called long-period comets because their trajectory orbit is 200 years or more.

Now, here is the thing. The Oort Cloud has never actually been observed. It is simply a theoretical area where we know the long-period comets travel to make their U-turn back to the sun. Evolutionary scientists estimate that there are trillions of comets out there. But that is simply conjecture. In reality, we have no idea how many long-term comets are circling the sun, because we have only been watching and recording them for a very short time. Thus far, we have only managed to catalog less than 3,000 comets. No one knows for sure how many comets actually exist.

So far, the empirical data show that a very small minority of all the comets contain the same chemical signature of the Earth's water. That makes it quite impossible for the evolutionary-imagined heavy bombardment event, which is estimated to be in the range of 30 million comets required to fill our oceans. Not only is the mechanism highly suspect since such a bombardment required within a relatively short period of time would, in fact, melt and sterilize the Earth and release all the water into steam, but we also cannot know if there are even 30 million comets in our entire solar system. It is simply all speculation based on wish projection.

I suspect that the few comets found with water of a similar composition to Earth's may even be fossils of the Great Flood. When the fountains of the deep were uncorked by the cracking of our crust, it created a curtain of jetting water that spanned our globe and shot water up at supersonic speeds into the stratosphere. Some fell back to Earth. But some water and some rocks were hurled into space and became Earth-borne asteroids and comets. It is more likely that the comets did not seed our oceans, but we might have seeded space with some terrestrial comets and asteroids.

Terrestrial Comets and Asteroids

As we previously stated, it would be impossible to explain the vast underground deposits of water had our crust been melted, and all water would have evaporated into the atmosphere. Much of it would have escaped into space. Instead of a dry planet, what we do find are telltale signs of a deluge. Evidence that the entire Earth was once covered by water abound, including the following:

1. Huge salt dome deposits
2. Huge areas of mineral deposits
3. Vast fields of trees washed onto the ocean floors in the Arctic as well as the Antarctic regions

4. Large coal and oil deposits made from the plants and animals killed by the flood and turned, by the deep pressure of the sedimentation above them, to the carbon fuels on which our world runs today.

5. Sedimentary rocks at the peaks of every one of the very highest mountains, replete with shells and marine fossils. This intimates that the present mountain formations were created after the Great Flood as the continents ground to a halt. (See Book 4 of this series, *The Descent of Man*, for a more exhaustive treatment of this topic and for the physical evidence to support it.)

Evidence for at least some comets of terrestrial origin may be suggested from the examination of their mineral content. Some comets contain crystalline dust, most of it being magnesium-rich olivine. Of the more than 2,000 known minerals, olivine is perhaps the most common terrestrial mineral. This is not so on the other planets. Therefore, it may point to the Earth as the place of origin for at least some of them.

Perhaps we have the cart before the horse? Could it be that some of the comets were created by mass hydro-ejections from our own planet? As the water from these fountains of the deep became frozen in the cold of space, they may have brought some of our water to the moon as terrestrial comets. If, in future lunar missions, we are able to test the chemical signature of the water on the moon, we may be able to identify our planet as its source rather than deep space comets. We are now able to study the chemical composition of water in comets, and in the future, I predict that some will be found to be voyagers, which were expulsed from our very own Earth. But these will be the exception to the rule.

On the other hand, the opposite idea championed by evolutionists that comets emanating from deep space are the harbingers of the water in our oceans and the seeding of life in our planet is simply not possible. This evolutionary imagination does not correlate with the true data we now know from the study of the chemical signature of most comets. Nevertheless, a small remnant will show that their origins may be terrestrial and may contain some organic material peculiar to Earth, which will not be found in comets that were created in deep space.

So we see that at the other end of the spectrum, some evolutionists, putting the cart before the horse, have theorized that comets were the agents that seeded life into our planet. The idea that water and life were brought here from outer space has become a very popular evolutionary claim. However, the evidence garnered from the study of many comets clearly shows the majority of these comets to be composed of two to three times the ratio of heavy water from Earth's oceans. The chemical signature of Earth's water is not that of most comets.

If our model of the cause of the Great Flood is correct, several giant asteroids might have simultaneously struck Earth, causing the cracking of our crust. The ripping crust would have opened like a zipper all along the Mid-Oceanic Ridge and released the subterranean aquifers. These reservoirs, such as the one in the transition zone, would have had carbon dioxide gas emulsified in it under the high pressure created by the 10 miles or so of continental granite above them.

The violent pressure wave created by the multiple strikes would have not only caused the violent shaking of the underground water but would have also suddenly removed the weight of the continental rocks above it by the cracks created through the powerful impacts. This sudden change in pressure, coupled with the violent pressure waves, created a sudden outgassing as the cracks in the crust neared the depth of the aquifers.

Like a shaken soda bottle suddenly uncorked, the underground aquifers, rich in mineral content, would have violently erupted all along the rift, creating a curtain of water shooting upward at supersonic speed. These were, in effect, giant geysers like the ones we see on Enceladus, but on steroids. Water flowed out of the cracks in the crust and ended our First Earth in a hydraulic cataclysm.

It may very well be that some of the asteroids revolving around the sun between Mars and Jupiter were literally thrust there from Earth. These asteroids would have been created not only by the initial ejecta of the striking meteors but also by the expulsion of rocks eroded by the powerful jetting waters. Large chunks of rocks as well as copious amounts of minerals and water would have escaped into space to become asteroids and comets. Some blasted the moon's surface; others blasted Venus, Mercury, and Mars on the other side of our planet. Others flew past Mars and began to orbit around the sun in the asteroid belt.

These comets and asteroids would therefore have the same chemical signature of water from our planet as well as similar rock composition containing nickel, iron, titanium, granite, olivine, and salt. A few of these terrestrial comets may have extended as far as the Kuiper belt, and fewer yet would have extended to the much further Oort Cloud. But we would probably expect that most would be concentrated around the much nearer asteroid belt.

This asteroid belt extends from only about 2.2 AU to 3.2 AU between the orbits of Mars and Jupiter. But this asteroid belt is not like in the movies, where the *Millennial Falcon* from *Star Wars* is dodging huge rocks and zigzagging through a dense minefield of asteroids. We know of only about 7,000 asteroids, and they are spread across a very large area of space. If you stood on an asteroid, you could look in every direction and not see another one. It is for this reason that we can safely send satellites through them to the outer planets. All the asteroids in the belt together would not even make an object as large as our moon.

Some of them may be pieces of the ejecta from the very meteors that struck Earth. These would have the telltale signs of shocked crystals associated with such energetic impacts. Others may be ejecta from the erosion of the continental shelves by the jetting waters of the fountains of the deep. Many are solid chunks of stone, but some are loose aggregates of broken stones that must have been shot in the same direction and at the same speed in order for them to be able to loosely aggregate. Had they been gathered randomly, there would have been high-energy collisions that would have tended to further disperse them rather than allow them to fly in tandem.

Of these objects in the asteroid belt, the largest are Ceres, Vesta, Pallas, and Hygeia. These four asteroids alone contain half the mass of the entire asteroid belt. Temperatures in Ceres, which is about the size of Texas, range between 130 Kelvin and 200 Kelvin, compared to 300 Kelvin on Earth. Any water present in these asteroids would therefore be in the form of ice since there is no giant planet close enough to them to create the gravitational havoc that generates heat.

Contrary to the evolutionary claim, the evidence seems to point to a recent origin for comets. Comets move in elliptical orbits around the sun. As they come nearest the sun (perihelion), their speed accelerates due to the pull of the sun's gravity to its maximum energy. The comet turns the corner and then begins to move away from the sun until it reaches its farthest point in the ellipse (aphelion), where it slows to its minimum speed by the drag of the sun's gravity before returning to the sun.

All the while, the solar wind, or cosmic rays released from the sun, blasts into the comet as it approaches its vicinity. This causes it to melt or disintegrate and release particles, which we can observe as the tail of the comet. For this reason, the tail is always on the opposite side from the sun. The continuous bombardment of solar wind creates an opposing force to the forward momentum of comets, which not only causes their tails to always point away from the sun but also slows down their speed in the same manner that air slows down a baseball, until gravity overtakes its forward momentum and brings it to the ground. Therefore, this loss of momentum inevitably alters a comet's trajectory.

Eventually, given enough time, all meteors and comets will be drawn into the sun; that is, if they manage to escape hitting another heavenly body during their lifetime, or if they avoid being ejected altogether from the solar system by the gravitational perturbation of a planet in its path.

The solar wind also causes the ice to vaporize; thus, it continually reduces the size of the comet as it orbits. Given enough time, all comets will completely vaporize. The fact that so many comets exist seems to point to a not very long age for our solar system; otherwise, they would have long ago disappeared. It is highly unlikely that they could have been created billions of years ago during the supposed gradualist formation of our solar system that evolutionists propose. Unless there is some hidden cosmic water faucet (i.e., a renewable source that continues to produce comets and asteroids, which no scientist has observed), comets would no longer be circling the sun today. It is for this statistical reason that evolutionary scientists so like to exaggerate the number of comets that actually exist in the Oort Cloud.

Moreover, the predominant number of comets with a short-period orbit compared to the number of long-period comets, suggests that their origins may be from some recent event that took place within or near the center of our solar system.

In addition, the fact that planets such as Mercury still contain some traces of water also points to a recent event, not an event that transpired millions of years ago. This tiny rock being so close to the sun and having no atmosphere to contain the water vapor from escaping into space also has a surface temperature that melts lead. And yet it still contains some traces of water. That points to a recent event, not an event that transpired millions of years ago. Otherwise, it would have long ago become completely dry.

The Evolution of Our Moon

Much that our evolutionary scientists believed about the Earth's moon has been proved wrong. Due to the extreme ages they claim for Earth and its moon, they confidently declared that the moon was a solid rock without a molten interior. It would be quite difficult to reconcile such long evolutionary ages for the moon with a molten interior. Well, they were dead wrong. In a NASA website article, the truth regarding the molten core of the moon was revealed from the data gathered by the *Apollo* seismic experiments.

> *State-of-the-art seismological techniques applied to Apollo-era data suggest our moon has a core similar to Earth's. Uncovering details about the lunar core is critical for developing accurate models of the moon's formation. The data sheds light on the evolution of a lunar dynamo—a natural process by which our moon may have generated and maintained its own strong magnetic field. The team's findings suggest the moon possesses a solid, iron-rich inner core with a radius of nearly 150 miles and a fluid, primarily liquid-iron outer core with a radius of roughly 205 miles. Where it differs from Earth is a partially molten boundary layer around the core estimated to have a radius of nearly 300 miles. The research indicates the core contains a small percentage of light elements such as sulfur, echoing new seismology research on Earth that suggests the presence of light elements—such as sulfur and oxygen—in a layer around our own core. The researchers used extensive data gathered during the Apollo-era moon missions. The Apollo Passive Seismic Experiment consisted of four seismometers deployed between 1969 and 1972, which recorded continuous lunar seismic activity until late-1977 (Boen).*

It is doubtful that such a small body could retain a molten core if it had the many billions of years evolutionists claim. Recent evidence brought back from lunar manned exploration teams has shown that the vast majority of the lunar bombardments, which left our moon's far face so deformed with craters, took place during a singular cataclysmic event. This was a complete surprise to scientists who thought these craters had been formed through singular meteors, striking one at a time, throughout millions of years. Evolutionists always have a proclivity to see things in long ages in order to justify their imagined gradualist model.

We can see by observing the surface of our own moon that a catastrophic event sent a massive meteor shower that struck, in particular, the far-side area of the moon and cratered it much more than the side of the moon that now faces us. The statistics of these meteors striking mostly the far side of the moon while the moon is rotating is quite unlikely. That event was much more likely to have happened in a relatively short period of time.

It is my suspicion that the moon during the First Earth, as was originally designed by God, revolved around our Earth in a perfectly circular orbit and rotated at a greater speed so people viewing from the Earth could see both sides of the moon every month. But all of that ended with the death of the First Earth during the Great Flood.

Far side of the moon heavily cratered.   Near side of the moon with huge lava beds.

Gradualism has ever been the preferred method of evolutionists, but the factual evidence speaks otherwise. Evidence for this singular cataclysmic event that impacted the far side of the moon is also found in the asteroid Vesta. This huge asteroid, measuring some 326 miles across, is the third-largest asteroid in volume and is reported to contain about 9 percent of the volume of the asteroid belt. It is within the relative orbit of Cera around the sun, which is considered the largest in the belt.

This should not surprise us. It is the old uniformitarian dogma that continues to thrive in the evolutionary community, in spite of the fact that it has been thoroughly discredited. They call this event, which caused the vast majority of the craters on the moon, the Lunar Cataclysm. Of course, they date this event at millions of years in our past in order to fit within their evolutionary hypothesis. The evolutionary magic wand only works in two modes: either millions of years or billions of years.

Nevertheless, the fact that some planets and moons are cratered only on one side seems to point to at least one cataclysmic event that would have thrust an enormous number of comets and asteroids into space. Why is it that the surface of Mars has 73 percent of its craters on one side and only 7 percent on the opposite side? In a rotating planet, the chances of one hemisphere being pelted by almost three-quarters of the total asteroid strikes makes the gradualist model of the evolutionists quite improbable. It is more likely that this is a result of a single catastrophic event that impacted the moon and Mars and other heavenly bodies.

It would also explain some of the similarities between asteroids and comets. I am not saying that all comets and asteroids come from Earth. Most asteroids and comets have probably come from collisions of other heavenly bodies. Perhaps some comets were formed by collisions of icy planetoids such as Pluto. Nevertheless, the future will one day show that some of them may have come from Earth.

The cataclysmic meteor impacts that cracked the surface of the Earth into seven continents also sent an enormous amount of material outward into space. Many of those ejected rocks slammed into what is now the far side of the moon. That impact shattered one side, scarring the surface with craters, and subsequently slowed down its rotation so now it only rotates once a month. Now only one side faces us all the time. Moreover, it also changed its perfect, stable orbit, slinging it farther away from Earth by 3.8 centimeters every year.

We can see the evidence of this catastrophic event on both sides of the moon. For years, scientists have wondered how the moon's crust on the far side could have become 30 miles thicker and much denser than on the visible side. The formation of a spinning object (such as the evolutionary model that claims our moon has been reconstituted from scattered debris caused by a planetary impact with Earth) would result in an even distribution of the melted material throughout the surface.

Yet this is not what we observe. The far side of the moon is 30 percent denser than the near side. That denser side in a rotating system would cause it to be pushed outward in the circumference of the rotation, thus changing the rotation of the moon from two turns every month to one turn, always having one side (the heavy side) facing away from Earth in the center of the circumference.

Additionally, they have wondered why one side of the moon looks so different from the other. The visible side is dominated by a preponderance of giant dark lakebeds of now-hardened lava and relatively few meteor craters. These large dark lakes were long ago thought by some scientists to be oceans on the moon. The far side, on the other hand, is riddled with craters of many sizes and no giant lava flows. I suggest that these are, in fact, the impact craters created by the material shot out in the powerful ejecta plumes from the multiple asteroid strikes that caused the Great Flood on the First Earth.

These ejected rocks, bursting outward from our planet, smashed into the far side of the moon, which at that time was facing Earth during its normal bimonthly rotation. The impact was so great that the pressure waves traveled through the interior of the moon and impinged on the opposite side (now the visible side), causing it to thrust outward with massive lava flows bursting from the molten interior core. As the angular momentum of the spin brought stability to the shape of the crust, the area that faces Earth was left with a much less dense crust.

This is the very same thing that happened on Earth as the giant meteors' impact pressure wave traveled through the center of Earth's core and impinged on the opposite side of the Earth. These massive pressure waves caused the western Pacific side to erupt, forming the Deccan Traps and Siberian Traps at the opposite ends of their impacts. For this reason, the areas in the western Pacific show a greatly reduced density and therefore lower gravity as the expulsed material left behind the deepest ocean trenches on the entire planet.

On the moon, the slowing of the rotation to its present rate may have caused the molten core to slow down its churning speed and subsequently lose its magnetosphere. A magnetic field around the moon would have greatly aided in protecting Earth against cosmic radiation. It is highly probable that all these catastrophic events recorded in the rocks of the moon and Earth are interconnected to the very mechanism that initiated the global flood and forever changed Earth and moon from the perfect order in which they were previsioned and intended.

Our Conclusion

Science is not in opposition to the Genesis account or to our Judeo-Christian cosmological model. Naturalism, in spite of its rhetoric, is not the absolute measure of science. It is an atheistic worldview that tries to explain the origins of the cosmos without God. It begins with the premise that there is no God and then attempts to put together scientific data to support a mechanism for our origins. But we have documented here that scientific data do not fit well with the naturalistic theory.

As we have seen so far, the more scientific knowledge we get about our universe, the more the Genesis claims are verified. We have seen that contrary to the evolutionary theory, the universe did have a beginning. The universe is not infinite.

Sudden cataclysms are very much a part of our history. The uniformitarian bias of gradualism has been dealt a lethal blow by the Big Bang. Catastrophism has been vindicated by the acknowledgment of past meteor strikes. The age of the universe does not contradict the Genesis account.

Our universe could not have evolved by random ordering. The fine-tuned parameters that governed every aspect of the formation of our universe can only be explained if they were, in fact, designed by an intelligent mind. Our Earth was specially created and designed by God to inhabit life. The account of the six days of creation in Genesis is not mythological but rather a space-time historical reality.

God made it thus so we could not deny it. Yet incredibly, we still deny it. Naturalists wanted desperately to believe that the universe just was—and that it would always be. But empirical evidence has blown both of those possibilities right out of the water, at least for those who have eyes to see and ears to hear. They wanted desperately to believe that all events are randomly ordered. But the entire universe screams of ordered conformity to a specific design from the first jiffy of existence even until now.

The idea that life could have evolved in the harsh conditions of deep space, unprotected from the lethal radiation of the sun, is even more preposterous than the idea of it evolving in muck here on Earth. But before we can appreciate the immense improbability of such a claim, we must first examine the very complexity of what they term "simple" life forms and the nature of "life" itself. We will examine this in Book 3 of this series, *Codes: God's Indelible Fingerprint*. The code of life proves that there was an intelligent designer who previsioned our universe and the living creatures that inhabit it.

REFERENCES

• • •

Adler, Mortimer J. 1990. *Truth in Religion*. New York: Macmillan Publishing.

Alinsky, Saul D. 1971. *Rules for Radicals*. New York: Vintage Books.

Allen, Richard Hinckley. 1963. *Star Names: Their Lore and Meaning*. New York: Dover Publication.

Asphaug, Erik. 2000. "The Small Planets." *Scientific American 282*, May 2000.

Barron, Robert. 2011. "Why I Loved to Listen to Christopher Hitchens." *Word on Fire* (blog), December 28, 2011, https://www.wordonfire.org/resources/article/why-i-loved-to-listen-to-christopher-hitchens/430/.

Barrow, John D., and Frank J. Tipler. 1996. *The Anthropic Cosmological Principal*. New York: Oxford University Press.

Boen, Brooke. 2011. "NASA Chat: The Moon's Earth-like Core." January 19, 2011. https://www.nasa.gov/connect/chat/moon_core_chat.html.

Brundage, Burr Cartwright. 1963. *Empire of the Inca*. Norman, OK: Oklahoma Press.

Budge, Sir Ernest A. Wallis. 1967. *The Egyptian Book of the Dead*. New York: Dover Publications.

Chesterton, G. K. 1955. *The Everlasting Man*. New York: Image Books.

Church, George M., and Ed Regis. 2012. *Regenesis: How Synthetic Biology Will Reinvent Nature and Ourselves*. New York: Basic Books.

Clavin, Whitney, and Alan Buis. 2011. "Astronomers Find Largest, Most Distant Reservoir of Water." *NASA*, July 22, 2011. https://www.nasa.gov/topics/universe/features/universe20110722.html.

Copi, Craig J., et al. 2010. "Large-Angle Anomalies in the CMB." *Advances in Astronomy 2010* 2010. https://www.hindawi.com/journals/aa/2010/847541/.

Corliss, William R. 1987. *Stars, Galaxies, Cosmos.* Glen Arm, MD; The Sourcebook Project.

Davies, Paul. *Other Worlds.* 1980. New York: Simon and Shuster.

Egnor, Michael. 2016. "Your Deluded Brain Thinks It's Conscious." *Evolution News & Science Today*, July 11, 2016. https://evolution-news.org/2016/07/your_deluded_br/.

Einstein, Albert. 1952. *The Principle of Relativity.* New York: Dover Publications.

Fazekas, Andrew. 2011. "Star Found Shooting Water "Bullets." *National Geographic News*, June 13, 2011. http://news.nationalgeographic.com/news/2011/06/110613-space-science-star-water-bullets-kristensen/.

Fichte, J. G. 1982. *The Science of Knowledge.* trans. P. Heath and J. Lachs. Cambridge: Cambridge University Press.

Fröböse, Rolf. 2012. *The Secret Physics of Coincidence.* Urheberrechtlich geschütztes Material.

Greene, Brian. 2003. *The Elegant Universe.* New York: Vintage Books.

———. *The Hidden Reality.* 2011. New York: Vintage Books.

Guillen, Michael. 2015. *Amazing Truths.* Grand Rapids, MI: Zondervan.

Hawking, Stephen. 1998. *A Brief History of Time.* New York: Bantam Books.

Hawking, Stephen, and Leonard Mlodinow. 2012. *The Grand Design.* New York: Bantam Books.

Humphreys, D. Russell. 1994. *Starlight and Time.* Green Forest, AR: Master Books.

Johnston, George Sim. 1999. "Designed for Living." *The Wall Street Journal.* October 15, 1999.

Kaku, Michio. 2004. *Einstein's Cosmos*. New York: Atlas Books.

Kane, Gordon. 2013. *Supersymmetry and Beyond*. Cambridge, MA: Basic Books.

Kerr, Richard A. 1992. "Theoreticians Are Putting a New Spin on the Planets." *Science 258*, October 23, 1992.

Krause, Lawrence M. 2016. "Freedom from Religion Foundation Advertisement." *Scientific American*, Nov. 2016: 19.

Lemonick, Michael D. 2012. "Dark-Matter Mystery: Why Are 400 Stars Moving As If There's Nothing There?" *Time*, May 1, 2012.

Lewis, C. S. 1943. *The Pilgrim's Regress*. Grand Rapids, MI: William B. Eerdmans Publishing.

Maldacena, Juan. 2006. "Black Holes and Wormholes and the Secrets of Quantum Spacetime." *Scientific American 315*, October 18, 2006:28–30.

McDonald, Marianne. 2000. *Mythology of the Zodiac*. New York: Metro Books.

Mercatante, Anthony S. 1995. *Who's Who in Egyptian Mythology*. New York: Metro Books.

Meyer, Stephen C. 2009. *Signature in the Cell*. New York: HarperCollins Publishers.

Moskowitz, Clara. 2014. "Gravitational Waves from Big Bang Detected." *Scientific American 315,* March 17, 2014.

NASA. 2010. "NASA's Hubble Captures First Images of Aftermath of Possible Asteroid Collision." October 13, 2010. https://www.nasa.gov/mission_pages/hubble/news/asteroid-collision.html.

Newton, Isaac. 1962. *Sir Isaac Newton's Mathematical Principles of Natural Philosophy and His System of the World, Vol. 1*, trans. A. Motte and Florian Cajori. Berkeley: Berkeley University of California Press.

Nobel Foundation. 2013. "2013 Nobel Prize in Physica: Higgs Particle and the Origin of Mass." *Science Daily* (October 8, 2013). https://www.sciencedaily.com/releases/2013/10/131008075834.htm.

Parrinder, Geoffrey. 1983. *World Religions*. New York: Barnes and Noble.

Peratt, Anthony. 1992, 2015. *Physics of the Plasma Universe*. New York: Springer.

Pilipenko, S. V. 2007. "The Space Distribution of Quasars." *Astronomy Reports 51:10*, October 2007: 820–829.

Prigogine, Ilya, and Isabelle Stengers. 1984. *Order Out of Chaos*. New York: Bantam Books.

Planck, Max. 1949. *Scientific Autobiography and Other Papers*. New York: Philosophical Library.

Quirke, Stephen. 1992. *Ancient Egyptian Religion*. New York: Dover Publications.

Rehwinkel, Alfred. 1951. *The Flood*. St. Louis, MO: Concordia Publishing House.

Ross, Hugh. 1989. *The Fingerprint of God*. Orange, CA: Promise Publishing.

Russell, Bertrand. 1957. *Why I Am Not a Christian*. New York: George Allen & Unwin.

Schroeder, Gerald S. 1997. *The Science of God*. New York: Free Press.

Seiss, Joseph A. 1999. *The Gospel in the Stars*. Grand Rapids, MI: Kregel Publications.

Sternglass, Ernest J. 1997. *Before the Big Bang*. New York: Four Walls Eight Windows.

Suh, Christine. 2003. "NASA Unveils New Picture of Universe." *UPI Science News* (February 12, 2003). Accessed 8 November

8, 2017 at https://www.upi.com/NASA-unveils-new-picture-of-universe/74471045099658/.

Sungenis, Robert A., and Robert J. Bennett. 2014a. *Galileo Was Wrong: The Church Was Right, Volume I*. State Line, PA: Catholic Apologetics International Publishing.

———. 2014b. *Galileo Was Wrong: The Church Was Right, Volume II*. State Line, PA: Catholic Apologetics International Publishing.

Susskind, Leonard. 2006. *The Cosmic Landscape*. New York: Back Bay Books.

———. 2008. *The Black Hole War*. New York: Back Bay Books.

Tarnas, Richard. 2007. *Cosmos and Psyche*. New York: Plume.

Thorne, Kip S. 1994. *Black Holes & Time Warps*. New York: W.W. Norton and Co.

Tifft, W. G., and W. J. Cocke. 1984. "Global Redshift Quantization." *The Astrophysical Journal 287*, December 15, 1984: 492–502.

Trefil, James. 1988. *The Dark Side of the Universe*. New York: Charles Scribner's Sons.

Wakker, Bart P., and Philip Richter. 2004. "Our Growing, Breathing Galaxy." *Scientific American 290:1*, January 2004.

Weinberg, Steven. 1977. *The First Three Minutes*. New York: Basic Books.

Whitehead, Alfred North. 1967. *Science and the Modern World*. New York: The Free Press.

Zacharias, Ravi. 2008. *The End of Reason*. Grand Rapids, MI: Zondervan.

INDEX

• • •

A

G

Q

R

www.ingramcontent.com/pod-product
Lightning Source LLC
Chambersburg PA
CBHW031531210326
41599CB00015B/1859